Lecture Notes in Civil Engineering

Volume 110

Lecture Notes in Civil Engineering (LNCE) publishes the latest developments in Civil Engineering - quickly, informally and in top quality. Though original research reported in proceedings and post-proceedings represents the core of LNCE, edited volumes of exceptionally high quality and interest may also be considered for publication. Volumes published in LNCE embrace all aspects and subfields of, as well as new challenges in, Civil Engineering. Topics in the series include:

- Construction and Structural Mechanics
- Building Materials
- Concrete, Steel and Timber Structures
- Geotechnical Engineering
- Earthquake Engineering
- Coastal Engineering
- Ocean and Offshore Engineering; Ships and Floating Structures
- Hydraulics, Hydrology and Water Resources Engineering
- Environmental Engineering and Sustainability
- Structural Health and Monitoring
- Surveying and Geographical Information Systems
- Indoor Environments
- Transportation and Traffic
- Risk Analysis
- Safety and Security

To submit a proposal or request further information, please contact the appropriate Springer Editor:

- Mr. Pierpaolo Riva at pierpaolo.riva@springer.com (Europe and Americas);
- Ms. Swati Meherishi at swati.meherishi@springer.com (Asia - except China, and Australia, New Zealand);
- Dr. Mengchu Huang at mengchu.huang@springer.com (China).

All books in the series now indexed by Scopus and EI Compendex database!

More information about this series at http://www.springer.com/series/15087

Magd Abdel Wahab
Editor

Proceedings of 1st International Conference on Structural Damage Modelling and Assessment

SDMA 2020, 4–5 August 2020, Ghent University, Belgium

Springer

Editor
Magd Abdel Wahab
Faculty of Engineering and Architecture
Ghent University
Ghent, Belgium

ISSN 2366-2557 ISSN 2366-2565 (electronic)
Lecture Notes in Civil Engineering
ISBN 978-981-15-9123-5 ISBN 978-981-15-9121-1 (eBook)
https://doi.org/10.1007/978-981-15-9121-1

This Springer imprint is published by the registered company Springer Nature Singapore Pte Ltd.
The registered company address is: 152 Beach Road, #21-01/04 Gateway East, Singapore 189721, Singapore

Organising Committee

Chairman

Prof. Magd Abdel Wahab, Ghent University, Belgium

International Scientific Committee

Prof. S. Abdullah, Universiti Kebangsaan Malaysia, Malaysia
Dr. H. T. Ali, University of Bristol, UK
Dr. I. Hilmy, International Islamic University Malaysia
Prof. G.-R. Gillich, Eftimie Murgu University of Resita, Romania
Dr. S. Khatir, Ghent University, Belgium
Prof. M. Korovina, Lomonosov Moscow State University, Russia
Dr. C. Le Thanh, Open University Ho Chi Minh City, Vietnam
Prof. H. Matevossian, Russian Academy of Sciences, Russia
Prof. N.-A. Noda, Kyushu Institute of Technology, Japan
Prof. K. Oda, Oita University, Japan
Prof. R. V. Prakash, Indian Institute of Technology, India
Prof. T. Rabczuk, Bauhaus University Weimar, Germany
Prof. A. Rudawska, Lublin University of Technology, Poland
Prof. J. Toribio, University of Salamanca, Spain
Dr. L. V. Tran, Sejong University, South Korea
Prof. L. Vanegas Useche, Universidad Tecnológica de Pereira, Colombia
Dr. C. Wang, Liaocheng University, China
Prof. H.-N. Xuan, Hutech University, Vietnam
Dr. Y.-L. Zhou, National University of Singapore
Dr. X. Zhuang, Leibniz Unversität Hannover, Germany
Prof. Yongtao Bai, Chongqing University, China

Preface

This volume contains the proceedings of the 1st International Conference on Structural Damage Modelling and Assessment (SDMA 2020), August 4–5, 2020, Online conference. The conference is a major international forum for research topics relevant to damage modelling and assessment of engineering structures and systems including numerical simulations, signal processing of sensor measurements and theoretical techniques, as well as, experimental case studies. The presentations of SDMA 2020 are divided into two main parts, namely: (1) Damage in Civil Engineering and (2) Damage in Damage in Mechanical and Materials Engineering.

The organising committee is grateful to keynote speakers: Prof. Luca Susmel, The University of Sheffield, UK, for his presentation entitled 'An advanced elasto-plastic approach to design notched metals against variable amplitude multiaxial fatigue loading' and Prof. Gilbert-Rainer Gillich, Faculty of Engineering and Management at "Eftimie Murgu" University of Resita, Romania, for his presentation entitled 'A model to predict the evolution of the natural frequencies of a beam due to damage: the equivalent healthy beam'.

Special thanks go to members of the Scientific Committee of SDMA 2020 for reviewing the articles published in this volume and for judging their scientific merits. Based on the comments of reviewers and the scientific merits of the submitted manuscripts, the articles were accepted for publication in the conference proceedings and for presentation at the conference venue. The accepted papers are of a very high scientific quality and contribute to advancement of knowledge in all research topics relevant to SDMA conference.

Finally, the organising committee would like to thank all authors, who have contributed to this volume and presented their research work at SDMA 2020.

Ghent, Belgium

Prof. Magd Abdel Wahab
Chairman of SDMA 2020

Contents

Damage in Civil Engineering

Monitoring and Quantifying Crack-Based Light Damage in Masonry Walls with Digital Image Correlation

Paul A. Korswagen and Jan G. Rots

Abstract Recent, induced earthquakes in the north of the Netherlands have led to a large number of damage claims. Many claims can be considered to fall into the category of 'light damage' to the ubiquitous, unreinforced masonry structures in the region. To evaluate and predict the behaviour of cracks, characteristic of light masonry damage, caused by seismic or other actions, an experimental campaign, linked to the validation of computational models, has been pursued. To accurately capture the initiation of visible cracks, wider than 0.1 mm, Digital Image Correlation (DIC) was applied to monitor the entire surface of full-scale wall panels and smaller specimens. Moreover, an optimised speckle pattern and solving algorithm was developed to be able to monitor not only the initiation, but also the propagation of the cracks during subsequent (repeating) loading cycles. In this approach, the crack data is then used to characterise the intensity of damage with a single scalar; the parameter, denoted Ψ and comprising the number, width and length of the cracks, is used to evaluate the progression of light damage in experiments and finite element models. A description of the DIC technique applied and of the development and usage of the damage parameter for masonry is presented herein.

Keywords Masonry · Cracks · Light damage · DIC

1 Introduction

Structures of all kinds are subjected to actions that have the potential of causing damage whenever these lead to undesired conditions in the structures. In general, any state deviating from the original state or the intended state of a structure can be categorised as damage. The deviating state can be a direct result of a particular action on a structure or its components. Loss of strength due to cracks in walls, crushing of bricks or loss of elements, as well as loss of section due to chemical action or

P. A. Korswagen (✉) · J. G. Rots
Materials, Mechanics, Management & Design (3MD), Delft University of Technology, Delft, The Netherlands
e-mail: p.a.korswageneguren@tudelft.nl

M. Abdel Wahab (ed.), *Proceedings of 1st International Conference on Structural Damage Modelling and Assessment*, Lecture Notes in Civil Engineering 110,
https://doi.org/10.1007/978-981-15-9121-1_1

freezing, or high distortions due to creep behaviour or overloading, are all examples of damage in masonry structures.

When observing minor levels of damage, strategies developed for near-collapse or ultimate limit states (like surveying if there is a severe loss of strength or excessive distortion of the structure) may fail to accurately describe minor damage. Additionally, these expressions of damage may not be readily detectable or are difficult to assess by an inexperienced observer or inspector. Cracking, on the other hand, the occurrence of cracks and fissures on or through walls and other elements in a masonry structure, is easily observable and directly relatable to some degree of damage. Moreover, many phenomena such as earthquakes or settlements are commonly observed to cause cracking. Consequently, many studies [e.g. 1–6 have opted for the consideration of cracks as a measure for the evaluation of (minor) damage; where, put simply: the absence of cracks corresponds to no damage, a few small cracks to very light damage, and larger cracks to more intense damage.

Nonetheless, even if a clear expression of damage such as cracking, is set, the quantitative evaluation of the damage remains unclear. It is thus necessary to categorise the diverse expressions and intensities of cracking into damage categories [7] and be able to quickly and objectively quantify and compare the intensity of damage between different cases or for different actions. For example, if only the crack width is used as a measure of damage, then the increase of the crack width can be related to the increase in damage; but, what happens if a second crack develops?

Consequently, a strict, mathematical definition of crack-based damage is herein proposed to objectively quantify damage in masonry structures. The goal of this definition is to assess the initiation, and most importantly, the progression of damage over time or during laboratory or computational experiments. Moreover, this is complemented with a high-resolution implementation of Digital Image Correlation [8] tailored to the evaluation of masonry cracks in laboratory tests. The usage of this definition allows for a more precise evaluation of light damage potentially caused by seismic events in the north of the Netherlands.

2 The Ψ Damage Parameter

2.1 Definition

Cracks in a masonry structure appear when the masonry tensile strength is exceeded [9]. Since masonry has a very low tensile strength, cracks are reasonably likely to appear in any structure; however, for a crack to be detectable it also needs to widen. Hence, deformations also need to take place. In Fig. 1, an example of a wall with an opening subjected to a lateral load in its plane is presented. As the lateral force increases, so does the displacement measured at the top of the wall. This elastic relationship starts to degrade as the displacement increases at a larger rate than the force. This is linked to the appearance of crack(s) somewhere in the wall. Damage

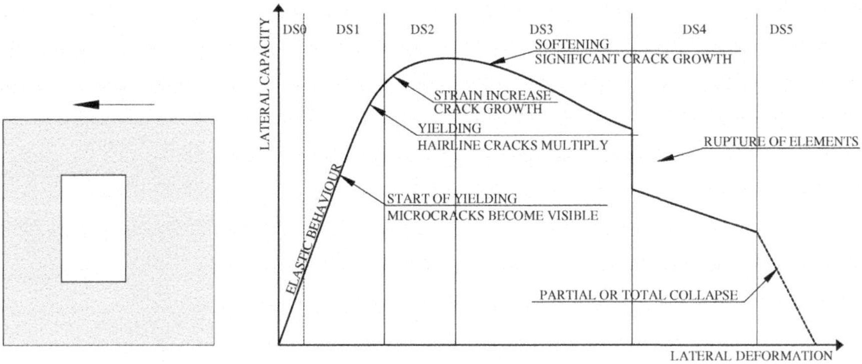

Fig. 1 Typical initiation and propagation of cracks illustrated on a lateral force–displacement curve of a masonry wall

States, as defined by [7, 10], are usually employed to categorise the damage in the wall but these are not directly related to the cracks. After the wall reaches its maximum force capacity, cracks will become wider and failure will ultimately occur. Yet, the aggravation of cracks within the states up to DS2 is difficult to assess from force–displacement graphs, while also problematic to quantify using qualitative definitions for DS1 and DS2.

Therefore, the summative crack pattern is better characterised with a parameter directly computed from the properties of the cracks. As mentioned before, the width of the cracks can be related to the intensity of the damage; however, the number of cracks is also influential. The assessment of laboratory specimens and inspection of real-world damage cases [7, 11] led to the realisation that cracks narrower than 0.1 mm were difficult to see with the naked eye. In fact, from an anatomical perspective, the normal human eye can detect differences of down to 30 μm in ideal light and contrast conditions (see for instance, [12]). Since cracks in masonry walls do not satisfy these ideal conditions even during rigorous inspections of plastered walls, a limit of 100 μm was deemed reasonable, especially considering that the outer walls in Groningen masonry are mainly unplastered [13]; and, since DS1 is related to aesthetic damage, damage that cannot be observed, is thus not relevant. This boundary is also employed as a cut-off value when measuring the length of cracks. Hence, a width of 0.1 mm was set as the lower boundary, below which no damage could be assumed.

The parameter that determines the damage intensity is herein based on the number, width, and length of the cracks following Eq. 1. The damage parameter Psi (Ψ) is based on a scale that defines the ease of repair of the cracks (adapted from Boscardin and Cording [14], Burland and Wroth [15], and, at its latest, Giardina et al. [16]); see Table 1. Here, the total of visible cracks is expressed in one number such that the narrowest visible cracks with a width of 0.1 mm result in a value of around one ($\Psi = 1$), slightly larger cracks of close to 1 mm width correspond to two ($\Psi = 2$) and cracks of approximately 4 mm in width give a value of three ($\Psi = 3$). This range of

Table 1 Definition of the Ψ parameter

Category of damage	Damage		Description of typical damage and ease of repair	Approx. crack width (mm)
Aesthetic damage (DS1)	Negligible	DL1	Hairline cracks	up to 0.1 mm
	Very slight	DL2	Fine cracks which can easily be treated during normal decoration. Perhaps isolated slight fracturing in building. Cracks in external brickwork visible on close inspection	up to 1 mm
	Slight	DL3	Cracks easily filled. Redecoration probably required. Several slight fractures showing inside of building. Cracks are visible externally and some repainting may be required externally to ensure water tightness	up to 5 mm

visible cracks from $\Psi = 1$ to $\Psi = 3$ is herein described as light damage (DS1). In this manner, Psi (Ψ) can be computed from both DIC and FEM data analogously: in the former by differentiating the displacement fields to obtain the crack width, and in the latter by employing the crack width data directly produced by finite element models with cracking material models.

$$\Psi = 2 \cdot n_c^{0.15} \cdot \hat{c}_w^{0.3}\ with\ \hat{c}_w = \frac{\sum_{i=1}^{n_c} c_{w,i}^2 \cdot c_{L,i}}{\sum_{i=1}^{n_c} c_{w,i} \cdot c_{L,i}} \tag{1}$$

where:

n_c is the number of cracks in the wall/specimen
$\hat{c}_w$ is the width-weighted and length-averaged crack width (in mm) calculated with:
c_w is the maximum crack width along each crack in mm
c_L is the crack length in mm
For $n_c = 1$, $\hat{c}_w = c_w$. In this expression, the crack width of each crack is measured at their widest point.

The parameter equation is graphically shown in Fig. 2 for some values of 'n_c'. The exponents (0.15 and 0.30) and coefficient (2) in the expression (Eq. 1) are tuned such that the relationship to the damage levels shown in Table 1 is maintained. Since these are qualitative descriptions, the defining expression of Ψ can be made to fit

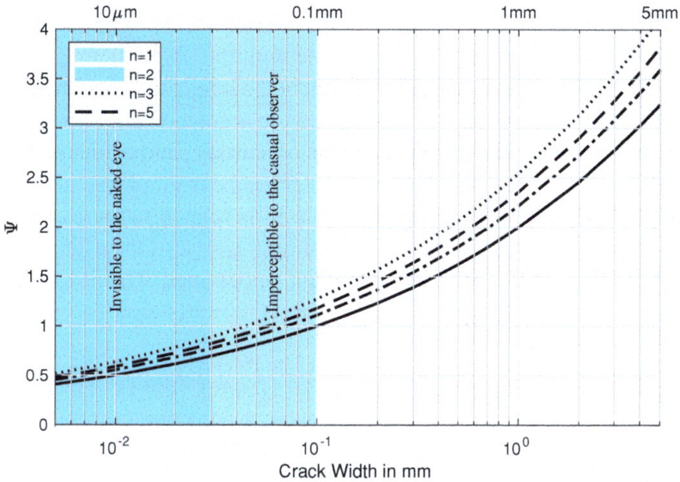

Fig. 2 Illustration of the Ψ damage parameter against crack width and for different numbers of cracks

nicely. It is evident that a specimen or wall with multiple cracks is more damaged than one with a single crack. Moreover, Fig. 3 gives a few examples of the usage of this parameter; one of the illustrated cases exceeds light damage and would probably better evaluated using a different kind of damage metric [17].

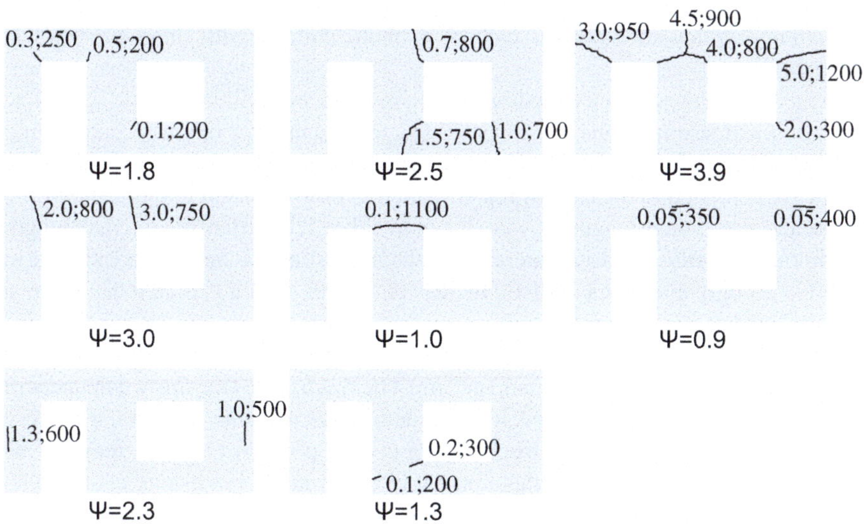

Fig. 3 Example of a façade with various crack patterns identified with 'width; length' in millimetres, and labeled with the computed value of Ψ

Nonetheless, the Ψ parameter allows for the comparison of the intensity of damage regardless of the specimen size. This is in-line with the parent damage scale (DS1-DS5), where the damage states are independent of the size of the structure and only the importance of the damage to each specific structure is considered. This is particularly advantageous when observing the progression of damage, and comparing it between samples of different dimensions. This parameter has been employed so far in several linked studies [11, 18–20]. Furthermore, since Ψ is related to the ease of repair of the damage, when the parameter is multiplied by the area of the affected wall, then a direct relationship to the cost of the repair can be obtained. This is treated later on.

2.2 *Limitations*

Since damage is directly evaluated independently of the crack configuration, the progress (or intensification) of damage can be observed throughout experiments. Nevertheless, the use of one parameter to characterise the entire damage picture of a specimen is accompanied by certain limitations.

First, there is the loss of cause as the mechanism observed in a crack pattern cannot be captured in the value of one parameter. Second, in some cases, there is a loss of veridicality: Some changes such as the increase in length of one narrow crack while observing no changes in any other cracks, will produce an unexpected change in the value of Ψ. This is an unrealistic situation for which the parameter has not been calibrated. Such changes, however, have a small influence in the value of Psi and will be limited to a centesimal change. This leads to a loss of precision, but helps to realise that attempting to capture aesthetic damage with a high precision is not sensible.

In this light, the parameter needs always to be evaluated within realistic scenarios. For example, masonry walls are subjected to a limited number of cracks: attempting to evaluate Ψ with a high number of cracks is hence unrealistic. Moreover, since the parameter is related to the ease of repair, which in turn is related to the width of the cracks and not to their length, as was shown in Table 1, an extension in crack length will not necessarily lead to an increase in Ψ; in masonry, a significant increase in length is accompanied by a realistic increase in width, which is then reflected by a higher Ψ value.

Furthermore, when considering the definition of damage based on an observable measure of damage, "transitory damage" must be discussed. Transitory damage will differ from the actual damage level of a structure. As its name suggests, transitory damage reflects the state of a structure at a certain point in time and may not be the same as the damage state at the moment of observation. For example, during an earthquake, a structure may deform such that cracks of 1 mm appear on the walls at the moment of maximum deformation; but, by the end of the earthquake, the cracks may have partially closed. If a picture had been taken at the moment of maximum deformation, the damage level would likely appear higher than the residual damage state. In this study, damage is analogous to transitory damage. This is a conservative approach yet especially suited for light damage due to the following reasons:

Firstly, unlike larger cracks exceeding DS1 which may close significantly compared to their maximum transitory state, narrower cracks corresponding to DS1 are not able to close once open because of the roughness of the newly developed crack interface. Moreover, cracks through bricks or in finished walls are irreversible: once a crack appears it will remain visible. Thus, when observing light damage, transitory and residual damage are more alike.

Secondly, since unreinforced masonry is designed without taking into account its tensile strength, it is usually not subjected to forces that would keep the cracks open once they have formed. However, when additional tensile stresses in more directions are present, such as those generated by hygro-thermal expansion or settlement actions, small cracks are more likely to remain open after they have formed. Additionally, unlike laboratory experiments where a restitutory force may exert the work required to close the cracks, in real cases, such a force may not be present; in fact, in real cases, forces might be present that keep the cracks open [11].

Thirdly, in contrast to field cases, observing transitory damage in laboratory experiments is possible and easier than the sometimes hardly-noticeable residual damage of lightly damaged cases subjected to only one action. Here, transitory damage will be higher than residual damage, and it is not possible to determine what the actual residual damage would have been, had the laboratory case been a real case in the field with complex interactions with other structural or non-structural elements and finishings. Furthermore, the maximum (transitory) and residual damage can also be obtained easily from computational models, which provides an additional point of comparison.

Fourthly, when analysing the effect of a combination of actions or of repeated actions, the true damaged state of the structure is that revealed by the maximum transitory state. The transitory state will be a more accurate representation of the loss of strength experienced by the structure and hence its response to subsequent excitations. The analysis of the behaviour of the structure to subsequent damage causes or events should be performed with the maximum damage and not the (perhaps inapparent) residual damage. Thus, when multiple damaging causes are considered, the residual damage would not be adequately suited.

Fifthly, when considering the structural design of a structure and whether it adheres to regulations, drift limits are specified towards the maximum displacement of the structure and not the residual displacement. It is thus common practice to look at the maximum transitory state of a structure when assessing its final damage state.

Therefore, transitory damage was used consistently in this study when referring to damage, and while it is expected that for light damage, the transitory damage will be close to the final damage, it must be noted that the final damage is bound to be slightly lower than the transitory damage measure employed.

2.3 Perception of Damage

Two identical houses subjected to similar actions will be similarly damaged; however, if one house has walls covered in a plaster that is old and stiff while the other has walls covered in flexible wallpaper, the former will display any crack prominently, while the latter will hide cracks. The first house will be perceived to be more damaged than the second one. This is the perceived damage state which may differ from the actual damage state.

This work focuses on real damage and observes the physical processes that lead to it, but it is still important to acknowledge that damage can be subjective and that certain combinations of architectural building parameters will lead to more reported damage. Understanding how damage is likely to be perceived also gives insight into overall damage conditions in the region.

The following is an empirical proposal of how and which parameters relating to the aesthetic and architectural disposition of the structure, as well as the situation in which damage was observed, may affect the way in which damage is perceived.

$$\psi = \Psi^{\frac{k}{3}} \tag{2}$$

where:

ψ is the perceived damage intensity (lowercase Ψ)
Ψ is the damage parameter (uppercase Ψ)
k is the average of the empirically-determined influence parameters for each category as shown in Table 2.

It is possible to include as many categories as deemed relevant into the evaluation of the perception of damage. Additionally, the relation can be inverted to try to estimate the actual damage of a structure from a study case report. In the former case, an estimation of how damage will be perceived can be inferred from a numerical model, while in the latter, a more accurate estimation of damage can be registered from an uncertain field report.

2.4 Agglomeration of Psi

The Ψ parameter has been mainly developed to assess the progression of damage on a certain specimen or between identical specimens. However, sometimes different walls or structures would like to be compared to each other to determine which presents lower or higher damage. In this case, it is convenient to express a relative version of Ψ based on the surface area of the masonry:

$$\Psi_i' = \Psi_i \cdot \frac{A_i}{\overline{A}} \tag{3}$$

where A_i is the surface are of the wall i, $\overline{A}$ is the mean area of the walls considered and Ψ' is the relative Ψ. Furthermore, if a structure where each wall is monitored separately wants to be characterised with a single value of Ψ, the damage in the N walls can be accounted as:

$$\overline{\Psi} = \frac{\sum_{i=1}^{N} \Psi_i \cdot A_i}{A_T} \tag{4}$$

where $\overline{\Psi}$ is the mean Ψ value and A_T is the sum of the surface areas.

3 Digital Image Correlation for Cracks

Photogrammetry techniques can be used to automatically detect cracks [21–23]. Digital Image Correlation (DIC) [24] is widely used in laboratories to measure displacement and strains on small samples or large surfaces where the use of multiple sensors is inconvenient [3, 8]. However, strains are not easily linked to cracks. Usually, measured strains are smoothed out over a certain surface to correct for noise and the relatively low resolution of DIC. The smooth strains are thus not representative of the discrete cracks appearing on masonry specimens.

As part of the experimental validation of the usage of the Ψ parameter, cracks had to be automatically detected to assess the progression of damage over hundreds of loading cycles [20]. For this purpose, a 51 Mpx DSLR camera and a 35 mm lens stopped down to f/9.0 were used to acquire high resolution images. Shots were illuminated with a flash at a speed of 1/63000 s to produce even lighting conditions and eliminate image blur. The setup, in combination with an optimised speckle pattern [25] and the DIC-algorithm developed for these kind of tests, allowed for the observation and, most importantly, the progression (in width and in length) of cracks invisible to the naked eye over the entire surface of a full-scale wall. Figure 4 presents an example of a 2.7 m-tall masonry wall covered in a speckle pattern where dots are randomly positioned and randomly vary in diameter between 4 and 8 pixels. To produce such a pattern, a stencil was laser-cut from a flexible acrylic plate and was applied on the masonry by spraying black paint with a compressed-air nozzle, similar to the approach of Ghorbani et al. [5]. The random pattern was generated with a script that allowed changing the sizes and distances of the dots. Multiple patterns were tested on small specimens to determine the best set of parameters. The pattern allows a (standard) DIC-algorithm to detect the relative displacements between an initial image and a later image. In Fig. 4, the displacement field reveals the presence of cracks. The DIC pattern on the wall in combination with the camera allowed for the monitoring of the full displacement field of the wall in a grid with a spacing of 2.6 × 2.6 mm and a precision of 20 μm, comprising over 1.2 million measurement (or grid) points. Images were taken at precise time-points throughout the test. The

Fig. 4 Laboratory wall specimen (3.1 m × 2.7 m) overlaid with the horizontal displacement field obtained with DIC during testing of a lateral top displacement in-plane. Right, zoomed-in corner of the wall depicting the speckle pattern used for DIC

accuracy of the setup allows to detect displacements as small as a fifth of the threshold set for visible cracks; this has herein been considered as high-resolution.

The raw or unmodified displacement field acquired by a DIC-algorithm using small subsets (of approximately 10 pixels) to avoid smoothing the displacement values, can be scanned for discontinuities above a certain threshold; large groups of discontinuities are likely to correspond to cracks. The relative displacement between one side of the discontinuity and the other corresponds to the width of the crack. Figure 5 presents the result of this operation where each crack is automatically characterised in width and length. Note that the wall specimens of Figs. 4 and 5 are not the same. While many authors utilise strains to map cracks [5, 6, 23, 26], strains would need to be integrated over a crackbandwidth to be able to output the crack width and length, and can thus only be used in an illustrative manner. The detection and characterisation of discontinuities, on the other hand, is much better tailored to obtain the crack kinematics. Gehri et al. [27] specify this approach in detail. The resulting data can provide an in-depth look at crack progression; Fig. 6 presents the

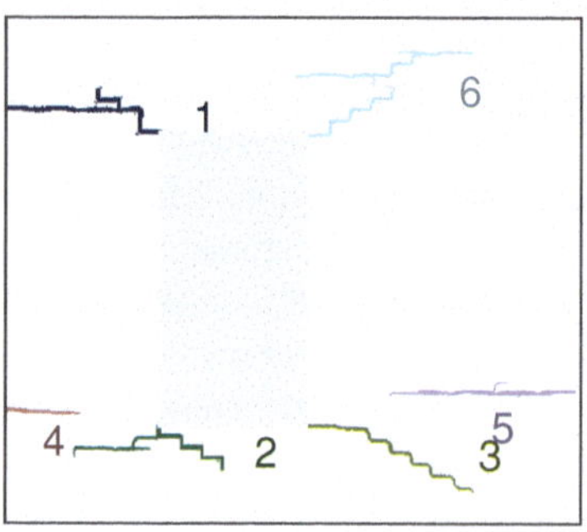

Crack	Width (μm)	Length (mm)
1	750	980
2	285	866
3	361	1278
4	375	445
5	676	1052
6	482	1330

Fig. 5 Detected cracks in DIC data of a laboratory masonry wall

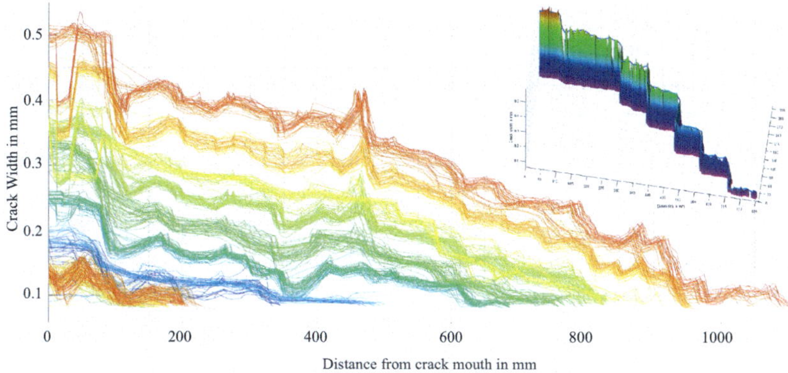

Fig. 6 Crack width against crack length as measured by the centreline of the crack, for numerous test cycles

crack width at the centreline of the crack over multiple experimental cycles. It can be observed that the crack grows in width and in length throughout the experiment. The centreline is captured automatically by following the trend of the maximum width over the crack. It can also be seen that the DIC output is not free from noise; however, with thousands of measurement points over the crack, reasonable values can be computed.

For each frame in the experiment, with crack widths and lengths determined, the Ψ value is computed as in Fig. 7. This illustrates how Ψ increases throughout the experiment, even during cycles of equal amplitude.

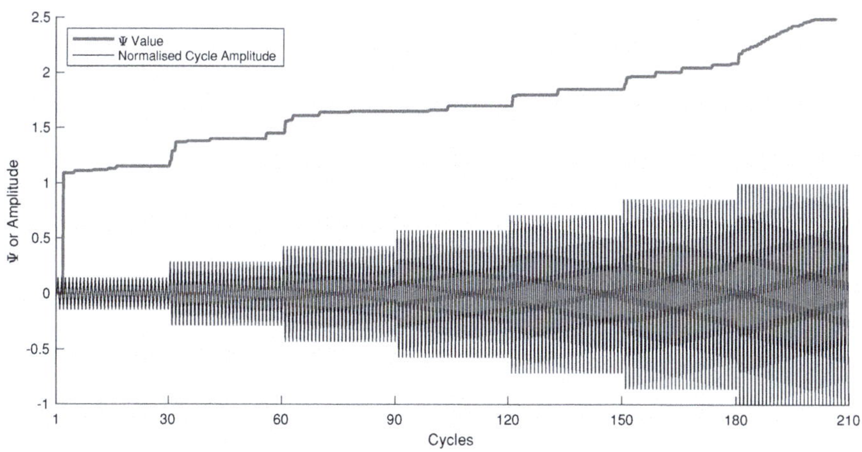

Fig. 7 Development of the Ψ value throughout testing of a wall indicating also the progression of the amplitude of the applied lateral top displacement

While the algorithm employed here was custom-written to detect cracks with the highest accuracy possible, commercial DIC tools have started to implement crack-oriented solutions [28] and it remains convenient to tailor solutions to specific experiments [27]. These analysis tools will allow for an easier characterisation of light damage in masonry.

4 Conclusions

Digital Image Correlation can be used successfully for the crack characterisation of laboratory masonry walls. The resolution achieved is sufficient for the accurate assessment of crack propagation even during repeating cycles of equal amplitude. The aggravation of damage can then be measured using a purposely-developed parameter that considers the number of cracks on the wall specimen, their width and their length. The parameter can be used to compare the damage progression even when cracks increase in number.

Acknowledgements This research was funded by Nederlandse Aardolie Maatschappij (NAM) under contract number UI67339 'Damage sensitivity of Groningen masonry building structures—Experimental and computational studies', contract holders: Jan van Elk and Jeroen Uilenreef. This cooperation is gratefully acknowledged.

Appendix

See Table 2.

Table 2 Influence values for various non-structural aspects influential in damage perception. Empiric exemplary values

Category	Subcategories	Influence value (k)	Description
Age	Very old	2	Older than 1970
	Older	3	Between 1970 and 2000
	New	4	Newer than 2000
Material	Baked clay	3	
	Calcium silicate	4	
Wall type	Slim	4	Less than 120 mm
	Thick	3	Greater than 120 mm
	Double	2	More than one layer of bricks
Cavity	Without cavity	3	One single leaf

(continued)

Table 2 (continued)

Category	Subcategories	Influence value (k)	Description
	Cavity and aesthetic	4	Two leaves, where only one is structural
	Cavity and structural	3	Two leaves, both structural
Brick Type	Regular bricks	3	Units with a height smaller than 150 mm
	Large blocks	4	Units with a height larger than 150 mm
	Hollow units	3	Units that are not solid
Mortar	Slim	4	The joints are around 3 mm according to EC
	Free verticals	2	Vertical joints between the bricks are not filled
	Normal	3	All joint are filled and greater than 3 mm
Finish	Exposed	2	The bricks and joints can be seen
	Plaster + paint	3	The wall is covered with plaster and painted
	Mortar + paint	4	The wall is covered with mortar and painted
	Elastomeric paint	2	Wall is (covered and) painted with flexible paint
	Wall paper	1	The wall is plastered and covered with paper

1—Reduces the perception of damage significantly; 2—Reduces the perception of damage; 3—Does not influence; 4—Increases the perception of damage; 5—Increases the perception of damage significantly

References

1. Didier M, Abbiati G, Broccardo M, Beyer K, Danciu L, Petrović M, Mojsilović N, Stojadinović B (2017) Quantification of non-structural damage in unreinforced masonry walls induced by geothermal reservoir exploration using quasi-static cyclic tests. In: Proceedings of the 13th Canadian masonry symposium, Halifax, Canada
2. Didier M, Abbiati G, Hefti F, Broccardo M, Stojadinovic B (2018) Damage quantification in plastered unreinforced masonry walls using digital image correlation. In: 10th Australasian masonry conference, 14–18 Feb 2018
3. Ramos T, Furtado A, Eslami S, Alves S, Rodrigues H, Arêde A, Tavares P, Moreira P (2015) 2D and 3D digital image correlation in civil engineering—measurements in a masonry wall. Proc Eng 114(2015):215–222
4. Bosiljkov V, Page AW, Bokan-Bosiljkov V, Zarnić R (2008) Evaluation of the seismic performance of brick masonry walls. Struct Control Health Monit 17:100–118. Published online 22 Dec 2008 in Wiley InterScience (www.interscience.wiley.com). https://doi.org/10.1002/stc.299

5. Ghorbani R, Matta F, Sutton MA (2015) Full-field deformation measurement and crack mapping on confined masonry walls using digital image correlation. Exp Mech 2015(55):227–243
6. Kumar SL, Aravind HB, Hossiney N (2019) Digital image correlation (DIC) for measuring strain in brick masonry specimen using Ncorr open source 2D MATLAB program. Results Eng 4(2019):100061
7. de Vent I, Rots JG, van Hees RPJ (2011) Structural damage in masonry—developing diagnostic decision support. TU Delft
8. Laurin F, Charrier JS, Lévêque D, Maire JF, Mavel A, Nuñez P (2012) Determination of the properties of composite materials thanks to digital image correlation measurements. Proc IUTAM 4(2012):106–115
9. Gattesco N, Macorini L, Dudine A (2016) Experimental response of brick-masonry spandrels under in-plane cyclic loading. ASCE J Struct Eng 142(2)
10. Grünthal G (1998) European Macroseismic Scale 1998 (EMS-98). European Seismological Commission, sub commission on engineering seismology, working group Macroseismic scales. Conseil de l'Europe, Cahiers du Centre Européen de Géodynamique et de Séismologie, vol. 15, Luxembourg
11. Van Staalduinen P, Terwel K, Rots JG (2018) Onderzoek naar de oorzaken van bouwkundige schade in Groningen Methodologie en case studies ter duiding van de oorzaken. Delft University of Technology. Report number CM-2018-01, Downloadable from www.NationaalCoordinatorGroningen.nl
12. Österberg G (1935) Topography of the layer of rods and cones in the human retina. Acta Ophthalmol [Suppl] 13(6):1–102
13. Jafari S, Esposito R, Rots JG, Messali F (2017) Characterizing the material properties of Dutch unreinforced masonry. Proc Eng 193:250–257
14. Boscardin MD, Cording EJ (1989) Building response to excavation-induced settlement. J Geotech Eng 115(1):1–21
15. Burland JB, Wroth CP (1974) Settlement of buildings and associated damage. In: Proceedings of conference on settlement of structures. Pentech Press, Cambridge, pp 611–654
16. Giardina G, van de Graaf AV, Hendriks MAN, Rots JG, Marini A (2013) Numerical analysis of a masonry façade subject to tunnelling-induced settlements. Eng Struct 54(2013):234–247
17. Korswagen PA, Jonkman SN, Terwel K (2019) Probabilistic assessment of structural damage from coupled multi-hazards. Struct Safety 76:135–148. ISSN 0167-4730. https://doi.org/10.1016/j.strusafe.2018.08.001
18. Korswagen PA, Longo M, Meulman E, Rots JG (2019) Crack initiation and propagation in unreinforced masonry specimens subjected to repeated in-plane loading during light damage. Bull Earthq Eng. https://doi.org/10.1007/s10518-018-00553-5
19. Korswagen PA, Longo M, Rots JG (2020a) High-resolution monitoring of the initial development of cracks in experimental masonry shear walls and their reproduction in finite element models. Eng Struct 211(2020):110365
20. Korswagen PA, Longo M, Rots JG (2020b) Calcium silicate against clay brick masonry: an experimental comparison of the in-plane behaviour during light damage. Bull Earthq Eng 18:2759–2781. https://doi.org/10.1007/s10518-020-00803-5
21. Mojsilović N, Salmanpour AH (2016) Masonry walls subjected to in-plane cyclic loading: application of digital image correlation for deformation field measurement. Institute of Structural Engineering, Department of Civil, Environmental and Geomatic Engineering, ETH Zurich, 8093, Switzerland
22. Dhanasekar M, Prasad P, Dorji J, Zahra T (2019) Serviceability assessment of masonry arch bridges using digital image correlation. J Bridge Eng 24(2):04018120
23. Tung S, Shih M, Sung W (2008) Development of digital image correlation method to analyse crack variations of masonry wall. Sadhana 33:767–779, Part 6
24. Blaber J, Adair B, Antoniou A (2015) Ncorr-open-source 2D digital image correlation Matlab software. Exp Mech. https://doi.org/10.1007/s11340-015-0009-1

25. Crammond G, Boyd SW, Dulieu-Barton JM (2013) Speckle pattern quality assessment for digital image correlation. Opt Lasers Eng 51(2013):1368–1378
26. Smrkić MF, Košćak J, Damjanović D (2018) Application of 2D digital image correlation for displacement and crack width measurement on RC elements. Građevinar 70(9):771–781
27. Gehri N, Mata-Falcón J, Kaufmann W (2020) Automated crack detection and measurement based on digital image correlation. Constr Build Mater 256(2020):119383
28. GOM Correlate—visited June 2020. www.gom.com/3d-software/gom-correlate/gom-correlate-features.html#video51438

Model Updating for a Railway Bridge Using a Hybrid Optimization Algorithm Combined with Experimental Data

H. Tran-Ngoc, H. Ho-Khac, T. Le-Xuan, Hieu Nguyen-Tran, Guido De Roeck, Thanh Bui-Tien, and Magd Abdel Wahab

Abstract This paper proposes a hybrid optimization algorithm combining particle swarm optimization (PSO) with genetic algorithm (GA) to update a railway bridge. PSO is an evolutionary optimization algorithm based on global search techniques to look for the best solution. Nevertheless, since PSO relies crucially on the quality of initial particles, it may reduce its effectiveness and robustness in tacking optimization issues. If the positions of initial populations are too far from the global best, it is challenging to determine the best solution. To overcome these shortcomings, we propose a hybrid optimization algorithm applying the advantages of both PSO and GA. GA is first used to generate the most elite populations based on its crossover

H. Tran-Ngoc · H. Nguyen-Tran · M. Abdel Wahab
Department of Electrical Energy, Metals, Mechanical Constructions, and Systems, Faculty of Engineering and Architecture, Ghent University, 9000 Ghent, Belgium
e-mail: hoa.tran@ugent.be

H. Nguyen-Tran
e-mail: nthieu@utc.edu.vn

M. Abdel Wahab
e-mail: magd.abdelwahab@ugent.be

H. Tran-Ngoc · T. Le-Xuan · T. Bui-Tien (✉)
Department of Bridge and Tunnel Engineering, Faculty of Civil Engineering, University of Transport and Communications, Hanoi, Vietnam
e-mail: btthanh@utc.edu.vn

T. Le-Xuan
e-mail: thang151113016@st.utc.edu.vn

H. Ho-Khac
Department of Transportation Nghe An, Le Hong Phong street, Vinh city, Nghe An, Vietnam
e-mail: hokhachanh@gmail.com

H. Nguyen-Tran
Department of Network and Information Systems, Faculty of Information Technology, University of Transport and Communications, Hanoi, Vietnam

G. De Roeck
Department of Civil Engineering, KU Leuven, B-3001 Leuven, Belgium
e-mail: guido.deroeck@kuleuven.be

M. Abdel Wahab (ed.), *Proceedings of 1st International Conference on Structural Damage Modelling and Assessment*, Lecture Notes in Civil Engineering 110,
https://doi.org/10.1007/978-981-15-9121-1_2

and mutation characteristics. Those populations are then employed to seek the best solution based on the global search capacity of PSO. The experimental measurements of the railway bridge are carried out under ambient vibrations used to validate the proposed algorithm (PSO-GA). The result demonstrates that PSO-GA, GA, and PSO possibly determine uncertain parameters of the bridge exactly and PSO-GA surpasses GA alone and PSO alone in terms of convergence level and accuracy.

Keywords Model updating · Railway bridge · Hybrid optimization algorithm

1 Introduction

During operation, structures may be damaged due to many different reasons e.g., natural hazards (storm, flood, earthquake…) or human impacts (overload, collision). Those adverse influences not only reduce operational effectiveness but also shorten the lifespan of structures. Vibration-based structural health monitoring (VBSHM) is a non-destructive structural health monitoring (SHM) method that can determine uncertain parameters changing over time such as boundary conditions, or material properties, and possibly detects damages happening in the structures as soon as possible. SHM has become common over the past decades [1–7]. Tran-Ngoc et al. [8] analyzed different connection conditions (rigid, pin, and semi-rigid) of truss members of a railway bridge in Vietnam to determine the stiffness of truss joints. They demonstrated that the semi-rigid connection could present structural dynamic behavior with high accuracy. Wu et al. [9] applied model updating to a highway bridge to identify uncertain parameters of the bridge based on measured results utilizing optical fiber sensors. Uncertain parameters of a railway bridge were determined and updated in Ref. [10] applying structural dynamic characteristics with the ambient excitation sources of a trainload and wind. El-Borgi et al. [11] conducted SHM of a reinforced concrete bridge using both theoretical analysis and an improved frequency decomposition method. The objective function comprised natural frequencies and mode shapes. Bayraktar et al. [12] dealt with the inverse problems to determine unknown parameters of a large-scale concrete bridge. The author used natural frequencies of the first 10 modes (the objective function) to reduce the discrepancies be-tween calculated and measured results.

Evolutionary optimization algorithms have been developed using global search techniques that assist in avoiding local minima problems [13, 14]. PSO also belongs to evolutionary optimization algorithms applied successfully in many fields [15]. This algorithm is based on the evolutionary rules of nature's creatures such as birds, fish, even humans looking for food. Kaveh and Maniat [16] employed PSO to detect damages in beam and truss structures. The results showed that PSO could accurately determine different damage scenarios occurring in the tested structures although noise and incomplete data were fully considered. Seyedpoor [17] used PSO combined with a strain energy function for damage identification in structures. Khatir and Wahab [18] employed PSO coupled with eXtended IsoGeometric Analysis (XIGA) to identify damages in plate structures. The result showed that the proposed combination not only detected crack locations of the considered structures exactly but also was superior to PSO alone and XIGA in terms of accuracy and computational time.

However, PSO has considerable disadvantages, because it relies crucially on the quality of initial particles. If those populations produce bad results, it can lead to the failure of looking for the best solution. Therefore, in this paper, GA with mutation and crossover capacities is coupled PSO to deal with optimization issues. To consider the effectiveness of PSO-GA, a railway bridge in Vietnam is employed. PSO-GA, GA together with PSO are used to determine unknown parameters consisting of the stiffness of truss joints, the stiffness of bearings, and Young's modulus of truss members of the bridge. The results obtained from the three algorithms are compared with each other in terms of accuracy and computational cost.

Apart from the introduction part, the structure of this paper is split into 5 main sections. Section 2 introduces PSO, GA, and PSO-GA. After that, Sects. 3 and 4 describe the bridge and how the measurement was carried out. Model updating is presented in Sect. 5. Finally, the most vital conclusions are summarized.

2 A Hybrid Algorithm (PSO-GA)

2.1 PSO

PSO is an optimization algorithm that operates based on global search techniques. This algorithm is developed by Kenedy and Maniat [16] in 1905 that imitates the strategy of animals such as fishes or birds looking for foods. Those animals often work in a group and frequently communicate with each other in the process of seeking food. This strategy helps them to know where to contain the most food, ignoring places without potential. Thanks to that, the opportunity for looking for food is higher and the moving time is reduced. The working principle of PSO is based on two main equations.

The first equation is applied to determine the position of particles:

$$P_{i+1} = P_i + V_{i+1} \tag{1}$$

The second equation is applied to identify the suitable velocity for each individual. With individuals far away from the best solution, their velocity has to increase, whereas individuals having positions near the best solution will keep a stable velocity.

$$V_{i+1} = w * V_i + C_1 * t_1 * (LB_i - P_i) + C_2 * t_2 * (GB_i - P_i) \tag{2}$$

Here, $P_{i,} V_i$ and $P_{i+1,} V_{i+1}$ are positions and velocities of elements at step i and i+1, respectively. C_1 and C_2 are learning coefficients, whereas t_1, t_2 are random numbers.

Fig. 1 PSO

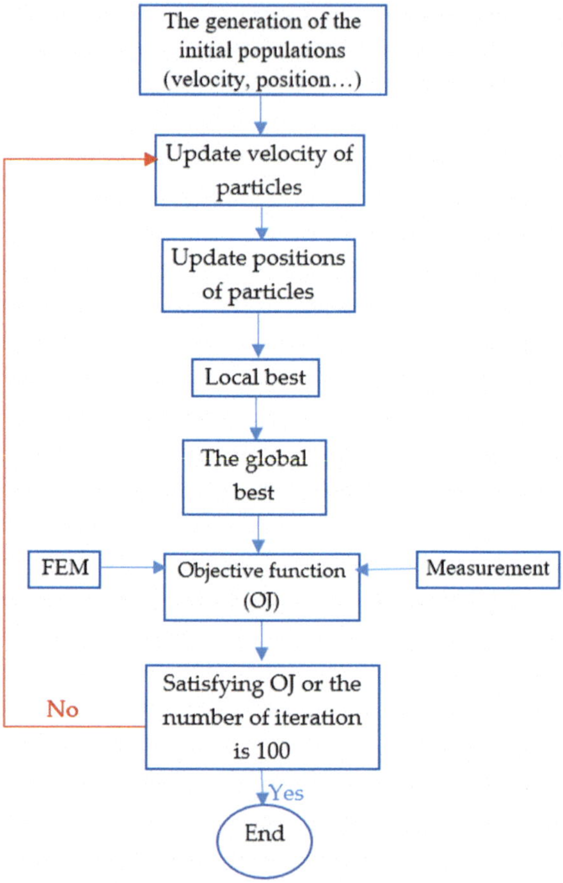

LB_i and GB_i are the best solutions in the local and global areas. w indicates the inertia weight parameter.

The main steps applying PSO to deal with optimization problems are described as Fig. 1.

2.2 GA

As natural law, to survive and thrive under the competitive pressure for food and shelter, creatures need to evolve increasingly better and be more in tune with their habitats. This means that the quality of the next generations is usually better than the previous ones. Only the advantages of the previous generations will be kept for the next ones and the disadvantages will be excluded. This will bring a higher chance for them to grow and survive. Inspired by the natural evolution theory of Darwin,

Mitchell et al. [19] proposed the first GA in 1975. After that other researchers have developed GA and made it become more effective and possibly apply for a wide range of fields. The operation of GA is based on three main features as follows:

1. Selection: This process is conducted to select the best populations for the next generations, and individuals with inferior quality will be removed.
2. Crossover and mutation: advantaged of parents are selected and exchanged to generate new offspring's and populations can mutate to improve quality.
3. Validation: if targets are achieved, the process will be finished. By contrast, if the quality of the new generations does not guarantee requirements, steps 1 and 2 are repeated.

The main steps applying GA to tackle optimization problems are described as Fig. 2.

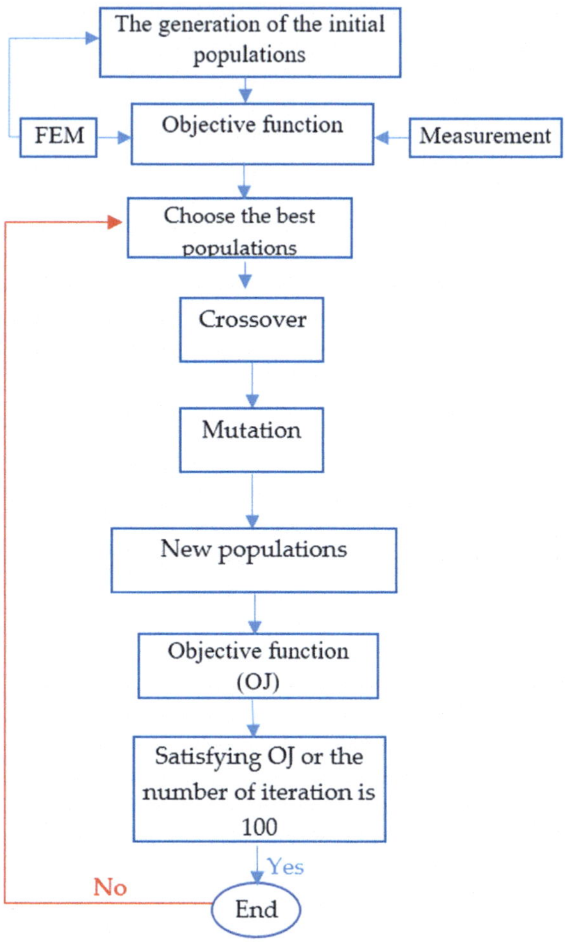

Fig. 2 GA

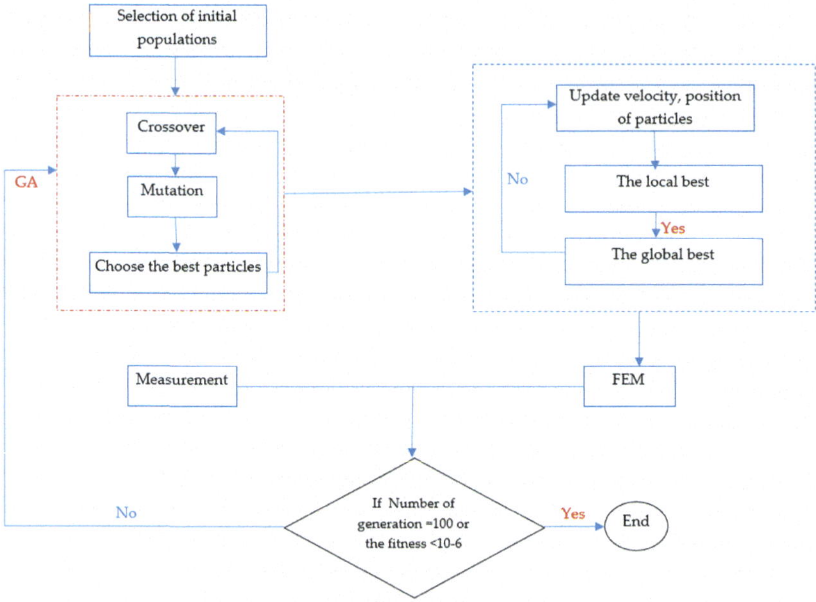

Fig. 3 PSO-GA

2.3 *PSO-GA*

PSO has disadvantages of depending too much on the quality of the initial population. If initial populations are far away from the best solutions, it may be difficult for them to look for the global best. To remedy those shortcomings, in this paper, GA is proposed to combine with PSO to deal with optimization problems. GA with crossover and mutation characteristics may improve the quality of populations after iterations. PSO with global search capacity, and fewer parameters that have to change, not only generates higher opportunities to look for the best solution but also reduces computational time. The combination of PSO and GA is illustrated in Fig. 3.

3 Description of the Bridge

Nam O bridge is located in the middle of Vietnam playing an essential role in connecting the railway traffic from the South to the North. The bridge was not only built and operated in a long time (since 1950), but also has been carrying the heavy loads of trains with a high frequency (at the moment, there is only one railway line connecting from South to North in Vietnam). Therefore, the Nam O bridge has undergone degradation (the reduction in stiffness caused by some defects such as rust appearing in some truss members). Fortunately, the bridge can still operate

Fig. 4 Nam O bridge

according to designed loads. The bridge includes four spans with almost the same length (75 m). Two abutments are U-shaped abutments, whereas piers are solid shaft piers commonly used for railway bridges. Roller and pin bearings are used to support spans. The layout of the bridge is shown in Fig. 4.

The main structure of Nam O bridge consists of upper chords, lower chords, vertical chords, diagonal chords, upper wind bracings, lower wind bracings, struts, and stringers. For more detail of truss member's dimensions, the readers can refer to Ref. [8].

4 Experimental Measurements

Since the bridge has been operated in a long time under the effect of repeated heavy loads of trains with a high frequency, a measurement campaign for monitoring the bridge was conducted in 2015. In the measurement process, the trains still operated normally. The movement of the trains and the effect of wind played a role as vibration excitation sources. The time for one setup lasted approximately 30 min, including the time for relocating sensors and other equipment. The sampling frequency is 1651 Hz. To obtain the mode shapes of many modes as possible, sensors are arranged to cover the dynamic behavior of the whole bridge. However, because Nam O is a large-scale bridge, the number of available sensors (ten sensors) is not enough. Six sensors had to use as reference sensors (roving sensors), and only four sensors were fixed. Totally, there are eight setups. Considering the characteristics of a truss bridge in general,

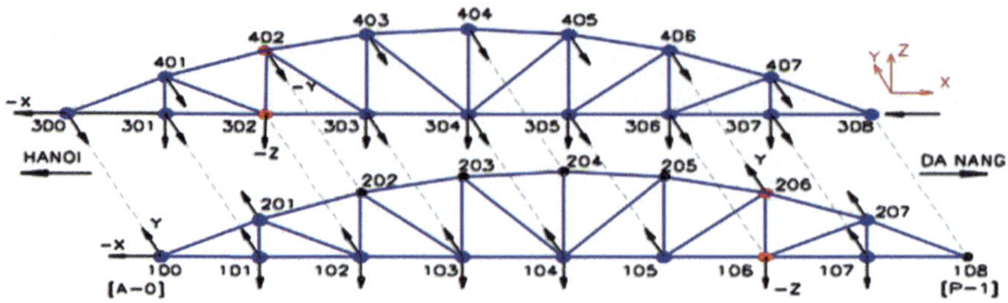

Fig. 5 The diagram of equipment arrangement

and Nam O in particular (without handrail at the upper chords), sensors could only be arranged at nodes of lower chords. Figure 5 describes the equipment arrangement.

For more information about equipment setup, data collection, and data processing, readers can find detail in Ref. [8].

5 Model Updating

To monitor the structural health, determine physical behavior, and identify uncertain parameters of the bridge, model updating was conducted. Uncertain parameters include Young's modulus of truss members, the stiffness of truss joints and the stiffness of bearings. The objective function consists of natural frequencies and mode shapes of the first five modes as shown in Eq. 3.

$$\text{Fitness} = \sum_{i=1}^{5}\left[1 - \frac{\left(\widetilde{\varphi_i^T}.\varphi_i\right)2}{(\varphi_i^T.\varphi_i).(\widetilde{\varphi_i^T.\varphi_i})}\right] + \sum_{i=1}^{5}\left(\frac{f_i - \widetilde{f}_i)^2}{\widetilde{f}_i^2}\right. \tag{3}$$

$((f_i, \varphi_i), (\widetilde{f}_i, \widetilde{\varphi}_i))$ are analytical and experimental natural frequencies, mode shapes, respectively, "*i*" is the modal order, and *T* denotes a transposed matrix. To deal with model updating problems of the bridge, PSO-GA, PSO, and GA are applied. For PSO-GA, learning coefficients C_1, and C_2 are 2, the inertia weight parameter (w) is 0.3. Crossover and mutation coefficients are 0.8 and 0.1, respectively. For PSO and GS, parameters are selected the same as those of PSO-GA. The number of population of all three algorithms is 50.

Figure 6 shows that the convergence level of PSO-GA is the highest, at 0.045, whereas those of PSO and GA are nearly 0.06 and 1.2, respectively. It can be understandable because PSO-GA applies both the advantages of PSO and GA to deal with optimization issues. Selection, crossover, and mutation features of GA are employed to generate the best populations and then the global search technique of PSO is applied to determine the global best. Table 1 shows the natural frequencies of modes before model updating, after model updating using PSO-GA, PSO, and GA.

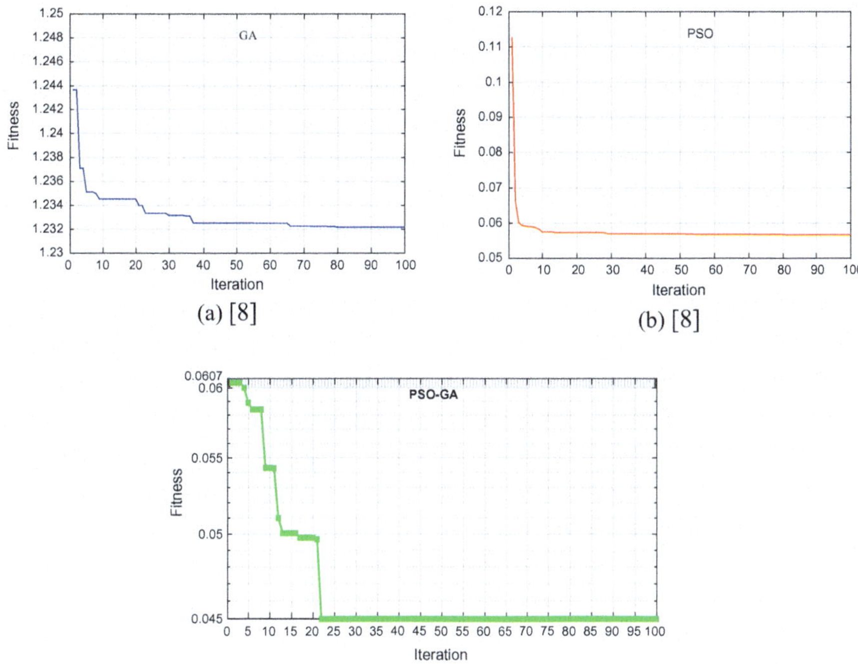

(a) [8] (b) [8]

(c)

Fig. 6 Fitness toleration **a** GA; **b** PSO; **c** PSO-GA

Table 1 Natural frequencies of ten modes before and after model updating

Mode	Before model updating (Hz)	GA (Hz)	PSO (Hz)	PSO-GA (Hz)	Measurement (Hz)
1	1.47 (1.4%)	1.47 (1.3%)	1.450 (0%)	1.45 (0%)	1.45
2	3.14 (1%)	3.06 (1.6%)	3.106 (0.3%)	3.10 (0.3%)	3.11
3	3.32 (1.2%)	3.29 (0.3%)	3.267 (0.32%)	3.27 (0.3%)	3.28
4	4.80 (3.7%)	4.70 (1.7%)	4.669 (0.8%)	4.66 (0.8%)	4.62
5	6.96 (13%)	6.53 (7.3%)	6.55 (7.6%)	6.47 (6.9%)	6.05
6	7.21 (1.35%)	7.11 (0.2%)	7.15 (0.4%)	7.13 (0.14%)	7.12
7	7.50 (2.74%)	7.35 (0.7%)	7.33 (0.5%)	7.31 (0.13%)	7.30
8	8.33 (14,1)	8.21 (10%)	8.10 (8.57%)	8.05 (7.91%)	7.46
9	9.18 (10.81%)	9.05 (9.2%)	9.00 (7.94%)	8.89 (7.23%)	8.29
10	9.79 (10.16%)	9.64(8.5%)	9.57 (7.10%)	9.42 (5.96%)	8.89

Table 1 shows that after updating the model using all three algorithms GA, PSO, and PSO-GA, the difference between the measured and calculated natural frequencies is reduced to less than 10%. Results from PSO-GA are superior to those of GA and PSO. The MAC values in Fig. 7 are also higher than 0.9. That shows a close correlation between the numerical model and the experimental measurements. Uncertain parameters are determined as in Table 2.

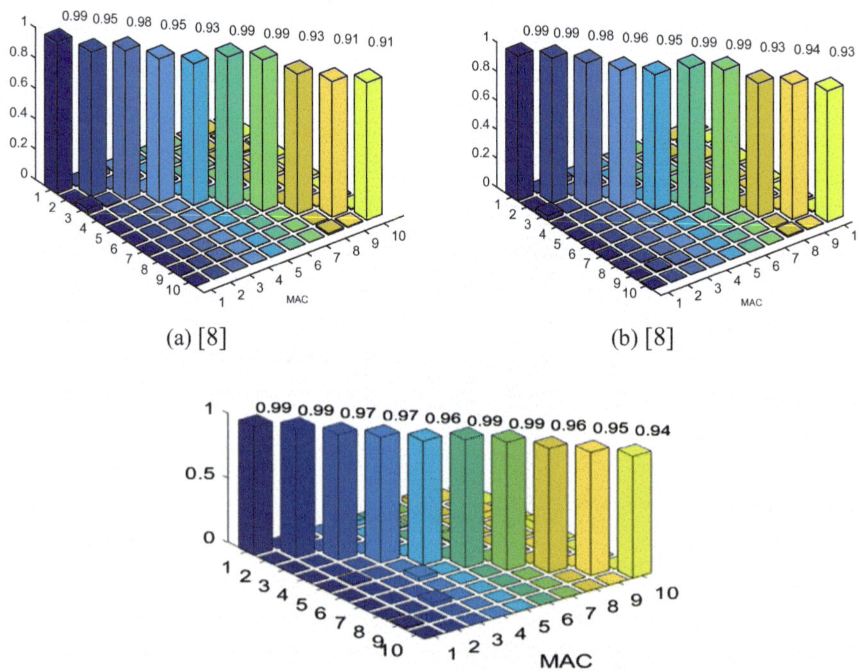

(a) [8] (b) [8]

(c)

Fig. 7 Mac values; **a** GA; **b** PSO; **c** PSO-GA

Table 2 Uncertain parameters

	K_1	K_2	K_3	K_4	K_5	K_6	K_7	E
Lower bound	1.0	1.0	1.0	1.0	1.0	1.0	7	190
GA	1.27	1.22	1.19	1.21	1.45	1.38	7.8	199
PSO	1.20	1.16	1.12	1.16	1.40	1.33	7.6	198
PSO-GA	1.18	1.14	1.10	1.14	1.38	1.31	7.5	197
Upper bound	2.0	2.0	2.0	2.0	2.0	2.0	9	220

Note unit of k_1, k_2, k_3, k_4, is 10^{10} N/m, unit of k_5, k_6 is 10^7 N/m, unit of k_7 is 10^5 N.m/rad, unit of E is GPA

6 Conclusion

The paper proposes a hybrid algorithm to monitor the health of a railway bridge. This algorithm is a combination of PSO and GA algorithms. PSO is an algorithm that works on the principle of global search, but its disadvantage is the dependence on the quality of the original populations. Moreover, populations work independently of each other, as well as no crossover, and mutation features can be applied to improve quality. Therefore, in this paper, GA with selection, crossover, and mutation features is proposed to improve PSO. To evaluate the effectiveness of PSO-GA, PSO and GA are also applied to update the model and identify uncertain parameters of the bridge. From the results, some conclusions have been drawn as follows:

- GA, PSO, and PSO-GA provide a good convergence and the difference between analysis and measurement results is significantly reduced after model updating.
- PSO-GA outperforms GA and PSO in terms of convergence and accuracy.
- To validate the robustness of PSO-GA, it is necessary to apply it to deal with other optimization issues.

Acknowledgements The authors acknowledge the financial support of VLIR-UOS TEAM Project, VN2018TEA479A103, 'Damage assessment tools for Structural Health Monitoring of Vietnamese infrastructures' funded by the Flemish Government. Moreover, the first author needs to acknowledge the financial supports from Ministry of Education and Training (MOET) under the project research "B2020—GHA—02" and Bijzonder Onderzoeksfonds (BOF) of Ghent University.

With the support of THE BELGIAN DEVELOPMENT COOPERATION

References

1. Nguyen DH, Bui TT, De Roeck G, Wahab MA (2019) Damage detection in Ca-Non Bridge using transmissibility and artificial neural networks. Struct Eng Mech 71:175–183
2. Tran-Ngoc H, Khatir S, De Roeck G, Bui-Tien T, Nguyen-Ngoc L, Wahab MA (2019) Stiffness identification of truss joints of the nam o bridge based on vibration measurements and model updating. In: International conference on arch bridges. Springer, Cham, pp 264–272
3. Hoa TN, Khatir S, De Roeck G, Long NN, Thanh BT, Wahab MA (2020) An efficient approach for model updating of a large-scale cable-stayed bridge using ambient vibration measurements combined with a hybrid metaheuristic search algorithm. Smart Struct Syst 25(4):487–499
4. Tran-Ngoc H, He L, Reynders E, Khatir S, Le-Xuan T, De Roeck G, Bui-Tien T, Wahab MA (2020) An efficient approach to model updating for a multispan railway bridge using orthogonal diagonalization combined with improved particle swarm optimization. J Sound Vibration 11531
5. Khatir S, Boutchicha D, Le Thanh C, Tran-Ngoc H, Nguyen TN, Abdel-Wahab M (2020) Improved ANN technique combined with Jaya algorithm for crack identification in plates using XIGA and experimental analysis. Theor Appl Fracture Mech 102554
6. Ho VL, Tran NH, De Roeck G, Bui TT and Wahab MA (2018) System identification based on vibration testing of a steel I-beam. In: International conference on numerical modelling in engineering. Springer, Singapore, pp 254–268

7. Hoa TN, Thanh BT (2019) Damage detection in a steel beam structure using particle swarm optimization and experimentally measured results
8. Tran-Ngoc H, Khatir S, De Roeck G, Bui-Tien T, Nguyen-Ngoc L, Abdel Wahab M (2018) Model updating for Nam O bridge using particle swarm optimization algorithm and genetic algorithm. Sensors 18(12):4131
9. Wu B, Lu H, Chen B, Gao Z (2017) Study on finite element model updating in highway bridge static loading test using spatially-distributed optical fiber sensors. Sensors 17(7):1657
10. Feng D, Feng MQ (2015) Model updating of railway bridge using in situ dynamic displacement measurement under trainloads. J Bridge Eng 20(12):04015019
11. El-Borgi S, Smaoui H, Cherif F, Bahlous S, Ghrairi A (2004) Modal identification and finite element model updating of a reinforced concrete bridge. Emir J Eng Res 9:29–34
12. Bayraktar A, Altunisik AC, Sevim B, Turker T (2010) Finite element model updating of Kömürhan highway bridge based on experimental measurements. Smart Struct Syst 6(4):373–388
13. Tran-Ngoc H, Khatir S, De Roeck G, Bui-Tien T, Wahab MA (2020) Damage assessment in beam-like structures using cuckoo search algorithm and experimentally measured data. In: Proceedings of the 13th international conference on damage assessment of structures. Springer, Singapore, pp 380–385
14. Tran-Ngoc H, Khatir S, De Roeck G, Bui-Tien T, Wahab MA (2019) An efficient artificial neural network for damage detection in bridges and beam-like structures by improving training parameters using cuckoo search algorithm. Eng Struct 199:109637
15. Kennedy J, Eberhart R (1995) Particle swarm optimization. In: Proceedings of ICNN'95-international conference on neural networks, vol 4. IEEE, pp 1942–1948
16. Kaveh A, Maniat M (2015) Damage detection based on MCSS and PSO using modal data. Smart Struct Syst 15(5):1253–1270
17. Seyedpoor SM (2012) A two stage method for structural damage detection using a modal strain energy based index and particle swarm optimization. Int J Non-Linear Mech 47(1):1–8
18. Khatir S, Wahab MA (2019) A computational approach for crack identification in plate structures using XFEM, XIGA, PSO and Jaya algorithm. Theor Appl Fract Mech 103:102240
19. Mitchell M, Holland JH, Forrest S (1994) When will a genetic algorithm outperform hill climbing. Adv Neural Inf Process Syst 51–58

Degradation of Concrete Resistance: Analysis of a Homogeneous Area. The City of Caserta

Ersilia Biondi and Giorgio Frunzio

Abstract The concept that concrete degrades over time has now been acquired and this leads to a reduction of the resistance. In the case under consideration, an attempt is made to establish a relationship between the time elapsed from construction to the test date with the same environmental conditions, as all samples from the same geographical area were examined. A few dozen results are analyzed, in terms of compressive strength and an attempt is made to draw a relationship between time and reduction of resistance. Where the level of carbonation penetration is also available, an attempt is also made to find a correlation between carbonatation and reduction of resistance in the time. Resistance is measured on samples from the extraction of cores from existing public and private buildings and built over different periods of time. The investigations were carried out in the time interval between 2009 and today for various purposes requested by the owners at the TecnoLab laboratory in Naples which allowed us to access their archive, not disclosing personal customer data. The resistance was assessed both through the direct crushing of the extracted samples and through evaluations with non-destructive methods such as the combined SonReb tests.

Keywords Concrete · Carbonatation · Degradation · Inspection · Mechanical resistance

1 Introduction

For a long time, scholars and researchers of various nationalities and backgrounds, academics and engineers have been dealing with the problem of the deterioration of reinforced concrete and its durability over time [1–7]. Various international and Italian regulations have followed [8–12], following disasters and consequences to the emergence of structural problems that have become objects of study [13–18].

E. Biondi (✉) · G. Frunzio
Università Degli Studi Della Campania Luigi Vanvitelli, Caserta, Italy
e-mail: ersilia.biondi.a@gmail.com

M. Abdel Wahab (ed.), *Proceedings of 1st International Conference on Structural Damage Modelling and Assessment*, Lecture Notes in Civil Engineering 110,
https://doi.org/10.1007/978-981-15-9121-1_3

The article focuses on the buildings being studied in reinforced concrete, subjected to compression testing and to the thickness of the carbonation to evaluate the performance of the material over time, starting from the date of construction as the first key element, to the test. Structural carried out after a few decades. Data were acquired on public buildings, in particular on schools and other services, since they can be found at public administration offices. It was noted, first of all, that most of the buildings that could have been studied did not present documentation necessary for the purpose of structural knowledge, as well as other tests. The search for school buildings in the town built over several generations is highlighted.

1.1 History of the City

The city of Caserta, located in southern Italy a few kilometers from the famous city of Naples, boasts a history worthy of attention as a symbol of the monarchical power exercised by the Bourbon domination at the time of the Kingdom of the Two Sicilies, when Italy still it was not a united nation. While in France the bourgeois Enlightenment of Voltaire and Montesquieu under Louis XIV was advancing, in England the foundations were laid for the Industrial Revolution, in Germany there had been the Peace of Westphalia—on the Italian peninsula there were several dominations, among which the most powerful was the Spanish who conquered the south. In Naples and Caserta, in particular, the focus of attention deserves the urbanization and the urban layout that have materialized since this period thanks to the rulers who had the ambition to enhance the splendor and the artistic avant-garde of the two cities.

The credit for the identity that distinguishes the city of Caserta goes to the revolutionary and tireless father of neoclassical architecture, the architect and engineer Luigi Vanvitelli. The Royal Palace was born as a still primitive neoclassical style, in step with the enlightenment concept in vogue in France. Vanvitelli was responsible for the birth of a more modern urban concept starting from 1751 when the Royal

Fig. 1 **a** Aerial view of the Royal Palace of Caserta and, **b** aerial view of Garden of The Royal Palace

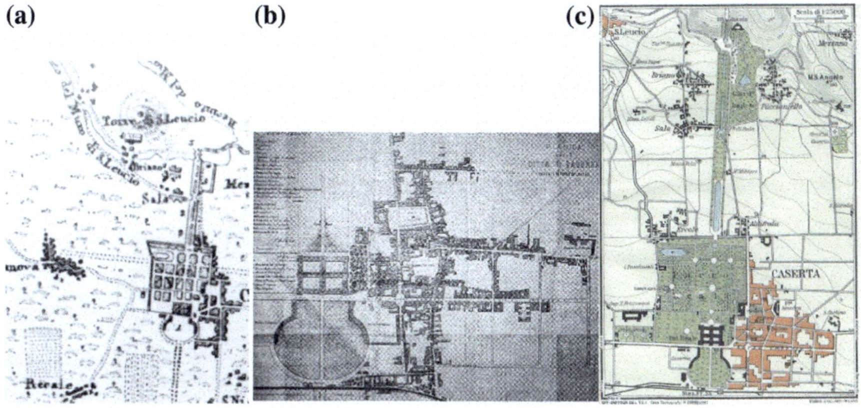

Fig. 2 **a** Caserta. City map_1780, **b** Caserta. City map_1879 and **c** Caserta. City map_1924

Palace of Caserta was built, the largest Royal Palace in the world that changed architecture. Figures 1a, b represent aerial view of Royal Palace and the Garden of The Royal Palace, respectively.

The construction of the Royal Palace had started a few years when the first settlements adjacent to it began to be born, along the right side of Palace and along the backbone that culminates with the waterfall at the end of the huge garden—the so-called districts of San Leucio, Sala and Briano, famous for the presence of the Royal Seteria of the Monumental Complex associated with the entire Bourbon complex, a UNESCO heritage site. In Fig. 2 you can see the expansion of the city on different dates.

1.2 History of Buildings

The expansion slowly proceeds until the early 1800s, when the right side of the palace to be urbanized is always preferred, becoming the ancient heart of the city, still a restricted monumental area. In the mid-800s the road layouts begin to be glimpsed, up to the map of 1879 which outlines a now identified city, constant until 1924, during the World War, thanks to the map owned by the Allies. Figure 3a, b show an example of reinforced concrete construction of the time.

The buildings considered in this work began to stand up after the Second World War. A particular building already appears on the 1956 map. This is the Sant'Anna Clinic, historic in the city also for its central location, seen in Fig. 4. In 1960 the urbanization of the city of Caserta grew dramatically: the most important residential districts were built, which today are still a reference point for citizenship. Figure 5 shows the buildings built before 1960. All the case studies of the city are highlighted in yellow on Fig. 6, mapped by google earth.

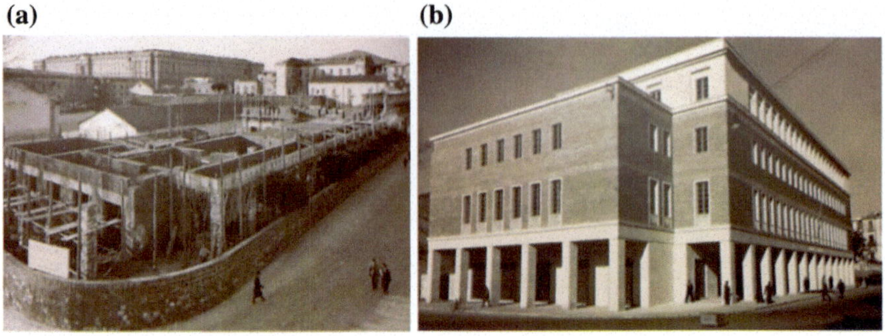

Fig. 3 a. Genio Civile Palce: example of work in progress of a reinforced concrete building. **b** Genio Civile palace; 60s

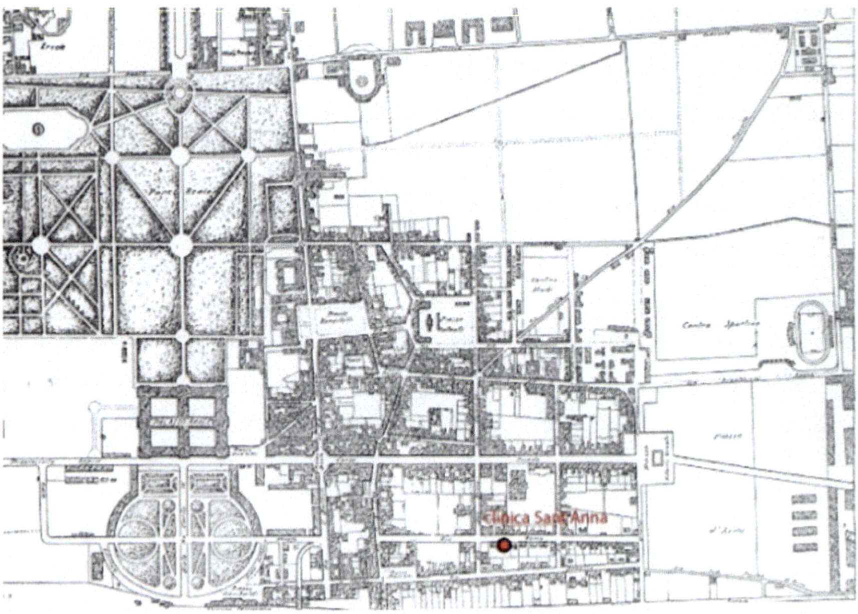

Fig. 4 Caserta. City map_1956. Clinic in red (first case study existing)

2 Study Cases

The following are the buildings studied and mapped in Fig. 7:

- Clinic Sant'Anna: 50s
- Primary School Tuoro: 50s
- Municipal Police Headquarter St. Gobain: 50s
- Ex former St. Gobain: 50s

Fig. 5 Caserta. City map_1960. Some cases study in red

Fig. 6 Caserta. City map_2020 (Google earth) and all cases study in yellow

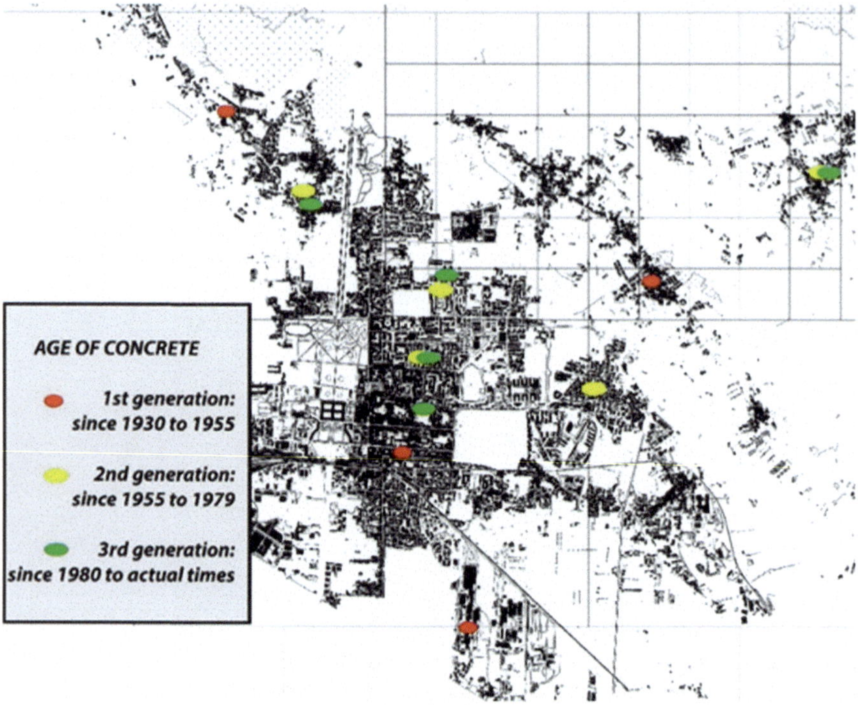

Fig. 7 Caserta. City map¬_2020. 3 generations of concrete with classification of buildings by their age

- Primary School Patturelli: 50s
- Nursery and Primary School Lorenzini: 1960
- Nursery School S. Barbara: 1962/1965
- Art Institute San Leucio_build. A: 1962/1964; B: early 90s; C: early 90s; D: mid 90s
- Nursery and Primary School Lorenzini plexus Casola: 1973
- High School Manzoni: 1973/1974
- Surveyor Institute Buonarroti: 1980/1982
- Nursery School Rione Tescione: 1990/1992
- Parking S. Carlo: 2012.

The buildings have been mapped on the city territory: the oldest ones can already be seen on the maps of the time; others are located on the recent map. Several city neighborhoods are involved. The buildings were built hand in hand with the expansion of the neighborhood in which they are located, forming a complete fabric of public and private buildings: it can be assumed that the materials of origin and, consequently, their own pretensions, are consistent with each other. Taking these buildings as a sample allows us to study how the mechanical performance of concrete varies over time and therefore estimate its durability.

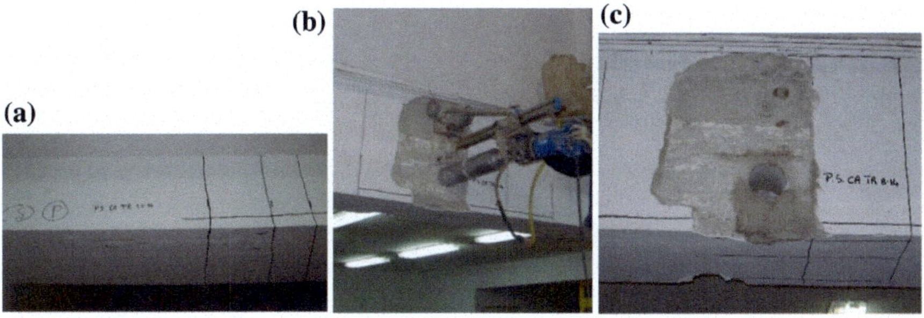

Fig. 8 The **a** Pacometric test to find reinforcement, **b** Sample extraction and **c** The baem after sample extraction

3 Materials and Methods

The tests on the cementitious material were carried out through various destructive and non-destructive tests[1] [19].

The survey methodology common to all the structures studied is coring. The execution of the destructive test is regulated by the UNI EN12504-1 standard [20]. Through it, a direct evaluation of the strength of the concrete is obtained through the site extraction of cylindrical elements. The extraction of the coring was carried out after pacometric investigation which identified the reinforcement in order to indicate the areas to be avoided in the extraction. It is to be avoided the presence of the reinforcement for simple construction safety reasons, to avoid reducing its resistance and altering the values of the compression test of the specimen itself. According to UNI standards, coring has a diameter equal to 3 times the maximum diameter of the aggregates specific to the material. It should be noted that the sample of coring taken in situ, subjected to a axial-load test in laboratory, has a lower resistance than the resistance of samples cured in laboratory at standard conditions.[2] This both depends on the difference in conditions of the installation and subsequent maturation, and on the disturbances to which the samples are subjected while they are extracted. In Fig. 8 are shown three phases of extraction of cylindrical samples from a beam.

Pacometric test is a non-destructive test, regulated by BS 4408 standard [21] and by Eurocode2, based on the "eddy currents". The pacometer detects the location of the reinforcement, the size of the concrete cover and the diameter of the bars. This investigation is preparatory to any other type of subsequent diagnostic methodology used, such as coring, sclerometric and ultrasonic. In Fig. 9 is shown the compression test in laboratory.

The SONic REBound (SONREB) method is a non-destructive test, characterized by the combination of the data of the sclerometric and ultrasonic tests, regulated

[1] Bossio A. (2014) Corrosione e diagnostica delle strutture in calcestruzzo armato. Criteri di progettazione di intervento di ripristino conservativo. Wolters kluwer Italia 2014.

[2] UNI EN 12,504–1, Prove sul calcestruzzo nelle strutture. Carote, prelievo, esame e prova di compressione. 2002.

(a) (b) (c) (d)

Fig. 9 The **a, b, c** and **d** Core in the laboratory: treatment, compression test and final result

by UNI EN 12504-2:2001 [22] and UNI EN 12504-4:2005 [23] standards. This allowed to determine the resistance of the concrete to outline the picture of the deterioration that occurs. This is a somewhat advantageous method, since it neglects the humidity and degree of maturation of the cement conglomerate; compared to the ultrasonic test, it allows the reduction of the granulometry of aggregates, dosages and any additives and, compared to the sclerometric method, it differentiates the quality between surface and depth in a lesser way. The ultrasonic test (UNI EN 12504-4) identifies defects, degradations and physical-performance characteristics. Waves propagate through the examined object that investigate elasticity and resistance, consequently detecting cracks, cavities and inhomogeneities. The sclerometric test (UNI EN 12504-2) allows dynamic tests in the site to assess the surface quality of the material according to the extent of rebound after a stop body. The method therefore combines, through empirical methods and mathematical algorithms, two parameters which are the propagation speed of the ultrasound waves and the rebound index. SONREB investigations were carried out according to RILEM 043-CN Recommendations [24]: the values correlate with the compressive strength according to law of variation:

$$Rc = A \times IB \times VC(MPa) \quad (1)$$

where Rc represent the compressive strength of concrete, I = average rebound index; V = average speed of ultrasound, $A = 7.695 * 10^{-11}$; B = 1.4; C = 2.6.

The resistances obtained with the SONREB method are compared with those obtained from the axial-load tests on the samples and, from this, it is possible to extend the results to other points of the structure, avoiding to extract other cylindrical samples. Carbonation penetration on concrete specimens need to be determined according to UNI EN 14630:2007 standard [25]. A solution of phenlphthalein with 1% ethyl alcohol was sprayed on the surface immediately after collection. The substance colored purple the surfaces where the pH was greater than 9.2;

remained colorless for lower pHs. To be specified, the well proportioned concrete has an alkaline pH (about 12.5) which undergoes the oxidation reactions of the reinforcement.

3.1 Results

Table 1 shows the results obtained by testing about seventy cylindrical samples.

4 Degradation of Concrete

Considered structures, despite being dated in several decades of the twentieth century and, therefore, having different ages at the time of the resistance and carbonation test, all show more or less evident and dangerous forms of degradation following the diagnostic investigations carried out in the site. Precise information on the characteristics of the cementitious material used in the initial construction phase is not available, but from the inspection of the coring we note how they are carbonated and, during the inspection, how the buildings present physical situations of evident degradation [26]. They depend on physical, chemical, accidental, technological and design causes[3]; overall, they affect the durability of the material over time. Concrete is porous and not homogeneous: it is not immediately clear how and when it will collapse or lose its mechanical characteristics. Of course, what primarily affects is the thermo-hygrometric, and therefore climatic, variation based on the geographical location and morphological stock. During their life, these buildings have been subjected to both climatic and geothermal and accidental weather. All are united by the phenomena of degradation of the corrosion of the reinforcements, evoked by inspections and photographic reports, and of carbonation, also examined in the laboratory.

4.1 Carbonation and Corrosion

The alkaline environment of a concrete properly packaged and proportioned in its "ingredients" inhibits the oxidation reactions of the reinforcement. On the other hand, concrete is permeable, therefore carbon dioxide penetrates its pores and starts the carbonation phenomenon. This phenomenon generates consequences such as steel dimensional variations, progressively leading to cracks. It happens that the cement pH

[3]Biondi E., Frunzio G. (2019) Analysis of degradation for the conservation of reinforced Concrete. WORLD heritage and legacy. Culture, Creativity, Contamination" Gangemi Edito-re, ISBN 978-88-492-3751-1.

Table 1 Test values

Test date	Case study building	Concrete core	Internal carbonation [cm]	Strength of concrete [Mpa]	σ compression [Mpa]	Breaking strength [KN]	Strength of concrete SONREB [Mpa]	Date of construction
2015	Clinic S.Anna	c1		33.35	30.24			Late 50s
		c2		34.00	30.03			
		c3		34.49	30.09			
2017	Primary school, Tuoro	c1		22.17			20.1; v (m/s) 20.1	Late 50s
		c2		14,96			16,06; v (m/s) 34,07	
2015	Municipal police headquarters, St. Gobain	C1	1.2	26.31	23.57		25.65	Late 50s
		C2	1	30.22	27.18		29.67	
		C3	1,1	27.70	24.82		25.51	
		C4	1,3	25.34	23.03		27.70	
		C5	1,4	25.80	23.52		26.63	
		C6	1.2	25.04	22.82		25.43	
2019	Ex Former St. Gobain	C1	2.5	30.69			30.8	Late 50s
		C2	2.7	29.05			30.01	
		C3	2.2	19.07				
2017	Primary school "Patturelli"	c1	3.25	25.21		179		Late 50s
		c2	7.82	26.82		190		
		c3	11.42	9.21		40		
		c4	10.23	14.16		61		
2017	Nursery and Primary School, Piazza Cavour	c1		17.46				1960

(continued)

Table 1 (continued)

Test date	Case study building	Concrete core	Internal carbonation [cm]	Strength of concrete [Mpa]	σ compression [Mpa]	Breaking strength [KN]	Strength of concrete SONREB [Mpa]	Date of construction
		c2		15.06				
		c3		16.05				
		c4		16.98				
		c5		17.81				
2017	Nursery school, S. Barbara	c1		28.29				1962/1965
		c2		27.81				
2010	Art Institute, San Leucio Building A	PI CA P22	1.3	31.70	25.28	198.52	30.27	1962/1964
		Parete 1	3.6	22.45	18.17	148.72		
		PS CA Tr8_14	3.6	43.32	37.63	295.53	42.35	
		PR CA P3	5.8	11.00	9.81	77.01	12.48	
		P1 CA Tr3_27	5.3	21.80	17.90	140.59	19.28	
		P2 CA P18	6.2	12.21	9.63	75.65	13.68	
2017	Nursery Primary School Lorenzini, Casola	c1	2.5	18.8		130		1973
		c2	12	16.97		118		
		c3	15	10.46		73		
		c4		18.13		126		
2010	High School, Manzoni	M1	5.5	13.5		55		1973/1974
		M2	5.8	13		53		
		M3	5.5	13.2		54		
		M6	4.8	13.5		55		

(continued)

Table 1 (continued)

Test date	Case study building	Concrete core	Internal carbonation [cm]	Strength of concrete [Mpa]	σ compression [Mpa]	Breaking strength [KN]	Strength of concrete SONREB [Mpa]	Date of construction
		M8	3	13.8		56		
		M9	4	14.2		58		
		M10	3.5	14.2		58		
		M11	4.5	13.1		53		
		M12	6.5	13.4		55		
		M13	2	12.9		53		
		M14	2.5	14.5		59		
		M15	5.5	14.4		59		
		M16	2	13.2		54		
		M17	3	12,9		53		
		M18	3	12.8		52		
		M19	3.5	12.9		53		
		M20	0	13		53		
		M21	5	13.5		55		
		M22	1.5	13,3		54		
		M4		13.8		172		
		M5		17.8		215		
		M7		20.4		246		
2010	Surveyor Institute, Buonarroti	C1		9.08		40.12	13.52	1980/1982
		C2		10.47		46.28	15.59	

(continued)

Table 1 (continued)

Test date	Case study building	Concrete core	Internal carbonation [cm]	Strength of concrete [Mpa]	σ compression [Mpa]	Breaking strength [KN]	Strength of concrete SONREB [Mpa]	Date of construction
		C3		9.52		42.06	14.18	
2017	Nursery school, Rione Tescione	c1	0.5	26.1		182.7		1990/1992
		c2	0.5	22.77		159.4		
		c3	0.5	22.46		157.2		
		c4	0.5					
2010	Art Instit, San Leucio Building B	PS CB Tr2*-3*	2.6	23.02	17.85	140.21	23.73	Early 90s
		PR CB P10*	2.5	33.66	26.37	207.11	32.20	
		P1 CB P11*	3	21.05	16.16	126.94	22.13	
		P2 CB P11*	3.3	12.34	9.69	76.14	12.49	
2010	Art Instit, San Leucio Building C	PT CC P4	1.1	15.06	13.17	103.46	15.53	Early 90s
		PTCC Tr40-41	0	30.61	25.02	196.42	30.17	
2010	Art Instit, San Leucio Building D	PT CD P48	1.6	17.91	15.48	121.61	17.91	Mid 90s
2016	Parking S. Carlo	C1	0	21.65	17.34			2012
		C2	0	24.26	20.65			
		C3	0	22.77	18.64			

lowers (in a healthy state the alkaline pH of the concrete is 13), up to 8.5. Gradually the carbonated layer descends into the cortical layer, penetrating deeper and deeper, until it meets the reinforcements. Here the corrosion of the armatures is triggered, when the aggressive agents come into contact following the formation of a corrosive cell, mainly responsible for the propagation. At the time of the inspections for subsequent diagnostic tests on the buildings in question, the load-bearing structures presented situations of this type.

Corrosion is dangerous as it reduces the section of the bars, increasing their deformations and decreasing their breaking strength and fatigue strength.[4] On the other hand, the cracking of the concrete occurs, the main danger of the anchoring of the reinforcement and consequently the cause of the total detachment of the material and the main accelerator of the corrosion rate [27].

Carbonation is triggered when calcium carbonate reacts with carbonic acid to form calcium bicarbonate. If it rains on a carbonated surface, the material produces diffuse erosion, losing cohesion and forming spongy incrustations of a color tending to yellow. This reduces the porosity of the concrete and increases the mechanical resistance. The negative consequences of carbonation affect the reinforcement because there is a change in pH and the steel is in contact with pure water with a pH of 11.5, the minimum value that ensures the conditions of passivity. It follows that the steel that oxidizes by carbonation increases its volume up to 5 times: there will be lateral pressures that damage the concrete, until the reinforcements are exposed towards the outside in contact with the atmosphere and aggressive agents. Rust appears instead of the resistant section of the material.

These are the cases involving coring examined by the buildings studied. The compressive strength properties are altered, therefore, by the problem of degradation that the structures present and this is demonstrated by the purple color on the surface of the coring after spraying the phenolphthalein and measured, with millimeter precision, the thickness concerned. Figure. 10 show the state of degradation of the elements.

5 Data Analysis

The data collected, as far as it relates to a single municipality, are not satisfactory as it is not possible to reconstruct the actual level of exposure of each sample in the time elapsed between the construction of the work and the test carried out.

Wanting to look for relationships between the carbonation parameter and the resistance seems to have greater significance to evaluate those buildings in which more data are available, regardless of the knowledge of the point where the same samples

[4]"[...] corrosion represents the most important form of degradation for materials and structures, both for wide diffusion and the amount of danger it presents". Bossio A., Fabbrocino F., GP DI Lignola, Monetta T., Bellucci F., Manfredi G., Prota A. (2016) Corrosion effects on seismic capacity of reinforced concrete structures.

https://www.degruyter.com/view/journals/corrrev/37/1/article-p45.xml

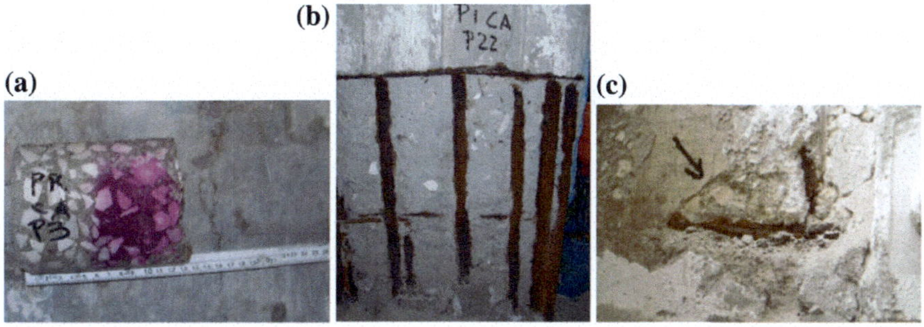

Fig. 10 **a** Carbonate thickness measurement, **b**. and **c** Corroded steel of the pillar

were taken, the hypothesis however acceptable, and that in the same construction homogeneous type of concrete was used during construction, without being able to better investigate the causes of different levels of carbonation found.

The following buildings are therefore examined:

- Nursery and Primary School Lorenzini plexus Casola: 1973,
- Primary School Patturelli: 50s,
- Municipal Police Headquarter St. Gobain: 50s,
- Art Institute San Leucio, build. A: 1962/1964.

For which the graphs that report on the abscissa axis the level of penetration of the carbonation in cm and on the ordinate axis the resistance assessed following the extraction of cores are summarized.

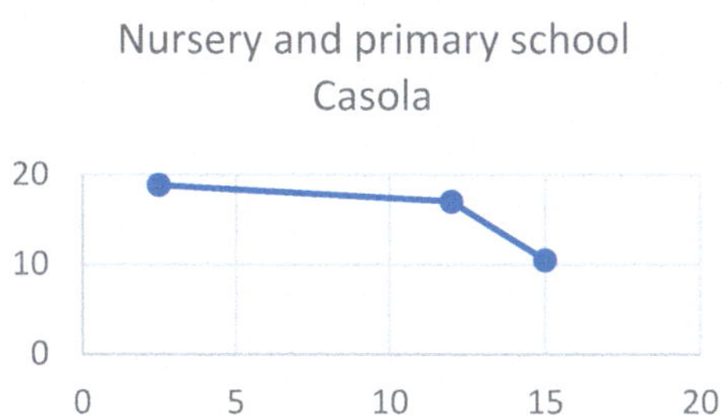

About 45 years after construction.

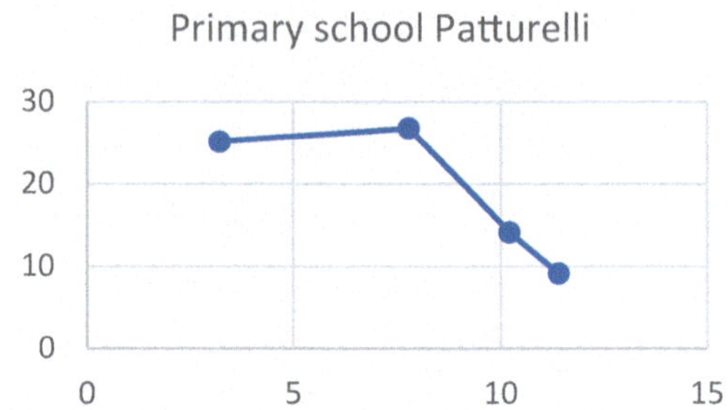

About 60 years after construction

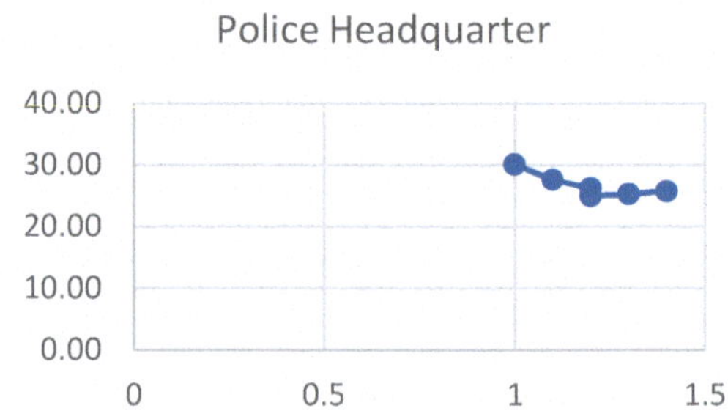

Over 60 years of construction

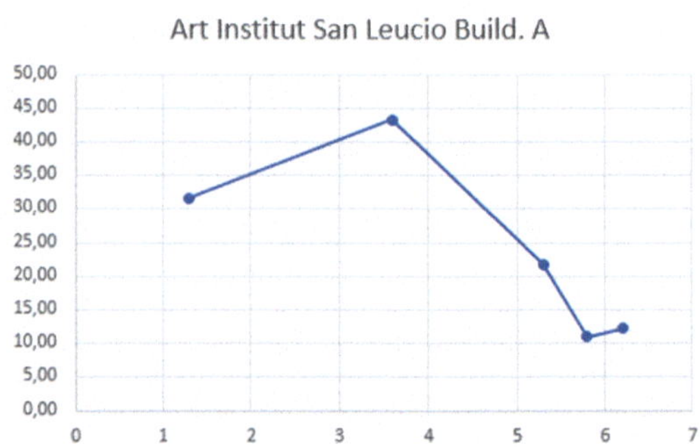

About 50 years after construction

6 Conclusions

In order to find a concrete relationship that regulates the phenomena analyzed, there are not too many data and, moreover, certain aspects are missing which are known to have fundamental importance in the phenomenon examined:

- Initial resistance value
- Protection of the structure over time or more generally the actual environmental conditions to which the concrete has been subjected during its life.

These are constructions referring to the second generation of the age breakdown of concrete. They are spaced a few years, someone closer to the first generation and someone close to the third. The data are heterogeneous but they, on a large scale, are not available in the documentation that has been obtained, so the aim is to examine, even with new test campaigns, which unfortunately in this period have not been possible, others and more numerous buildings in concrete.

References

1. Basile A, Frunzio G, Mattiello G (2015) Diagnosis of the critical state of concrete. In: Heritage and technology mind knowledge experience, vol 56, pp 1783–1791, Aversa
2. Faella E, Martinelli E, Salerno N (2008) Some considerations on non-destructive methods for de-termining the mechanical properties of concrete. Extract of the acts of the seventeenth century C.E. Rome
3. Collepardi (2002) The new concrete. Edizioni Tintoretto
4. Marcelli (2007) Complaints under construction: causes and solutions for damage and loss in the works
5. Pucinotti (2013) Assessment of the strength of the characteristic concrete in situ. In: Building and construction materials
6. Rossetti (2003) Concrete, materials and technology. McGraw-Hill
7. Ruggerone (2005) Structural diagnostics—analysis of the state of conservation of the structures, theory and practice of rehabilitation interventions
8. American Society for Testing and Materials ASTM, A. S. (2002) Test method for pulse velocity through concrete, ASTM C 597-02
9. Eurocode2, C (2003) General rules and rules for buildings (final draft). In: Design of concrete structures. Bruxelles, Belgium
10. Eurocode8, C (2004) Assessment and retrofitting of buildings (draft n. 6). In: Design of structures for earthquake resistance. Bruxelles, Belgium
11. Comité Euro-international du Béton CE-I (1989) Diagnosis and assessment of structures: state of the art report. Bulletin D´Information n° 192. Copenhagen
12. Primi elementi in materia di criteri generali per la classificazione sismica del territorio, 3274 (OPCM Ordinanza del Presidente del Consiglio dei Ministri Marzo 20, 2003)
13. Augenti (2003) The resistance of concrete in existing buildings. II National Conference on collapses and reliability of structures. Naples
14. Gonçalves (2015) Study of pathologies and their causes in reinforced concrete structures and construction works. Monograph, UFRJ. Rio de Janeiro
15. Masi (1991) Analysis of construction techniques and materials, in the manual for the as-sessment of earthquake safety and for seismic adaptation (coord. F. Braga), Ordine Ingegneri Potenza
16. Masi. Estimating the strength of concrete in situ by means of destructive and non-destructive tests. DiSGG, Università degli Studi di Basilicata, Potenza
17. Pucinotti (2005) Pathology and diagnostics of reinforced concrete. In: Non-destructive in-vestigations and coring in the works to be consolidated
18. Viola E., Pascale G, Di Leo A (1984) Core sampling size in N.D.T. of concrete structures. In: International conference on in-situ non-destructive testing of concrete. Ottawa

19. Bossio A (2014) Corrosione e diagnostica delle strutture in calcestruzzo armato. criteri e interventi di ripristino. Wolters kluwer Italia
20. UNI EN 12504-1 (2002) Prove sul calcestruzzo nelle strutture. Carote, prelievo, esame e prova di compressione
21. BS (1974) British standard 4408: part 5, non-destructive methods of test for concrete measurement of the velocity of ultrasonic pulses in concrete, London
22. UNI EN 12504-2:2001 Prove sul calcestruzzo nelle strutture - Prove non distruttive - Determinazione dell'indice sclerometrico.
23. UNI EN 12504-4:2005 Prove sul calcestruzzo nelle strutture - Parte 4: Determinazione della velocità di propagazione degli impulsi ultrasonici
24. RILEM 043-CND: Combined non-destructive testing of concrete
25. UNI EN 14630:2007 "Prodotti e sistemi per la protezione e la riparazione delle strutture in calcestruzzo - Metodi di prova - Determinazione della profondità di carbonatazione di un cal-cestruzzo indurito con il metodo della fenolftaleina".
26. Biondi E, Frunzio G Analysis of degradation for the conservation of reinforced Concrete. WORLD heritage and legacy. Culture, Creativity, Contamination, Gangemi Editore, ISBN 978-88-492-3751-1
27. Bossio A, Fabbrocino F, Lignola GP, Monetta T, Bellucci F, Manfredi G, Prota A (2016) Effects of corrosion on reinforced concrete structures, XIV International Forum World Heritage and Degradation, Capri, Italy

Strenght Tests of Two Reinforced Concrete Plinths of the Former Saint Gobain Factory in Caserta Dating Back to the 1960s

Ersilia Biondi and Giorgio Frunzio

Abstract The article highlights the current status and the physical conditions in which they pour two reinforced concrete plinths, belonging to the foundations of the former glass factory Saint Gobain of Caserta (Italy). They are about half a century old and, thanks to the destructive and non-destructive analyzes carried out on the plinths, followed by structural tests in the laboratory, it was possible to establish how resistant the structural elements were and how much they affected environment and degradation factors on quality and relative durability of the same. The old structure is incorporated into the new one on the south side. The two examples of plinths: they are identical in constitution and material but the physical and chemical state in which they are found today is different. The old structure is often clearly visible, although it is based on in more than disastrous situations; the second, a few meters away, has always been left at the mercy of the passing of years, exposed to atmospheric and related agents, surrounded by rubble. The tests performed are: Pacometric; Sclerometric; Ultrasonic; No. 2 samples of cylindrical sample (coring); Test on compression on the extracted carrots. Furthermore, the sclerometric and ultrasonic tests were processed with the SONREB method, to obtain a more reliable estimate of the strength of the concrete on site. Comparing the values obtained and given the passage of time and the advanced age of the materials, it can be said that the result was surprising and unexpected.

Keywords Concrete · Plinths · Degradation · Destructive and non-destructive analyzes

1 Introduction

The Pisani Saint Gobain glassworks landed in Southern Italy in the second half of the last century and covered an entire industrial area of about 500,000 m^2, with a glass and reinforced concrete plant (Fig. 1). It had a relatively short duration, given

E. Biondi (✉) · G. Frunzio
Università Degli Studi Della Campania Luigi Vanvitelli, Caserta, Italy
e-mail: ersilia.biondi.a@gmail.com

M. Abdel Wahab (ed.), *Proceedings of 1st International Conference on Structural Damage Modelling and Assessment*, Lecture Notes in Civil Engineering 110,
https://doi.org/10.1007/978-981-15-9121-1_4

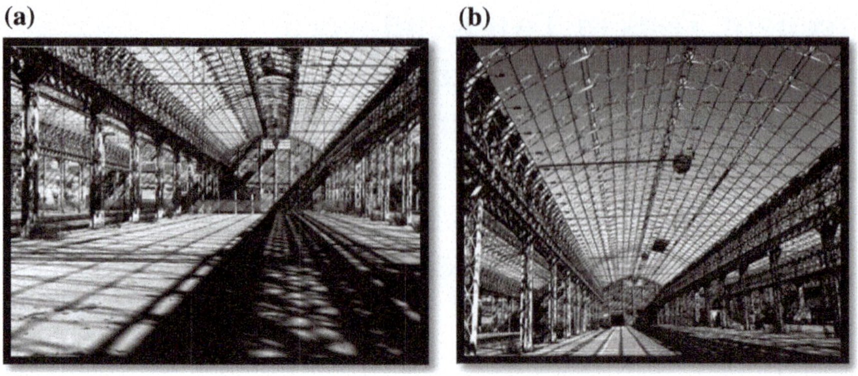

Fig. 1 **a** and **b**. Current state of the former Saint Gobain factory in Caserta

that for about twenty years the factory has been decommissioned and abandoned and has been replaced by a subsequent urbanization of the area, regulated by the Caserta PRG which evolved over the years, where residential neighborhoods were born, services available to citizens, hotels, sports centers, palace of health and much more. The factory was never demolished, but at times it was incorporated into new buildings and, for a large portion, it is periodically used by movie productions which find ample space for mounting sets and scenographies.

The area covered by the former factory is located in the city of Caserta, on the border with the municipality of San Nicola la Strada (Fig. 2) with which it is connected in all respects, both thanks to the infrastructure and the services available, between inhabited centers, schools and the tertiary sector. Currently, a large part of

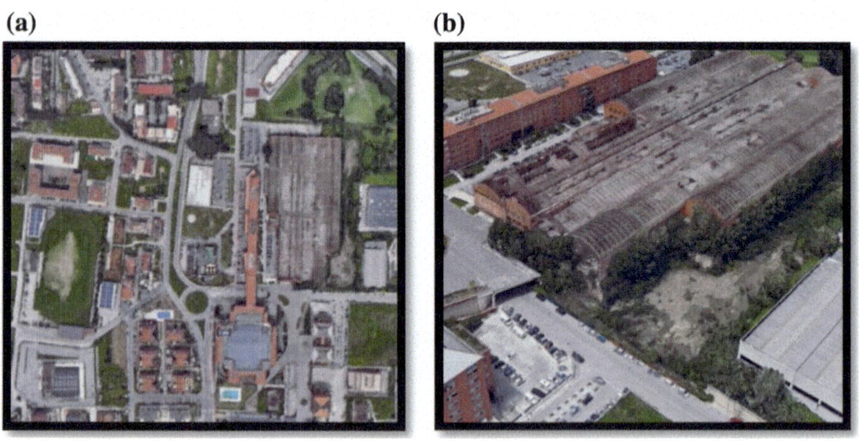

Fig. 2 **a** Satellite image of the former area and **b**. South East view of the old factory currently contextualized

the area has been incorporated into the large hotel of the international chain Golden Tulip. It is an expanding area, the new crucial point of the two municipalities.

The object of study is represented by two examples of reinforced concrete plinths of the factory foundation. It was possible to carry out direct analyzes on site because the factory was never demolished or considered an obsolete object to be removed. Excavations for new buildings were carried out adjacent to the old foundations and, for this reason, it was possible to carry out an on-site inspection [1] where the foundation soil is also visible. The old structure is incorporated in the new one to the south. On the other three sides, the former factory is no longer accessible, both for bureaucratic reasons and because it has not withstood the weather to which it has been subjected over the years. Two examples of plinths have been examined: they are identical in constitution and material but the physical and chemical state in which they are today is different. The former has endured more thanks to the protective barrier provided by the new hotel construction, resting on land currently unbolted in order to rebuild the area and serve it through buildings with different uses; the second plinth, a few meters away, has always been exposed to atmospheric and related agents, surrounded by rubble (Figs. 3 and 4).

Fig. 3 Case study 1—Foundation plinth incorporated in the new building

Fig. 4 Case study 2—Extreme South East external plinth

The survey conducted offers a starting point to study and evaluate the state of conservation of the cementitious material, its duration over time, the comparison between them and an overview of its durability [2–5].

2 Case Study 1. Covered Plinth and Incorporated in the New Structure

The first plinth, shown in Fig. 5, represents the starting point of the study. Tests performed are:

- a. Pacometric test on the structural element
- b. Sclerometric tests on the plinth
- c. Ultrasonic tests on the plinth
- d. Nr. 2 sampling of cylindrical sample (coring)
- e. Compression test on extracted coring.

In addition, the sclerometric and ultrasonic tests were processed with the SONREB method, in order to obtain a more reliable estimate of the strength of the concrete on site. The study is conducted through non-destructive tests in the first phase, which provided results directly in situ [6, 7].

2.1 Test with a Pacometer

The Profoscope pacometer (Fig. 6), which uses the electromagnetic pulse induction method, was used. The coils in the probe are cyclically loaded with electrical impulses that generate a magnetic field. A parasitic current (or eddy current) is created on

Fig. 5 Foundation plinth of the former factory and bare land with adjacent new construction

Fig. 6 **a** and **b**. Tests with pacometer and determination of the position of the reinforcing bars

the surface of each conductive material within the magnetic field, which induces a magnetic field in the opposite direction. The resulting voltage change can therefore be used for measurement. Profoscope uses different coil arrangements in order to generate different types of magnetic fields. Advanced signal processing allows to:

- Locate a steel bar
- Locate the midpoint between the bars
- Determine the thickness of the concrete cover
- Estimate the diameter of the bar.

This method is not affected by any non-conductive material such as concrete, wood, plastic, brick etc. Instead any conductive material present in the magnetic field (in a radius of about 400 mm/16) will influence the measurement. The pulse induction principle used by the Profoscope involves a predefined range and accuracy. The expected accuracy of the size of the concrete cover is indicated in the following graph and complies with BS 1881 part 204 [8] and Eurocode 2 [9] for a single bar with sufficient spacing and known diameter.

The results:

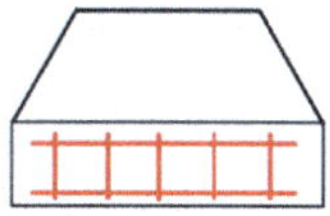

Concrete cover: ~4.1–5.5 cm
Horizontal bars: diameter = 12/14
Brackets: diameter = 8 (pace 17–22 cm).

2.2 Test with a Sclerometer

A type N sclerometer was used, typical for carrying out non-destructive tests on the material in use. Provides an immediate indication of the concrete's compressive strength. Before using the tool, it is advisable to clean the surface of the pillar to be examined with hard brushes or grindstones to eliminate any roughness left by the wooden formwork or the hardened cement dust deposited during the vibration of the concrete. It is advisable to remove a surface thickness of a few millimeters with the help of an abrasive stone. The structural element to be examined must have a minimum thickness of 15 cm. Subsequently, the measuring points are identified (at least 9) by drawing a regular grid of lines from 25 to 50 mm apart; the stop points are constituted by the intersection points of the grid lines. Then, position the instrument perpendicular to the surface gradually increasing the pressure on the hammer until the impact is produced. After the impact, the rebound index is read on a graduated scale, shown in Fig. 7. Once the measurements are completed, the sclerometric rebound index is determined by calculating the average of all the measurements made. Then, using appropriate correlation curves, the cubic compressive strength of the concrete is obtained.

In addition to determining the compressive strength of concrete, the elastic modulus Ec of the same was determined through the use of the following formula:

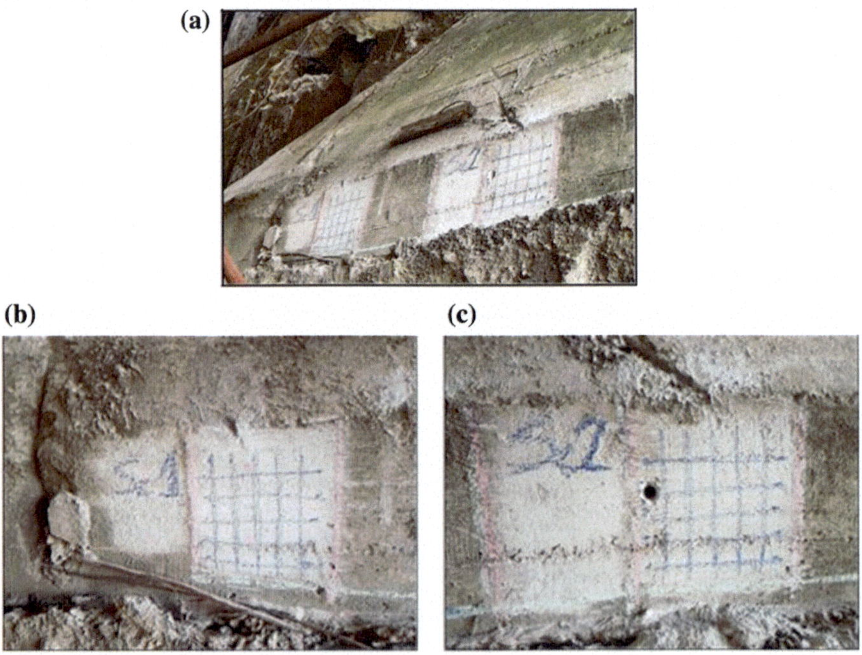

Fig. 7 **a**, **b** and **c**. Tests with sclerometer and subdivision of two surfaces, with relative study area

Table 1 Results of sclerometric test

Station 1		Station 2	
Foundation plinth		Foundation plinth	
Rebound value	Deviation from the average	Rebound value	Deviation from the average
36	0.2	34	−1.0
36	0.2	32	−3.0
36	0.2	36	1.0
36	0.2	36	1.0
36	0.2	36	1.0
34	−1.8	34	−1.0
36	0.2	38	3.0
38	2.2	34	−1.0
34	−1.8	36	1.0
36	0.2	34	−1.0
Rebound average		Rebound average	
35.8		35.0	
Stop angle		Stop angle	
0°		0°	
Resistance read on the calibration curves		Resistance read on the calibration curves	
(MPa)		(MPa)	
32.7		31.3	
E (elastic module)		E (elastic module)	
(MPa)		(MPa)	

$$EC = 18000\sqrt{Rck} \quad (1)$$

The procedure carried out in the investigation complies with UNI EN12504-2 [10]. Table 1 shows the results.

2.3 *Ultrasonic Test*

A portable ultrasound detector was used to measure the characteristic values of the material by ultrasonic pulses [11]. The probe, placed in contact with the surface under examination and coupled thanks to special materials, seen in Fig. 8, generates ultrasonic pulses that propagate in the medium according to spherical wave fronts. A digital timer records the travel time. The instrument made it possible to measure:

- The speed of the impulse in the material, given the distance between the probes
- The distance between the probes, given the speed of sound in the material

(a) (b)

Fig. 8 **a** and **b**. Ultrasonic tests on two surfaces

- The time taken by the impulse to cross the material
- Young's modulus, given the distance between the probes and the density of the material.

The measurements are obtained by direct transmission, i.e. by applying the two probes on the two opposite faces of the element in question; in case of inaccessibility of one of the two faces, as in the case study, there are two types of transmission (Fig. 9): semi-direct (transducers are applied at points belonging to two adjacent faces); indirect (by placing the probes on the same face at a known distance).

The procedure carried out in the survey complies with the UNI EN 12504-4 standard [12].

The numerical values of the test are shown in Tables 2 and 3.

(a) (b) (c)

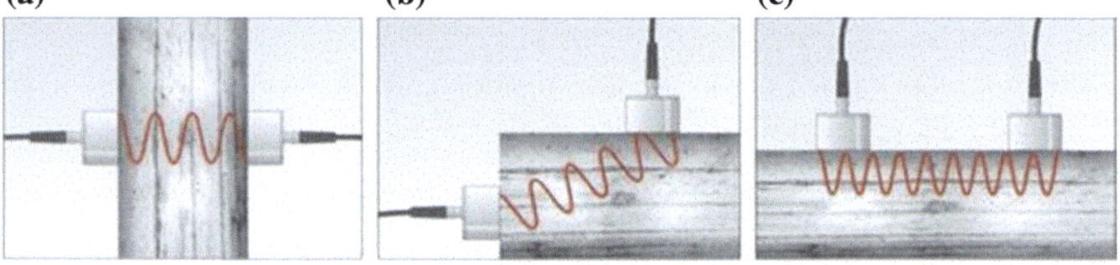

Fig. 9 **a** Direct transmission; **b** semi-direct transmission; **c** indirect transmission

Table 2 Numerical results of the ultrasonic test

Nr. stations	Structural element	Type of measure	Distance of the probes	Transit time (microsec.)	Speed (m/s)	Average speed	Dynamic modulus of elasticity (Mpa)
1	Foundation plinth	Direct	0.20	48.7	4107	4223	35.673
		Direct	0.20	46.1	4338		
1b	Foundation plinth	Direct	0.10	23.6	4237	4265	36.386
		Direct	0.10	23.3	4292		
2	Foundation plinth	Direct	0.20	46.9	4264	4237	35.924
		Direct	0.20	47.5	4211		
2b	Foundation plinth	Direct	0.10	23.5	4255	4160	34.626
		Direct	0.10	24.6	4065		

Table 3 Other numerical results of the ultrasonic test

Speed US (m/s)	Concrete quality
>4.500	Excellent
4.500–3.500	Good
3.500–3.000	Medium
3.000–2.000	Moderate
<2.000	Poor

2.4 SONic REBound (SONREB)

The combination of non-destructive methods using sclerometer and ultrasound allows to obtain more reliable results on the strength of the concrete on site. The advantages of the method can be summarized as follows:

- Cancellation of the influence of humidity and of the degree of maturation of the concrete on the results of the analysis, since they have, with the same effective breaking strength, the opposite effect on the measurements of the speed of propagation of the ultrasound and the rebound index;
- Reduction, compared to the ultrasonic method, of the influence of the granulometry of the aggregate, the dosage and type of cement and any additive used for pouring the concrete;
- Decrease, compared to the sclerometric method, of the importance of quality variations between surface and deep layers of concrete.

The method is applied by determining two pairs of values for each test area:

- Average propagation speed of ultrasonic pulses
- Average rebound index.

Table 4 Sonreb test results

Nr. station	Structural element	Sclerometric index	Average speed (m/s)	Rc (MPa)
1	Foundation plinth	35.8	4223	30.8
1b	Foundation plinth	35.8	4265	31.6
2	Foundation plinth	35.0	4237	30.1
2b	Foundation plinth	35.0	4160	28.7

This investigation is carried out according to RILEM 43 CND Recommendations [13]. The two values are correlated with the compressive strength of concrete according to the law of variation:

$$Rc = A \times I^{B} \times V^{C} (MPa) \quad (2)$$

- Rc = compressive strength of c.l.s.;
- I = average rebound index;
- V = average speed of ultrasound;
- $A = 7695 * 10^{-11}$; B = 1,4; C = 26 (GIACCHETTI).

Results in Table 4.

2.5 Coring

It is the destructive method used to obtain a direct assessment of the strength of concrete and to evaluate the internal state of pathologies due to degradation [14, 15]; 2 cores with a diameter of about 100 mm were taken from the structural element constituting a reinforced concrete foundation plinth (Fig. 10). The sampling operations were carried out by means of continuous coring with water circulation, with a diamond crown, after identifying the reinforcement bars inside the concrete mass with a pacometer. After taking the sample, the structural element was restored using compensated shrinkage cement mortar. It's regulated by UNI 12504-1 standard [16].

Fig. 10 **a**, **b**, **c** and **d**. Coring extraction C1 and C2

Coring 1 Carbonated thickness, 2.5 cm

Coring 2 Carbonated thickness, 2.7 cm

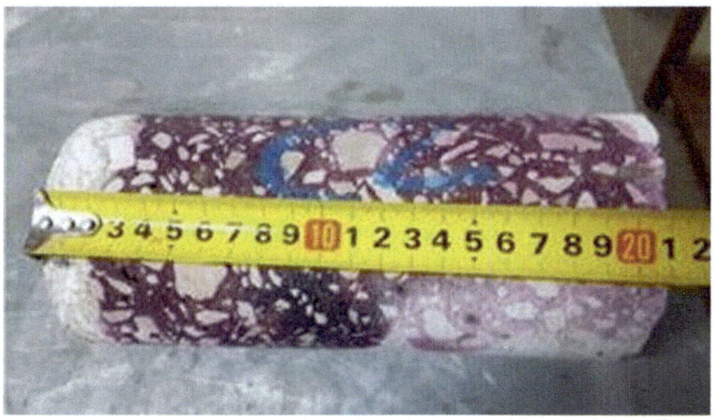

Carbonation test

Correctly proportioned concrete has a strongly alkaline environment (pH13) which inhibits the oxidation reactions of the reinforcement. Concrete is permeable, so that carbon dioxide can spread inside it, reacting with the substances it encounters, giving rise to the phenomenon of "carbonation" and to dimensional variations that cause cracks. This process modifies the alkaline properties of the cement conglomerate since the consequent reduction of calcium hydroxide causes a lowering of the pH of the cement paste (for a fully carbonated concrete, it drops to about 8.5). The phenomenon is triggered by the cortical layer and is gradually affecting the deeper layers of the concrete. If the carbonated layer reaches the concrete that surrounds the reinforcement, with the lowering of the pH, the anti-corrosive protection of the cement paste is lost and the reinforcement iron is no longer "passivated": in the presence of humidity and oxygen it oxidizes and corrodes, resulting in rust.

The measure of the carbonation depth, whose regulatory reference is given by the UNI 9944: 1992 standard [17, 18], can be determined by spraying a 1% solution of ethyl alcohol on the surface of the concrete. The test can be performed on a cylindrical specimen extracted by coring from the element itself, by spraying the solution on the external surface of the coring itself immediately after its removal (to avoid carbonation by direct contact with the air).

Phenolphthalein turns violet on contact with material whose pH is greater than about 9.2 and remains colorless for lower pH values. If only a weak color appears, the treatment with the phenolphthalein solution is repeated. The measurement of the carbonation depth must be measured with millimeter precision.

2.6 Compression Tests

In the laboratory it was possible to trace the value of the mechanical strength of the concrete by crushing the coring through the press. Before subjecting them to this test, it is advisable to adapt them geometrically so that they have perfect flatness between the faces, according to UNI EN 12390 [19]. The coring undergoes mechanical grinding through the use of a disc cutter and flattener. The tolerances allowed for the aforementioned geometric characteristics are:

- Perpendicularity of the generator of the cylinder with respect to the bases: ±0,5 nm1;
- Height, which is equal to 1 time the diameter "d": 5%;
- Latness of the surfaces subjected to load: ±0,0006 d.

At the end, planarity and perpendicularity occur with a dial protractor.

Before proceeding with the crushing (Fig. 11), the specimen is weighed and its height and diameter are measured, and having noted these characteristics, it is positioned between the plates of the machine.

Once the specimen is brought into constant contact with the upper plate, the load is applied without strokes and with a constant gradient, i.e. continuously increasing until no further load can be supported [20] point the maximum load that causes the specimen to break is recorded.

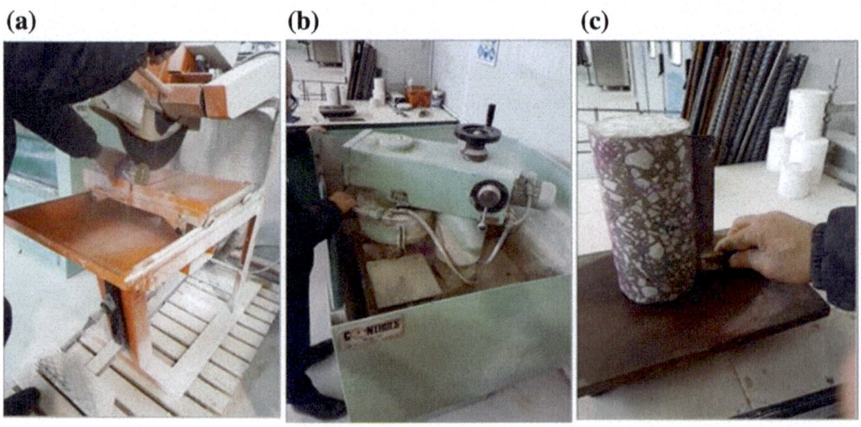

Fig. 11 **a**, **b** and **c**. Laboratory procedure to geometrically uniform the coring extracted on site

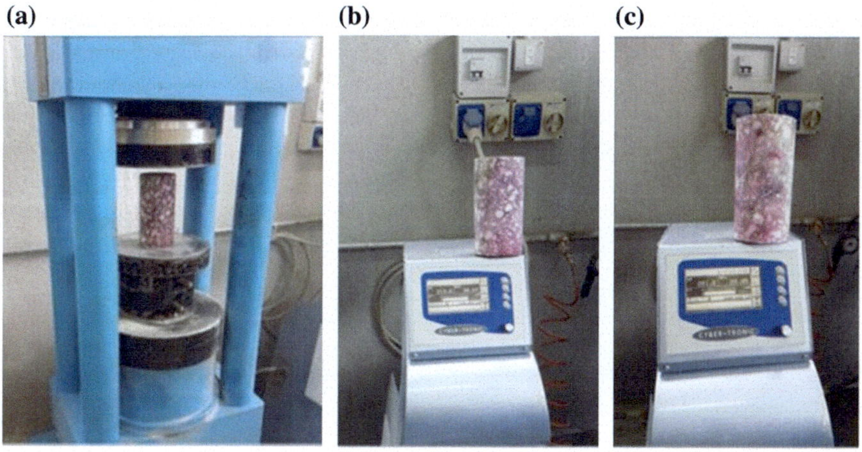

Fig. 12 a, b and c. Compression tests in the laboratory of corings C1 and C2

Table 5 In situ core test

Coring abbreviation	Sample dimension (mm)	MASS (g)	R coring (N/mm^2)
C1	d = 94 h = 182	2882	30,69
C2	d = 94 h = 187	2958	29,05

Table 5 shows each specimen:

- The geometric characteristics;
- The mass;
- The resistance of the coring to compression corresponding to the ratio between the value of the maximum load and the cross-sectional area.

The evaluation of the characteristics of the concrete in situ [21, 22], based on the resistance detected on the core, must take into account the conversion factors that compensate for the effect of slenderness and the transition from cylindrical resistance to cubic resistance. The type of failure ascertained for the samples is of type "1" or satisfactory. The results of test (Fig. 12), are shown in Table 5:

3 Study Case 2. External Plinth

The second test was carried out on the plinth located at the southeast end of the foundation of the old factory (Fig. 13). It presents itself to the eye in a more degraded manner than in the first case study. It was also difficult to carry out the inspection due to the conditions in which it poured the state of the place. Leaving aside the non-destructive tests, since they would have negligibly differed from the first case, we concentrated on the extraction of the coring in order to carry out, in the laboratory, the compression test to evaluate its resistance to be compared with the first, by virtue of the different physical conditions. The procedure was completely similar to the previous case (Fig. 14).

3.1 Carbonation Test

In line with the first procedure, the results of the current case study emerged as follows (coring C3 and Fig. 15):

Fig. 13 Study case 2. External plinth

Fig. 14 **a**, **b**, **c** and **d**. Coring C3 extraction

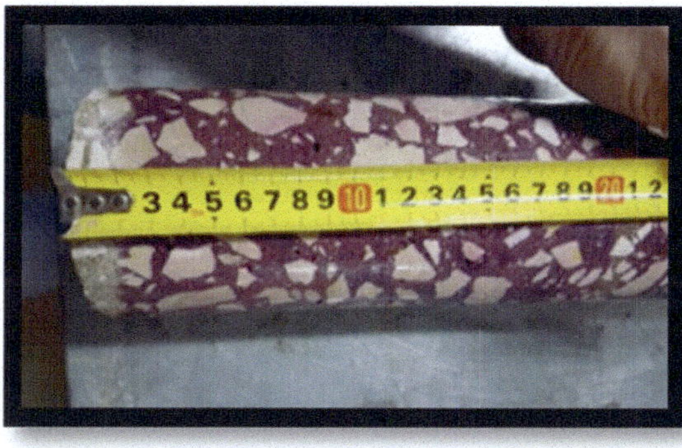

Coring 3 Carbonated thickness, 2.2 cm

(a) **(b)**

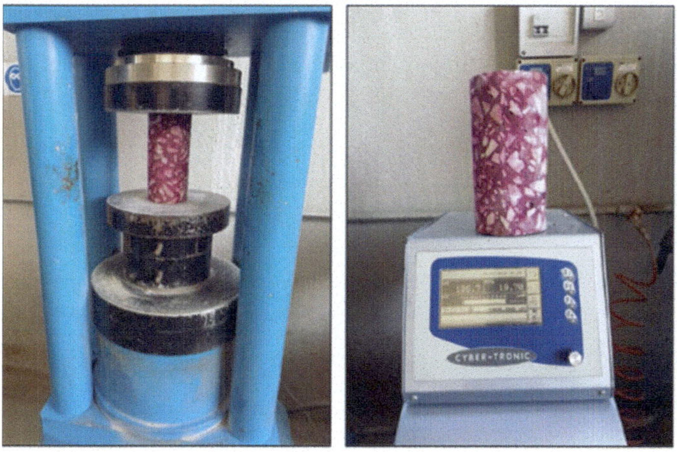

Fig. 15 **a** and **b**. Coring C3 compression tests in the laboratory

3.2 Compression Test

The same working phases were repeated, with the following results in Table 6.

Table 6 Compression test results C3

Coring abbreviation	Sample dimension (mm)	MASS (g)	R coring (N/mm^2)
C3	d = 94 h = 186	3033	19,70

4 Comparison

The first extraction of two coring from the same plinth quantified a compressive strength of 30.69 and 29.05 N/mm^2 respectively. Quite good as a result, considering the structure of a certain age. The second case quantified a lower resistance, i.e. of 19.70 N/mm^2.

5 Conclusions

The hypothesis, which is certainly acceptable, is that the plinths were cast with concrete of similar quality, the results of the tests present a significant difference which is clearly attributable to the different conditions [23–25] of exposure to atmospheric agents that the two elements have had over time, confirming, if still necessary, that the protective conditions of the concrete determine both the level of carbonation and the residual strength of the concrete.

References

1. Cianfrone F, Facaoaru J (1979) Study on the introduction into Italy of the combined non-destructive method for the determination of in-situ concrete strenght. Mater Struct RILEM 12(71):1979
2. Augenti M (2003) The resistance of concrete in existing buildings. II National Conference on collapses and reliability of structures. Naples
3. Pederferri P, Bertolini L (2000) The durability of reinforced concrete. McGraw-Hill
4. Rossetti V (2003) Concrete, materials and technology. McGraw-Hill
5. Siviero E, Cantoni R, Forin M Durability of concrete works. Francoangeli Editore
6. Faella E, Martinelli E, Salerno N (2008) Extract of the acts of the seventeenth century C.E. Rome,—Some considerations on non-destructive methods for determining the mechanical properties of concrete
7. Pucinotti R, Nucera F (2002) Non-destructive tests on concrete: critical review of some experimental investigations, Il giornale delle prove non distruttive e monitoraggio e diagnostica, n. 1/2002
8. British Standard BS 1881-204:1988. Testing concrete. Recommendations on the use of electromagnetic covermeters
9. Eurocode2 C (2003) General rules and rules for buildings (final draft). In: Design of concrete structures. Bruxelles, Belgium.
10. UNI EN 12504-2, Prove sul calcestruzzo nelle strutture – Prove non distruttive – Determinazione dell'indice sclerometrico, 2002
11. Cianfrone F (1993) Micro-seismic and ultrasonic investigations, seminar proceedings, experimentation on structures, Venice
12. UNI EN 12504-4:2005 Prove sul calcestruzzo nelle strutture - Parte 4: Determinazione della velocità di propagazione degli impulsi ultrasonici
13. RILEM 043-CND: Combined non-destructive testing of concrete
14. Pucinotti R (2005) Pathology and diagnostics of reinforced concrete. Non-destructive investigations and coring in the works to be consolidated

15. UNI 6131 (2000)Prelevamento di campioni di calcestruzzo indurito
16. UNI EN 12504-1: 2009. Prove sul calcestruzzo nelle strutture - Parte 1: Carote - Prelievo, esame e prova di compressione
17. UNI 9944:1992 - Corrosione e protezione dell'armatura del calcestruzzo. Determinazione della profondità di carbonatazione e del profilo di penetrazione degli ioni cloruro nel calcestruzzo
18. UNI EN 14630 (2007) Prodotti e sistemi per la protezione e la riparazione delle strutture in calcestruzzo - Metodi di prova - Determinazione della profondità di carbonatazione di un calcestruzzo indurito con il metodo della fenolftaleina
19. UNI EN 12390-1 (2012) Prova sul calcestruzzo indurito - Parte 1: Forma, dimensioni ed altri requisiti per provini e per casseforme
20. UNI EN 12390-3 (2009) Prove sul calcestruzzo indurito - Parte 3: Resistenza alla compressione dei provini
21. Masi. Estimating the strength of concrete in situ by means of destructive and non-destructive tests. DiSGG, Università degli Studi di Basilicata, Potenza
22. Pucinotti R (2013) Assessment of the strength of the characteristic concrete in situ. Build Constr Mater 44:63–73
23. Basile A, Frunzio G, Mattiello G (2015) Diagnosis of the critical state of concrete. In: Heritage and technology mind knowledge experience, vol 56, pp 1783–1791, Aversa
24. Biondi E, Frunzio G Analysis of degradation for the conservation of reinforced Concrete. WORLD heritage and legacy. Culture, Creativity, Contamination, Gangemi Editore, ISBN 978-88-492-3751-1
25. Gonçalves (2015) Study of pathologies and their causes in reinforced concrete structures and construction works. Monograph, UFRJ. Rio de Janeiro

Improved Hydrostatic-Season-Time Model for Dam Monitoring: Inclusion of a Thermal Analytical Solution

Ahmed Belmokre, David Santillan, and Mustapha Kamel Mihoubi

Abstract Hydrostatic-season-time model (HST) are commonly used by engineers to assess dam safety. The model predicts deformations by considering the effects of temperatures in terms of seasonal function, water level fluctuation and time. However, the non-necessity of real temperature measurements and the non-linear relationship between dependent and influencing variables represent a limit. Recently, researchers developed a new one-dimensional analytical model which gives more accurate estimation of thermal field in concrete dams than those obtained through numerical models or the Stucky and Derron method. From that, in this work we tried to improve HST model by correcting the thermal component. Two terms extracted from the analytical model were added to the HST model making it more sensitive to the daily variation of air and water temperatures variations, the first one simulates deformations induced by solar radiation, rainfall and long wave radiation, while the second term assume water temperature effect. The performance of the proposed model was tested using displacements data recorded on six pendulums installed at Tichy Haf arch dam located in north Algeria during five years. The root means square error (RMSE) and the mean absolute error (MAE) indices indicated that the improved model provided better results than the traditional one.

Keywords Arch dam · HST model · Concrete temperature · Statistical model · Analytical model · Solar radiation

A. Belmokre (✉) · M. K. Mihoubi
Mobilisation et Valorisation des Ressources en Eau (MVRE), Ecole Nationale Supérieure d'Hydraulique (ENSH), Blida, Algeria
e-mail: a.belmokre@ensh.dz

D. Santillan
Departamento de Ingeniería Civil: Hidráulica, Energía y Medio Ambiente, Universidad Politécnica de Madrid, 28040 Madrid, Spain

M. Abdel Wahab (ed.), *Proceedings of 1st International Conference on Structural Damage Modelling and Assessment*, Lecture Notes in Civil Engineering 110,
https://doi.org/10.1007/978-981-15-9121-1_5

1 Introduction

The socioeconomic interest of concrete dams requires high levels of safety [1]. Displacements data can describe dam safety [2] by analyzing collected information, detecting anomalies and making the appropriate repairs [3]. Statistical models are extensively utilized with success for dam monitoring [4], for that, hydrostatic-season-time model (HST) are commonly used for predicting dam displacements [5]. Based on multiple linear regression method, HST model assumes that displacements are mainly influenced by water level variation in the reservoir, concrete temperature and aging [6].

However, HST model contain some drawbacks that may affect the accuracy of results [7]. The model assumes that environmental variables and displacements are linearly related, whereas in reality they are strongly nonlinear [8, 9]. Furthermore, thermal component is approximated by harmonic seasonal function without real measurements of air and concrete temperatures [10, 11]. Hence, for days when temperatures are higher or colder than seasonal values, the model performs poorly.

Strenuous efforts were made to enhance the prediction of thermal deformations. Léger et al. [12] established Hydrostatic-Temperature-Time model (HTT) where thermal displacements are estimated from the measurements of thermometers installed in dam's body. Nevertheless, the inadequate position of thermometers or their insufficiency may provide incorrect estimates. A hybrid model that possessed more physical sense by taking into account thermal gradient produced between dam faces was suggested by Tatin et al. [7]. Although the authors have supposed that thermal loads are uniform on dam faces, several works have shown that solar radiation, shading have a significant effect on dam displacements [13].

Santillan et al. [14] proposed a new analytical solution for heat transfer problems in concrete dams. The authors computed one-dimensional thermal field whose ends are subject to variations in the temperatures of air and water as well as heat flux created by solar radiation, rainfall and night cooling as an equivalent temperature.

Herein, an improved HST model was proposed. Thermal component was rectified by adding two terms extracted from thermal analytical solution, the first term simulates displacements generated by solar radiation, night cooling, evaporative cooling and the difference between the daily air temperature and the seasonal function. On the other side, the second term consider deformations caused by water temperature variation. The performances of the new model are validated by comparing its results with those of the traditional one using recorded data from a case study. In addition, the evolution of thermal displacements along cross sections was investigated. This paper is structured as follows. A short description of HST model and the thermal analytical solution was presented in Sect. 2. In Sects. 3 and 4, the case study and the obtained results are illustrated and discussed. Lastly, we summarize the conclusions in Sect. 5.

2 Methodology

2.1 Hydrostatic-Season-Time Model

HST model is composed of functions multiplied by coefficients estimated using multiple linear regression method. The general expression of HST model is described by Eq. (1). Where a_0 is a constant term, $f(h)$ is the function that models the elastic effect hydrostatic pressure, $f(s)$ is the thermal component, and $f(t)$ are the irreversible displacements [15].

$$f(h, s, t) = a_0 + f(h) + f(s) + f(t) \tag{1}$$

Thermal component is determined by a sinusoidal function, it is expressed by Eq. 2, where J is the number of days in the year [5].

$$f(s) = a_7 \cos\left(\frac{2\pi J}{365}\right) + a_8 \sin\left(\frac{2\pi J}{365}\right) + a_9 \cos\left(2\frac{2\pi J}{365}\right) + a_{10} \sin\left(2\frac{2\pi J}{365}\right) \tag{2}$$

Equation 2 represents displacements provided by seasonal temperature with annual amplitude A_{air} and mean annual temperature T_{mean}. The annual temperature series T_a that can be deduced using Eq. 3 [16]:

$$T_a = A_{air} \sin\left(\frac{2\pi}{365} J\right) + T_{mean} \tag{3}$$

2.2 Thermal Analytical Model

The new analytical model considers that for a point x in a cross section of a length L the temperature of the concrete T is the sum of three thermal responses: u, v and w [14].

u is the temperature in section L when its ends are subjected to constant temperature: the mean water temperature $\overline{T}_w$ at upstream end and the mean air temperature $\overline{T}_{wair}$ at downstream end. Equation 4 is the stationary solution to the problem u, where h_{air},h_w are the convection coefficients at respectively downstream and upstream faces.

$$u(x, t) = \frac{(\overline{T}_w - \overline{T}_{air})h_{air}h_w x + \overline{T}_{air}(h_{air} + h_{air}h_w L) + \overline{T}_w h_w}{h_{air} + h_w + h_{air}h_w L} \tag{4}$$

The second term v is given by Eq. 5, it is the solution of a heat transfer problem whose downstream end is subjected to air temperature equal $A_{air}\sin\omega t$ while a temperature equal to zero is imposed at the upstream end.

$$v(x,t) = \frac{A_{air}h_{air}}{2}\left[S_p e^{i\omega t}.\frac{\sqrt{\frac{\omega}{2\chi}}(1+i)\cosh\left\{\sqrt{\frac{\omega}{2\chi}}(1+i)(L-x)\right\} + h_w \sinh\left\{\sqrt{\frac{\omega}{2\chi}}(1+i)(L-x)\right\}}{\left(\sqrt{\frac{\omega}{2\chi}}(1+i)S_p + h_{air}C_p\right)\left(\sqrt{\frac{\omega}{2\chi}}(1+i)S_p + h_w C_p\right) - h_{air}h_w}\right.$$
$$\left. + S_n e^{i\omega t}.\frac{\sqrt{\frac{\omega}{2\chi}}(1+i)\cos h\left\{\sqrt{\frac{\omega}{2\chi}}(1+i)(L-x)\right\} + h_w \sinh\left\{\sqrt{\frac{\omega}{2\chi}}(1+i)(L-x)\right\}}{\left(\sqrt{\frac{\omega}{2\chi}}(1+i)S_n + h_{air}C_p\right)\left(\sqrt{\frac{\omega}{2\chi}}(1+i)S_n + h_w C_p\right) - h_{air}h_w}\right] \quad (5)$$

The third term w is the response to a thermal problem in section L whose upstream end is subjected to a temperature equal to $A_w\sin\omega t$ and a zero temperature is imposed at downstream end. It is expressed by the following equation, where A_w is annual amplitude of water temperature.

$$w = \frac{A_w h_u}{2}\left[S_p e^{i\omega t}.\frac{\sqrt{\frac{\omega}{2\chi}}(1+i)\cosh\left\{\sqrt{\frac{\omega}{2\chi}}(1+i)x\right\} + h_{air} \sinh\left\{\sqrt{\frac{\omega}{2\chi}}(1+i)x\right\}}{\left(\sqrt{\frac{\omega}{2\chi}}(1+i)S_p + h_{air}C_p\right)\left(\sqrt{\frac{\omega}{2\chi}}(1+i)S_p + h_w C_p\right) - h_{air}h_w}\right.$$
$$\left. + S_n e^{i\omega t}.\frac{\sqrt{\frac{\omega}{2\chi}}(1+i)\cosh\left\{\sqrt{\frac{\omega}{2\chi}}(1+i)x\right\} + h_{air} \sinh\left\{\sqrt{\frac{\omega}{2\chi}}(1+i)x\right\}}{\left(\sqrt{\frac{\omega}{2\chi}}(1+i)S_n + h_{air}C_p\right)\left(\sqrt{\frac{\omega}{2\chi}}(1+i)S_n + h_w C_p\right) - h_{air}h_w}\right] \quad (6)$$

During service life, concrete temperature is influenced by heat fluxes generated on faces in contact with air in the form of an equivalent temperature T_{eq}, and water temperature on areas in contact with reservoir. Heat fluxes are created by mechanisms of convection, short-wave radiation and solar radiation [17].

The difference between T_{eq} and T_a temperatures T_{dif} and water temperature T_w are not considered by HST model. To overcome this defect, we thought to improve HST model by adding the terms v and w to the thermal component (Eq. 7) and maintaining it simple use at the same time. To include the effect of T_{dif}, we compute v at the location of the pendulum reading tables by imposing T_{dif} as boundary conditions at downstream face. We consider that displacements induced by the term u are assumed by the sinusoidal function of the model. The new model is called "improved hydrostatic-season-time" model (IHST).

$$f(s,v,w) = a_7\cos\left(\frac{2\pi J}{365}\right) + a_8\sin\left(\frac{2\pi J}{365}\right) + a_9\cos\left(2\frac{2\pi J}{365}\right)$$
$$+ a_{10}\sin\left(2\frac{2\pi J}{365}\right) + v + w \quad (7)$$

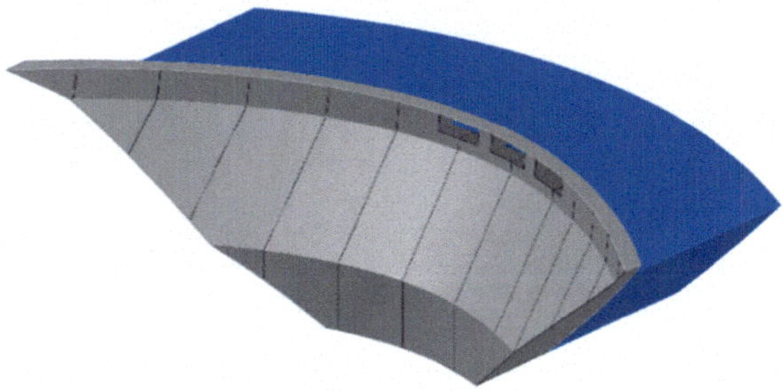

Fig. 1 Three-dimensional geometry of Tichy Haf dam

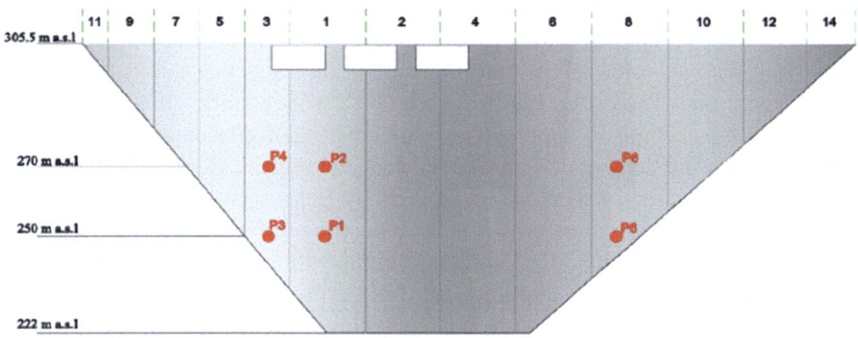

Fig. 2 Location of reading tables of selected pendulums

3 Case Study

We use monitoring data collected from Tichy Haf arch dam located in north Algeria that range between January 1, 2010 and July 30, 2014, specially water level variation and displacements data. Figure 1 shows a 3D view of the geometry of the dam. The performance of the new model compared to that of the HST model are evaluated based on displacement data recorded on six pendulums installed at different elevation and at several blocks of the case study as shown in Fig. 2. The length of cross section where reading tables are positioned varies between 25 m (for P1, P3 and P5) and 15 m for (P2, P4 and P6). Data before June 4, 2013 are used to train the model whereas the rest are served to investigate its accuracy.

4 Results and Discussion

Heat fluxes induced on dam faces in the form of an equivalent temperature are quantified applying the methodology described in detail by Belmokre et al. [18] and

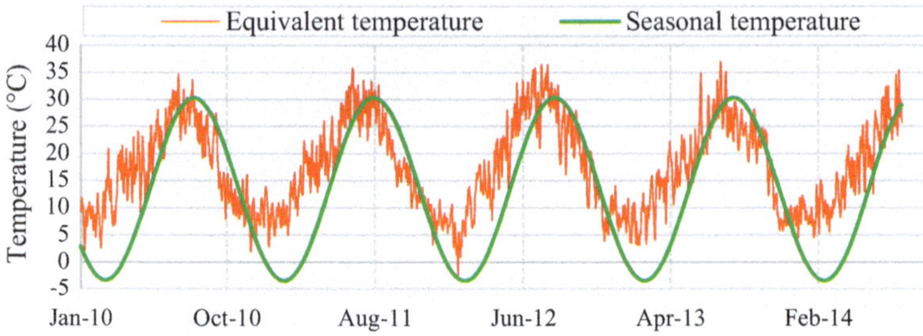

Fig. 3 Assessment of weather conditions acting on Tichy Haf dam: comparison between equivalent and seasonal temperatures at downstream face of the central block

Santillan et al. [17]. The equivalent temperature variation is illustrated in Fig. 3. The comparison with seasonal temperature fluctuation indicates that this is lower than the equivalent one. The difference can reach 6 °C, what confirms that HST model fails to take into account an important source of thermal displacements.

Concerning water temperature, it is calculated adopting the model of Bofang [19], Fig. 4 depicts water temperature evolution at Tichy Haf reservoir for several depths. Adopting the fast Fourier transform, we decompose the difference between seasonal and equivalent temperatures T_{dif}, and water temperature to a sum of cosine and sine functions defined by an amplitude and frequency. v and w are calculated at 6 points representing the position of the reading tables of the selected pendulums. The coefficients of the HST and IHST models are determined by the stepwise regression method. The mean absolute error (MAE) and the root mean square error (RMSE) are used to verify the accuracy of each model.

The RMSE and MAE values regrouped in Table 1 reveal that "Improved HST" is more accurate prediction than HST model. We notice an RMSE and MAE values are respectively equal to 0.524 mm and 0.453 mm for P4, while a slight improvement is reported at P1 with RMSE = 0.800 mm. Figure 5 illustrates measured and computed displacements by HST and IHST models at the six pendulums. It is obvious that the added terms enhance the prediction quality. Moreover, we observe that IHST model supply more reliable results for the shortest cross sections.

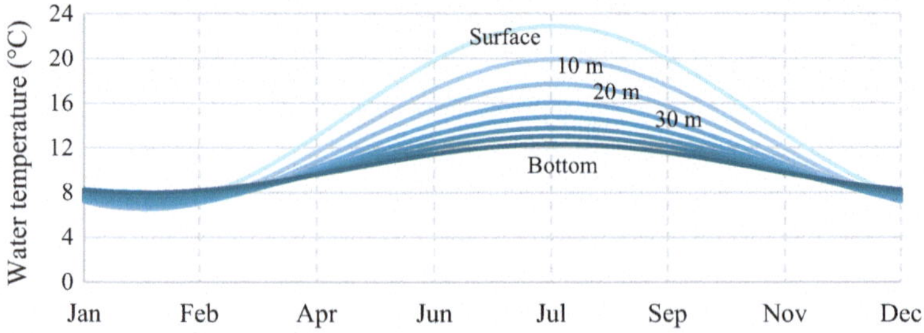

Fig. 4 Water temperature at several depths estimated by the model of Bofang

Table 1 Performance of HST and IHST models

Pendulum	HST		Improved HST	
	RMSE (mm)	MAE (mm)	RMSE (mm)	MAE (mm)
P1	0.813	0.712	0.800	0.693
P2	0.810	0.710	0.781	0.672
P3	0.728	0.642	0.686	0.609
P4	0.620	0.549	0.524	0.453
P5	0.852	0.380	0.822	0.665
P6	0.680	0.578	0.655	0.540

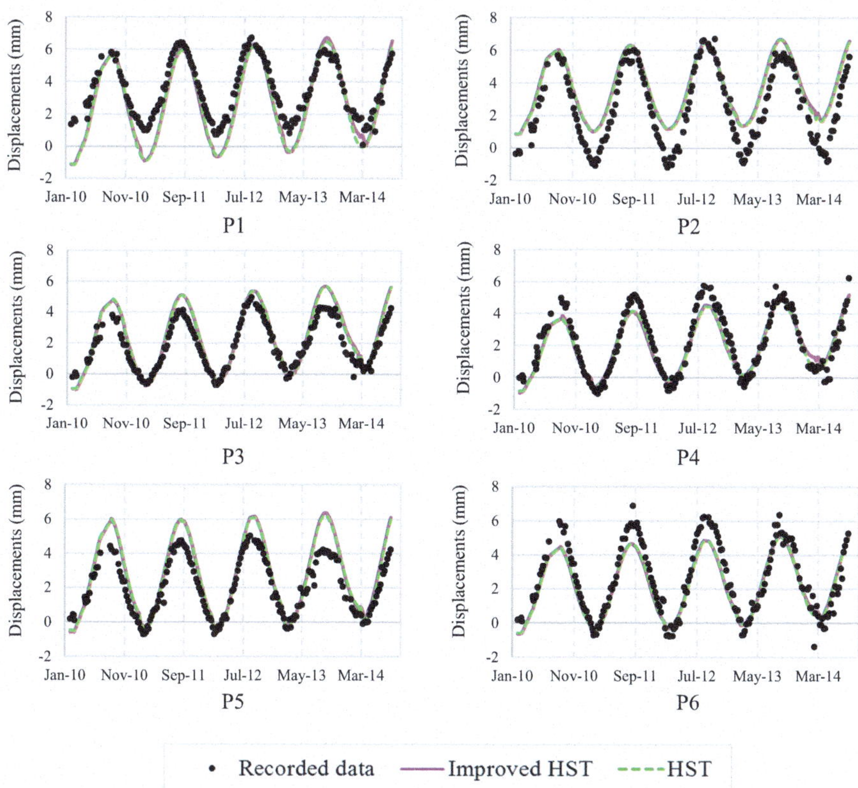

Fig. 5 Recorded and predicted deformations by HST and Improved HST models for six pendulums located at different dam elevations

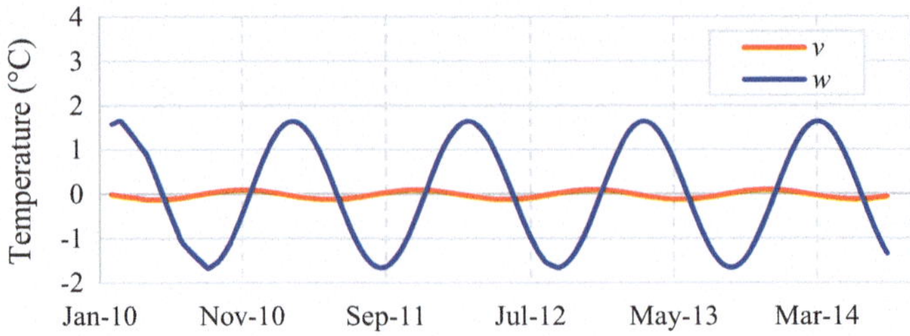

Fig. 6 Concrete temperature at the reading table P1 generated by water temperature (*w*) and solar radiation, night cooling and evaporative cooling (*v*)

Figure 6 presents the time evolution of the variables *v* and *w* at the pendulum P1. v is almost zero while w has an annual amplitude of 1.5 °C. Due to the close position of the pendulums to the upstream face, water temperature affects dam displacements more than T_{dif} in this zone.

Belmokre et al. [18] found that the thermal wave generated by the equivalent temperature has a penetration depth of five meters inward downstream face. Therefore, at a depth greater than the latter, the thermal displacements become less sensitive to equivalent temperature variation and can be simulated by the seasonal function of the model and also by the term *w* for the areas near the upstream facing (less than 5 m). This explains the small difference between the HST and IHST models in the obtained results. Also, several researchers admit that a part of displacements induced by the variation of water temperature are assumed by the seasonal function [7].

To understand how *v* and *w* contribute to the prediction of thermal displacements, we study the effect of the term *v* separately. First, a numerical study to determine concrete deformations in the central block of the dam using the finite element method is carried out. The central block was discretized into a structured finite element mesh. The mesh is made up of 1275 quadrilateral elements and 4026 nodes. The average mesh size is 0.7 m (Fig. 7).

The numerical outcomes are depicted in Fig. 8, negative and positive signs show respectively displacements to upstream and downstream direction. It appeared that regions near to the downstream face are in permanent deformation towards upstream.

The results of numerical modeling are then considered as data to validate the IHST model at different point of the dam. We tried to predict displacements for the point located at the height 270 m a.s.l at two meters inward downstream side through IHST model. Figure 9 shows that IHST model yields excellent results with RMSE equal to 0.487 mm and MAE up to 0.409 mm.

The variation of *v* and *w* are depicted in Fig. 10, it is obvious that displacements estimated by IHST model near downstream are mainly caused by thermal sources presented by *v*. On the other hand, *w* is almost zero because the effect of water temperature on thermal field is limited to two or three meters from upstream face. This approach confirm that thermal deformations do not occur in the same manner along the cross sections and prove that adding *v* and *w* not only improves the results but also gives more physical sense to the model.

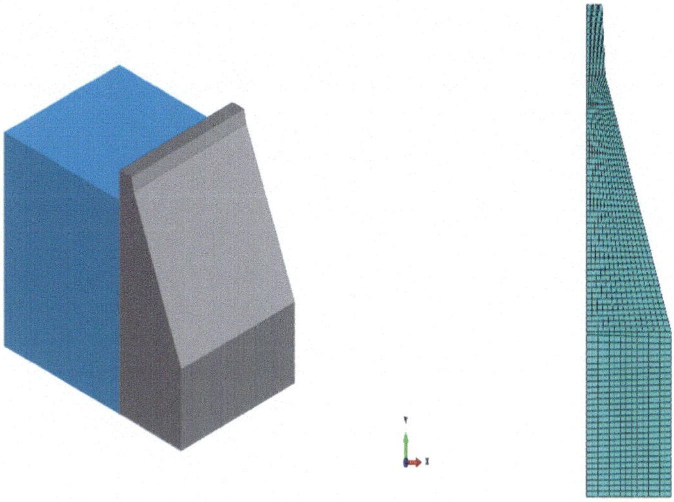

Fig. 7 Geometry and adopted mesh for the central block

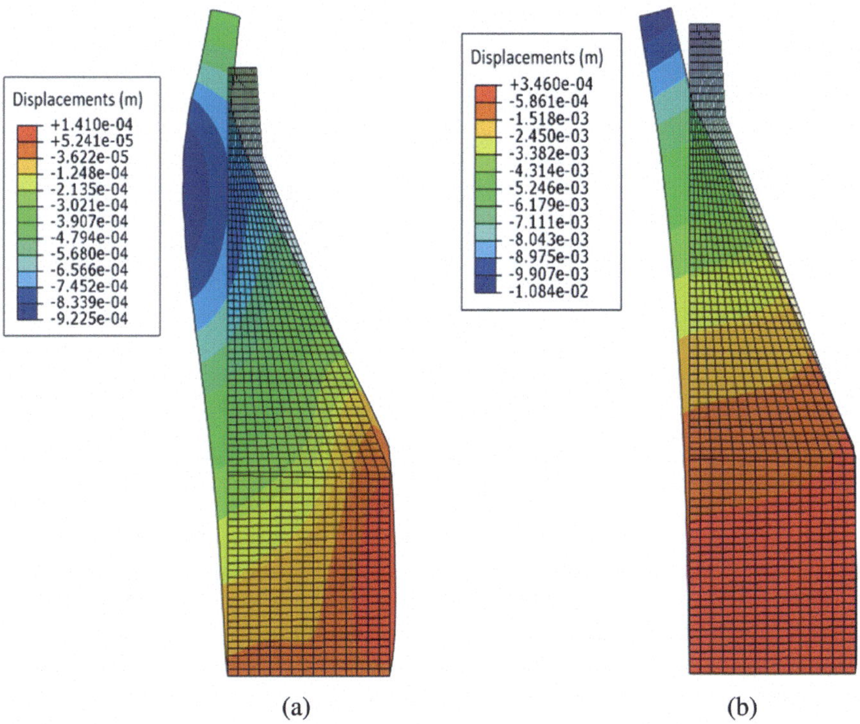

Fig. 8 Displacements evolution of the central block for the hottest (**a**) and coldest (**b**) days during the studied period

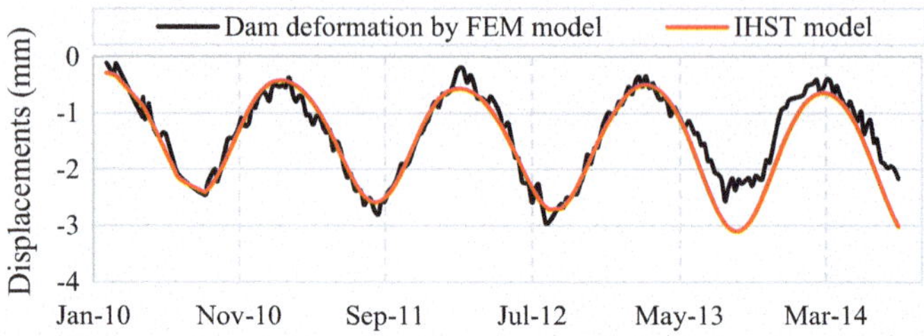

Fig. 9 Deformations estimated by the IHST model at two meters inward downstream face of the central block at 270 m a.s.l

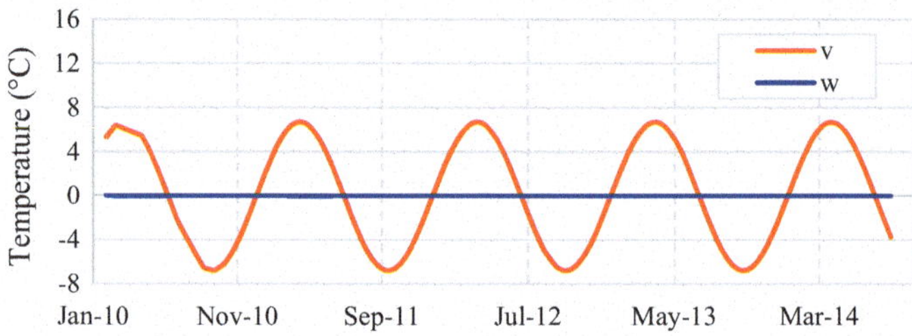

Fig. 10 Concrete temperatures at 2 m inward downstream face generated by water temperature (w) and solar radiation, night cooling and evaporative cooling (v)

5 Conclusion

In order to give a more physical interpretation to its results, an improvement of the HST model to predict arch dam displacements was discussed in this study. We tried to correct the thermal component by adding two terms that take into account the effect of different heat sources influencing the concrete temperature. Unlike the HST model, which considers that displacements are uniform along the cross sections, the improved one can predict deformation at each point of dam. The approach is validated using meteorological data and displacements recorded by six pendulums installed in Tichy Haf dam during 5 years. For all pendulums, results show that the improved model gives better estimates than the traditional one. However, the more the pendulum is near the upstream facing, the more the performances of the existed and the new HST models are similar.

Finite element method is then used to obtain daily deformations in the central block of the case study. We employ the numerical results to assess the performance of the improved model near the downstream face. It turned out that the developed approach provides better results in these areas, due to the significant effect of equivalent temperature on concrete temperature. We can conclude that to properly capture thermal displacements, it is more convenient to install pendulums closer to the downstream face and monitor permanently concrete temperature. The efficiency and the simple use of the new model can make it a good alternative for dam monitoring.

References

1. Li Z (2020) Global sensitivity analysis of the static performance of concrete gravity dam from the viewpoint of structural health monitoring. Arch Comput Methods Eng. https://doi.org/10.1007/s11831-020-09434-0
2. Chen S, Gu C, Lin C, Zhang K, Zhu Y (2020) Multi-kernel optimized relevance vector machine for probabilistic prediction of concrete dam displacement. Eng Comput. https://doi.org/10.1007/s00366-019-00924-9
3. Li Y, Bao T, Gong J, Shu X, Zhang K (2020) The prediction of dam displacement time series using STL, extra-trees, and stacked LSTM neural network. IEEE Access 8:94440–94452. https://doi.org/10.1109/ACCESS.2020.2995592
4. Kang F, Liu X, Li J (2019) Temperature effect modeling in structural health monitoring of concrete dams using kernel extreme learning machines. Struct Heal Monit. https://doi.org/10.1177/1475921719872939
5. Salazar F, Toledo MÁ, Oñate E, Suárez B (2016) Interpretation of dam deformation and leakage with boosted regression trees. Eng Struct 119:230–251. https://doi.org/10.1016/j.engstruct.2016.04.012
6. Gamse S, Henriques MJ, Oberguggenberger M, Mata JT (2020) Analysis of periodicities in long-term displacement time series in concrete dams. Struct Control Heal Monit 27. https://doi.org/10.1002/stc.2477
7. Tatin M, Briffaut M, Dufour F, Simon A, Fabre J-P (2015) Thermal displacements of concrete dams: accounting for water temperature in statistical models. Eng Struct 91:26–39. https://doi.org/10.1016/j.engstruct.2015.01.047
8. Han B, Geng F, Dai S, Gan G, Liu S, Yao L (2020) Statistically optimized back-propagation neural-network model and its application for deformation monitoring and prediction of Concrete-Face Rockfill Dams. J Perform Constr Facil 34:04020071. https://doi.org/10.1061/(ASCE)CF.1943-5509.0001485
9. Li B, Yang J, Hu D (2020) Dam monitoring data analysis methods: a literature review. Struct Control Health Monit 27. https://doi.org/10.1002/stc.2501
10. Gamse S, Zhou W, Tan F, Yuen K, Oberguggenberger M (2018) Hydrostatic-season-time model updating using Bayesian model class selection. Reliab Eng Syst Saf 169:40–50. https://doi.org/10.1016/j.ress.2017.07.018
11. Tatin M, Bri M, Dufour F, Simon A, Fabre J (2018) Statistical modelling of thermal displacements for concrete dams: influence of water temperature profile and dam thickness profile. Eng Struct 165:63–75. https://doi.org/10.1016/j.engstruct.2018.03.010
12. Léger P, Leclerc M (2007) Hydrostatic, temperature, time-displacement model for concrete dams. J Eng Mech 133:267–277. https://doi.org/10.1061/(ASCE)0733-9399(2007)133:3(267)
13. Santillán D, Salete E, Vicente DJ, Toledo MÁ (2013) Treatment of solar radiation by spatial and temporal discretization for modeling the thermal response of arch dams. J Eng Mech 140:1–18. https://doi.org/10.1061/(ASCE)EM.1943-7889.0000801
14. Santillán D, Salete E, Toledo MÁ (2015) A new 1D analytical model for computing the thermal field of concrete dams due to the environmental actions. Appl Therm Eng 85:160–171. https://doi.org/10.1016/j.applthermaleng.2015.04.023
15. Mata J, De Castro AT, Sá J (2013) Constructing statistical models for arch dam deformation. Struct Control Health Monit, 1–15. https://doi.org/10.1002/stc.1575
16. Sheibany F, Ghaemian M (2006) Effects of environmental action on thermal stress analysis of Karaj concrete arch dam. J Eng Mech 132:532–544. https://doi.org/10.1061/(ASCE)0733-9399(2006)132:5(532)
17. Santillán D, Salete E, Toledo MÁ, Granados A (2015) An improved 1D-model for computing the thermal behaviour of concrete dams during operation . Comparison with other approaches. Comput Concr 1:103–126. https://doi.org/10.12989/cac.2015.15.1.103

18. Belmokre A, Mihoubi MK, Santillán D (2019) Analysis of dam behavior by statistical models: application of the random forest approach. KSCE J Civ Eng. https://doi.org/10.1007/s12205-019-0339-0
19. Bofang Z (2014) Thermal stresses and temperature control of mass concrete. Elsevier Inc., Oxford, UK

Estimation of Bending Collapse Load for Triangle Tubes

Kenichi Masuda

Abstract Estimation method of bending collapse load for rectangular tubes was studied by Kecman. In the study, flange buckling at the compression side or plastic yielding was collapse type. Recently, we indicated that the buckling at the compression web may also occur. In order to estimate the bending maximum moment of rectangular tubes, the effective width which shows the post buckling strength was used. In this paper, the purpose is to investigate whether the previous proposal can be used in estimation method with triangle tubes under pure bending. It is found that the collapse type is the same as rectangular tubes. However, in the triangle tubes, buckling width of the web is smaller than the actual one due to the condition of pure bending. Therefore, the boundary condition of tension side is assumed to be free in contrast to rectangular tubes. Its validity is checked by comparing with the results of FEM.

Keywords FEM · Pure bending · Triangle tube · Buckling · Effective width

1 Introduction

Hollow section structures are used from a standpoint of lightweight and high rigidity. For example, automobiles, trains, vehicles, aircraft have many such structural members. Therefore, it is important to understand the collapse load of thin-walled structures under various loads. In case of the cylindrical tubes, representative examples are as follows. Alexander [1] proposed an approximate analysis of the collapse under axial loading. Abramowicz [2] studied dynamic axial crushing. Wierzbicki [3] proposed an approximate analysis of Brazier effect under bending.

In case of the rectangular tubes under bending, representative example is Kecman's study [4]. In the study, two collapse types were indicated. First is a collapse due to flange buckling at the compression side, and second is a collapse due to plastic yielding. Recently, we indicated that the buckling at the compression web may also

K. Masuda (✉)
University of Toyama, 3190 Gofuku, Toyama-shi, Toyama, Japan
e-mail: masuda@eng.u-toyama.ac.jp

M. Abdel Wahab (ed.), *Proceedings of 1st International Conference on Structural Damage Modelling and Assessment*, Lecture Notes in Civil Engineering 110,
https://doi.org/10.1007/978-981-15-9121-1_6

occur [5]. In order to estimate the bending maximum moment of rectangular tubes, the effective width which shows the post buckling strength was used.

In this paper, our purpose is to study whether the previous method can be extended for triangle tubes. It is found that the collapse type is the same as rectangular tubes. However, in the triangle tubes, buckling width of the web is smaller than the actual one due to the condition of pure bending.

2 Numerical Analysis Method

In this paper, MSC. MARC [6] which is the FEM software is used. Figure 1 shows the analyzed triangle tubes. Tube thickness is shown by t, flange width is shown by a, and web width is shown by b. The tube material is a homogeneous and isotropic elasto-plastic material. Von Mises yield criterion is adapted and stress-strain relation is elastic perfect plastic. Young's modulus is 210 GPa, the yield stress is 420 MPa, and Poisson's ratio is 0.3. The yield stress is described as σ_y.

In order to solve the nonlinear equation, the updated Lagrange method, Newton-Raphson method and return-mapping method are used. Four-node quadrilateral thickness shell element is used for the triangle tubes. The element type is No. 75 for MARC User's Guide [6]. The elements divide the thin-walled width a and b into more than 20 sub lengths.

In order to neglect the influence of the boundary conditions, the ratio of the length and widths L/a, L/b is set to L/a, $L/b > 6$.

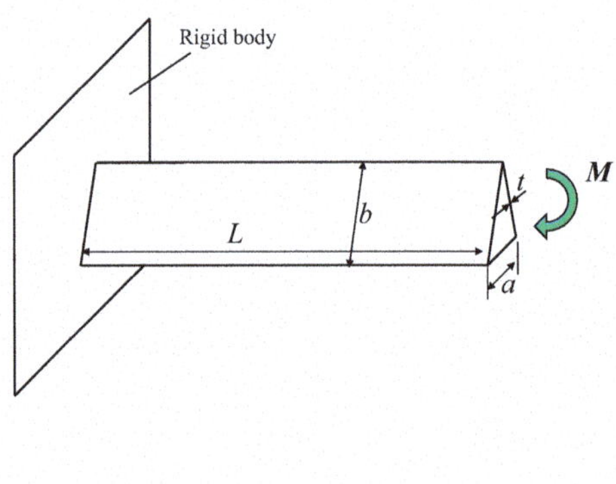

Fig. 1 Analysis model of triangle tube under pure bending

3 Investigation

3.1 *Proposal of Extended Kecman's Method [4] for Triangle Tubes*

By using an extended Kecman's method, this section is described for predicting the maximum bending moment of triangle tubes. The value of buckling stress $\sigma_{\text{buc-}a}$ of the compression flange is given by

$$\sigma_{buc-a} = \frac{k_a \pi^2 E}{12(1-\nu^2)}\left(\frac{t}{a}\right)^2 \quad (1)$$

where k_a is the buckling coefficient. The value is given by

$$k_a = 5.23 + 0.16\frac{a}{b}. \quad (2)$$

By using Eq. (1), three collapse cases are derived as shown in Fig. 2. Case 1 is buckling at the compression flange. Case 2 is plastic yielding at the flange or plastic yielding at the web. Case 3 is the cross-sectional fully plastic yielding. For Case 1, the maximum moment $M_{\max}$ is given by

$$M_{\max} = \frac{1}{3}\sigma_y t\sqrt{1-\frac{a^2}{4b^2}}\left(b^2 + 2a_e b - \frac{a_e^2}{2}\right). \quad (3)$$

For Case 2, the maximum moment $M_{\max}$ is given by

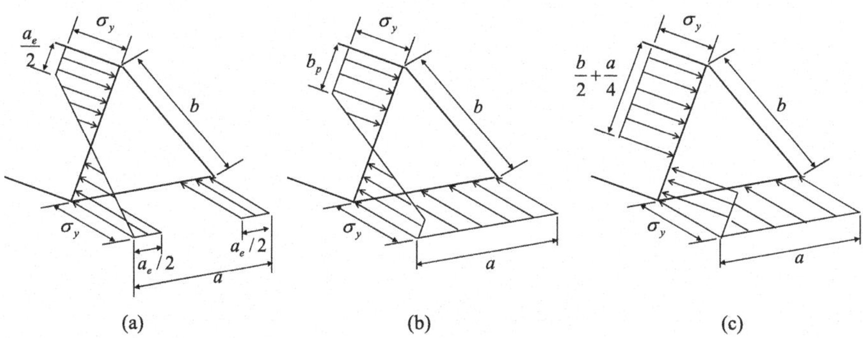

Fig. 2 Axial stress distribution in the maximum bending moment: **a** Case 1: $\sigma_{\text{buc-}a} < \sigma_y$; **b** Case 2: $\sigma_y < \sigma_{\text{buc-}a} < 3\sigma_y$; and **c** Case 3: $\sigma_{\text{buc-}a} > 3\sigma_y$

$$M_{\max} = M\big|_{\sigma_{buc-a}=\sigma_y} + \left(M_{pl} - M\big|_{\sigma_{buc-a}=\sigma_y}\right)\frac{\sigma_{buc-a} - \sigma_y}{2\sigma_y}. \tag{4}$$

For Case 3, the maximum moment $M_{\max}$ is given by

$$M_{\max} = M_{pl}. \tag{5}$$

In the above equations

$$a_e = a\left(0.7\frac{\sigma_{buc-a}}{\sigma_y} + 0.3\right) \tag{6}$$

and $M|_{\sigma \text{buc-}a} = \sigma_y$ is the moment in which $\sigma_{\text{buc-}a}$ is equal to σ_y

$$M\big|_{\sigma_{buc-a}=\sigma_y} = \frac{1}{3}\sigma_y t\sqrt{1 - \frac{a^2}{4b^2}}\left(b^2 + 2ab - \frac{a^2}{2}\right) \tag{7}$$

and M_{pl} is the cross-sectional fully plastic bending moment.

$$M_{pl} = \frac{1}{2}\sigma_y t\sqrt{1 - \frac{a^2}{4b^2}}\left(b^2 + ab - \frac{a^2}{4}\right). \tag{8}$$

3.2 Proposal of Extended Authors' Method [5] for Triangle Tubes

By using an extended authors' method, this section is described for predicting the maximum bending moment of triangle tubes. An extended authors' method means considering web buckling and boundary condition. In the triangle tubes, buckling width of the web is smaller than the actual one due to the condition of pure bending. Therefore, the boundary condition of tension side is assumed to be free in contrast to rectangular tubes. Linear strain is applied to the web of tube. The problem of linear stress buckling assumed to be web buckling is expressed in Fig. 3. In Fig. 3a, one of the longitudinal edges DA is fixed in the out-of-plane direction and the other edge BC is free. The edge BC is tension side. In a plate ABCD, width is shown by b' and thickness is shown by t. The loading condition is displacement control. It means that a linearly distributed on both edges (AB and CD) is applied through displacement control. Figure 3b shows the maximum loading after buckling. Two effective widths are shown by b_{e1} and b_{e2}. The distribution of compressive stress σ_x is described as a positive value.

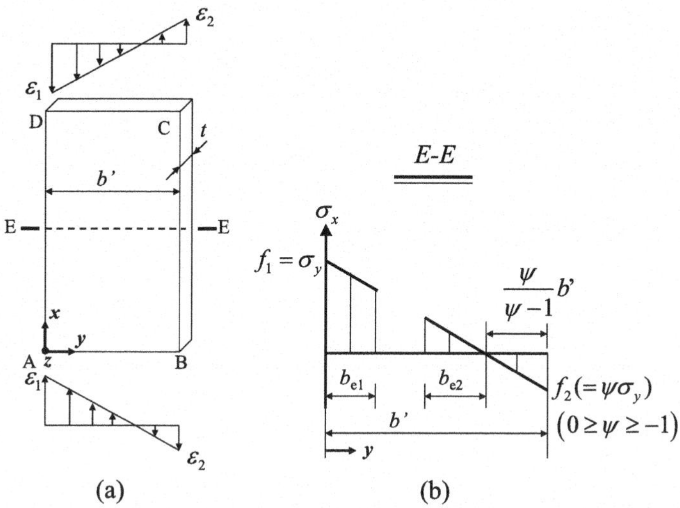

Fig. 3 Plate subjected to linear distribution through displacement control: **a** analysis model and **b** axial stress distribution on E-E cross section in maximum loading

Rusch and Lindner [7] proposed the values of effective widths b_{e1} and b_{e2}. The equations are given as follows:

$$\begin{cases} b_{e1} = b_e - b_{e2} \\ \dfrac{b_{e2}}{b'} = \dfrac{0.226}{\lambda^2} \end{cases} \tag{9}$$

where

$$\frac{b_e}{b'} = \frac{\rho}{1 - \Psi} \tag{10}$$

In addition, b_e cannot exceed the compression portion of the plate. λ and ρ are given in Eqs. (11) and (12), respectively. Ψ is ratio of f_1 and f_2. f_1 and f_2 are edge stresses shown in Fig. 3b

$$\Psi = \frac{f_2}{f_1} \tag{11}$$

λ is given by

$$\lambda = \sqrt{\frac{\sigma_y}{\sigma_{buc-b}}} \tag{12}$$

ρ is a reduction factor and the equation is given by

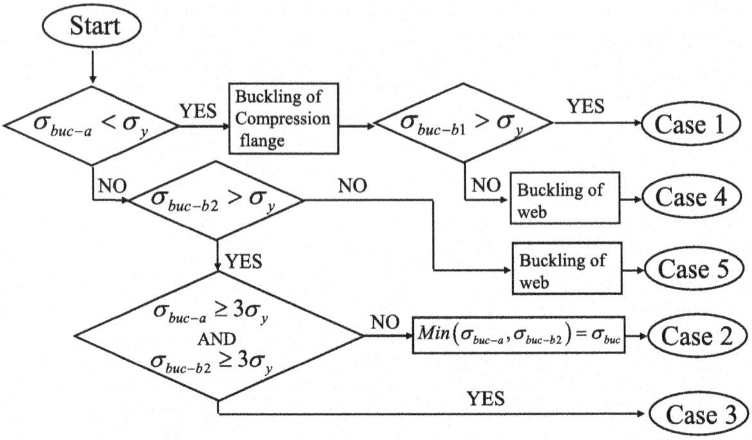

Fig. 4 Flow chart for predicting the maximum moment of triangle tubes under bending

$$\rho = \frac{1}{\lambda} \tag{13}$$

the buckling stress $\sigma_{\text{buc-}b}$ is given by

$$\sigma_{buc-b} = \frac{k_b \pi^2 E}{12(1 - \nu^2)} \left(\frac{t}{b'}\right)^2 \tag{14}$$

where k_b is the buckling coefficient and the value is given by

$$k_b = 1.7 - 5\Psi + 17.1\Psi^2 \tag{15}$$

Figure 4 shows a flow chart for predicting the collapse load of triangle tubes. In the flow chart $\sigma_{\text{buc-b1}}$ and $\sigma_{\text{buc-b2}}$ are the buckling stress of web but the widths are not b. The widths b' are given by

$$b' = b - \frac{a_e}{2} \text{ and } b' = b - \frac{a}{2} \tag{16}$$

respectively, and $\Psi = -1$. Moreover, in Case 4 and Case 5, the stress ratio Ψ is not -1 but unknown. By using the pure bending condition, the value of Ψ can be determined through trial and error. The cross-sectional stress distributions for Case 4 and Case 5 are schematically represented by Fig. 5a, b, respectively.

Figure 6a shows the proposal value of the maximum moment. This figure also shows the FEM results for the sake of comparison. It is seen from the figure that the smaller predicted value is in good agreement with the FEM analysis. Figure 6b shows the axial stress distribution at the maximum moment in $a = 50$ mm, $b = 100$ mm

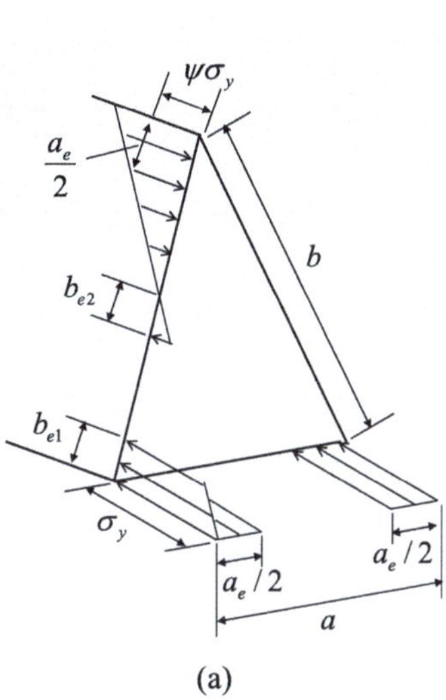

Fig. 5 Axial stress distribution in the maximum bending moment: **a** Case 4: $\sigma_{\text{buc-}a} < \sigma_y$ and **b** Case 5: $\sigma_{\text{buc-}a} > \sigma_y$

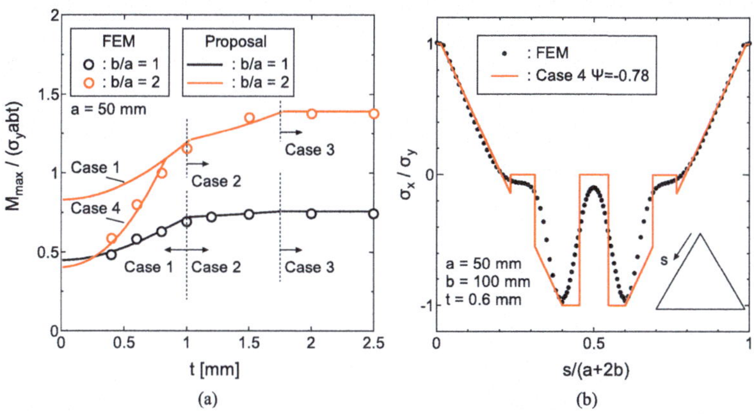

Fig. 6 Comparison of proposal value and FEM results for: **a** relations of $M_{\max}$ and t with $b/a = 1$ and 2; **b** axial stress distribution with $a = 50$ mm, $b = 100$ mm and $t = 0.6$ mm

and $t = 0.6$ mm. FEM result is black mark and our proposal value is red line. As shown in the figure, proposal value is in good agreement with the FEM results.

4 Conclusion

In this paper, the purpose is to investigate whether the previous proposal can be used in estimation method with triangle tubes under pure bending. The conclusion is as follows.

1. The collapse type is the same as rectangular tubes. Case 1 is buckling at the compression flange. Case 2 is plastic yielding at the flange or plastic yielding at the web. Case 3 is the cross-sectional fully plastic yielding. Case 4 is buckling at the compression flange and web. Case 5 is buckling at the compression web.
2. In the triangle tubes, buckling width of the web is smaller than the actual one due to the condition of pure bending. Therefore, the boundary condition of tension side is assumed to be free in contrast to rectangular tubes.
3. The present predicted maximum moment and stress distribution agree with the FEM results.

References

1. Alexander JM (1960) An approximate analysis of the collapse of thin cylindrical shells under axial loading. J Mech Appl Math 13–1:10–15
2. Abramowicz W, Jones N (1984) Dynamic axial crushing of circular tubes. Int J Impact Eng 2–3:263–281
3. Wierzbicki T, Sinmao MV (1997) A simplified model of Brazier effect in plastic bending of cylindrical tubes. Int J Press Vessels Pip 71:19–28
4. Kecman D (1983) Bending collapse of rectangular and square section tubes. Int J Mech Sci 25:623–636
5. Chen DH, Masuda K (2016) Estimation of collapse load for thin-walled rectangular tubes under bending. J Appl Mech 83
6. MSC (2013) MARC 2013 User's guide. MSC Software, Newport Beach, CA
7. Rusch A, Lindner J (2004) Application of level 1 interaction formulae to class 4 sections. Thin-Walled Struct 42–2:279–293

Intersection of Convex Cones as Stress Range for Plane Normal Elastic Bodies

Massimiliano Lucchesi, Barbara Pintucchi, and Nicola Zani

Abstract In this paper, the constitutive model of the normal elastic material has been generalized in order to describe stress states accounting for different strength characteristics in different directions. With this aim, suitable stress ranges can be defined by appropriately varying the relative position between different cones, each of which expresses a particular stress constraint. The model has been formulated and implemented in the Mady code. Then, it has been applied to the study of some test cases.

Keywords Orthotropic materials · Masonry panels · Stress constraints

1 Introduction

Moving on from past experience, the authors have recently formulated a constitutive model for materials with a constrain on the stress that limits their resistance to traction, compression and shear [1, 2]. This constraint implies that the stress belongs to the stress range $\mathcal{K}$, a closed and convex subset of the space of the second-order symmetric tensors. The material is completely characterized by both $\mathcal{K}$ and the tensor $\mathbb{C}$ of the elastic modules, which is hypothesized to be symmetric and positive-definite. Once a strain E has been assigned, the stress T is obtained by projecting $\mathbb{C}E$ onto $\mathcal{K}$ with

To the memory of Giampietro Del Piero

M. Lucchesi (✉) · B. Pintucchi · N. Zani
Department of Civil and Environmental Engineering, University of Florence, via S. Marta 3, Florence, Italy
e-mail: massimiliano.lucchesi@unifi.it
URL: https://www.dicea.unifi.it/p-doc2-2017-200002-Z-3f2a3d30332c2e-0.html

B. Pintucchi
e-mail: barbara.pintucchi@unifi.it

N. Zani
e-mail: nicola.zani@unifi.it

M. Abdel Wahab (ed.), *Proceedings of 1st International Conference on Structural Damage Modelling and Assessment*, Lecture Notes in Civil Engineering 110,
https://doi.org/10.1007/978-981-15-9121-1_7

respect to a suitably defined inner product. The strain is thus additively divided into an elastic part on which T depends linearly and into an inelastic part E^a which is characterized by belonging to the normal cone of $\mathcal{K}$ at T.

In [3], the model has been generalized admitting that $\mathbb{C}$ is orthotropic, in order to better modeling materials with different stiffness in different directions. Nevertheless, the characteristic of $\mathcal{K}$ of being a spherical set was maintained. Such a property implies that T and E^a are coaxial tensors, greatly simplifying the constitutive equation with the consequent benefits in implementing the model in a computer code. On the other hand, modeling materials having different strength characteristics in different directions was not allowed. Obviously, this fact may result a shortcoming in some circumstances, as in case of structures with strength characteristics significantly dependent on masonry texture or when a damage process induces a strength's decay which depends on the direction [4, 5].

The purpose of this paper is to overcome this limitation by developing a constitutive equation in which the stress range is made by the intersection of several closed and convex cones, each one expressing a particular constrain. The vertices of the cones are not necessarily spherical tensors and, by varying their position, it is possible to vary the stress range of the material in a very general way. The model has been implemented for plane bodies in the finite element code Mady [6].

In the following, the model is firstly presented. Then, with reference to a simple masonry panel, a parametric analysis is provided in order to highlight how variations in value of the strengths parameters only in one direction may influence the wall's global response. By fixing the strength parameters of a reference case, values of the tensile, compressive and shear strengths are then opportunely varied and the shear-displacement curves compared. Lastly, some numerical results are compared to some experimental data available in the literature [10].

2 Convex Cones

Let $\mathbb{C}$ be a symmetric and positive-definite fourth order tensor, interpreted as tensor of the elastic modules of the material, and Sym be the space of the symmetric second order tensors with the inner product

$$A \cdot B = \mathrm{tr}(AB) \quad A, B \in \mathrm{Sym} \tag{1}$$

and the corresponding Euclidean norm $\| \|$.

The energetic inner product $(A, B)_E = A \cdot \mathbb{C}^{-1} B$ and the corresponding norm denoted by $\| \|_E$ will also be considered. Moreover, let $\mathcal{K} \subset \mathrm{Sym}$ be a non empty, closed and convex set made by all the admissible stresses. For each $T \in \mathcal{K}$, the normal cone of $\mathcal{K}$ at T is the set

$$\mathcal{N}(\mathcal{K}, T) = \{A \in \mathrm{Sym} : (S - T) \cdot A \leq 0, \text{ for each } S \in \mathcal{K}\}, \tag{2}$$

which when $T \in \partial\mathcal{K}$ always contains some non-null elements. Note that if $\mathcal{K}$ is a spherical set, i.e. such that

$$S \in \mathcal{K} \text{ if and only if } QSQ^T \in \mathcal{K}, \tag{3}$$

for every rotation Q, then it also turns out that

$$Q\mathcal{N}(\mathcal{K}, T)Q^T = \mathcal{N}(\mathcal{K}, QTQ^T). \tag{4}$$

T is a *regular point* if there exists a neighborhood $\mathcal{U}$ of T in Sym and a differentiable function $f : \mathcal{U} \to \mathbb{R}$, such that

$$\begin{aligned} f(T) &< 0, \text{ if } T \in \mathcal{U} \cap \mathcal{K}^\circ, \\ f(T) &= 0, \text{ if } T \in \mathcal{U} \cap \partial\mathcal{K}, \\ f(T) &> 0, \text{ if } T \in \mathcal{U} \cap \mathcal{K}^c, \end{aligned} \tag{5}$$

where $\mathcal{K}^\circ$, $\partial\mathcal{K}$ and $\mathcal{K}^c$ are the interior, the boundary and the complement of $\mathcal{K}$, respectively. In this case, denoted as Df the derivative of f,

$$N(T) = \frac{Df(T)}{\|Df(T)\|} \tag{6}$$

is the outward unit normal vector to $\partial\mathcal{K}$ at T (In the following, the dependence of N on T will be omitted for the sake of brevity).

It is known that assigned a strain tensor E, the minimum norm theorem guarantees the existence and uniqueness of $T \in \mathcal{K}$ having the minimum energy distance from $\mathbb{C}E$, i.e. such that

$$\|\mathbb{C}E - T\|_E = \min_{S \in \mathcal{K}} \|\mathbb{C}E - S\|_E \tag{7}$$

or, equivalently, such that $E - \mathbb{C}^{-1}T \in \mathcal{N}(\mathcal{K}, T)$, i.e.

$$(S - T) \cdot E - (\mathbb{C}^{-1}T) \leq 0, \text{ for each } S \in \mathcal{K}. \tag{8}$$

Once set

$$E^e = \mathbb{C}^{-1}T, \quad E^a = E - \mathbb{C}^{-1}T, \tag{9}$$

it turns out that

$$T = \mathbb{C}(E - E^a) \quad E^a \in \mathcal{N}(\mathcal{K}, T). \tag{10}$$

Relations (10) define the constitutive equation of a non linear hyperelastic material that is said *normal elastic material* [7]. E^e and E^a are said *elastic* and *inelastic part of the deformation*, respectively.

In the particular case where T is a regular point of $\partial\mathcal{K}$, relations (6) and $(10)_2$ implies

$$E^a = \alpha N(T) \qquad \alpha \geq 0. \tag{11}$$

and from $(9)_2$

$$\mathbb{C}E - T = \alpha \mathbb{C}N \tag{12}$$

follows. Scalarly multiplying both members of this equality by N, it follows

$$\alpha = \frac{(\mathbb{C}E - V) \cdot N}{N \cdot \mathbb{C}N} \tag{13}$$

and, again from (12), the equation

$$T = \mathbb{C}E - \frac{(\mathbb{C}N \otimes N)}{N \cdot \mathbb{C}N}(\mathbb{C}E - V). \tag{14}$$

is obtained.

Below is shown how constitutive equation (21) is simplified if $\mathcal{K}$ is the intersection of a certain numbers of convex cones, a case frequently encountered in applications.

A *convex cone* with vertex V is a non-empty, closed set $\mathcal{C}$ containing the origin O of Sym, such that

$$V + a(S - V) + b(T - V) \in \mathcal{C}, \text{ for each } S, T \in \mathcal{C} \text{ and for each } a \geq 0 \text{ and } b \geq 0. \tag{15}$$

Proceeding in a similar way to what was done in [8], it can be verified that for this set the normal cone $\mathcal{N}(\mathcal{C}, T)$ is made up of all the elements $A \in$ Sym such that

$$\text{(i) } (T - V) \cdot A = 0, \tag{16}$$

$$\text{(ii) } (S - V) \cdot A \leq 0, \text{ for each } S \in \mathcal{C}. \tag{17}$$

Let $T \in \partial\mathcal{C}$ be a regular point and $N(T)$ the corresponding outward unit normal. Then, in view of (16) it holds

$$(T - V) \cdot N = 0 \tag{18}$$

that implies

$$(\mathbb{C}E - T) \cdot N = (\mathbb{C}E - T + T - V) \cdot N = (\mathbb{C}E - V) \cdot N. \tag{19}$$

Hence, in substitution of (9) and (11), the relations

$$\alpha = \frac{(\mathbb{C}E - V) \cdot N}{N \cdot \mathbb{C}N} \tag{20}$$

and

$$T = \mathbb{C}E - \frac{(\mathbb{C}N \otimes N)}{N \cdot \mathbb{C}N}(\mathbb{C}E - V). \tag{21}$$

are obtained, which are much more convenient as V is known in applications.

In the next three sections constrains are considered which imply that the stress belongs to convex cones. It will be used the theorem of Hamilton-Caley, i.e. the relation

$$T^2 - (\mathrm{tr}T)T + (\det T)I = 0. \tag{22}$$

From this, scalarly multiplying by the identity tensor I, equation

$$\|T\|^2 - (\mathrm{tr}T)^2 - 2\det T = 0 \tag{23}$$

is obtained.

3 Materials with Limited Tensile Strength

Let Sym$^-$ and Sym$^+$ be the subsets of Sym consisting of the negative and positive semi-definite tensors, respectively, and in a Cartesian reference system $O;x,y$ let

$$T_t = \sigma_{tx} e_1 \otimes e_1 + \sigma_{ty} e_2 \otimes e_2 \tag{24}$$

be a positive semi-definite tensor. A material is said to have *limited tensile strength* if the stress is constrained to belongs to the closed and convex *tensile cone* with vertex T_t

$$\mathscr{T} = \{T \in \mathrm{Sym} : T - T_t \in \mathrm{Sym}^-\} =$$

$$\{T \in \mathrm{Sym} : \mathrm{tr}(T - T_t) \leq 0, \det(T - T_t) \geq 0\}, \tag{25}$$

whose boundary is made up, in addition to $\{T_t\}$ of the set

$$\{T \in \mathrm{Sym} : \mathrm{tr}(T - T_t) < 0, \det(T - T_t) = 0\}. \tag{26}$$

It comes down to the usual case when T_t is a spherical tensor, i.e. when $\sigma_{tx} = \sigma_{ty} = \sigma_t$ [9].

From (23) it follows that the conditions $\det(T - T_t) = 0$ and $\mathrm{tr}(T - T_t) < 0$ are equivalent to

$$f_t(T) = \|\bar{T}\| + \mathrm{tr}\,\bar{T} = 0 \tag{27}$$

where $\bar{T}$ denotes $T - T_t$. Then the unit outward normal to $\partial\mathscr{T}$ at T is

$$N_t = I - \frac{\bar{T}}{\mathrm{tr}\bar{T}}. \tag{28}$$

Let $S = \mathbb{C}E$ be a tensor whose projection T belongs to $\partial\mathscr{T}T_t$. From (21) the equation

$$T = \mathbb{C}E - \frac{(\mathbb{C}N_t \otimes N_t)}{N_t \cdot \mathbb{C}N_t}(\mathbb{C}E - T_t) \tag{29}$$

is obtained and, by setting

$$\mathbb{C}E - T_t = \bar{S}, \tag{30}$$

it follows

$$\bar{T} = \bar{S} - \frac{(\mathbb{C}N_t \otimes N_t)}{N_t \cdot \mathbb{C}N_t}\bar{S}, \tag{31}$$

with N_t given by (28). In other words, $\bar{T}$ can be determined with the same procedure used in the case of a no-tension material, i.e. when $T_t = 0$.

4 Materials with Limited Compressive Strength

Let

$$T_c = \sigma_{cx} e_1 \otimes e_1 + \sigma_{cy} e_2 \otimes e_2 \tag{32}$$

be a positive semi-definite tensor. A material is said to have limited compressive strength if the stress T is constrained to belong to the *compressive cone* with vertex $-T_c$,

$$\mathscr{C} = \{T \in \mathrm{Sym} : T + T_c \in \mathrm{Sym}^+\} =$$

$$\{T \in \mathrm{Sym} : \mathrm{tr}(T + T_c) \geq 0, \det(T + T_c) \geq 0\} \tag{33}$$

whose boundary is made up of $\{T_c\}$ and the set

$$\{T \in \mathrm{Sym} : \mathrm{tr}(T + T_c) > 0, \det(T + T_c) = 0\}. \tag{34}$$

It comes down to the usual case when T_c is a spherical tensor, i.e. when $\sigma_{cx} = \sigma_{cy} = \sigma_c$.

Analogously to the case of limited tensile strength, we can obtain, respectively in place of (27) and (28),

$$f_c(T) = \left\|\tilde{T}\right\| - \mathrm{tr}\tilde{T} = 0 \tag{35}$$

and

$$N_c = -I + \frac{\tilde{T}}{\mathrm{tr}\tilde{T}} \tag{36}$$

where $T + T_c$ has been denoted by $\tilde{T}$.

Similarly to the previous case, by setting $\mathbb{C}E + T_c = \tilde{S}$,

$$\tilde{T} = \tilde{S} - \frac{(\mathbb{C}N_c \otimes N_c)}{N_c \cdot \mathbb{C}N_c}\tilde{S} \tag{37}$$

is obtained, and therefore the same considerations given at the end of the previous Section hold.

5 Mohr-Coulomb Materials

For each unit vector n let σ and τ be the normal and tangential components of the stress T, i.e.,

$$\sigma = n \cdot Tn, \quad |\tau| = \sqrt{n \cdot T^2 n - (n \cdot Tn)^2}. \tag{38}$$

A Mohr-Coulomb material is characterized by the stress constrain $|\tau| \leq f(\sigma)$ where f is an experimentally determined function. Here the hypothesis is made that the stress range is the *shear cone*

$$\mathscr{S} = \left\{T \in \mathrm{Sym} : |\tau| \leq c - \sigma \tan\phi\right\} \tag{39}$$

where $c \geq 0$ and $\phi > 0$ are material parameters usually named *cohesion* and *friction angle*, respectively. In view of (38), from (39) it is obtained

$$n \cdot T^2 n - (n \cdot Tn)^2 - c^2 - (n \cdot Tn)^2 \tan^2\phi + 2c(n \cdot Tn)\tan\phi \leq 0, \tag{40}$$

and from the Hamilton-Caley theorem it follows

$$T^2 = (\mathrm{tr}T)T + \frac{1}{2}(\|T\|^2 - (\mathrm{tr}T)^2)I, \tag{41}$$

so that (40) becomes

$$-2(1 + \tan^2\phi)\chi^2 + 2(\mathrm{tr}T + 2c\tan\phi)\chi + \|T\|^2 - (\mathrm{tr}T)^2 - 2c^2 \leq 0, \tag{42}$$

that is

$$-2\chi^2 + 2(\mathrm{tr}T\cos^2\phi + 2\kappa\sin^2\phi)\chi + \left(\|T\|^2 - (\mathrm{tr}T)^2\right)\cos^2\phi - 2\kappa^2\sin^2\phi \leq 0, \tag{43}$$

where $\chi = n \cdot Tn$ and $\kappa = c \operatorname{cotan}\phi$.

This is a quadratic equation with respect to χ, with a negative first coefficient. Hence, a sufficient condition for (40) to be satisfied for every n is that the discriminant of equation (43) is negative, i.e.,

$$(\text{tr}T\cos^2\phi + 2\kappa\sin^2\phi)^2 + 2[\left(\|T\|^2 - (\text{tr}T)^2\right)\cos^2\phi - 2\kappa^2\sin^2\phi] \leq 0 \tag{44}$$

from which, dividing by $cos^2\phi$,

$$-(1+\sin^2\phi)(\text{tr}T)^2 - 4\kappa\sin^2\phi(\kappa - \text{tr}T) + 2\|T\|^2 \leq 0 \tag{45}$$

follows. It is now easy to verify that T must belong to the set

$$\mathscr{S} = \{T \in \text{Sym} : \|T - \kappa I\|^2 - \frac{1+\sin^2\phi}{2}(\text{tr}(T - \kappa I))^2 \leq 0, \text{tr}(T - \kappa I) \leq 0\} \tag{46}$$

which is a convex cone with vertex κI.

Note that in the particular case when it is $\sin\phi = 1$, the constrain that is expressed by (46) coincides with that of a material with limited tensile strength having $T_t = \kappa I$.

The boundary of $\mathscr{S}$, which is regular everywhere except the vertex $\{\kappa I\}$, is the set where function

$$f_s(T) = 2\|T - \kappa I\|^2 - (1+\sin^2\phi)\text{tr}(T - \kappa I)^2 \tag{47}$$

vanishes and

$$Df_s(T) = 4(T - \kappa I) - 2(1+\sin^2\phi)\text{tr}(T - \kappa I)I \tag{48}$$

holds. As

$$\|Df_s(T)\|^2 = 8\sin^2\phi(1+\sin^2\phi)\text{tr}(T - \kappa I)^2, \tag{49}$$

the unit outward normal to $\partial\mathscr{S}\backslash\{\kappa I\}$ is

$$N_s = \frac{2(T - \kappa I) - (1+\sin^2\phi)\text{tr}(T - \kappa I)I}{\sin\phi\,\text{tr}(T - \kappa I)\sqrt{2(1+\sin^2\phi)}}. \tag{50}$$

The previous relations suggest to generalize this constrain, admitting that the vertex of $\mathscr{S}$ can be any tensor T_s symmetric and positive semi-definite,

$$T_s = \kappa_x e_1 \otimes e_1 + \kappa_y e_2 \otimes e_2. \tag{51}$$

Then, instead of (47) and (50) it is obtained

$$\bar{f}_s(T) = 2\|T - T_s\|^2 - (1+\sin^2\phi)(\text{tr}(T - T_s))^2 \tag{52}$$

and

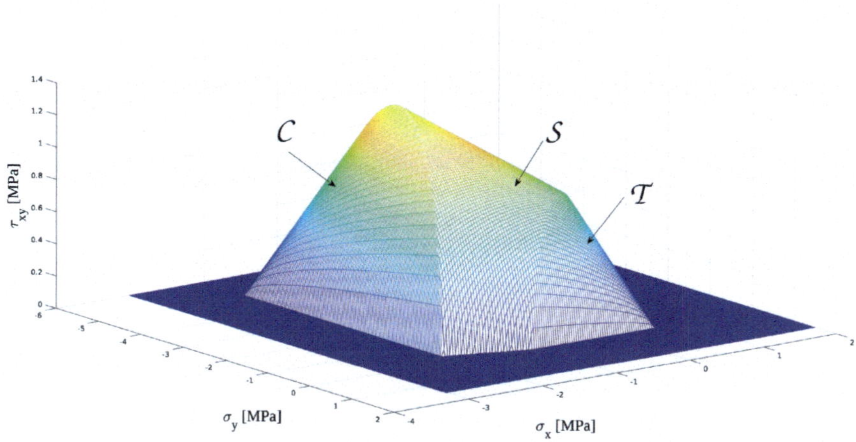

Fig. 1 Stress range as intersection of the three cones: $\mathscr{T}$ tensile, $\mathscr{C}$ compressive and $\mathscr{S}$ shear cone

$$N_s = \frac{2(T - T_s) - (1 + \sin^2\phi)\mathrm{tr}(T - T_s)I}{\sin\phi\, \mathrm{tr}(T - T_s)\sqrt{2(1 + \sin^2\phi)}}. \tag{53}$$

The normal cone to $\mathscr{S}$ at T_s is the set

$$\mathcal{N}(\mathscr{S}, T_s) = \{A \in \mathrm{Sym} : \|A\|^2 - \frac{2}{1 + \sin^2\phi}(\mathrm{tr}A)^2 \leq 0\}. \tag{54}$$

Therefore, for $\mathbb{C}E$ to belong to the region that is projected into the vertex T_s it must be

$$\|E - \mathbb{C}^{-1}T_s\|^2 - \frac{2}{1 + \sin^2\phi}(\mathrm{tr}(E - \mathbb{C}^{-1}T_s))^2 \leq 0. \tag{55}$$

For the sake of example, Fig. 1 shows the stress range in the half-space σ_x, σ_y, $\tau_{xy} \geq 0$, obtained for the following values of the parameters: $\sigma_{tx} = 0.6$ MPa, $\sigma_{ty} = 0.1$ MPa, $\sigma_{cx} = 2$ MPa, $\sigma_{cy} = 5$ MPa, and assuming the two values of the cohesion $c_x = 0.8$ MPa, $c_y = 0.6$ MPa and the friction angle $\phi = 22^o$. Within realistic and reasonable ranges, the values of parameters have been selected to highlight the three cones that constitute the stress range. The stress range is symmetrical with respect to the plane $\tau_{xy} = 0$.

6 Numerical Results

In the following, some numerical results are provided. All the results presented have been obtained via the finite element code MADY [6], which implements the proposed model. All the analyses have been conducted using four-node isoparametric elements, in the framework of plane stress hypothesis.

6.1 *Parametric Analysis*

In this Section, some results of a parametric analysis are provided for the purposes of illustrating the model. The structure considered is a simple masonry panel, 2.0 m height and a rectangular cross-section 1 m in width 0.1 m in thickness. It is perfectly constrained at the base. Two possible constraints conditions have been considered at the top of the panel, which firstly has been assumed free (and denoted in the following as case P1); then a restraint to rotation at the top has been added (named thereafter as case P2).

In the numerical simulations, the panel has been discretized into 200 finite elements. It has first been subjected to a constant vertical load q equal to 10 kN/m^2. The self-weight has also been considered, assuming for the mass density a value = 1900 kg/m^3. Then, the analyses have been conducted by applying a monotonic increasing horizontal displacement at the top of the wall.

Taking into account a masonry texture with horizontal bed-joints, it is reasonable to assume that the horizontal and vertical directions (denoted as x and y) are those in which the panel exhibits the maximum and minimum strength under uniaxial stress. Thus, the x and y axes necessarily coincide with the principal directions of T_t, T_c, and for simplicity, it is assumed that T_s is also coaxial with them.

Moreover, a reference case has been defined by assuming equal values of strengths in each directions and the other following values of the mechanical parameters: Young's modulus $E_x = E_y = E = 1000$ MPa, Poisson's ratio $\nu = 0.1$, $G = \frac{E}{2(1+\nu)}$, tensile strength $\sigma_{tx} = \sigma_{ty} = 0$, compressive strength $\sigma_{cx} = \sigma_{cy} = 1$ MPa, cohesion $c_x = c_y = 0.1$ MPa and friction angle $\phi = 22^o$.

For each contraint conditions, i.e. for cases denoted in the foregoing as P1 and P2, six different analyses have been performed, by opportunely varying the tensile, compressive and shear strengths in each direction (x and y) one by one, whereas all the other values of the parameters have been kept unchanged and equal to the reference case. With the aim of highlighting the generality of the model, the parameters have been varied without accounting for possible relations among them.

With reference to the panel P1, Figs. 2, 3 and 4 show the graphs of the horizontal reaction at the base of the panel as a function of the top displacement, for the six cases described. Figures 5, 6 and 7 are the analogous of Figs. 2, 3 and 4 for the panel P2. The shear-displacement curves are always compared to that of the reference case.

As expected and confirmed herein by Fig. 2, the capability of the panel to sustain lateral loads increases as σ_{ty} increases while it is substantially independent of σ_{tx}. Moreover, the maximum lateral load grows with increasing the value of c_x, contrary to what happens with c_y whose variation does not induce significant effects (see Fig. 4). Lastly, changes in values of σ_{cx} and σ_{cy} does not lead to any variation on the shear-displacement curve (Fig. 3).

The trends exhibited by the results obtained for panel P2 differ significantly from the previous case P1. As shown in Fig. 5, the tensile parameter which mainly influences the panel's lateral capability is now σ_{tx} and not σ_{ty}, leading in this case to horizontal load values always higher with respect to the case P1.

Moreover, as the value of σ_{cx} increases the panel horizontal capability increases as well since, for this kind of constraint and load conditions, crushing occurs in the lower right corner of the panel (Fig. 6).

As regards to variation of c_x and c_y, the higher horizontal load is obtained for the lower value of c_y. The reason can be found in the fact that the intersection between

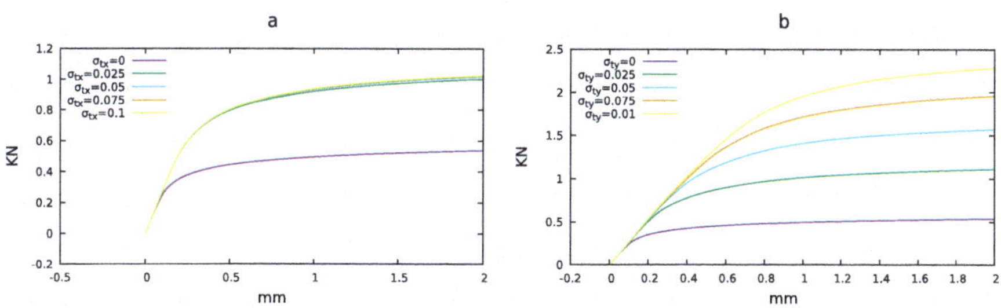

Fig. 2 Shear force versus displacement for case P1 and several values of **a** σ_{tx} and **b** σ_{ty}

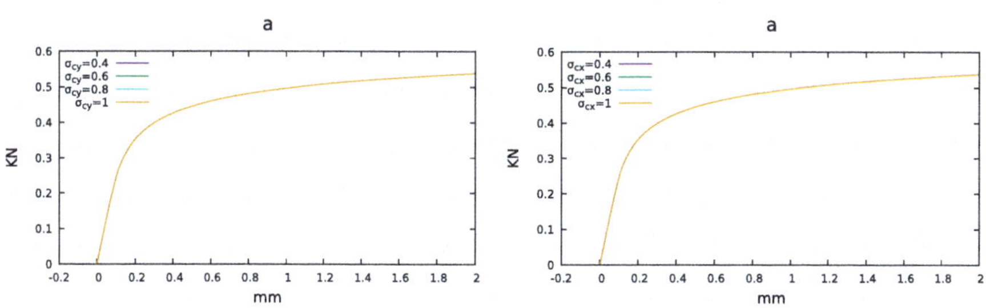

Fig. 3 Shear force versus displacement for case P1 and several values of **a** σ_{cx} and **b** σ_{cy}

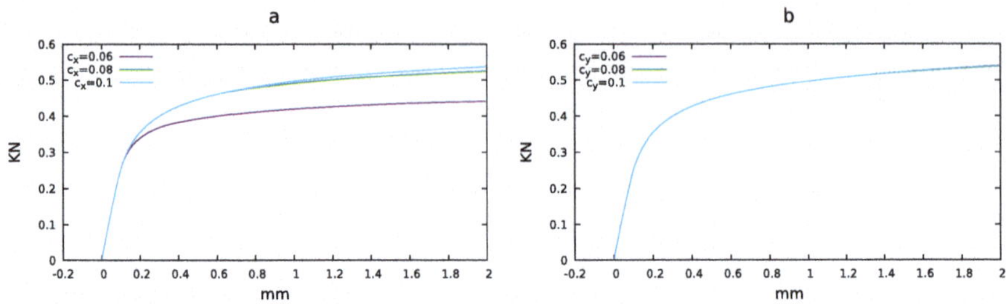

Fig. 4 Shear force versus displacement for case P1 and several values of **a** c_x and **b** c_y

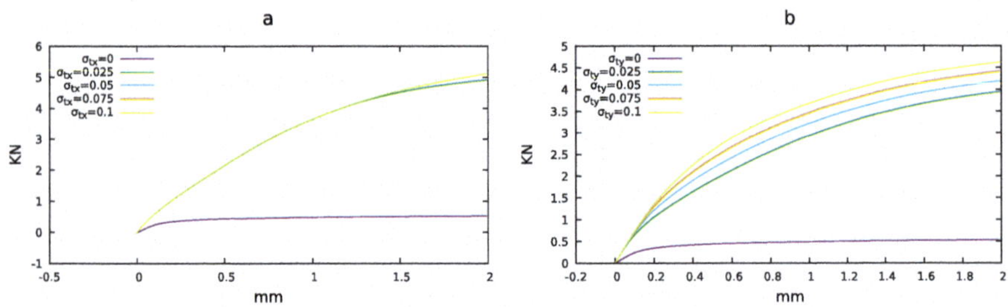

Fig. 5 Shear force versus displacement for case P2 and several values of **a** σ_{tx} and **b** σ_{ty}

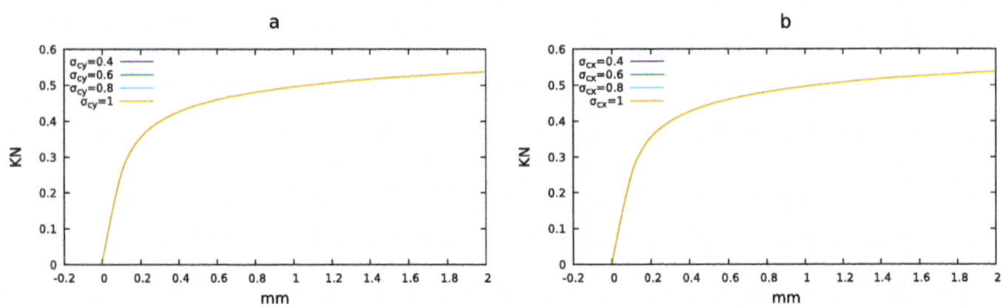

Fig. 6 Shear force versus displacement for case P2 and several values of **a** σ_{cx} and **b** σ_{cy}

the shear cone $\mathscr{S}$ and the tensile cone $\mathscr{T}$ for the lower value of c_y leads to the most favourable stress range for this kind of load process.

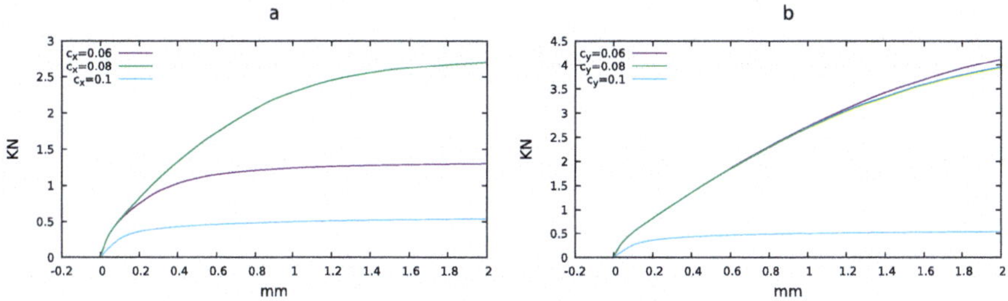

Fig. 7 Shear force versus displacement for case P2 and several values of **a** c_x and **b** c_y

6.2 *Validation of the Model*

The results provided in this section refer to the masonry wall, denoted as W1, tested in the experimental campaign conducted at ETH Zurich by [10]. To summarize briefly, this specimen consists of a panel placed between two lateral masonry flanges. Two concrete slabs are also placed at the top and at the bottom of the specimen (see Fig. 8). A uniformly distributed vertical load, which is equal to 0.77 MPa, is applied on the top [11].

In performing the numerical analyses, the values of the mechanical properties for masonry are assumed accordingly to literature [10, 12–14], i.e. $E_1 = 2460$ MPa, $E_2 = 5460$ MPa, $\nu = 0.18$, $G = 1130$ MPa, $\sigma_{tx} = 0.28$ MPa, $\sigma_{ty} = 0.05$ MPa, $\sigma_{cx} = 1.87$ MPa, $\sigma_{cy} = 7.61$ MPa, $c_x = c_y = 0.26$ MPa and $\phi = 43^o$. For these values, the stress range is the one depicted in Fig. 8b.

As can be seen from Fig. 8c, the graph obtained numerically fits quite well the experimental one.

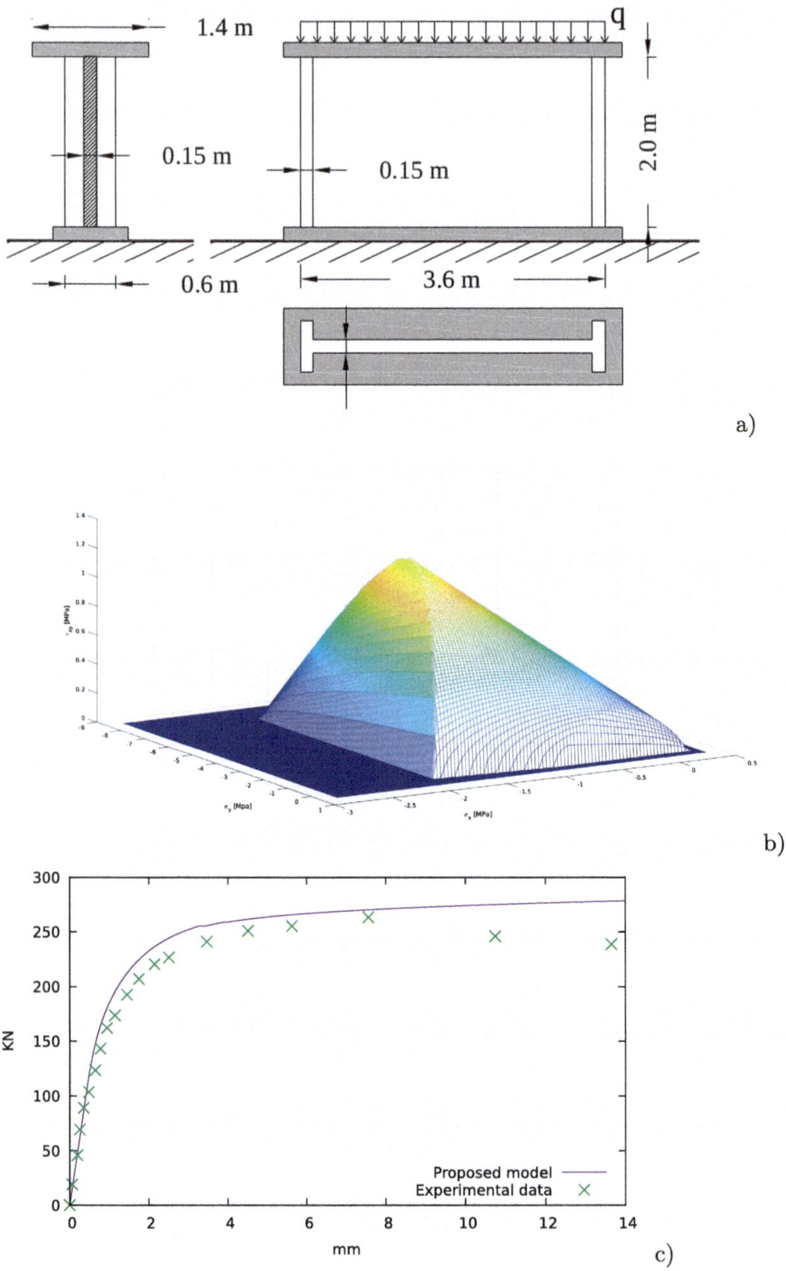

Fig. 8 **a** Geometry and load condition of the specimen; **b** Stress range for the assumed values of the parameters; **c** Shear force versus displacement: numerical and experimental results

7 Conclusions

Thanks to the formulation proposed, enhanced stress ranges for normal elastic materials can be considered, which suitably describe any material's strength characteristics. In particular, as the vertices of the cones whose intersection determines the overall constraint on the stress are no longer necessarily spherical tensors, the model can now well represent situations in which the resistance of the material depends on the direction. As shown by the examples, this feature has significantly increased the flexibility of the model.

References

1. Lucchesi M, Pintucchi B, Zani N (2018) Masonry-like material with bounded shear stress. Euro J Mech A Solids 72, 329:340
2. Lucchesi M, Pintucchi B, Zani N (2018) Bounded shear stress in masonry-like bodies. Meccanica 53:7, 1777:1791
3. Lucchesi M, Pintucchi B, Zani N (2018) Orthotropic plane bodies with bounded tensile and compressive strength. J Mech Mat Struct 13:5, 691:701. https://doi.org/10.2140/jomms.2018.13.691
4. Berto L, Saetta S, Scotta R, Vitaliani R (2002) An orthotropic damage model for masonry structures. Int J Numer Methods Eng 55:2, 127:157
5. Pelá L, Cervera M, Roca P (2011) Continuum damage model for orthotropic materials: application to masonry. Comput Methods Appl Mech Eng 200:9-12, 917:930
6. Lucchesi M, Pintucchi B, Zani N, Modelling masonry structures through the MADY code (in preparation)
7. Del Piero G (1989) Constitutive equation and compatibility of the external loads for linear elastic masonry-like materials. Meccanica 24:3, 150:162
8. Šilhavý M (2014) Mathematics of the masonry-like model and limit analysis. In: Angelillo M (ed) Machanics of masonry structures, CISM Courses and Lectures, vol 551
9. Lucchesi M, Padovani C, Pasquinelli G, Zani N (2008) Masonry constructions: mechanical models and numerical applications. Springer, Berlin
10. Ganz HR, Thrlimann B (1982) Tests on the biaxial strength of masonry (in German), Report No. 7502-3, Institute of Structural Engineering, ETH Zurich
11. Lurati F, Graf H, Thrlimann B (1990) Experimental determination of the strength parameters of concrete masonry (in German), Report No. 8401-2, Institute of Structural Engineering, ETH Zurich
12. Loureno PB, Rots JG (1997) A solution for the macro-modelling of masonry structures. In: Proceedings of the 11th International Brick/Block Masonry Conference, Shanghai, China, pp. 1239–1249
13. Ghiassi B, Soltani M, Tasnimi AA (2012) A simplified model for analysis of un reinforced masonry shear walls under combined axial, shear and flexural loading. Eng Struct 42:396–409
14. Addessi D, Marfia S, Sacco E, Toti J (2014) Modeling approaches for masonry structures. Open Civil Eng J 8, 288:300

Structural Health Monitoring Using Handcrafted Features and Convolution Neural Network

Dung Bui-Ngoc, Thanh Bui-Tien, Hieu Nguyen-Tran, Magd Abdel Wahab, and Guido De Roeck

Abstract Structural Health Monitoring is an important field that involves the continuous measuring of the structural status of infrastructures. In order to be able to detect the damage status, data collected from sensors have to be processed to identify the difference between the damaged and the undamaged states. There exist machine learning techniques attempting to extract features from vibration data, however, they require prior knowledge about the factors affecting the structure. In this paper, we propose a novel method of damage detection using a convolution neural network and a handcrafted feature extraction. This method uses a convolution neural network to extract deep features in time series and uses handcrafted features to find a statistically significant correlation of each lagged features in time series data. These two types of features are combined to increase discrimination ability compared to deep features only. Finally, the neural network will be used to classify the time series into normal and damaged states. The accuracy of damaged detection was tested on a benchmark dataset from Los Alamos National Laboratory and the result shows that hybrid features provided a highly accurate damage identification.

D. Bui-Ngoc · H. Nguyen-Tran
Faculty of Information Technology, University of Transport and Communications, Hanoi, Vietnam
e-mail: dnbui@utc.edu.vn

H. Nguyen-Tran
e-mail: nthieu@utc.edu.vn

T. Bui-Tien (✉)
Faculty of Civil Engineering, University of Transport and Communications, Hanoi, Vietnam
e-mail: btthanh@utc.edu.vn

M. Abdel Wahab
Department of Electrical Energy, Metals, Mechanical Constructions, and Systems, Ghent University, Ghent, Belgium
e-mail: magd.abdelwahab@ugent.be

G. De Roeck
Department of Civil Engineering, Katholieke Universiteit Leuven, Leuven, Belgium
e-mail: guido.deroeck@kuleuven.be

M. Abdel Wahab (ed.), *Proceedings of 1st International Conference on Structural Damage Modelling and Assessment*, Lecture Notes in Civil Engineering 110,
https://doi.org/10.1007/978-981-15-9121-1_8

Keywords Damage detection · Structural health monitoring · Convolution neural network · Handcrafted feature extraction

1 Introduction

In recent decades, structural health monitoring (SHM) systems have emerged as powerful solutions to detect critical damages to structures. The main objective of SHM is aimed at developing efficient methodologies to process the measured data and provide results associated with different levels of the damage identification process. There are a lot of methods to extract the characteristics of the vibration data to detect damage in real structures [1, 2]. There are two main approaches to the diagnostic phase of structural health monitoring: the former is based on the solution of inverse problems [3], and the latter is based on pattern recognition or machine learning [4]. Traditional methods rely on modal parameters such as natural frequencies, mode shapes to detect damage in structural health monitoring. However, they required prior knowledge about the factors that affect the structure.

With data getting larger in size and dimensionality, new approaches for structural health monitoring using machine learning have been proposed [4–6]. A key component of existing machine learning approaches for damage detection is to select and extract proper damaged features and learn optimal classifiers. Autoregressive model is fitted to the acceleration time series data and these model coefficients can be used as damaged sensitive features [7]. Support vector machine (SVM) is a popular machine learning method in damaged detection, which yields high accuracy with small data [5, 8]. In [8], the least square SVM with a combination of radial basis and wavelet kernel function gives higher accuracy in damage detection. Statistical features, also known as statistical indicators, were extracted from raw vibration data and are used as inputs to support vector machine and artificial neural network to classify the levels of the damage state [4].

Recently, with the rapid development of measurement devices leading to a large amount of data collected, as well as increasing computing power, a lot of deep learning techniques have been proposed for structural health monitoring [9, 10]. The common deep learning model is convolution neural network, which has successfully applied on image classification [11], speed recognition [12], object recognition [13]. In damage detection, the time series data was converted into images to take the advantaged of Convolution Neural Network (CNN) in image processing. CNN can automatically extract and select optimal features for damaged classification [10]. In [14], the 1D convolution neural network was applied to the raw acceleration data to detect the damage status in the structure. However, 1D CNN can only learn the internal features of time series and ignore the correlation of time series observations. In this paper, we propose a new method that combined CNN and handcrafted features to find the optimal features. We also examine correlation in time series data to increase the ability to discriminate between the damage and undamaged status from the structure.

2 Methodology

2.1 *Time Series Classification Using Convolution Neural Network*

The Convolution Neural Network (CNN) is a deep learning model that first was proposed by LeCun [11] and is used mainly for image processing or classification. A CNN comprises convolution and pooling layers [13], then these layers are connected to one or more fully-connected layers. From convolution and pooling layers, feature maps are extracted, which are two-dimensional matrices of CNN neurons. A key advantage of CNN is that they are able to learn relevant features from data and weight sharing. It can help to reduce computation complexity and to save memory compared to a traditional neural network. As CNN's input data has traditionally been two-dimensional, a modified model of CNN, called 1D CNN, can be developed to accept input data as one-dimensional time series, while keeping the existing advantages of CNN in image processing. Recent studies show that 1D CNN has certain advantages using time series as input in certain applications. Using 1D CNN, the forward propagation and backward propagation computation only implement array operations instead of matrix operations, which helps reduce computation complexity. Also, 1D CNN with shallow architecture can learn the task in time series problems. These architectures are much easier in training and implementing. The architecture of the 1D CNN for damaged detection is shown in Fig. 1.

The architecture in this paper includes two main parts: the convolution layers that concurrently implement both 1D convolution and pooling operations to extract features. The convolution layers are followed by fully connected layers operating as multi-layer perceptron that implements classification tasks.

In each layer, the forward propagation is calculated as below

$$x_k^l = b_k^l + \sum_{i=1}^{N_{l-1}} conv\left(w_{ik}^{l-1}, s_i^{l-1}\right)$$

where x_k^l is the input, b_k^l is the bias of the kth neural at layer l, s_k^{l-1} is the output of ith neural at layer $l-1$, w_{ik}^{l-1} is the kernel from i^{th} neural at layer $l-1$ to k^{th} neural at layer l. The operation *conv* performs the convolution by multiply the overlapping

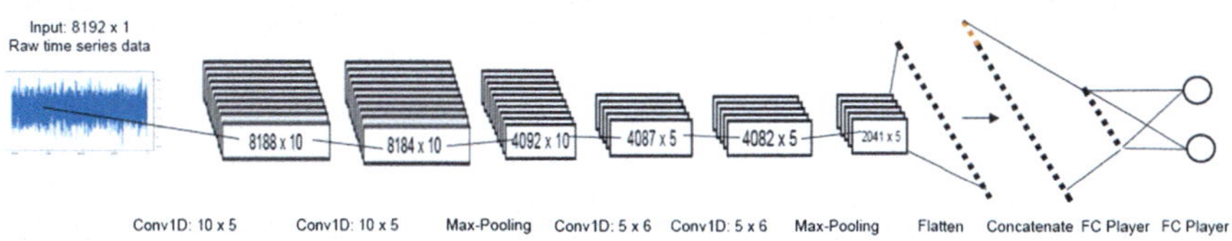

Fig. 1 Framework of the designed convolution neural network

values of the kernel and time series at each position of the kernel, and adds up the results. The activation function $f(.)$ can be used with the input x_k^l to produce the intermediate output $y_k^l = f(x_k^l)$.

To train the 1D CNN, the back propagation algorithm will be used to compute the gradient of the loss function $E(y)$ with respect to the weights of the network. The derivative of the error with respect to each weight is calculated by:

$$\frac{\partial E}{\partial \mathrm{w}_{i,k}^l} = \Delta \mathrm{w}_{i,k}^l$$

By using the chain rule, we can compute the gradient one layer at a time, iterating backward from the last layer, and update the weight as below:

$$\mathrm{w}_{i,k}^{l*} = \mathrm{w}_{i,k}^l + \eta \mathrm{w}_{i,k}^l$$

where $\mathrm{w}_{i,k}^{l*}$ is weights of the next iteration and η is the learning rate. The detail of the algorithm are presented in [14].

2.2 *Feature Extraction Using Autocorrelation*

Although convolution neural network is very effective in automatically extracting features in time series, this approach does not work well with historically dependent time series data, meaning every point in time series depends on previous time instances. To best capture these dependencies, autocorrelation can be used to characterize the relationship between lagged values of a time series. Technically, autocorrelation is a mathematical tool for finding features in the time series. Given time series x, the autocorrelation coefficient C_τ measures the relationship between x_t and $x_{t-\tau}$, the value of C_τ can be calculated as

$$C_\tau = \frac{\sum_{t=\tau+1}^{T}(x_t - \overline{x})(x_{t-\tau} - \overline{x})}{\sum_{t=1}^{T}(x_t - \overline{x})^2}$$

where T is the length and $\overline{x}$ is the mean of time series. The autocorrelation coefficients plot of a time series by lag is autocorrelation function, and useful for finding the features such as presence of periodic signal under noise in structural health monitoring [15].

2.3 Proposed Method

The proposed method includes two main steps, as shown in Fig. 2. The method takes the raw time series signals as an input to the framework. As explained above, in the case of the vibration time series data, CNN cannot capture the temporal relations from the beginning to the end of the time series. We use handcrafted features extraction with autocorrelation to capture the dependency and the coefficients are used as sensitive damaged features. In the first step, autocorrelation were used to extract the coefficients and these autocorrelation coefficients will be used as features for classification. In the second step, the time series were fed to the convolution neural network to automatically extract the features. We constructed the CNN by adjusting all the two-dimensional layers to one-dimensional layers for training and testing. The network settings are shown in Fig. 3. Here, the kernel moves in one direction from the beginning of a time series towards the end to perform convolution. The elements of the kernel get multiplied by the corresponding elements of the time series that they cover at a given point. Then the results of the multiplication are added together and a nonlinear activation function is applied to the value. Finally, the features from two steps are concatenated before going to the fully connected neural network.

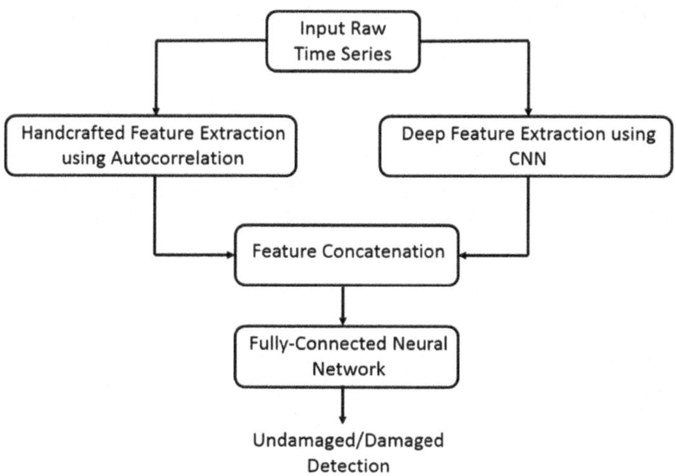

Fig. 2 Overview of the proposed method

```
(features): Sequential(
    (0): Conv1D(filters=10, kernel_size=(5,),stride=(1,))
    (1): ReLU
    (2): Conv1D(filters=10, kernel_size=(5,),stride=(1,))
    (3): ReLU
    (4): MaxPool(pool_size=2, stride=None, padding='valid')
    (5): Dropout(rate=0.5,seed=789)
    (6): Conv1D(filters=5, kernel_size=(6,),stride=(1,))
    (7): ReLU
    (8): Conv1D(filters=5, kernel_size=(6,),stride=(1,))
    (9): ReLU (10): MaxPool(pool_size=2, stride=None, padding='valid')
    (11): Dropout(rate=0.5,seed=789)
    (12): Flatten(in_features=(2041,5), out_features=10205) )
    (merge): Concatenate(int_features = [10205,19], out_features=10224)
(classifier): Sequential(
    (0): Dense(in_features=10224, out_features=512,bias=True)
    (1): ReLU
    (2): Dropout(rate=0.5,seed=789)
    (3): Dense(in_features=512, out_features=2,bias=True)
  (4): Softmax
```

Fig. 3 Network architecture of CNN

3 Results

3.1 Data

Data used in this paper is from Los Alamos National Laboratory (LANL) [15], which is a three-story building structure and is used as a damaged detection test-bed structure. The structure includes several aluminum columns and plates connected using bolted joints to form 3 floors, and slides on rails that allow movement in the horizontal direction. The damage was simulated using the center column on the top floor connected to the bumper, which is adjusted to variation. There are 5 sensors installed at each floor to capture the dynamic response during the excitation. The detail of the structure and data collected is described in [15]. The state conditions can be divided into two states, undamaged states (state 1–9) and damaged states (state 10–17), each state including 10 times testing. The samples of undamaged and damaged signals are shown in Fig. 4.

Data are divided into training and testing set randomly, which is 70% for training and 30% for testing. During the training phase, the features and labels are provided for all the time series in the training set. Afterward, the model has built to capture the relationship between features and class labels. Here, in our method, the features are extracted from deep neural network automatically and also estimated using autocorrelation method. Figure 5 (left) presents the visualization of autocorrelation function coefficients between states. The y-axis is coefficients, which is chosen 15 in this experiment. The x-axis shows the channels and states, where each channel shows two

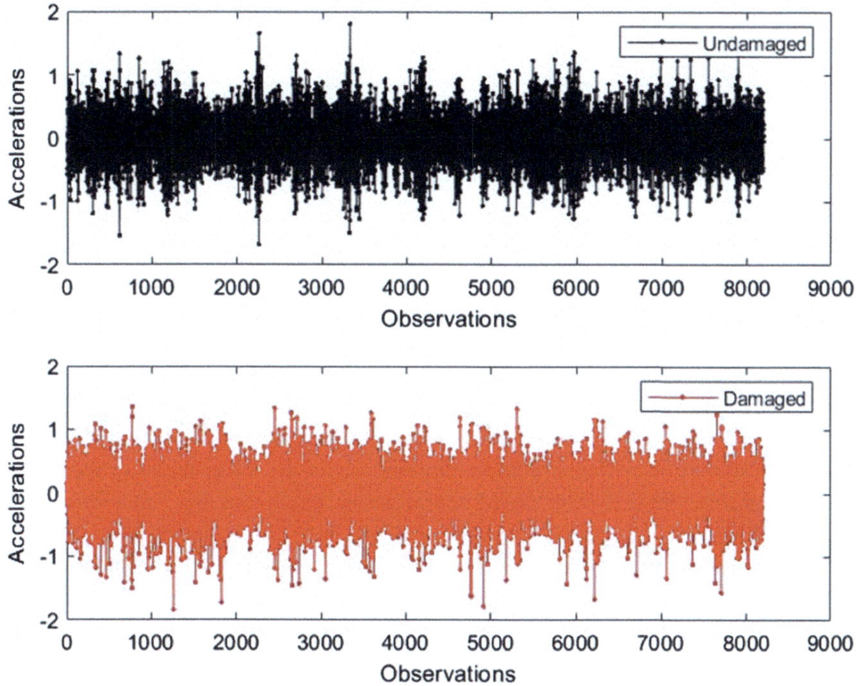

Fig. 4 Sample data in channel 5. Top (undamaged), Bottom (damaged)

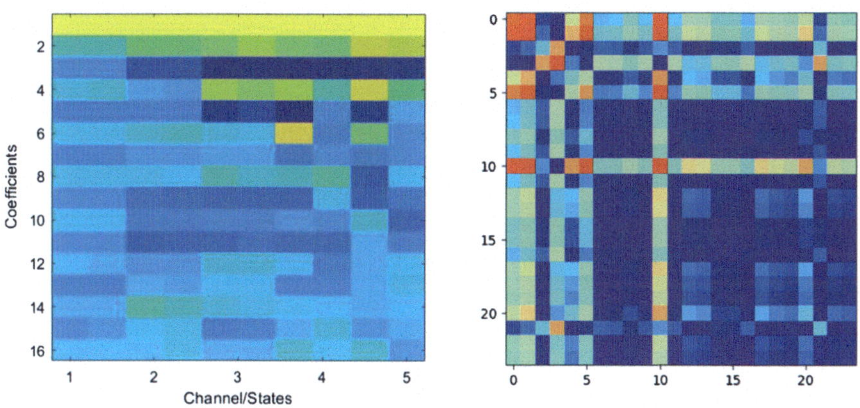

Fig. 5 Visualization of the feature map learned in autocorrelation (left) and CNN (right)

Table 1 Confusion matrix for CNN with handcrafted method (left) and CNN feature (right)

CNN with handcrafted	Predicted		CNN	Predicted	
Actual	109	20	Actual	122	7
	16	110		44	82

states for comparison. Using the autocorrelation function, the damaged and undamaged states do not distinguish because these sensors are located far from the damage source [15]. Channel 3–5 [last three columns in Fig. 5 (left)] show the difference between in coefficients, meaning that they can distinguish the damaged and undamaged states. It proves the feasibility that autocorrelation coefficients can be used as damaged sensitive features. The source of damage can make the change in these coefficients related to the level of damage. Figure (right) shows the features learned from CNN, visualization by Gramian Angular Summation Field (GASF) method [16]. The convolution extracts local features in x-axis. By concatenating these two kinds of features, the fully connected neural network can classify better than used only use CNN.

Prediction accuracy is evaluated on the testing set. We evaluate the accuracy of the methods using the ground truth notion of positive and negative detection. The confusion matrix for two methods CNN and CNN with handcrafted features is shown in Table 1. The accuracy of the method will be calculated as the percentage of correctly classified samples compared with the total number of samples.

$$Accuracy = \frac{(TP + TN)}{(TP + TN + FP + FN)}$$

where TP is true positive, TN is true negative, FP is false positive, FN is false negative. We also evaluate the results using F-measure [17], which is takes both false positive and false negative into account and useful when the data are unbalanced.

$$F\text{-}measure = \frac{2 * \text{Recall} * Precision}{\text{Recall} + Precision}$$

where Recall = TP/(TP + FN) and Precision = TP/(TP + FP).

Based on the matrix of CNN with the handcrafted features, we can see that 109 data samples of undamaged and 110 data sample damaged were correctly detected, as well as 16 data samples of undamaged and 20 data of damaged were misclassified. It means the accuracy of the method is 85% and f-measure is 86%. Similarly, the accuracy and f-measure of CNN method is 80% and 79%, respectively. In the CNN with handcrafted feature, the undamaged state is low accuracy than CNN method, while the damaged state is more accurate than the CNN method. This suggests the CNN method cannot learn the temporal relation from previous to present of the time series.

4 Conclusion

In this paper, we proposed the new damaged detection method using convolution neural network with autocorrelation. The handcrafted feature extraction combined with the features extracted from CNN help to collect richer features than using only CNN method. Experimental results show that the handcrafted features are suitable for damaged detection by capturing the dependence in time series, while feature from CNN is suitable for normal detection. Our results indicated that our proposed method outperforms CNN method in damaged detection.

Acknowledgements The authors acknowledge the financial support of VLIR-UOS TEAM Project, VN2018TEA479A103, 'Damage assessment tools for Structural Health Monitoring of Vietnamese infrastructures' funded by the Flemish Government.

References

1. Kong X, Cai C-S, Hu J (2017) The state-of-the-art on framework of vibration-based structural damage identification for decision making. Dimitrios G. Aggelis. Appl Sci 7:497
2. Sinou J-J (2013) A review of damage detection and health monitoring of mechanical systems from changes in the measurement of linear and non-linear vibrations. In: Sapri RC (ed) Mechanical vibrations: measurement, effects and control. Nova Science Publishers, Inc., pp 643–702, 2009, 978-1-60692-037-4. ffhal-00779322f
3. Deraemaeker A, Worden K (2010) Structural health monitoring using pattern recognition. New trends in vibration based structural health monitoring, CISM Courses and Lectures, vol 520, pp 183–246
4. Finottia RP, Cury AA, de Souza Barbosaa F (2019) An SHM approach using machine learning and statistical indicators extracted from raw dynamic measurements. Latin Am J Solids Struct 16(2):1–17
5. Gui G, Pan H, Lin Z, Li Y, Yuan Z (2016) Data-driven support vector machine with optimization techniques for structural health monitoring and damage detection. KSCE J Civ Eng 21(2):523–534
6. Noori MN, Cao Y, Hou Z et al (2010) Application of support vector machine for reliability assessment and structural health monitoring. Int J Eng Under Uncertain: Hazard Assess Mitig 2010(2):89–98
7. Da Silva S, Paixão J (2019) Extrapolation of autoregressive models for structural health monitoring. Compos Struct. https://doi.org/10.26678/ABCM.MECSOL2019.MSL19-0166
8. Ghiasi R, Torkzadeh P, Noori M (2016) A machine-learning approach for structural damage detection using least square support vector machine based on a new combinational kernel function. Struct Health Monit, 1–15
9. Tabian I, Fu H, Khodaei ZS (2019) A convolutional neural network for impact detection and characterization of complex composite structures, sensors, pp 1–25
10. Tang Z, Chen Z, Bao Y, Li H (2019) Convolutional neural network-based data anomaly detection method using multiple information for structural health monitoring. Struct Control Health Monit 26(1)
11. LeCun Y, Boser BE, Denker JS, Henderson D, Howard RE, Hubbard WE, Jackel LD (1990) Handwritten digit recognition with a back-propagation network. In: Advances in neural information processing systems. MIT Press, Cambridge, MA, USA, pp 396–404

12. Abdel-Hamid O, Mohamed A-r, Jiang H, Deng Li, Penn G, Dong Yu (2014) Convolutional neural networks for speech recognition. IEEE/ACM Trans Audio Speech Lang Process 22(10):1533–1545
13. Christian S, Alexander T, Dumitru E (2013) Deep neural networks for object detection. Adv Neural Inf Process Syst 26:1–9
14. Malek S, Melgani F, Bazi Y (2018) One-dimensional convolutional neural networks for spectroscopic signal regression. J Chemom 32:1–17
15. Figueiredo E, Park G, Figueiras J, Farrar C, Worden K (2009) Structural health monitoring algorithm comparisons using standard data sets. Los Alamos National Laboratory, LA-14393
16. Wang Z, Oates T (2015) Imaging time-series to improve classification and imputation. arXiv preprint arXiv:1506.00327
17. Tharwat A (2018) Classification assessment methods. In: Appl Comput Inf

A Versatile Interrogation-Free Magnetoelastic Resonator Design for Detecting Deterioration-Inducing Agents

Dimitrios G. Dimogianopoulos and Dionysios E. Mouzakis

Abstract The experimental design of a versatile interrogation-free magnetoelastic resonator suitable for detecting local accumulation of substances/agents potentially harmful to their environment, is presented. The concept uses a thin flexible polycarbonate slab with a Metglas® 2826MB magnetoelastic ribbon attached on its surface, and implements typical damage detection principles. The slab is fixed as a cantilever on one end, with a mobile talk-phone attached to the free end for providing vibration. Owing to the magnetoelastic principle, the vibrating ribbon creates variable magnetic flux, which induces voltage into a low-cost pick-up coil circuit placed in a contact-less manner above the slab. This voltage is recorded by a conventional oscilloscope. No extra interrogation coil or sophisticated hardware are necessary, thus enhancing versatility of the concept while minimizing its complexity and cost. Analysis of the vibration-dependent voltage signal provides specific slab eigenfrequencies, which depend upon (and characterize) its surface loading. The latter may result from environmental/biological/infectious agents or substances accumulating on the resonator's (suitably coated) surface, which may cause damage/deterioration to fragile or valuable objects placed close to the resonator. Eigenfrequency shifting may, then, allow for detecting such factors and triggering alerts. Tests with small loads on the vibrating slab confirm its potential for surface load detection.

Keywords Magnetoelastic materials · Interrogation-free resonator · Accumulating substance detection

---

D. G. Dimogianopoulos (✉)
Department of Industrial Design and Production Engineering, University of West Attica, 12241 Athens, Greece
e-mail: dimogian@uniwa.gr

D. E. Mouzakis
Sector of Mathematics and Engineering Applications, Mechanics Laboratory, Hellenic Army Academy, Vari, 16673 Attica, Greece

M. Abdel Wahab (ed.), *Proceedings of 1st International Conference on Structural Damage Modelling and Assessment*, Lecture Notes in Civil Engineering 110,
https://doi.org/10.1007/978-981-15-9121-1_9

1 Introduction

The arduous development of magnetoelastic resonators over the last 20 years, allowed for wireless monitoring of physical, chemical, or mechanical quantities in applications where conventional sensors could not be implemented. Such applications include environments affected by chemical/biological substances or infested by agents potentially hazardous to humans, or simply confined places hardly accessible for monitoring and maintenance. The magnetoelastic (magnetostrictive) property of some (ferromagnetic) materials which modify their shape due to external varying magnetic fields and vice versa, has been nicely reviewed in [1] and [2]. The applicability of magnetoelastic materials to sensing applications has been reviewed and classified according to the quantity to be measured (gas concentration, viscosity of liquids, mass, temperature) [1, 2], or to the sensing setup principle of the considered applications [3].

Of particular interest are magnetoelastic resonators detecting loading variations on the purposely coated resonator's surface, due to deposition of biomedical [4] or environmental agents [5], reaction of volatile organic compounds (i.e. sources of long-term health problems) with surface coating [6], and H_2O [7] or H_2O_2 [8] adsorption to name but a few. Early detection of such damage-inducing agents/substances is of outmost importance for preserving human health or valuable objects such as items in museum collections [9] from deterioration. Most of such resonators operate using an interrogation coil which, under suitable electrical excitation, produces variable magnetic flux towards the magnetoelastic material (usually in form of a ribbon), thus driving it to resonance. A pick up coil, which is also placed above the ribbon in a contact-less manner, is used for transforming the flux emitted by the vibrating ribbon into electrical signal of the corresponding frequency. The latter only changes with sensor surface loading. Effort has been invested in maximizing the resonator sensitivity to loading, for instance by optimizing the shape of resonator surface [10], or by allowing for optimal load placement [11] wherever possible. Nonetheless, similar effort has rarely been channeled towards making the resonator simpler to build and operate, and more versatile, hence, cost-effective.

The current work aims at addressing this issue by experimenting with a versatile and robust design, which operates on typical damage detection principles and minimizes use of equipment by eliminating the interrogation coil. The magnetoelastic material is vibrated via a conventional talk-phone, meaning that the signal generator and associated circuitry for the interrogation task is no longer required. The experimental design retains the wireless pick-up of the flux emitted by the vibrating ribbon via a coil placed at a distance above it. Its electrical output may be recorded either by a conventional oscilloscope, or by using the soundcard of the portable computer used for further processing this signal and evaluating any resonant shifts due to surface loading. The paper is organized as follows: Further details of the experimental setup are given in the following section, whereas in Sect. 3, results along with relevant discussion are shown. Finally, some concluding remarks are presented in Sect. 4.

2 Materials and Methods

The resonator design is based on a thin slab measuring 50 × 30 × 0.8 mm with a Metglas® 2826MB ribbon of 25 × 5 mm laterally centered on the slab surface and attached with cyanoacrylic glue. The longitudinal and lateral dimensions may be adjusted at will, meaning that the resulting resonator design may be easily adapted to the available space on site. This is useful when monitoring accumulation of damage-inducing factors in objects such as items in museum collections, since any monitoring setup must be of limited size and located out-of-view.

The slab shown in Fig. 1 (left) is fixed as a cantilever on one end, with a conventional mobile talk-phone fixed on the backside of the free end providing vibration sweeps of 4 secs at frequencies of 120–180 Hz [12]. The phone is fixed in place so that the slab shows no vertical deformation at rest. Thus nonlinear phenomena in the vibrating slab's response are kept to a minimum, with the principle of frequency shifts due to change surface load retained as in [1, 2]. According to the magnetoelastic phenomenon [1], the vibrating ribbon produces varying magnetic flux depending on the vibration characteristics of the slab. This in turn induces voltage in a low-cost pick-up coil (Vishay IWAS) placed 15 mm above the ribbon and, hence, bearing no contact with it. The voltage signal is recorded by a digital oscilloscope, and is forwarded (even wirelessly via the Local Area Network) for frequency analysis. Hence, specific slab eigenfrequencies characteristic of the changes in surface loading are obtained.

Contrary to the standard resonator design with a ribbon carefully suspended and vibrated at its resonance frequency by an interrogation coil, in the current setup the phone operates both as slab support and as exciter. Thus, no interrogation coil is required. The ribbon may be attached onto a less flexible supporting structure (thin slab), which may be vibrated for monitoring changes in its eigenfrequencies, similarly to traditional schemes used for fault detection in structures. The resulting

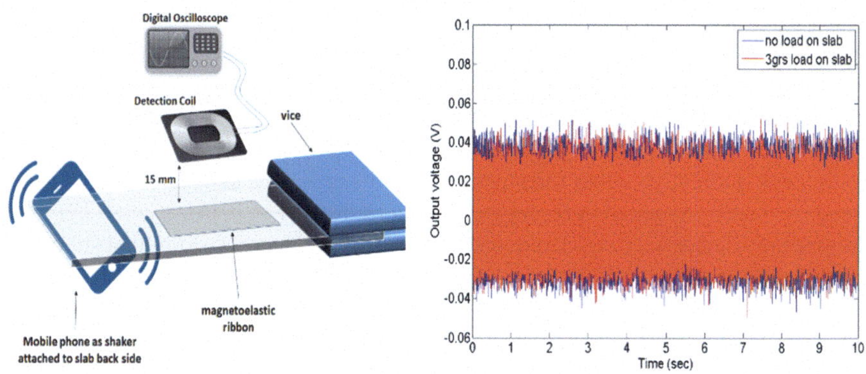

Fig. 1 Representation of the experimental setup (left) and coil output time histories from tests with different mass loads (0 and 3 g) on the slab (right)

Table 1 Tests and load on surface of resonator

Test	Load on surface (g)	Duration (s)
P1–P3	0	10
L1–L3	0.5	10
H1–H3	3.0	10

versatility and robustness is well suited to sensing applications in confined (or not safely accessible) environments, requiring low implementation costs.

Tests have a typical duration of 10 secs, with coil output (in V) recorded at 200 KHz. Nine series of tests have been carried out (see Table 1). Three tests (namely P1–P3) without load on slab surface, another three with tiny metallic fragments of approximately 0.5 g (L1–L3) on the slab surface at 10 mm from the fixed end, and finally three tests with more fragments weighting approximately 3 g (H1–H3) at the same position as for L1–L3 tests. The loading position was as feasibly close to the vice as possible, but ultimately the ribbon coating (used for trapping the agents/substances considered) may be placed for achieving maximum sensitivity to loading [11].

The power spectral density (PSD) of the recorded signals is estimated, via the Welch method, which separates the signal into segments, uses the discrete Fourier transform on each segment for computing the spectrum estimate and averages results. Two different settings are used for calculating the PSD: The first (low-resolution) one involves many small segments thus averaging out noise contribution, but allowing for identifying frequency bands where only principal eigenfrequencies lie. The second (high-resolution) setting involves few but large segments, thus providing a detailed picture of the identified eigenfrequencies (but also of the noise).

3 Results and Discussion

From Fig. 1 (right), it is obvious that no clear conclusions may be drawn from comparison of signal time histories of P-tests (no load on slab) even with the worst case of H-tests (significant load on slab). The mobile phone cannot create the level of vibration required for the ribbon to generate sufficient electrical power via the magnetoelastic phenomenon. Consequently, when such passive setups (without interrogation coil) are utilized, qualitative rather than quantitative changes may, at best, be found in the signals recorded. As noted in recent work of the authors [13], peaks as those in Fig. 2 indicate quite accurately eigenfrequencies, with this fact confirmed in that study via finite element analysis. Another finding in [13] regarded the fact that mainly eigenfrequencies at higher bands may be distinguishable when such passive setups are used. The PSD plots in Fig. 2 illustrate this fact, with the low resolution plot (left) indicating two high magnitude peaks (i.e. frequency bands) of interest for P1–P3 and L1–L3 tests (at 1.3 and 80 KHz) and again two for H1–H3 tests (at 1.3 and 45 KHz).

In turn, the high resolution PSD plot in Fig. 2 (right) allows for obtaining a quite clear description of the eigenfrequencies in those regions (1.3, 45 and 80 KHz)

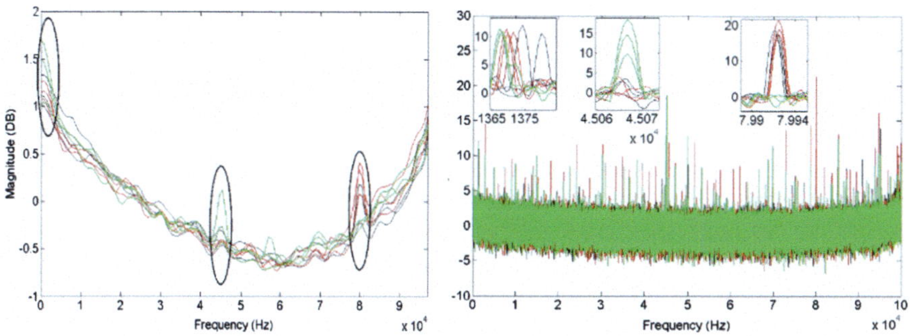

Fig. 2 Power spectral density of P1–P3 (all three in black), L1–L3 (all three in red) and H1–H3 (all three in green) in low (left) and high (right) resolution, with the frequency bands of interest equal to 1.3, 45 and 80 KHz [shown in ellipses (left) or insets (right)]

at the expense of also visualizing spectral contribution of the noise. It results that around 1360 Hz there seems to be an eigenfrequency shift to lower frequencies with growing load on the slab, which is reasonable. There also seems that the value of (approximately) 79,935 Hz corresponds to eigenfrequencies only for tests with unloaded (P1–P3) or lightly loaded (L1–L3) slabs. When significant load is placed on the slab (tests H1–H3) this eigenfrequency shifts to values around 45,060 Hz. Thus, obtaining a clear distinction between a quite loaded (tests H1–H3) and unloaded slabs (tests P1–P3) is definitely feasible. Detecting lightly loaded (tests L1–L3) from unloaded slabs seems more complicated, even if a definitive trend towards lower eigenfrequency values with growing load seems to be present.

Thus, in principle it is possible to use the current experimental design for monitoring those eigenfrequencies characterizing the behavior of an unloaded slab under vibration, and conclude on the existence of sensor loading when eigenfrequency shifting occurs. In the current case, a thin layer of sticky substance was created on the slab surface at approximately the ribbon's location, in order to fix the metallic fragments used as surface load when testing. The same principle applies to cases involving monitoring of the ambient space around fragile or valuable objects for detecting pest, insects or other damage-inducing agents [9], without physical human intervention. The mobile talk-phone may be used to wirelessly operate the resonator, with analysis of the response signal (also wirelessly recorded) in terms of eigenfrequencies being compared to those previously obtained. Any eigenfrequency shifts indicate that agents have been trapped onto the (suitable coated) surface, thus triggering an alert.

4 Conclusions

A versatile magnetoelastic resonator design for detecting environmental agents/substances potentially detrimental to fragile or valuable objects has been presented. The experimental design operates using typical damage detection principles. Unlike common resonators based on an interrogation coil to drive the magnetoelastic ribbons to an imposed frequency of vibration, the current experimental design innovates in that no interrogation coil is used and the ribbon along with the supporting structure (thin flexible slab) is not driven to such resonant. Instead, a conventional mobile talk-phone is used for vibrating the setup (thin slab and ribbon) at frequencies up to 180 Hz. A low-cost remote pick up coil is, then, used for transforming the vibration-dependent magnetic flux into voltage. Its frequency analysis provides a means for monitoring the slab eigenfrequencies with their shifted values pointing to changing load on the slab surface. Testing at unloaded, lightly or significantly loaded configurations indicates the resonator's ability for detecting changing surface load, thus illustrating its potential use for applications where versatility, robustness and low-cost operation are important.

References

1. Le Bras Y, Greneche J-M (2017) Magneto-elastic resonance: principles, modeling and applications. In: Awrejcewicz J (eds) Resonance, chap 2. IntechOpen, 51000 Croatia. https://doi.org/10.5772/intechopen.70523
2. Grimes CA, Roy S, Rani S, Cai Q (2011) Theory, instrumentation and applications of magnetoelastic resonance sensors: a review. Sensors 2011(11):2809–2844
3. Dimogianopoulos D (2012) Sensors and energy harvesters utilizing the magnetoelastic principle: review of characteristic applications and patents. Recent Pat Electr Electron Eng 5(2):103–119
4. Ren L, Yu K, Tan Y (2019) Applications and advances of magnetoelastic sensors in biomedical engineering: a review. Materials 2019(12):1135. https://doi.org/10.3390/ma12071135
5. Grimes CA, Jain MK, Singh RS, Cai Q, Maso A, Takahata K, Gianchandani Y (2001) Magnetoelastic microsensors for environmental monitoring. In: Proceedings IEEE international conference on micro electro mechanical systems (MEMS). Interlaken, Switzerland. https://doi.org/10.1109/MEMSYS.2001.906532
6. Baimpos T, Boutikos P, Nikolakis V, Kouzoudis D (2010) A polymer-metglas sensor used to detect volatile organic compounds. Sens Actuators, A 158:249–253
7. Atalay S, Izgi T, Kolat VS, Erdemoglu S, Inan OO (2020) Magnetoelastic humidity sensors with TiO_2 nanotube sensing layers. Sensors 2020(20):425. https://doi.org/10.3390/s20020425
8. Samourgkanidis G, Nikolaou P, Gkovosdis-Louvaris A, Sakellis E, Blana IM, Topoglidis E (2018) Hemin-modified SnO_2/Metglas electrodes for the simultaneous electrochemical and magnetoelastic sensing of H_2O_2. Coatings 2018(8):284. https://doi.org/10.3390/coatings8080284
9. Michalski S (2004) Care and preservation of collections. In: Boylan PJ (eds) Running a museum: a practical handbook. ICOM—International Council of Museums, Maison de l'UNESCO, Paris, France, pp 51–90
10. Saiz P, Gandia D, Lasheras A, Sagasti A, Quintana I, Fdez-Gubieda ML, Gutiérrez J, Arriortua MI, Lopes AC (2019) Enhanced mass sensitivity in novel magnetoelastic resonators geometries for advanced detection systems. Sens Actuators B Chem 296:126612

11. Zhang K, Chai Y (2016) Numerical study on mass sensitivity of magnetoelastic biosensors with concentrated mass load under different resonance modes. J Sens (8341656). https://doi.org/10.1155/2016/8341656
12. Baek Y, Myung R, Yim J (2006) Have you ever missed a call while moving? The optimal vibration frequency for perception in mobile environments. In: Proceedings 6th WSEAS international conference on applied informatics and communications. Elounda, Greece, pp 241–245
13. Dimogianopoulos DG, Charitidis PJ, Mouzakis DE (2020) Inducing damage diagnosis capabilities in carbon fiber reinforced polymer composites by magnetoelastic sensor integration via 3D printing. Appl Sci 2020(10):1029. https://doi.org/10.3390/app10031029

Seismic Retrofitting of Buildings with Damped Braces by Using a Computer-Aided Design Procedure

Fabio Mazza and Carlo Pasceri

Abstract Damped braces (DBs) can be an easy and not expensive retrofitting solution of reinforced concrete (r.c.) framed buildings, in order to attain a designated seismic performance for a given intensity of the ground motion. Most of the current displacement-based design (DBD) procedures of DBs only consider the capacity curve of an initially undamaged structure. However, structures may be initially damaged by previous earthquakes, so the proportioning of DBs should include the effects of the cumulative damage induced by previous loading cycles. In this study, a displacement-damage-based design (DDBD) procedure of hysteretic damped braces (HYDBs) is implemented in an automated application named DAMPERS. It is divided into six electronic sheets, each one having blank boxes editable by the user and grey boxes showing the outcome of the automated evaluation. A six-storey r.c. framed building, representative of the Italian residential housing stock constructed during the 1990s, is retrofitted with HYDBs designed through the proposed DDBD procedure with the support of DAMPERS tool. In order to assess the relevance of damage due to previous earthquakes, the retrofitting solutions are also obtained with a simplified DBD procedure. Three structural configurations are considered for the original and retrofitted structures: i.e. bare frame, with nonstructural MIs; infilled frame, with a uniform in-elevation distribution of structural MIs; pilotis frame, with no MIs at the ground floor and structural MIs at the other floors. To check the effectiveness of the HYDBs, nonlinear seismic analysis of the test structures is carried out using the computation platform OpenSees.

Keywords R.C. framed buildings · Previous earthquakes · Loss of capacity · Seismic retrofitting · Hysteretic dampers · Computer-aided design procedure

---

F. Mazza (✉) · C. Pasceri
Dipartimento di Ingegneria Civile, Università della Calabria, Rende, Cosenza, Italy
e-mail: fabio.mazza@unical.it

M. Abdel Wahab (ed.), *Proceedings of 1st International Conference on Structural Damage Modelling and Assessment*, Lecture Notes in Civil Engineering 110,
https://doi.org/10.1007/978-981-15-9121-1_10

1 Introduction

Passive energy dissipation technology is increasingly used for retrofitting of framed buildings, preferring to adopt displacement-dependent (e.g. hysteretic, HY) damped braces (DBs) to other typologies, because of their stable hysteretic behaviour and relatively low production and maintenance costs. A wide variety of hysteretic (metallic) dampers is currently available, utilizing different yielding methods based on shear, flexural and axial behaviour [1]. Aimed at obtaining different performance requirements for different seismic intensities, displacement-based design (DBD) procedures have been proposed in literature [2–7]. Static pushover analysis is frequently used for evaluating the extent of damage experienced at a target displacement, so implying an initially undamaged structure for which cyclic effects from previous earthquakes are not considered. In detail, a target displacement corresponding to a performance objective (e.g. avoid structural collapse and/or reduce non-structural damage) is assigned, combining static pushover analysis of the multi-degree-of-freedom (MDOF) model of the original structure with the capacity spectrum of an equivalent single-degree-of-freedom (SDOF) system. However, cyclic effect of the seismic loading is not modelled in conventional nonlinear static analysis, so that the initial backbone (i.e. the monotonic loading curve) and the cyclic envelope (i.e. the curve enveloping the hysteretic response) laws are usually different. Specifically, the first curve may lead to an overestimation of the lateral strength and stiffness and an underestimation of the displacement demand, while the latter may be heavily loading-history dependent [8]. In order to capture the degradation of the pushover properties, a DBD design procedure of the HYDBs, proposed in [5, 9] and extended in [10, 11] to take into account in-elevation irregularity, is further upgraded in a displacement-damage-based design (DDBD) approach, to evaluate the influence of the cumulative damage of reinforced concrete (r.c.) framed buildings resulting from repeated seismic loads with reverse in signs. A damage index (Ψ) weighs the energy dissipation related to deformations exceeding or not the previous maximum values and couples the energy dissipation under positive and negative forces, while a degradation parameter (ε) takes into account well or poor detailed and designed structures. For a wider promptness and attractivity of the proposed DDBD procedure for practitioners, an automatic application named DAMPERS is implemented.

The proposed DDBD procedure of the HYDBs and its automatic implementation DAMPERS are applied for the seismic retrofitting of a six-storey building in L'Aquila, representative of the Italian residential housing stock during the 1990s. Three in-elevation configurations of masonry infills (MIs) are assumed, simmetrically distributed in plan along the perimeter: (i) Bare Frame (BF), with nonstructural MIs; (ii) Infilled Frame (IF), with an uniform in-elevation distribution of structural MIs; (iii) Pilotis Frame (PF), with no MIs at the ground level and structural MIs at the other floors. The effects of previous seismic damage is investigated comparing cases without and with degradation of the r.c. frame members. The stiffness distribution of the HYDBs is designed to exclude soft storey behaviour for the PF structure, while a proportional stiffness criterion is applied for the BF and IF structures. Moreover, the strength distribution of the HYDBs is assumed so that their activation occurs before the attainment of the ultimate values of frame and infill shear forces for the PF structure, while the distribution of the yielding loads of the HYDBs is assumed proportional to the stiffness distribution for the BF and IF structures. Finally, nonlinear dynamic analyses of the unbraced and damped braced (DBBF, DBIF and DBPF) structures are carried out using the computation platform OpenSees [12], considering records scaled in line with the design hypotheses adopted. A lumped plasticity model is used for r.c. frame members, with flexure- or shear-controlled moment-chord rotation at critical end sections. The shear behaviour of the beam-column joints is modelled by means of a scissor model, while MIs are modelled with a simplified diagonal pin-jointed strut model.

2 DAMPERS Computer Aided Tool to Design Damped Braces

A computer-aided tool named DAMPERS is used for an automatic design of mechanical properties for the HYDBs, in accordance with the proposed displacement-damage-based design (DDBD) procedure. The DAMPERS software is divided into six electronic sheets shown in Fig. 1. Parameters related to previous seismic damage of the structure that needs to be retrofitted are preliminarily required (Fig. 1a), with the possibility to set an initial value of the damage index (Ψ) and its evolution law (ε). Arrangement of steel braces (i.e. diagonal, chevron or cross configuration) can be also selected in sheet 1 (Fig. 1a). In sheet 2, dynamic and geometric properties of the original structure need to be uploaded (Fig. 1b). A multilinear approximation of the base shear-top displacement (V_F-d) initial backbone curve of the original structure is also required, together with the performance displacement (d_p). Finally, effective properties of the equivalent SDOF system are given automatically as an output. In sheet 3 (Fig. 1c), elastic response spectra of acceleration (A) and displacement (D) and their ADRS combination are defined at the collapse prevention (CP) limit state. An artificial accelerogram is also uploaded, whose response spectrum should be compatible with the above defined design response spectrum. In sheet 4 (Fig. 1d), cyclic curve of the SDOF system is obtained by means of the button NTHA, carrying out the nonlinear seismic analysis of the equivalent system subjected to the previously defined loading history. As a comparison, the undamaged capacity curve and target displacement (d^*_p) are also plotted in the same graph. Then, a multilinear approximation of the envelope curve can be automatically drawn by the user. Finally, viscous damping ratio equivalent to the hysteretic energy dissipation ($\xi_{F,d}$) of the degraded SDOF system is evaluated. To this end, an iterative calculation is required to the user that have to enter a multiplication factor (α) of the loading history. In sheet 5 (Fig. 1e), the iterative part of the DDBD procedure is implemented by setting design parameters of the equivalent damped brace. Stiffness hardening ratio

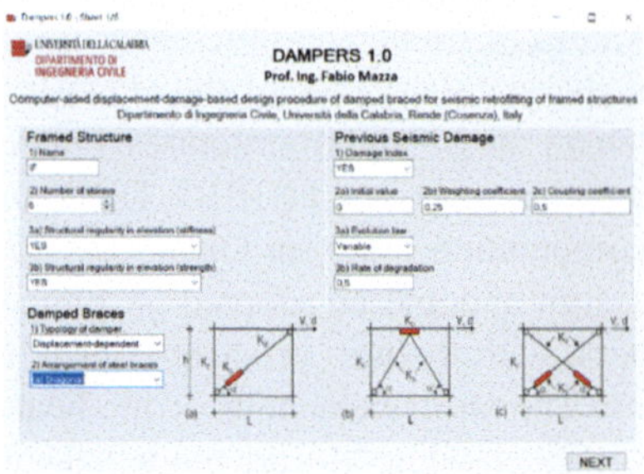

(a) First sheet.

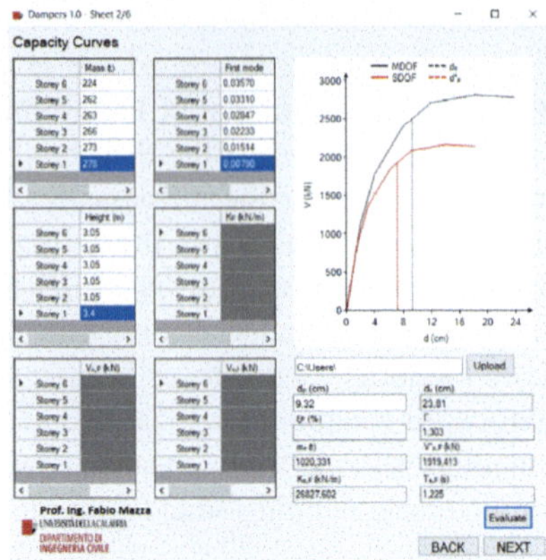

(b) Second sheet.

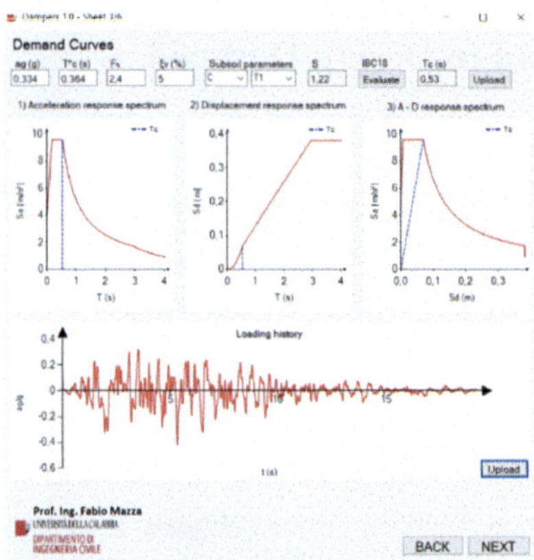

(c) Third sheet.

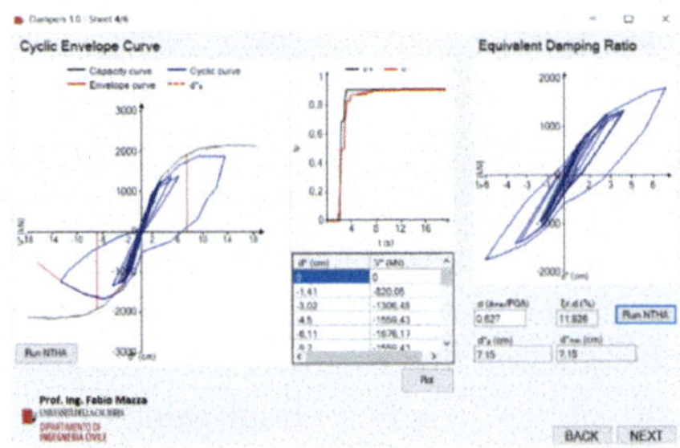

(d) Fourth sheet.

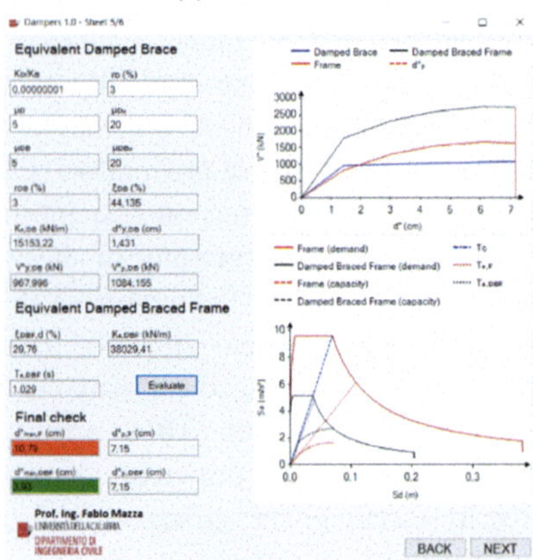

(e) Fifth sheet.

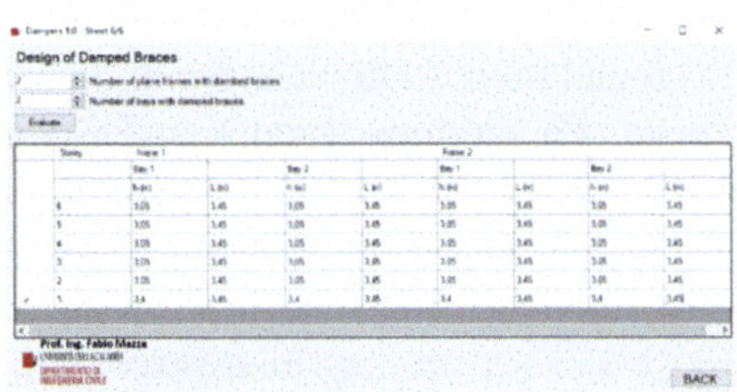

(f) Sixth sheet.

Fig. 1 Screenshots of the DAMPERS computer-aided design tool of HYDBs

(r_{DB}) and design (μ_{DB}) and ultimate (μ_{DBu}) ductility of the HYDB are automatically obtained as output together with viscous damping ratio of the degraded damped braced frame ($\xi_{DBF,d}$). Capacity curves (V^*-d^*) of the degraded frame, damped brace and their in-parallel combination are plotted. Finally, modified capacity curves in the acceleration (V^*/m_e)-displacement (d^*) format versus ADRS demand reduced by ($\xi_{DBF,d}$) are plotted together for the frame and damped braced frame, in order to verify that the displacement demand of the equivalent frame and damped braced frame are less than the target displacement. Geometric properties of the plane frames and bays with HYDBs are provided in sheet 6 (Fig. 1f). Consequently, design yielding force ($N_{y,DBij}$) and elastic stiffness (K_{DBij}) of the j-th damped brace at the i-th storey are evaluated and uploaded in an output file labelled with the name stated in sheet 1. Elastic stiffness of the supporting brace (K_{Bij}) and HYD (K_{Dij}), considered as elements acting in series, and corresponding cross-section areas (i.e. A_{Bij} and A_{Dij}) are provided to choose their size from commercial catalogues. Design of the HYDBs is carried out by using a proportional stiffness criterion with respect to the modal behaviour, in the case of in-elevation regular distribution of strength and stiffness for the original structure; while, a constant drift ratio criterion is assumed when an in-elevation irregular distribution of stiffness is selected in sheet 1.

3 Original and Retrofitted Test Structures

A case-study structure representative of the Italian residential building stock built during the 1990s is selected from the RINTC research project [13], financed by the Italian Department of Civil Protection. Specifically, the r.c. framed structure is designed for moderate seismic loads corresponding to L'Aquila (Italy), assuming deformed type of steel reinforcement. The six-storey building has a rectangular shape of the plan (Fig. 2a), with five and three bays along the principal X and Y directions (Fig. 2b), and a staircase with knee beams symmetrically placed along the Y direction. Double leaf (8 cm, internal layer, and 12 cm, external layer) masonry infills (MIs)

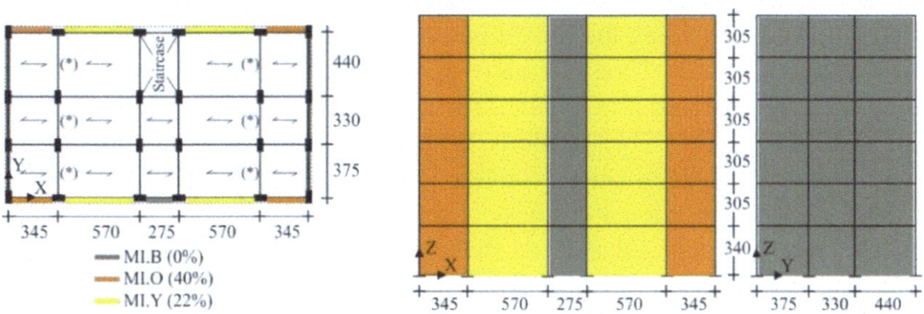

(a) Plan. (b) Bare frames (BFs).

Fig. 2 Original test structure (unit in cm)

of hollow clay bricks are placed along the perimeter. Deep beams are used in the perimeter frames and staircase, with cross section 30 cm × 60 cm at the first two levels and 30 cm × 50 cm at the other four levels, while all internal beams are flat with a cross-section 60 cm × 25 cm and 80 cm × 25 cm (the latest marked with an asterisk in Fig. 2a). Rectangular cross-section with the orientation shown in Fig. 2a is assumed for the columns, considering: 30 cm × 60 cm at the first two levels; 30 cm × 50 cm at the third and fourth levels; 30 cm × 40 cm at the fifth and roof levels. The percentage of openings varies depending on the architectural layout is assumed for the MIs, marked with different colours in Fig. 2a, b: i.e. 0%, MI.B in brown; 40%, MI.O in orange; 22%, MI.Y in yellow.

A simulated design of the bare frame (BF) is carried out in line the Italian codes IBC92 [14] and IBC86 [15] for the vertical and seismic loads, respectively, considering linear static analysis and the admissible tension method. Vertical gravity loads are represented by dead loads of 4.8 kN/m^2, on the roof, and 6.3 kN/m^2, on the other floors, and live loads of 2.2 kN/m^2, on the roof (snow), 2 kN/m^2, on the other floors, and 4 kN/m^2 for the staircase. Non-structural MIs are taken into account through an additional dead load of 3.5 kN/m^2, along the perimeter facades, with compressive and shear strength equal to 2 MPa and 0.4 MPa, respectively. Horizontal seismic loads are evaluated for a medium-risk zone (seismic coefficient, C = 0.07) and typical subsoil class (foundation coefficient, $\varepsilon = 1$). Concrete with cylindrical compressive strength of 25 MPa (maximum normal stress equal to 8.5 MPa) and steel reinforcement with yield strength of 430 MPa (maximum normal stress equal to 2600 MPa) are considered. Further details can be found in [13].

Three configurations of MIs are considered for the original test structure: i.e. bare frame (BF), with non-structural MIs designed so as not to affect structural deformability; (ii) infilled frame (IF), with structural MIs uniformly distributed along the height, in contact with the frame but not structurally connected to it; pilotis frame (PF), with a soft-storey at the ground level and the same MIs of the IF at the other levels.

The seismic retrofitting of the test structures is carried out by the insertion of diagonal steel braces equipped with HYDBs (see sheet 1 of DAMPERS). Attention is taken to avoid impeding the internal functioning of the building and only the corner bays of the perimeter frames are involved to ensure compatibility with the architectural layout providing the lowest percentage (X direction) and absence (Y direction) of balcony doors. Three alternative retrofits are designed, with reference to the bare (i.e. Damped Braced Bare Frame, DBBF in Fig. 3a), infilled (Damped Braced Infilled Frame, DBIF in Fig. 3b) and pilotis (i.e. Damped Braced Pilotis Frame, DBPF in Fig. 3c) structures. Two distributions of the HYDBs are adopted along the height, according to: (i) a proportional stiffness criterion [5], for the regular BF and IF original structures, assuming that mode shapes of the structures remain practically the same after the insertion of the HYDBs; (ii) a constant drift criterion [10], for the original irregular PF structure along the X and Y directions, to obtain a globally regular retrofit by balancing the soft-storey at the ground level. Moreover, the vertical distribution of the yield load is assumed proportional to the stiffness distribution in the X direction of the PF structure [5], while it is modified to be similar to that of

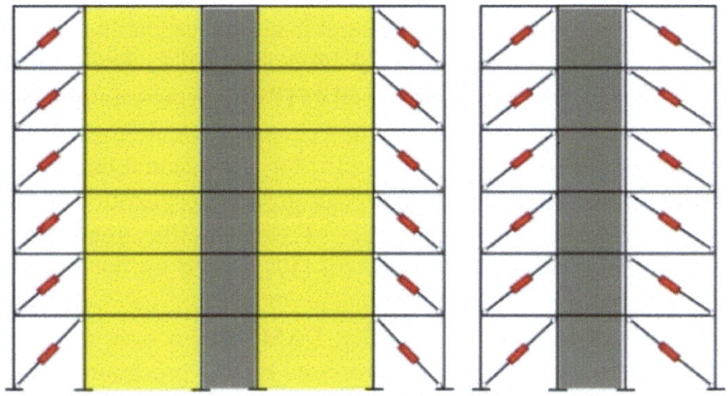

(a) DBBF structure.

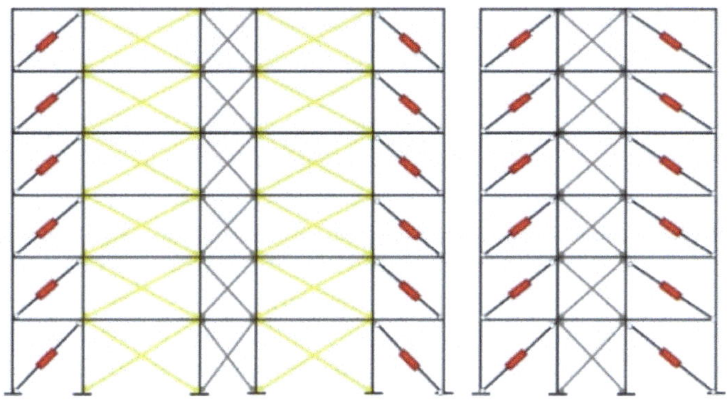

(b) DBIF structure.

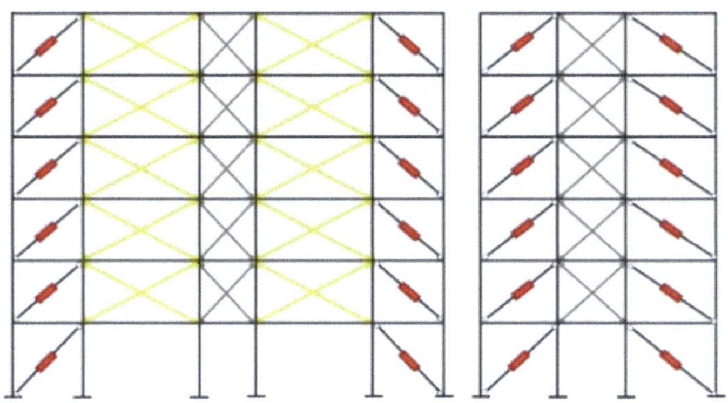

(c) DBPF structure.

Fig. 3 Alternative layouts of the retrofitted test structure the (elastic) shear force induced by the lateral loads corresponding to the inverted-triangular first vibration mode along the Y direction, where in-elevation strength irregularity is observed, so that activation of the HYDBs occurs simultaneously [10].

The displacement design spectrum at the collapse-prevention (CP) limit state is defined considering the IBC18 provisions [16] for a residential building located in L'Aquila. A high-risk seismic zone (i.e. peak ground acceleration on rock, a_g = 0.334 g) and moderately-soft subsoil (i.e. class C, site amplification factor S = 1.22) are assumed. The computer code SeismoArtif [17] is used for the generation of an artificial earthquake matching the CP design spectrum of acceleration provided by IBC18 (see sheet 4 of DAMPERS). Finally, DAMPERS is used to proportion the equivalent HYDBs in accordance to the proposed DDBD procedure, with reference to variable (i.e. degradation parameter $\varepsilon = 0.5$) values of the damage index. In order to assess the effects of previous seismic damage, the retrofitting solutions are also designed for $\Psi = 0$ considering the DBD proposed in a previous paper [5], where a reduction factor equal to 0.66 is assumed for the viscous damping equivalent to the hysteresis of the frame [16]. In particular, the performance displacement $d^*_p(=d_p/\Gamma =$ 7.15 cm), corresponding to a target displacement of the MDOF system d_p (=0.5%/H_{tot} = 9.32 cm, being H_{tot} = 1865 cm the total height of the building) and a participation factor $\Gamma(= 1.303)$, is combined with a constant design value of the damper ductility ($\mu_D = 5$); constant hardening ratios $r_F = 0\%$ and $r_D = 3\%$ are also assumed. The influence of the stiffness of the supporting brace is not considered (i.e. $K_{DB} = K_D$ and $r_{DB} = r_D = 3\%$). Main parameters of the equivalent HYDBs along the in-plan principal directions are reported in Table 1, with reference to the DBBF (Table 1), DBIF (Table 2) and DBPF (Table 3) structures: i.e. equivalent stiffness ($K_{DB,e}$) and shear force at the yielding ($V^*_{y,DB}$).

Table 1 Properties of the equivalent damped brace for DBBF (units in kN and m)

Previous damage	Damage index	X direction		Y direction	
		$K_{DB,e}$	$V^*_{y,DB}$	$K_{DB,e}$	$V^*_{y,DB}$
No	$\Psi = 0$	6844	636	12,946	897
Yes	Ψ = var., $\varepsilon = 0.5$	19,642	1220	16,094	992

Table 2 Properties of the equivalent damped brace for DBIF (units in kN and m)

Previous damage	Damage index	X direction		Y direction	
		$K_{DB,e}$	$V^*_{y,DB}$	$K_{DB,e}$	$V^*_{y,DB}$
No	$\Psi = 0$	12,477	797	5259	332
Yes	Ψ = var., $\varepsilon = 0.5$	15,205	971	17,520	1108

Table 3 Properties of the equivalent damped brace for DBPF (units in kN and m)

Previous damage	Damage index	X direction		Y direction	
		$K_{DB,e}$	$V^*_{y,DB}$	$K_{DB,e}$	$V^*_{y,DB}$
No	$\Psi = 0$	14,677	864	7572	445
Yes	$\Psi =$ var., $\varepsilon = 0.5$	16,273	958	14,629	861

4 Numerical Results

Nonlinear dynamic analysis of the original (i.e. BF, IF and PF) and retrofitted (i.e. DBBF, DBIF and DBPF) test structures are carried out using the OpenSees platform [12], taking into account the onset of shear failure in r.c. frame members and joints, prior to or following ductile ones, and the nonlinear in-plane behaviour of masonry infills considered as structural elements. Specifically, the failure mode of beams and columns is predetermined, by classifying each of them as brittle or ductile. Moreover, the HYDBs are simulated with truss elements through a bilinear axial force–displacement law, where yielding and buckling of the diagonal steel braces are assumed to be prevented. Seven earthquakes, reflecting the IBC18 provisions at the site in question [16], are selected from the Pacific Earthquake Engineering Research Center database [18] and scaled in order to match on average the life-safety (LS) design response spectrum, within lower (i.e. -10%) and upper (i.e. $+30\%$) bound tolerances, in a suitable range of vibration periods. The following results are obtained as a mean of the maximum values obtained for each of the seven pairs of accelerograms, at the final instant of simulation.

First, mean damage index of the rotational springs, lumped at the critical end sections of r.c. beams (i.e. $\Psi_{\theta,b} = \theta_{max,b}/\theta_{u,b}$) and columns (i.e. $\Psi_{\theta,c} = \theta_{max,c}/\theta_{u,c}$) and in the joints (i.e. $\Psi_{\theta,j} = \theta_{max,j}/\theta_{u,j}$), and truss elements, representing masonry infills (i.e. $\Psi_{\Delta,i} = \Delta_{max,i}/\Delta_{u,i}$), along the building height is reported in Figs. 4, 5 and 6. In particular, curves corresponding to the original and retrofitted structures are compared, the latter corresponding to the HYDBs designed with (e.g. $\varepsilon = 0.5$) and without (i.e. $\Psi = 0$) considering previous seismic damage of the structures. It is interesting to note that a "strong-beam weak-column" mechanism affects the original structures at the lowest two storeys (Figs. 4a, b, 5a, b and 6a, b), especially for the PF structure where an open ground storey is considered (Fig. 6a, b), leading to a significant decrease in the global ductility. Moreover, the shear demand in the beam-column joint panels, identified as the main cause of collapse of many buildings during recent earthquakes, is dominant at the third and fourth levels of all existing structures (Figs. 4c, 5c and 6c). From the second to the fourth level, masonry infills exhibit a medium–high level of damage, while little damage is noted at the first and top levels (Figs. 5d and 6d). Further results, omitted for the sake of brevity, highlight that the maximum damage index related to an earthquake may exceed the corresponding ultimate value, inducing shear failure of columns ($\Psi_{\theta,c} = 1$) and joints ($\Psi_{\theta,j} = 1$) of the staircase along the Y direction, for all the original structures, and brittle failure of masonry infills ($\Psi_{\Delta,i} = 1$), for the IF and PF. For all the examined case studies,

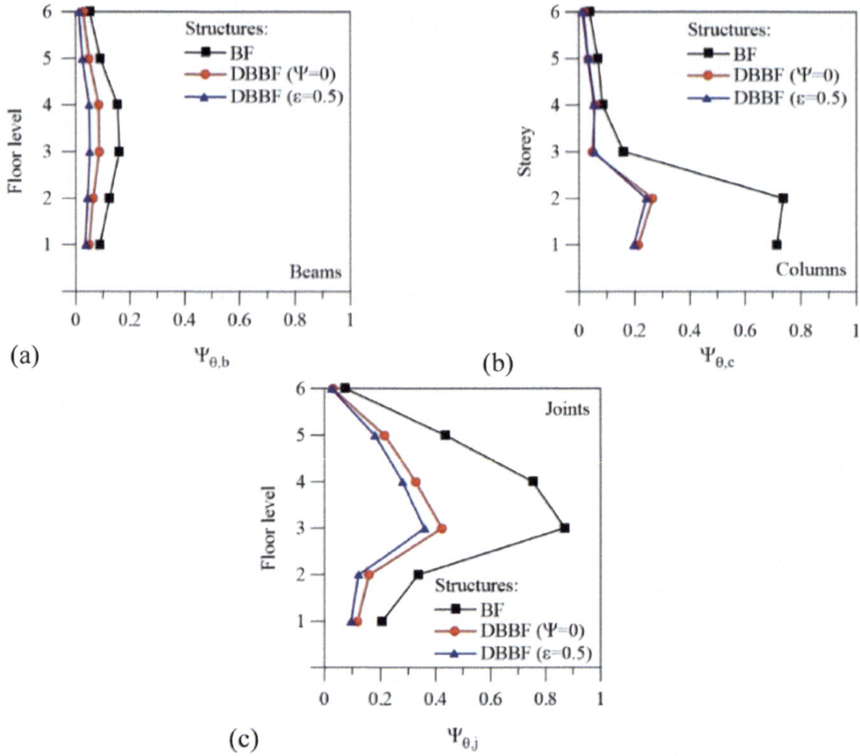

Fig. 4 Ductile and brittle damage indexes for the BF and DBBF structures

the effectiveness of the HYDBs is confirmed, ensuring a notable reduction of the seismic demand of both ductile and brittle failure modes.

Finally, mean values of the storey drift ratio at the LS limit state, defined as drift along the in-plan X (Δ_X) and Y (Δ_Y) directions normalized by the storey height (h), are shown in Fig. 7. As shown, the original structures exhibit higher deformability in the X (Fig. 7a, c, e) rather than in the Y (Fig. 7b, d, f) direction. An irregular distribution law of the drift ratio can be observed for the BF and IF structures, although the IBC18 criteria for regularity in elevation are satisfied. The proportional stiffness criterion, used for the HYDBs of DBBF (Fig. 7a, b) and DBIF (Fig. 7c, d), ensures a reduction of at least half of the drift demand when $\varepsilon = 0.5$ is assumed. Moreover, the constant drift criterion adopted for the DBPF structure improves the shape (almost uniform) and intensity of the drift ratio when previous damage is taken into account (Fig. 7e, f). For all examined cases, significant differences of drift ratio corresponding to $\Psi = 0$ and $\varepsilon = 0.5$ are observed at the intermediated levels.

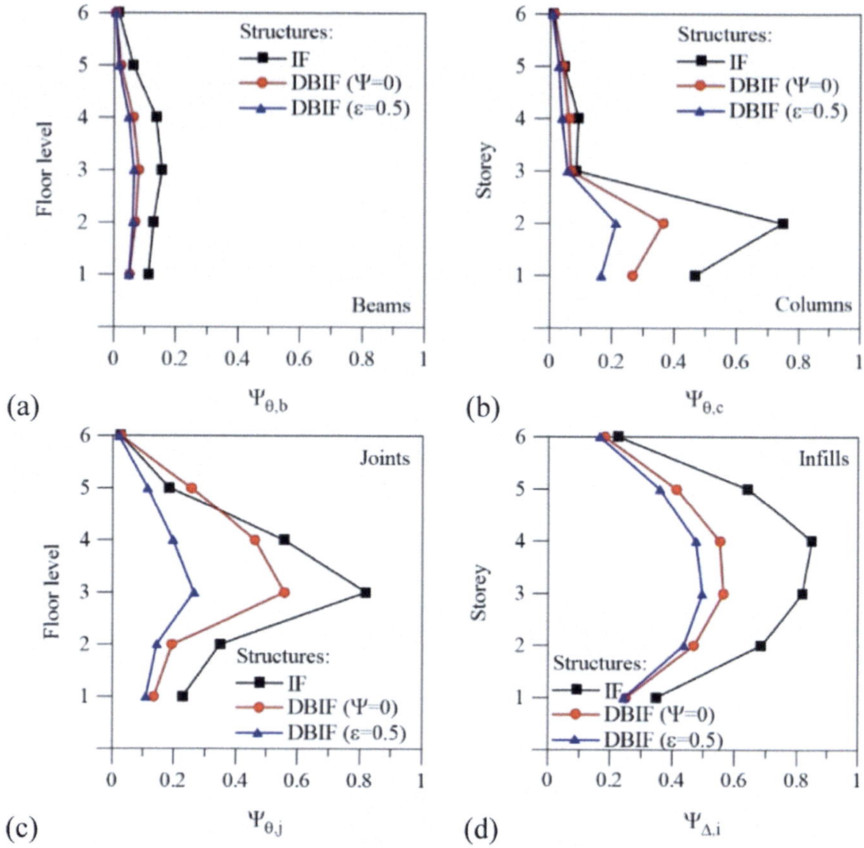

Fig. 5 Ductile and brittle damage indexes for the IF and DBIF structures

5 Conclusions

The present study is focused on the formulation of a displacement-damage-based design (DDBD) procedure of HYDBs, to consider the repeated cyclic loading in the inelastic range, and the use of a computer-aided software named DAMPERS, to automate its application to make it user friendly for practitioners. The effectiveness of the proposed DDBD procedure and implemented DAMPERS automated design tool are assessed with reference to an archetype representative of the Italian residential buildings built in the 1990s, assuming bare (BF), infilled (IF) and pilotis (PF) configurations.

In particular, original and retrofitted (DBBF, DBIF and DBPF) structures are compared, the latter corresponding to the HYDBs designed with ($\varepsilon = 0.5$) and without ($\Psi = 0$) considering previous seismic damage of the original structure. Based on the

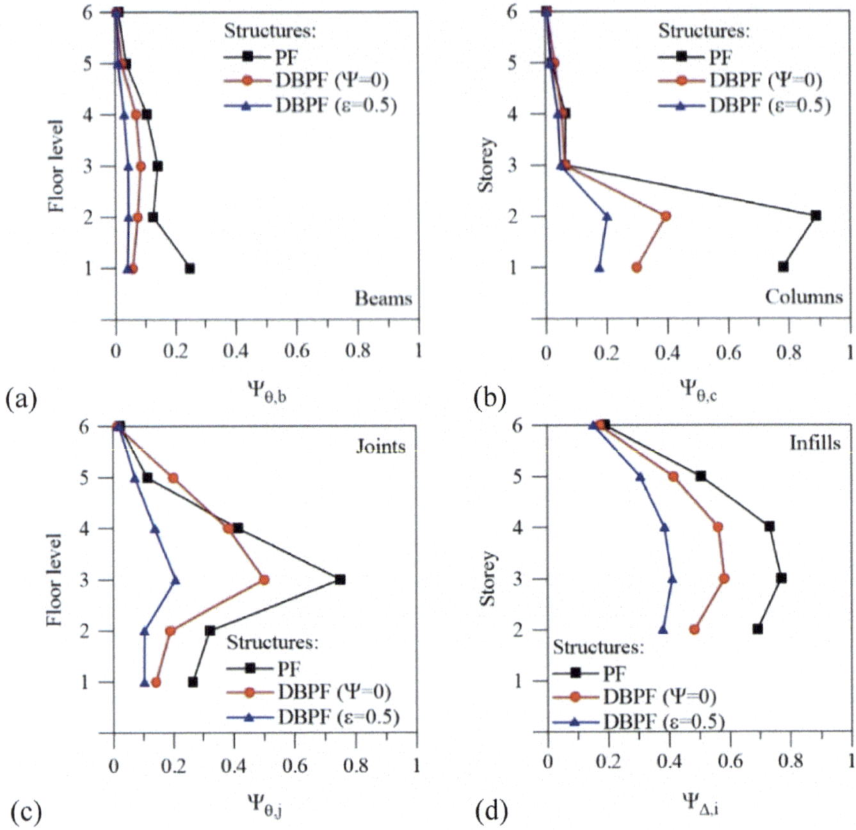

Fig. 6 Ductile and brittle damage indexes for the PF and DBPF structures

results of the nonlinear seismic analysis carried out on the OpenSees platform, the following conclusions can be drawn.

Brittle failure modes affect the behaviour of the original structures at the LS limit state, with a "strong-beam weak-column" mechanism at the lowest two storeys, high shear demand in the beam-column joints and columns of the staircase, at the third and fourth levels, and medium–high level of the in-plane damage for masonry infills placed from the second to the fourth level. The insertion of the HYDBs ensures a notable reduction of the post-retrofitting seismic demand, with an ever greater effectiveness when previous seismic degradation of r.c. frame members is considered. The irregular distribution law of the drift ratio observed for the original structures is avoided for the DBPF, while a reduction of at least half of the drift demand is obtained for $\varepsilon = 0.5$ also for the DBBF and DBIF where the proportional stiffness criterion is adopted. Moreover, ultimate ductility of some HYDBs designed assuming $\Psi = 0$ is unpredictable for the DBIF and DBPF.

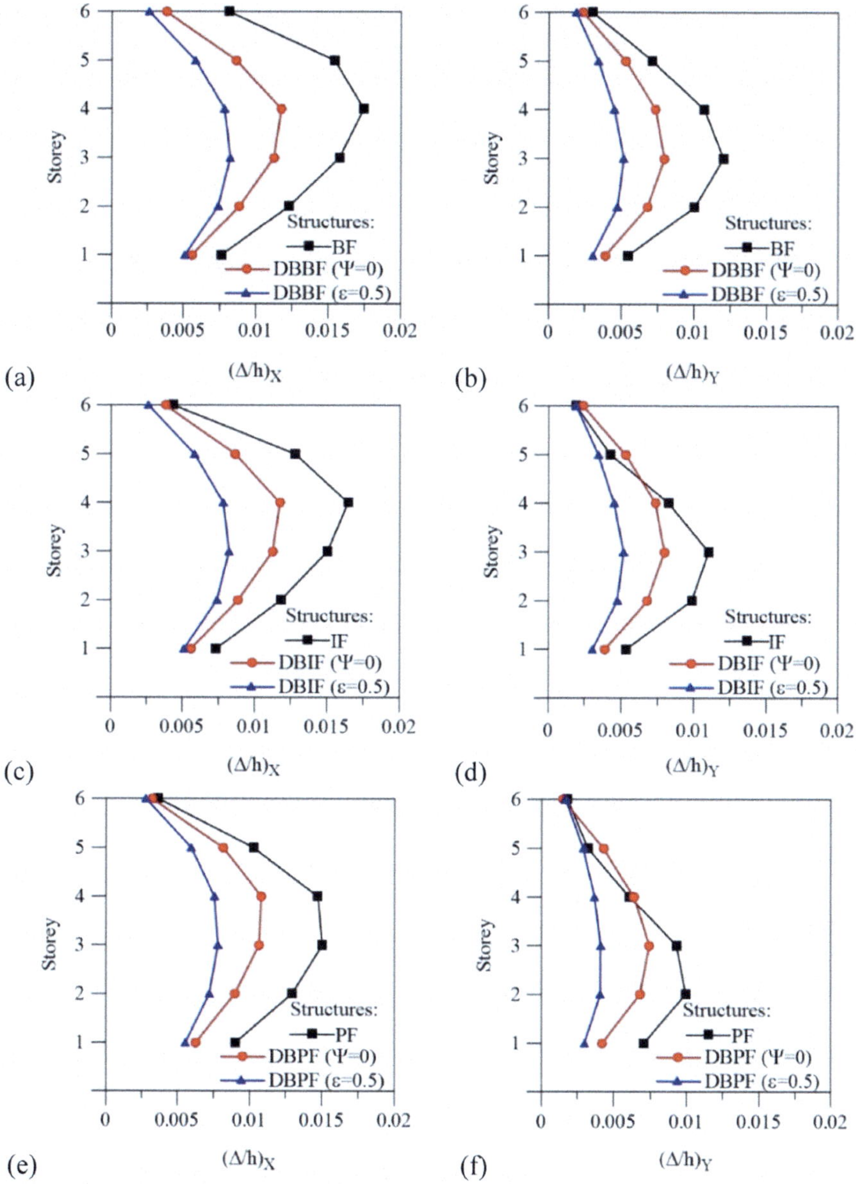

Fig. 7 Drift ratio for the original and retrofitted structures

Acknowledgements The present work was financed by Re.L.U.I.S. (Italian network of university laboratories of earthquake engineering), in accordance with the Convenzione D.P.C.-Re.L.U.I.S. 2019-2021, WP15, Code Revisions for Isolation and Dissipation.

References

1. Javanmardi A, Ibrahim Z, Ghaedi K, Benisi Ghadim H, Hanif MU (2019) State-of-the-art review of metallic dampers: testing, development and implementation. Arch Comput Methods Eng. https://doi.org/10.1007/s11831-019-09329-9
2. Lin Y-Y, Tsai M, Hwang J, Chang K (2003) Direct displacement-based design for building with passive energy dissipation systems. Eng Struct 25(1):25–37
3. Kim J, Choi H (2004) Behavior and design of structures with buckling-restrained braces. Eng Struct 26:693–706
4. Bergami AV, Nuti C (2013) A design procedure of dissipative braces for seismic upgrading. Earthq Struct 4(1):85–108
5. Mazza F, Vulcano A (2015) Displacement-based design procedure of damped braces for the seismic retrofitting of r.c. framed buildings. Bull Earthq Eng 13:2121–2143
6. Mazza F (2014) Displacement-based seismic design of hysteretic damped braces for retrofitting in-plan irregular r.c. framed structures. Soil Dyn Earthq Eng 66:231–240
7. Mazza F (2015) Seismic vulnerability and retrofitting by damped braces of fire-damaged r.c. framed buildings. Eng Struct 101:179–192
8. Mazza F (2019a) A plastic-damage hysteretic model to reproduce strength stiffness degradation. Bull Earthq Eng 17:3517–3544
9. Mazza F (2019b) A simplified retrofitting method based on seismic damage of a SDOF system equivalent to a damped braced building. Eng Struct 200:109712
10. Mazza F, Mazza M, Vulcano A (2015) Displacement-based seismic design of hysteretic damped braces for retrofitting in-elevation irregular r.c. framed structures. Bull Earthq Eng 69:115–124
11. Mazza F (2016) Nonlinear seismic analysis of r.c. framed buildings with setbacks retrofitted by damped braces. Eng Struct 126:559–570
12. McKenna F, Fenves GL, Scott MH (2000) Open system for earthquake engineering simulation. University of California, Berkeley, CA
13. Ricci P, Manfredi V, Noto F, Terrenzi M, De Risi MT, Di Domenico M, Camata G, Franchin P, Masi A, Mollaioli F, Spacone E, Verderame GM (2019) RINTC-E: towards seismic risk assessment of existing residential reinforced concrete buildings in Italy. In: COMPDYN 2019, 7th ECCOMAS thematic conference on computational methods in structural dynamics and earthquake engineering, Crete, Greece
14. Italian Building Code 92. Norme tecniche per le opere in c.a. normale e precompresso e per le strutture metalliche. DM 14-02-1992, Italian Ministry of Public Works, Rome, Italy
15. Italian Building Code 86. Norme tecniche relative alle costruzioni antisismiche. DM 24-01-1986, Italian Ministry of Public Works, Rome, Italy
16. Italian Building Code 18. Norme tecniche per le costruzioni. DM 17-01-2018, Italian Ministry of Infrastructures and Transports, Rome, Italy
17. Seismoartif (2019) A computer program for generation of artificial accelerograms. https://www.seismosoft.com
18. PEER. Pacific Earthquake Engineering Research Center database (2014). https://ngawest2.berkeley.edu

Multiscale Damage Modelling of Composite Materials Using Bayesian Network

Arvind Keprate and Ramin Moslemian

Abstract Fibre reinforced polymers or as they widely are known composite materials, have made it possible to develop and manufacture large wind turbine and tidal blades central to competitiveness of wind and tidal turbines against other energy sources. Composite materials are light, corrosion free and easy to manufacture into complex aerodynamic/hydrodynamic profiles of wind and tidal blades. Even though polymer-based materials like composites used in wind and tidal blades do not corrode the same way as metals do, they undergo environmental degradation. One of the main mechanisms triggering degradation of polymers is their material structure which is not as "tight" as metals with significant free space between the polymeric chains. Due to these characteristics, unlike metals, polymers tend to absorb elements of their surrounding environment such as liquids, gas and humidity. For glass reinforced epoxy which the wind and tidal blades are typically manufactured of, absorption of water and humidity causes change in both epoxy matrix and glass Fibres, consequently leading to degradation of the material properties. Due to multi-parameter complex and rather uncertain nature of the environmental degradation of composites, typically full-scale aging tests are done which are limited, costly and time consuming. The limited number of possible full-scale tests fail to generate enough data points for statistical evaluation of uncertainties. In this paper a methodology for evaluation of environmental degradation of composites in a probabilistic scheme using Bayesian Networks is presented. The developed scheme is used to develop a digital integrity assessment tools for blades in wind and tidal industries. An illustrative case study is performed to demonstrate the application of the developed tool.

Keywords Composite material · Multiscale damage modelling · Bayesian network

A. Keprate (✉)
Department of Mechanical, Electronics and Chemical Engineering, Oslo Metropolitan University, Oslo, Norway
e-mail: arvindke@oslomet.no

R. Moslemian
DNV GL, Høvik, Norway

M. Abdel Wahab (ed.), *Proceedings of 1st International Conference on Structural Damage Modelling and Assessment*, Lecture Notes in Civil Engineering 110,
https://doi.org/10.1007/978-981-15-9121-1_11

1 Introduction

Polymers and polymer-based composites are generally known for their low reactivity with their environments. They are widely used where corrosion has been a big problem with metals such as pipes used to transfer water. However, polymers and composites have their own environmental degradations. The structure of polymer-based materials is not as "tight" as metals and have significant free space between the polymeric chains. When polymers are exposed to a liquid or gas components from the environment, depending on their affinity to the polymer type and structure diffuse into the polymer. The diffusion mechanism is schematically shown in Fig. 1. Absorption of elements from the environment can potentially have two effects on the properties of the polymer:

1. **Swelling**: Depending on the type of polymer and its affinity to the environment, polymers may absorb different amount of the elements in their environment. For example, polyethylene at high temperature may absorb up to 15% of its weight if it is exposed to aromatic oil. For epoxy and water, up to 1.5% of the weight of epoxy water can be absorbed. The absorption of liquids and gases cause polymer to swell i.e. to expand and as a result of swelling polymers become soft and plasticized.
2. **Chemical degradation and brittleness**: The elements of the environment which have diffused into a polymer over time may react with polymer chains and cause chemical degradation and brittleness.

For Fibre reinforced polymers or composites the environmental degradation is more complex due to multi-material structure of composite materials. The process of diffusion into composite is similar to non-reinforced polymers, however the diffusion is anisotropic i.e. the diffusion rates is different in different directions. In Fibre reinforced polymers the effect of exposure to the environmental elements which diffuse into the material should be assessed for the three constituents of composite

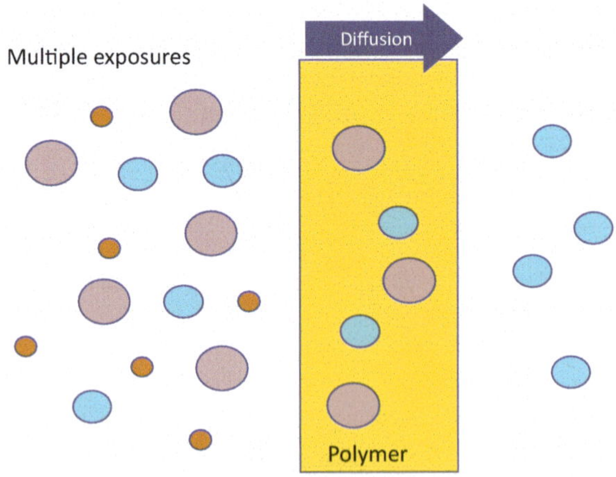

Fig. 1 Diffusion of environmental components into a polymer

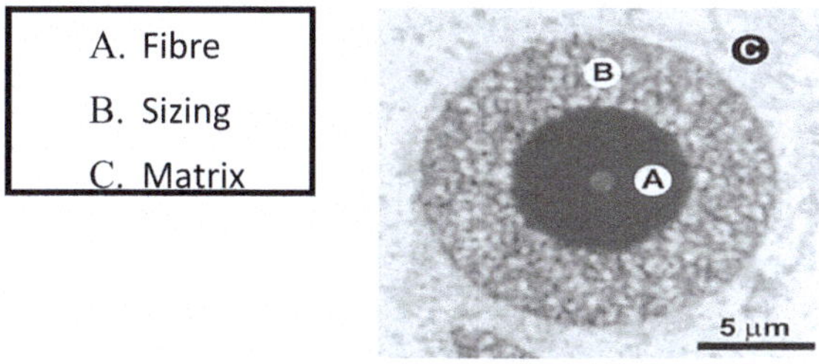

Fig. 2 A unit cell in composite materials

materials i.e. for the matrix, Fibre and Fibre/matrix interface as shown in Fig. 2, which makes the assessment much more complex.

In Fig. 3 the process of environmental degradation for composite materials is shown. After exposure to an environment, the process of diffusion starts, resulting in a concentration of environmental elements in the material and exposure of Fibre, matrix and their interface. Over time and depending on the temperature, chemical reactions may start in one of the three constituents of the material system, changing its properties including resistance to short-term and long-term static loads and cyclic loads. The process shown in Fig. 3 depends on characteristics such as different constituents of the material, the exposure environment, temperature, time, stress state and damage in the material and include various degradation mechanisms and strong anisotropy the material. Modelling such a complex process in a probabilistic framework (which is central for integrity assessment of structures made of composite materials for offshore application where significant environmental exposure is expected),

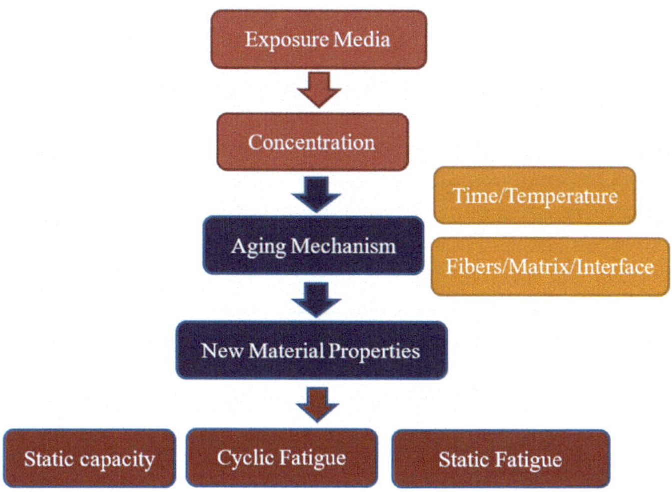

Fig. 3 The process of environmental degradation in composite materials

has been a big challenge, thereby requiring methods other than traditional structural reliability analysis (SRA).

2 Probabilistic Graphical Models

As discussed previously, modeling the progression of damage in composites materials is a challenging task mainly due to the uncertainty in the multi-scale physics of the damage process and the large variability in behavior that is observed, even for the tests of nominally identical specimens. As a result, there is much uncertainty related to the choice of the class of models among a set of possible candidates for predicting damage behavior. One tool for assessing risk and reliability for engineering structures is through structural reliability analysis (SRA). Let us first recall the basic principles of SRA.

In SRA, the Probability of Failure (PoF) is defined through a multidimensional integral [1]:

$$PoF = \int_{D_f} f_X(\boldsymbol{x}) d\boldsymbol{x}$$

where $\boldsymbol{x}$ is the vector containing all the basic random variables, i.e., all uncertain variables making up the system. The integrand $f_X(\boldsymbol{x})$ is the joint probability density function of all the variables$\boldsymbol{x}$, and the domain of integration is the failure domain $D_f = \{\boldsymbol{x} : g(\boldsymbol{x}) \leq 0\}$ which is given by the sign of the limit state function $g(\boldsymbol{x})$. The limit state function returns a negative value under system failure conditions, and a positive value under acceptable conditions. This function can be hard to formulate for complex problems, and it also depends on $\boldsymbol{x}$, that is, it depends on the integration variable. The fact that the domain of integration depends on the integration variable is a challenge in SRA.

To solve the aforementioned challenge, one may rely on Monte Carlo methods, i.e., sampling-based methods, to numerically approximate the integral. However, this is in practice is not applicable for real-life problems involving for example finite element models (due to the huge number of samples needed and the cost of running finite element models). Moreover, there are approximate methods such as first-order and second-order reliability methods (commonly abbreviated FORM or SORM, respectively), which have been very successfully applied in the industry. They are fast and efficient, but do not provide any guarantee in terms of the accuracy of the result. However, in order to use these methods, we have to formulate a limit state function and we must also assume that this function satisfies certain differentiability properties.

The above paragraph highlights some limitations of traditional SRA. Is there a different approach for assessing risk and reliability in complicated structural systems? In this manuscript, we have relied on the concept of Probabilistic Graphical Models

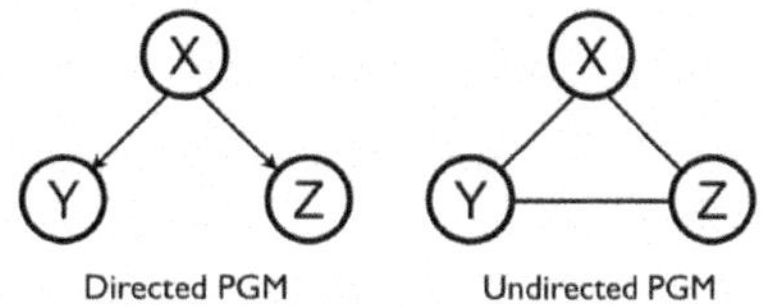

Fig. 4 A directed and an undirected PGM with nodes X, Y and Z

(PGM). PGMs are diagrammatic representations of probability distributions where each node represents a random variable, and edges express probabilistic relationships between random variables [2]. One may say that PGM is in a way a marriage between probability theory and graph theory: Probabilistic because they deal with chances and graphical because they express dependencies between variables on a graph. One can separate between two types of PGM, directed PGM and undirected PGM as shown in Fig. 4.

We will focus on a variant of directed PGM which is called Bayesian Network (BN). The Bayesian interpretation of probability represents probability as a degree or quantification of a personal belief or state of knowledge. Your initial (prior) belief changes in light of new information or evidence. Given an observation or measurement in an experiment, a revised probability of the desired outcome might change, i.e., we think in a conditional way. A BN is a tool to represent this Bayesian and conditional way of thinking via a Directed Acyclic Graph (DAG). BNs are therefore directed and contain no cycles. More information about BNs is discussed by the author in [3].

3 Bayesian Network Modelling of Composite Materials

To start building the Bayesian network of environmental degradation of composite materials, the degradation process must be broken down into its various mechanisms including the causal relationship between each mechanism. Once the problem is broken down, various nodes and edges for the BNs are identified. The environmental degradation of glass/epoxy composites which is the main material used in wind and tidal blades are shown in Fig. 5. Assume that the location of interest for us in the structure has the coordinates of x and y. When the structure is exposed to an environment (here water) it will diffuse into the material. Six parameters that control the amount and rate of absorption are maximum uptake in equilibrium condition M_{eq}, anisotropic diffusivity parameters D_{11}, D_{22} and D_{33}, matrix crack density and temperature. If these parameters are known, a diffusion analysis can be done using finite element method to determine the concentration of the exposure medium in the desired location at time t.

Once the concentration is known, the effect of concentration on matrix, fibres and interface is required to be evaluated. The primary effect of water on epoxy is softening and plasticization. However, exposure of glass fibres to water leads to reduction of their diameter, as components from fibres get dissolved in water

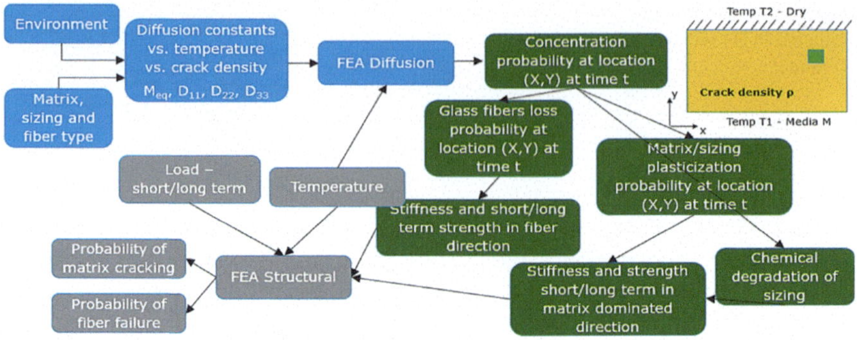

Fig. 5 Schematic representation of environmental aging process in glass/epoxy composite material

causing reduction in both stiffness and strength in fibre direction in the composite material. Furthermore, a coating-like layer on the surface of fibres, bonding them to matrix may undergo chemical degradation, cracking and depending from fibres. These three effects when combined change the mechanical properties in fibre and transverse to fibre direction. The degraded mechanical properties can then be used in the structure analysis to evaluate the possibility of failure in fibres, matrix and between the composite material's layers.

If the combined effect of degradation of mechanical properties and the applied loads is matrix cracking, it will accelerate the diffusion process (although this failure mechanisms is not considered to be critical to the structural integrity) that needs to be included in updating the diffusivity parameters.

4 Illustrative Case Study

4.1 Problem Description

The configuration of tidal turbine blade have many designs due to the existence of many competing technologies. For example, in addition to the shell and web structure similar to wind industry (Fig. 6 left), some devices have solid composite blade, such as OpenHydro turbine (Fig. 6 right). These blades are made of composite material (mainly glass fibre reinforced epoxy composite) due to its superior performance in corrosion resistance and weight saving. Tidal blades are often fully submerged in seawater condition. Hence, the environmental degradation due to moisture is the main challenge for the industry for ensuring long term integrity of the component. For this case study, OpenHydro's blade is chosen due to its simplicity in design and configuration.

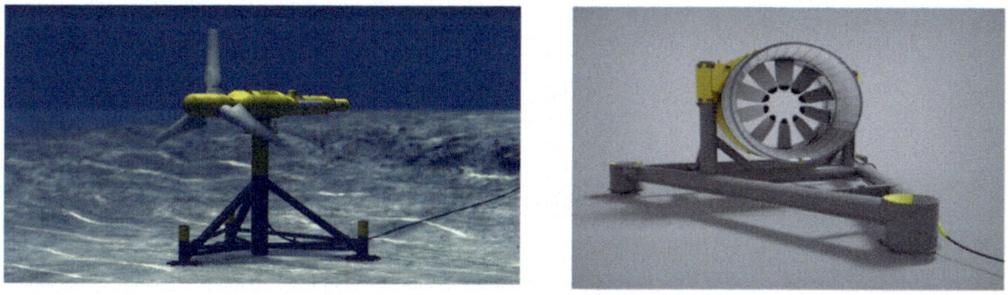

Fig. 6 Configuration of tidal turbine blade

4.2 Diffusion Data and Modeling

A diffusion model was created in Abaqus to resemble the OpenHydro blade. The aim of this model is to analyze the moisture content of the whole blade. The blade is modelled as a flat rectangle plate of 5000 mm × 2400 mm × 200 mm in length, width and thickness respectively. It is made up of 5 layers of Biax (±45°), 0° UD and 90° UD with the thickness depicted in Table 1. For the diffusion analysis, the critical parameters are the diffusivity constants: D_{ii}, where ii are 11, 22 and 33 which represent Fibre, transverse to Fibre and out of plane directions respectively, see Fig. 7.

Table 1 Fibre orientation and thickness in the model

Parameter	Ply thickness (mm)
Biax (±45°)	25
0° UD	65
90° UD	20
0° UD	65
Biax (±45°)	25

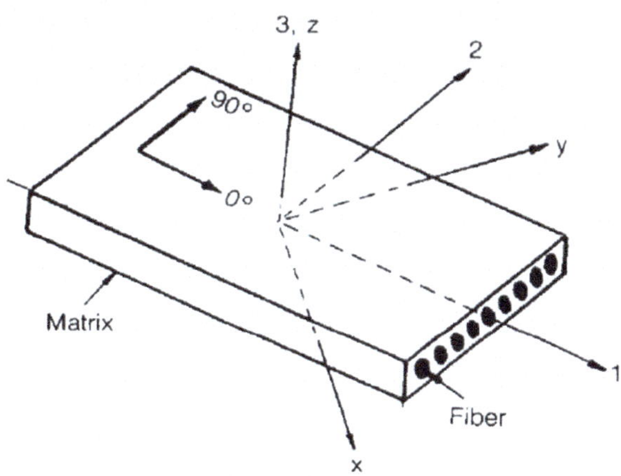

Fig. 7 Composite coordinate system (from DNV-DS-J102)

Table 2 Parameters for diffusion analysis

Parameter	Value
D_{11} (mm^2/h)	0.1598
D_{22} (mm^2/h)	0.036
D_{33} (mm^2/h)	0.036
Solubility[a] (g/cm^3)	0.01795

[a]Water mass solubility in Table 3 (0.93% = 0.0093) needs to be converted to volumetric by multiplying it for the composite density (1.93 g/cm^3). Hence, 0.01795 (=0.0093 × 1.93) is used

The diffusion coefficients are obtained from the experimental results from Norwegian University of Science and Technology (NTNU) [4] as part of the Joint Industry Project (JIP) affordable composite. The parameters used for the diffusion analysis are shown in Table 2. The diffusivity constants are artificially increased by 10 times the values in order to exaggerate the speed of diffusion for the demonstrator. In the model, it is assumed that the blade is exposed to water on all external surface except the root of the plate, which is assumed to be connected to blade root connection with limited exposure to water. Figure 8 shows the moisture content (concentration) distribution in the blade subjected to 20,000 h of exposure to water. Figure 9 shows the changes of concentration of moisture over time at different locations (nodes) along the thickness of the blade. The nodes nearer to the surface show the concentration reaching saturation quicker.

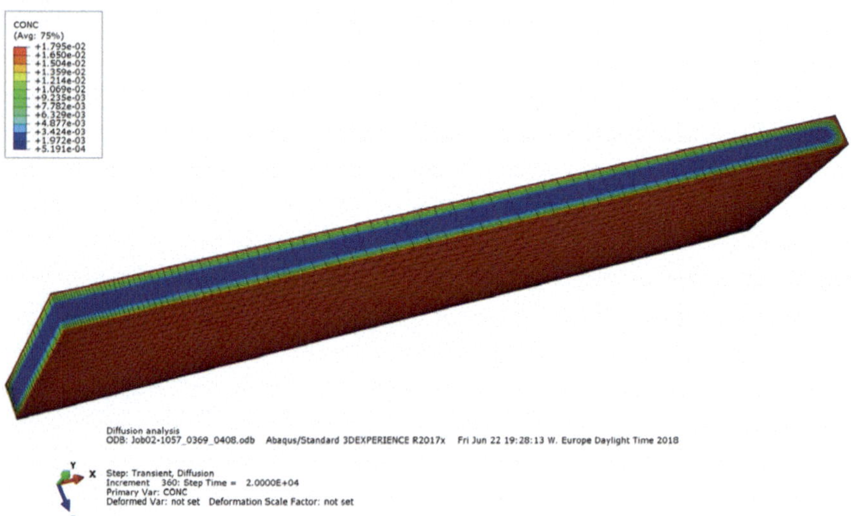

Fig. 8 Concentration of water in the blade subjected to 20,000 h of exposure

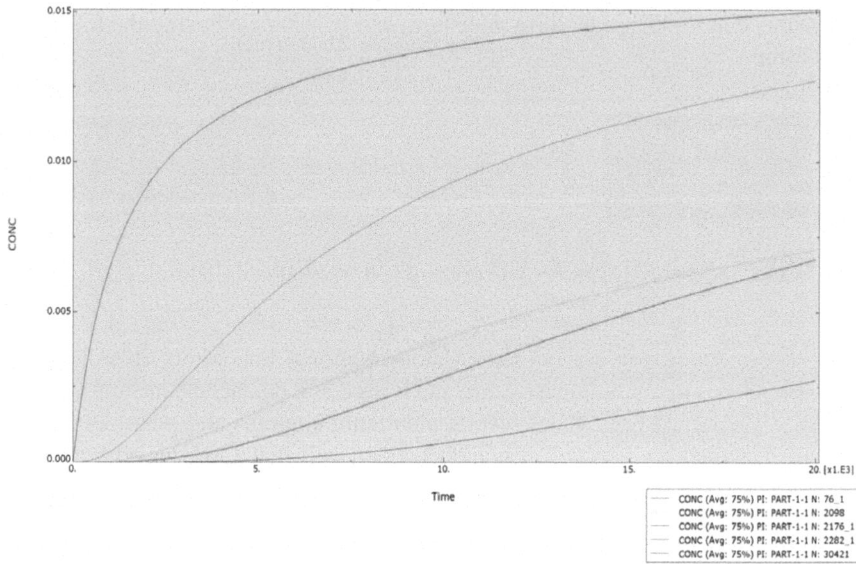

Fig. 9 Changes of concentration of moisture at different locations along the thickness of the blade

Table 3 Parameters for estimation of fibre radius subjected to moisture content

K_0^I (g/m^2 s)	K_0^{II} (g/m^2 s)	t_{st} (h)	R^2 ($t \leq t_{st}$)	R^2 ($t > t_{st}$)
3.00×10^{-9}	6.68×10^{-10}	166	0.8069	0.9653

4.3 Loss of Glass Fibres

One of the main degradation mechanisms is the loss of glass fibres due to corrosion (leeching of alkali metallic ions in the presence of water around the fibre). Due to the loss of glass fibre, the radius of the glass fibre will be reduced. As it is a chemically driven process, the following relationship was used to calculate the reduction in radius of the fibre follows the following relationship in:

$$\begin{cases} t \leq t_{st}: & r = r_0 - \dfrac{K_0^I}{\rho_{glass}} t \\ t > t_{st}: & r = r_{st} - \dfrac{K_0^{II}}{\rho_{glass}} (t - t_{st}) \end{cases}$$

and the following parameters (given in Table 3) are used to estimate the fibre radius after a certain period of exposure to the [5].

From the diffusion model, it is possible to estimate the average moisture content over the whole blade. Using the averaged moisture content in the blade, the new fibre radius due to degradation (by moisture) can be estimated using the following relationship:

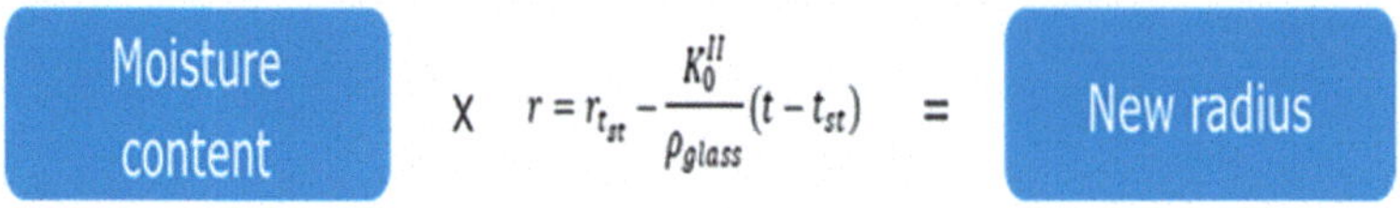

In DNVGL-ST-C501, [6], for UD plies, it can be assumed that:

$$E_1 = E_1^0 \cdot \frac{V_f}{V_f^0}$$

Hence, we can assume:

$$E_1(t) = \left[\frac{FVF_{new}(t)}{FVF_0}\right] = E_0 \Delta FVF(t) \quad (1)$$

where $E_1(t)$ is the young modulus after water exposure time t, $FVF_{new}(t)$ is the new fibre volume fraction at time t, with radius loss $r_0 - r(t)$. If we assume:

$$V_f \sim fibrecrosssection = \pi r^2$$

Then

$$\Delta FVF(t) = \frac{FVF_{new}(t)}{FVF_0} \approx \frac{\pi [r(t)]^2}{\pi {r_0}^2}$$

$$= \left(\left(r_0 - \frac{K_0^I}{\rho_{fibre}} t_{tst} - \frac{K_0^{II}}{\rho_{fibre}}(t - t_{tst})\right) / r_0\right)^2 \quad (2)$$

Hence, the degraded stiffness due to fibre loss can be estimated using Eqs. (1) and (2). Furthermore, due to the presence of moisture, matrix in the composite experiences plasticization. The impact of the aforementioned phenomenon is to cause reduction in the young modulus in the direction transverse to the fibre and in the out of plane direction. From the data provided by NTNU in [7] it was estimated that a reduction of approximately 11% in the matrix young modulus was seen, when the matrix is saturated in the moisture. Hence, in the finite element model, the matrix properties can be degraded by 11% to account for the plasticization.

Table 4 Mechanical properties of UD 0° and 90° and Biax 45°

Property	Value (UD 0° and 90°)	Value (Biax 45°)
E_{11}	26,700 MPa	26,700 MPa
E_{22}*	7132 MPa	26,700 MPa
E_{33}*	7132 MPa	7132 MPa
ν_{12}	0.26	0.26
ν_{13}	0.26	0.26
ν_{23}	0.26	0.26
G_{12}	3500 MPa	3500 MPa
G_{13}	2800 MPa	2800 MPa
G_{23}	2800 MPa	2800 MPa

4.4 Structural Finite Element Model

A structural finite element model was created to model the strain distribution in the blade under loading. For the sake of simplicity, the mechanical properties given in the DNVGL-ST-C501 are used in the finite element model for UD 0° and 90° and for Biax 45° stated in Table 4. In Table 4, E is the young modulus; ν is the Poisson's ratio; G is the shear modulus and subscripts of these properties indicate the direction following the coordinate in Fig. 7. For the loading, a uniformly distributed pressure loads of 0.164 MPa is applied on one main surface to represent the current loads, see Fig. 10.

4.5 Bayesian Model of Blade and Results

The discretized BN for the problem under consideration is shown in Fig. 11. A software program GeNIe [8] was used to build the BN.

Two vital steps of building a BN model of the tidal blade are to identify the nodes of the BN and to define various nodes and establish the relationships (between parent and child nodes). The former is done by using the expert judgement, while the latter is achieved either by experimental data or by obtaining a parametric fit to the FEA data (as shown in Fig. 12). In order to cater for the uncertainty in the experiments and FEA, nodes representing white noise (characterized by standard Gaussian distribution having mean of zero and some standard deviation) is added to the BN. These nodes are represented by the yellow color in Fig. 11. The uncertainty quantified in the first step of the BN is propagated through the next steps of the BN, as will be seen in the results.

Once the BN is established, it can be used in the forward direction to obtain the evolution of the probability density function (PDFs) with time for the parameter of interest i.e. moisture content, stiffness in the Fibre direction and strain in the Fibre direction. The PDFs of the parameters are shown in Figs. 13, 14 and 15 respectively,

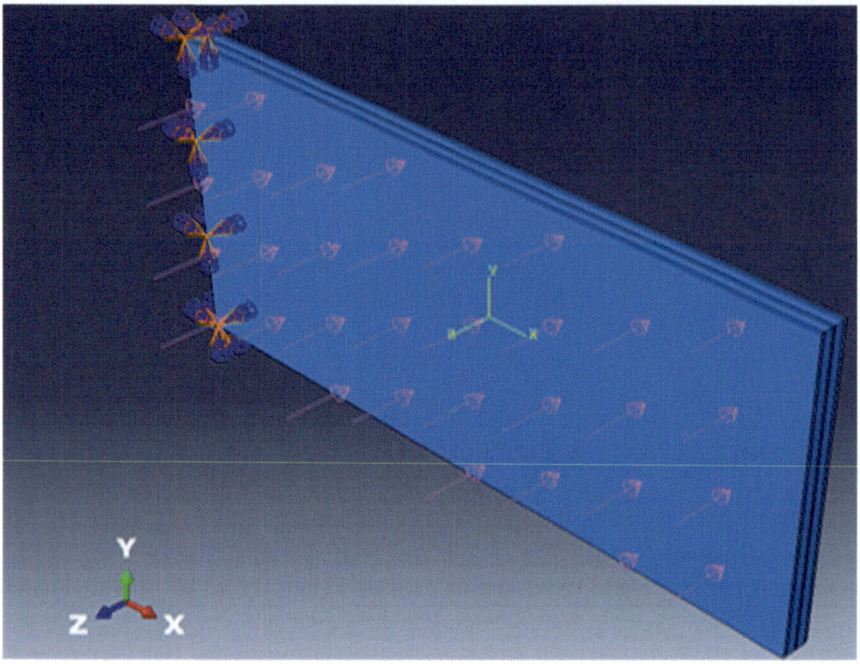

Fig. 10 Uniformly distributed pressure applied on the blade

and the trend of PDFs (i.e. shifting of mean and increasing of the standard deviation as the time progresses) is as expected.

Up until now, we demonstrated the use of the BN in the forward direction (i.e. causal reasoning) to predict the evolution of the output parameters (or parameters of interest), for a given set of input parameters. An added advantage of BN is that besides causal reasoning, it can also be used for inferential reasoning, i.e. given an evidence in one of the parameters, the probability distribution of the other parameters can be obtained. For e.g. Figure 16 indicates the probability distribution of Strain in Fibre direction, before and after evidence setting in Stiffness in the Fibre direction.

5 Conclusion

The main conclusion of the paper is as follows:

- Complex multi-parameter problems such as environmental aging are difficult to assess in the traditional SRA due to difficulty in determination of limit state function.
- To crack such problems Probabilistic Graphical Modelling (PGM) such as Bayesian Networks (BNs) can be utilized.

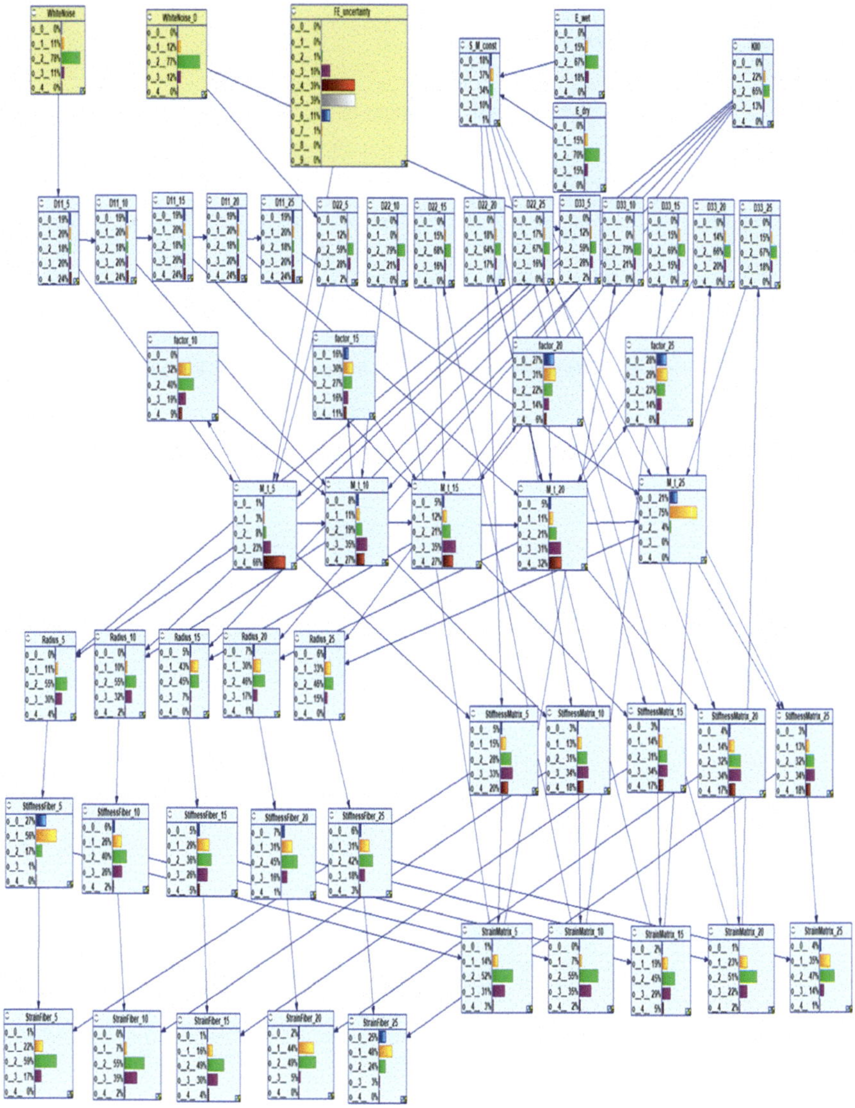

Fig. 11 Discretized Bayesian Network for the problem under consideration

- Bayesian Networks (BN) provides the possibility of adaption of real-time data and field measurements to update the models.
- BNs can be built in a transparent manner by a combination of deterministic physical models, expert knowledge and empirical data.

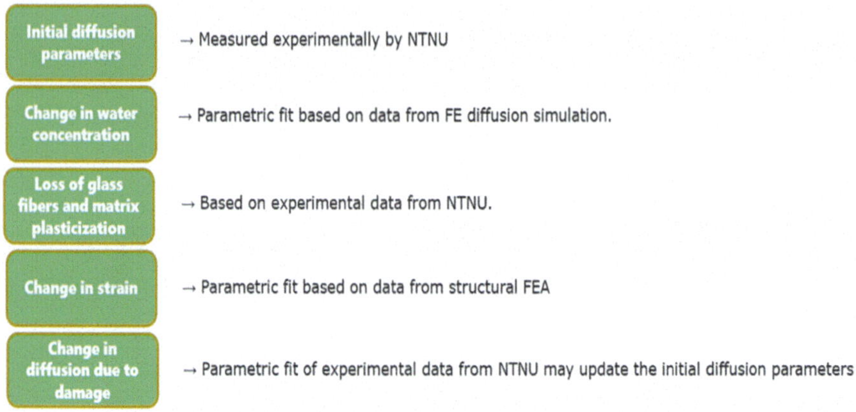

Fig. 12 Methods used to establish relationship between various nodes in the BN

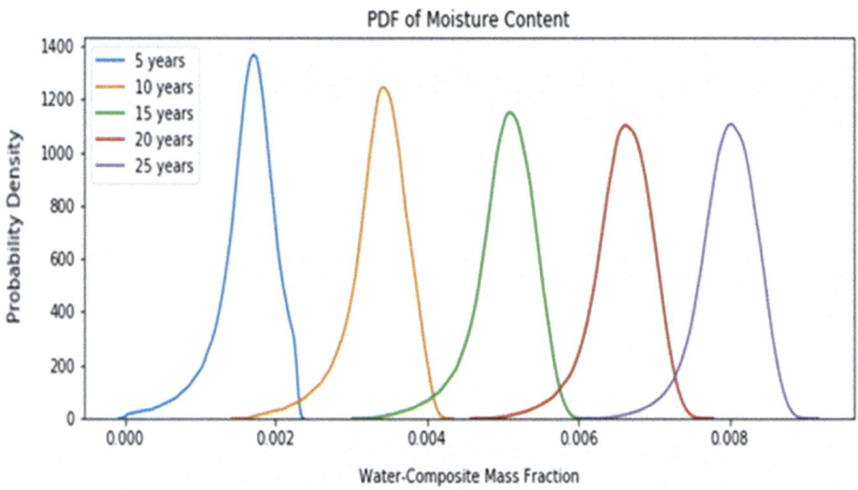

Fig. 13 PDF of moisture content for 5–25 years' time period

- The PDFs of the parameters (moisture content, stiffness and strain) and the trend of PDFs (i.e. shifting of mean and increasing of the standard deviation as the time progresses) is as expected.
- BN has also been used to demonstrate evidential reasoning by putting evidence of strain 0.85%. The PDFs of parameters (moisture content, stiffness and strain) thus obtained can be used for inspection planning of the tidal blades.

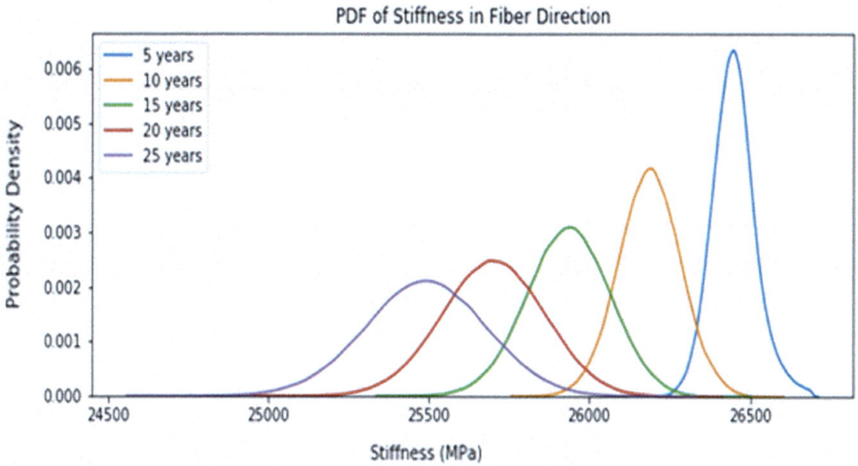

Fig. 14 PDF of stiffness for 5–25 years' time period

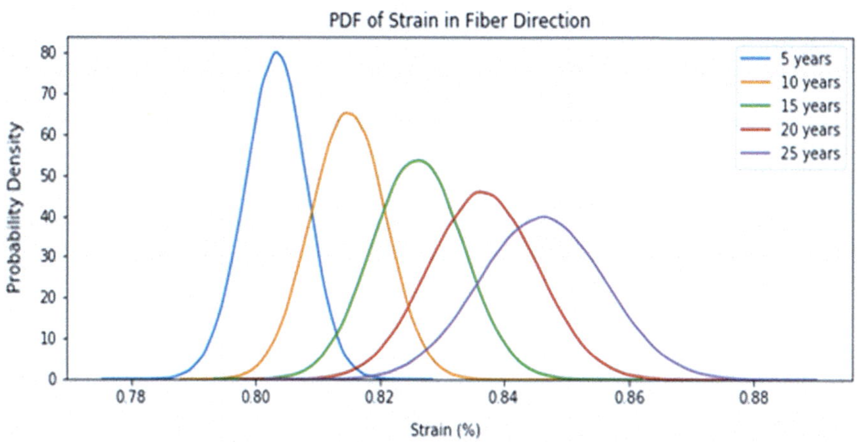

Fig. 15 PDF of strain for 5–25 years' time period

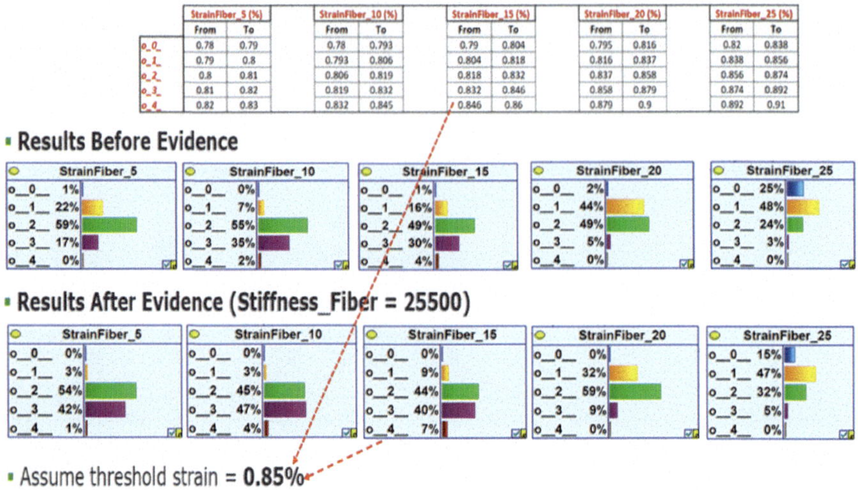

Fig. 16 PDF of strain in the Fibre direction for 5–25 years' time period

References

1. Maden HO, Krenk S, Lind NC (2006) Methods of structural safety. Dovert Publications, Mineola, NY, USA
2. Pearl J (1988) Probabilistic reasoning in intelligent systems. Morgan Kaufmann, San Francisco, USA
3. Keprate A, Ratnayake RMC (2019) Use of Bayesian Network for risk-based fatigue integrity assessment: application for topside piping in an arctic environment. Int J Offshore Polar Eng 29(4):421–428
4. Gagani A, Echtermeyer TA (2018) Fluid diffusion in cracked composite laminates—analytical, numerical and experimental study. Compos Sci Technol 160
5. Krauklis A, Echtermeyer TA (2016) Long-term dissolution of glass fibres in water described by dissolving cylinder zero-order kinetic model: mass loss and radius reduction. Open Chem 16
6. DNVGL-ST-C501 (2019) Composite Components. Det Norske Veritas, Høvik, Norway
7. Krauklis A (2018) Environmental degradation of constituents of composite materials. In: Presentation for JIP affordable composite meeting in April 2018
8. GeNIe Software, BayesFusion. https://www.bayesfusion.com/

Study on Fracture of Fiber-Reinforced Polymeric Composites Using Spiral Notch Torsion Test

Fen Du, Jy-An Wang, and Ting Tan

Abstract Fiber-reinforced polymer-matrix composites have been widely applied in various areas due to their excellent mechanical properties. In this study, carbon fiber-reinforced polymer-matrix composites and glass fiber-reinforced polymer-matrix composites failure tests under quasi static torsion have been demonstrated by using spiral notch torsion test methodology. Three types of cylindrical specimens with different fiber orientations were prepared and tested. For both composites, after increase stage, A type exhibited limited stress variations before the sudden failure, whereas B type showed gradual failure with torque drops. Displacement contours, stress contours and energy release rates were acquired using the anisotropic elastic finite element models. Microscopic characterization exhibited that the failure of A type occurred along interfaces between fibers and matrices, whereas circular cracks in B type propagated alternatively between polymeric matrices and interfaces. Type C failure is similar to type A whereas with a curvature observed on the fracture surface.

Keywords Fiber-reinforced composites · Fracture · Spiral notch torsion test

F. Du
Department of Mechanical Engineering, Vermont Technical College, Randolph Center, VT 05601, USA
e-mail: fen.du@vtc.edu

J.-A. Wang
Materials Science and Technology Division, Oak Ridge National Laboratory, Oak Ridge, TN 37830, USA

T. Tan (✉)
Department of Civil and Environmental Engineering, University of Vermont, Burlington, VT 05405, USA
e-mail: ting.tan@uvm.edu

M. Abdel Wahab (ed.), *Proceedings of 1st International Conference on Structural Damage Modelling and Assessment*, Lecture Notes in Civil Engineering 110,
https://doi.org/10.1007/978-981-15-9121-1_12

1 Introduction

Fiber-reinforced polymer-matrix (FRP) composites are made from polymeric matrices and different fibers. The carbon fiber-reinforced polymeric composites (CFRP) and glass fiber-reinforced polymeric composites (GFRP) are commercial FRP composites which the fiber is carbon and glass, respectively. Both CFRP and GFRP have been widely used in aerospace, renewable energy, automotive and construction industries, et al. due to their high-performance mechanical properties [1–4]. Study mechanical performance of unidirectional composites is critical because this would help further understand the more advanced composites structures. Substantial research has been performed to evaluate the fracture of unidirectional FRP composites using different methods [5, 6], such as compression [7, 8], double cantilever beam specimens [9, 10], Iosipescu tests [11, 12], edge crack torsional tests [13, 14]. Despite prior work, limited studies investigated how unidirectional CFRP and GFRP composites crack under pure torsion, or to quantify how torque drops evolve as cracks propagate.

The objective of this work is to find the fracture behavior of CFRP and GFRP composites using Spiral Notch Torsion Test (SNTT) [15]. As an innovative technology, SNTT used here can effectively obtain the failure performances of unidirectional CFRP and GFRP. First, cylindrical shaped specimens with cubic ends were fabricated. The V-shaped spiral notches were grooved in the gauge section of the specimens. Different types of specimens were prepared by varying the relative angle of the fiber direction and cylindrical sample longitudinal axis, i.e., the longitudinal axis of type A specimen was 90° to the fiber direction; type B specimen longitudinal axis was 0° to the fiber direction; and type C specimen longitudinal axis was 45° to the fiber direction. Then, SNTT torsional tests were conducted to fail all type specimens, respectively. The experimental system used to test CFRP composites was built with two load cells and two data acquisition cells. Finite element simulation results indicated characteristics of displacement, *von* Mises stress and energy release rates of different types of samples. Torque drops were examined for CFRP composites type B specimen after the peak load. Finally, microscopic characterization was conducted to further reveal distinctive failure mechanism of different types of composites.

2 Materials and Methods

2.1 Spiral Notch Torsion Test

A unique fracture testing methodology, Spiral Notch Torsion Test (SNTT), is used in this study to determine the fractural resistance of unidirectional CFRP and GFRP materials. The idea of this innovative method is applying quasistatic torsion forces on the two ends of the cylindrical test specimen. In Fig. 1a, when a cylindrical specimen is loaded in torsion, the spiral notch initiates the crack that propagate

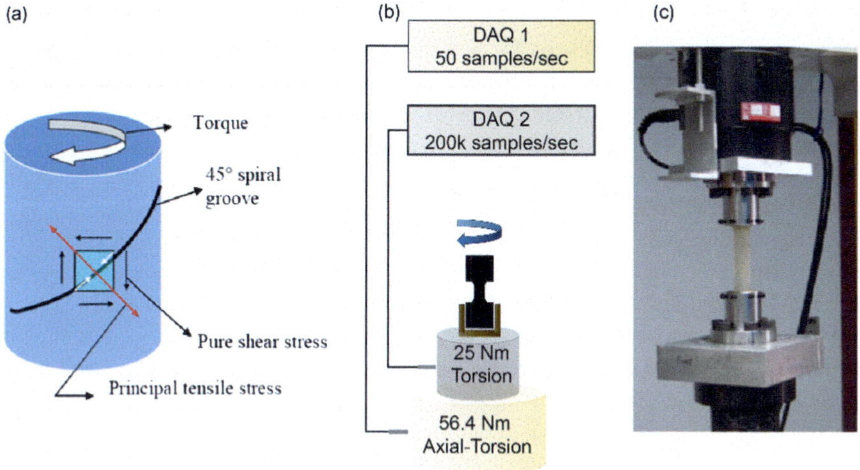

Fig. 1 **a** SNTT method schematic diagram. **b** Schematic diagram of experimental system for CRFP composite. Two load cells and two data acquisition cells are integrated in this system. **c** Experimental setup for GFRP SNTT test. **a**, **c** adapted from Tan et al. [19]; **b** adapted from Du et al. [20]

perpendicularly toward the central axis of the cylinder. More information about the SNTT methodology is included in prior studies [15–18].

In this study, SNTT is used and extended to study fracture of polymeric composites. The experimental set-up for CRFP composite was created with two load cells, one axially torsional load cell with a torque limit up to 25 Nm on the top, while the other load cell with a torque limit up to 56.4 Nm on the bottom, as shown in Fig. 1b. Two data acquisition units were adopted to collect testing date with frequencies at 200 kHz and 50 Hz for small and large load cells, respectively. The advantage of using an ultra-high sampling rate is to capture very small variations of the torque. This information could reveal composites' microstructural damages during the failure. SNTT torsional tests were conducted on type A and type B CFRP composite specimens to failure. This experimental system is displacement controlled. The displacement rate of 0.5°/s was applied.

The experimental system for GFRP composite is shown in Fig. 1c, the servo-hydraulic axially torsional testing machine was used to conduct SNTT testing [19]. Furthermore, the GFRP composite experimental system is load-controlled. The loadings rates of 0.56, 1.13 and 2.26 N m/s were applied for all three types of GFRP specimens to find whether mechanical performance varies on different loading rates.

2.2 *Specimen Preparations*

The unidirectional GFRP composite specimens were made from composite plates with 20 layers of unidirectional fabrics and resin [19]. The dimensions of these cylindrical specimens were 25.4 mm in diameter and a gauge length of 101.6 mm. Spiral notch with a pitch angle of 45° was grooved on each GFPR specimen surface. This V-shaped notch provided the root of crack initiations and 45° pitch angle made the failure into a Mode I loading in SNTT test due to applied pure torsion and cylindrical shaped specimens. The notch to diameter ratio was about 0.20.

Three different types of specimens were fabricated by varying the relative angle between the fiber orientation and cylindrical sample longitudinal axis (Fig. 2a). For type A specimen, the longitudinal axis was 90° to the fiber orientation (Fig. 2b), whereas this angle changed into 0° for type B specimen as the fibers were parallel to samples' longitudinal axis, and the angle was 45° for type C specimens (Fig. 2d).

Types A and B specimens were prepared following the same approach as shown in Fig. 2 using CFRP composites. The unidirectional CFRP composite specimens were obtained from composite plates of ~20 layers of unidirectional fabrics [20]. The properties of carbon fibers and epoxies of this composite are provided by the suppliers [21, 22]. These properties were later used to estimate the elastic properties of the composite. The dimensions of CFRP specimens were much smaller than those of GFRP. The diameter of cylinder was about 4.1 mm, the effective gauge length was

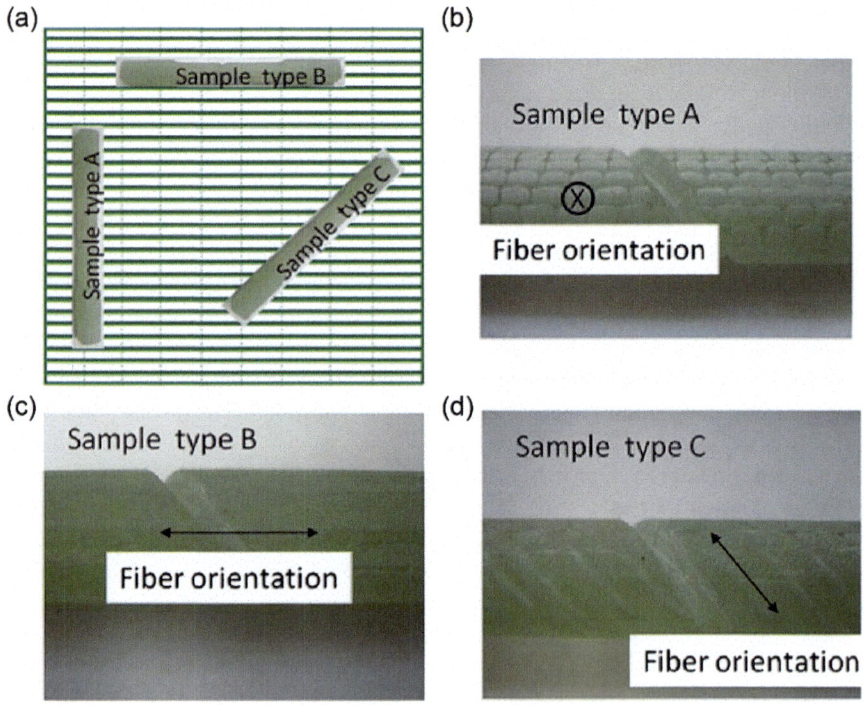

Fig. 2 **a** Fabrication directions on composite plate of different types of GFRP specimens, **b** type A specimen with spiral notch and fiber orientation, **c** type B, **d** type C specimens. Adapted from Tan et al. [19]. Type A and B CFRP specimens were prepared following the same approach

about 20.3 mm, and the entire specimen length was about 50.8 mm. Similar to GFRP specimens, spiral notches with 45° pitch angle were grooved along the specimen surface with a notch to dimeter ratio about 0.15 [20].

2.3 The Wiener Filter

In CFRP composite experiment, data collection rate of the load cell with up to 25 Nm torsion (Fig. 1b) was at 200 kHz. Due to the high frequency collection, noises could be a problem in data accuracy. The Wiener filter [23–25] was used to get high accuracy stress strain curves. The filter deconvolution details are described in Alghamdi et al. [26–29].

2.4 Fractographic Characterization

For GFRP composite type A, B and C specimens, both optical and scanning electron microscopic (SEM) techniques were used for fractural characterization. For CFRP composite type specimens, microCT images were captured besides SEM images. The detailed information about the equipment used in fractographic characterization can be found in Du et al. [20].

3 Finite Element Modeling

Finite element analyses were conducted using ABAQUS™ to reveal displacement contours and stress contours of all three types of specimens. The simulation focused on linear stage of load–displacement variation. Only a quarter of the entire cylindrical specimens were modeled because the spiral notches with 45° pitch angle on the cylindrical surface were axisymmetric. The notch was demonstrated as crack in the model. The length of the crack set in the model was consistent with the testing specimens, notch to diameter ratio about 0.15 of CFRP and 0.20 of GFRP composites, respectively. Two types of mesh elements were used. Around crack tip area wedge elements (C3D15) was applied, whereas the rest of the area reduced quadratic elements (C3D20R) was applied. The load added in the model was a concentrated torque [19, 20]. The elastic properties of unidirectional CFRP and GFRP composites were listed in Table 1. E_1, E_2, E_3 are Young's modulus in 1, 2 and 3 orientations; G_{12}, G_{23}, G_{13} are the shear modulus; ν_{12}, ν_{23}, ν_{13} are associated Poisson's ratios.

Estimations were calculated by using four mathematical models [30–33] due to limited access to accurate elastic properties of unidirectional CFRP composites. The predictions of Young's moduli and shear moduli along 1, 2, and 3 orientations from different models were agree with each other in general. The material properties of

Table 1 Mechanical properties of composites used in finite element modeling

	E_1 (GPa)	E_2 (GPa)	E_3 (GPa)	G_{12} (GPa)	G_{23} (GPa)	G_{13} (GPa)	ν_{12}	ν_{23}	ν_{13}
CFRP [20]	127.8	8.6	8.5	3.2	3.3	3.1	0.29	0.42	0.15
GFRP [34, 35]	44.6	17.0	16.7	3.5	3.5	3.8	0.26	0.35	0.26

GFRP composite were obtained from reference [34, 35]. The elastic properties of the unidirectional CFRP and GFRP composites used in this study were summarized in Table 1.

4 Results and Discussion

4.1 Load–Displacement Curves

Infrared thermography technique was used to capture the loading processes of GFRP composite samples during SNTT testing. Figure 3 showed the four stages of loading for a type A sample. Beginning state of the experiment was shown in Fig. 3a. In

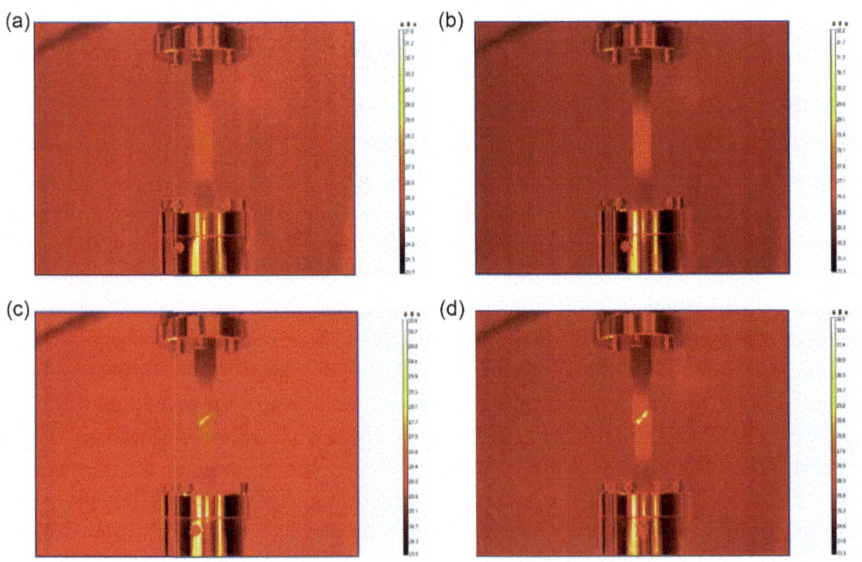

Fig. 3 Infrared images of a type A sample. **a** Image at the beginning of the testing. **b** Image at the crack initiation. **c** Image of the crack propagation. **d** Image after the completely separation of the sample. Adapted from Tan et al. [19]

Fig. 3b, the crack initiation was observed as a bright dot at the spiral notch indicating. This bright dot indicated a higher temperature than the surrounding area. The explanation of this temperature peak was likely to be the formation of a crack where the surface energy was released. This spot would likely to be the location of fracture initiation. In Fig. 3c, the spot region was grown into full length of the spiral notch as the applied torque during the test raised. The image revealed that in the SNTT test, crack propagation happened along the previously grooved notch. The failure of the specimen is shown in Fig. 3d, shiny bright area indicated the release of a large amount of surface energy due to the failure.

The torque-time loading curves were recorded during the SNTT tests for both CRFP and GRFP composite samples, as shown in Fig. 4. Load–displacement curves of type A CFRP composites were shown in Fig. 4a, whereas curves of type B CFRP composites were shown in Fig. 4b. Three specimens were tested of each type. The

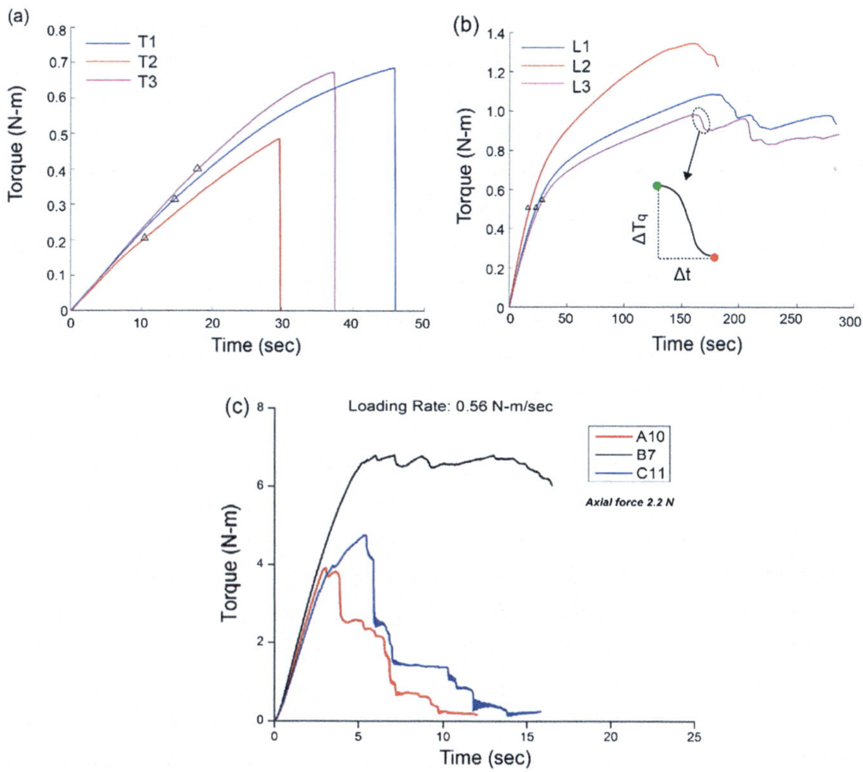

Fig. 4 **a** Loading curves of three type A CFRP composite specimens. **b** Loading curves of three type B CFRP composite specimens. **c** Loading curves of type A, B and C samples of glass fiber composite. **a**, **b** adapted from Du et al. [20]; **c** adapted from Tan et al. [19]

results demonstrated good consistency between different trials of same type of specimens. Load–displacement curves of types A, B and C of GFRP composite specimens were shown in Fig. 4c.

For type A specimens, two stages of the loading curves were observed (Fig. 4a), a smooth increasing stage followed by a sudden drop stage. During the increasing stage, two segments were indicated in the Fig. 4a, a linear increasing segment initially, and a nonlinear increasing segment until the sudden drop. Compare to type A specimens, type B specimens' failure behavior during SNTT test were quite different (Fig. 4b, c). At the beginning, a linear increasing stage was observed. Then a nonlinear increasing stage before drop was shown. Different than the sudden drop of type A, the failure of type B exhibited stepwise fluctuations over time after the maximum torque (Fig. 4b).

4.2 Finite Element Analysis

4.2.1 Displacement Contours

The finite element models of CFRP and GFRP composite specimens were built to simulate the loading process. CFRP composite type A displacement contours were presented in Fig. 5. The original mesh of a quarter of the specimen was shown in Fig. 5a, while Fig. 5b–d showed the deformed displacement contours in radial, tangential and longitudinal directions respectively. As can be observed in Fig. 5b, in the radial direction, the maximum displacements happened near the middle of the grooved notch. The contours of tangential direction were concentric rings grew from the center of the cylinder cross-section surface (Fig. 5c). Whereas in longitudinal direction, the contours were symmetric on the two sides of the grooved spiral notch. This can be explained as the tests were conducted in a torsion way, twisting the samples to the equal, opposite directions.

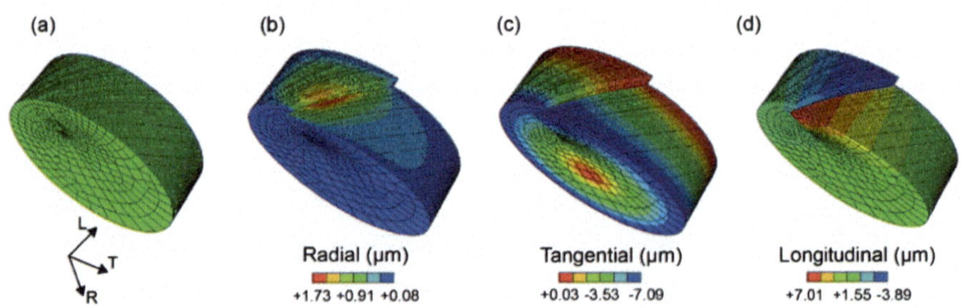

Fig. 5 CFRP composite type A specimen displacement simulation **a** mesh of a quarter of specimen before loading. Deformed contours of **b** radial direction, **c** tangential direction, and **d** longitudinal directions. Adapted from Du et al. [20]

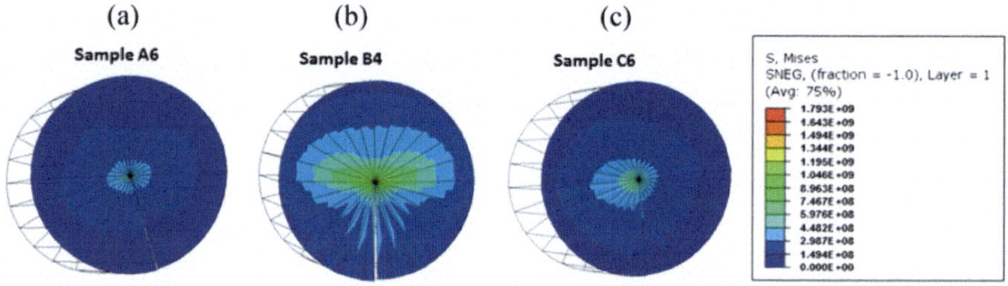

Fig. 6 GFRP composite **a** type A crake tip *von* Mises stress contour, **b** type B crack tip stress contour, and **c** type C crack tip stress contour. *Note* Scale deformed factor: 20; Unit: Pa. Adapted from Tan et al. [19]

4.2.2 *von* Mises Stress Contours

The *von* Mises stress distributions were investigated for both CFRP and GFRP composite specimens as well. Figure 6a–c demonstrated *von* Mises stress contours of the deformed GFRP composite specimens of type A, B and C respectively. As the fiber orientations are different in these three types, the stress contours exhibited different characteristics. Type A and C contours were asymmetric, whereas type B contours were symmetric. The magnitude of *von* Mises stress was higher of type B, while those of type A and C were very close. Similar contours of *von* Mises stress were obtained for CFRP composites.

From the crack tip *von* Mises stress contours information, energy release rates of three specimen types were obtained for both CFRP composite and GFRP composite, as shown in Figs. 7 and 8. Results showed that for CFRP composites, type A specimens had larger energy release rates than type B specimens (Fig. 7a). The stress

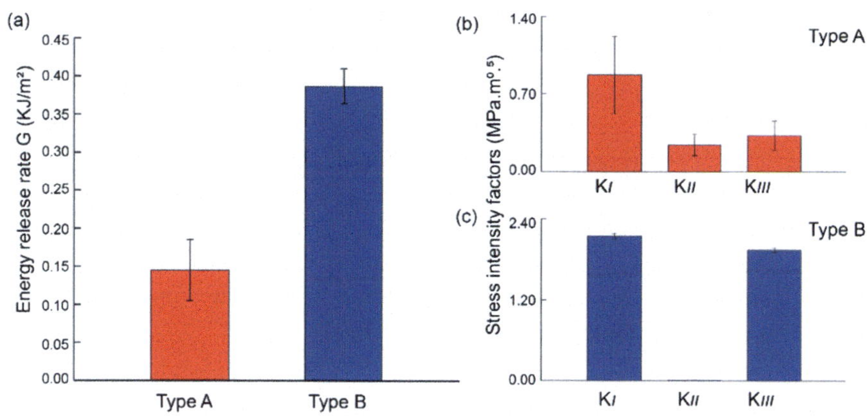

Fig. 7 **a** Energy release rates of type A and B CFRP composite specimens at the proportional limits. Stress intensity factors at different mode mixites of **b** Type A and **c** Type B specimens. Adapted from Du et al. [20]

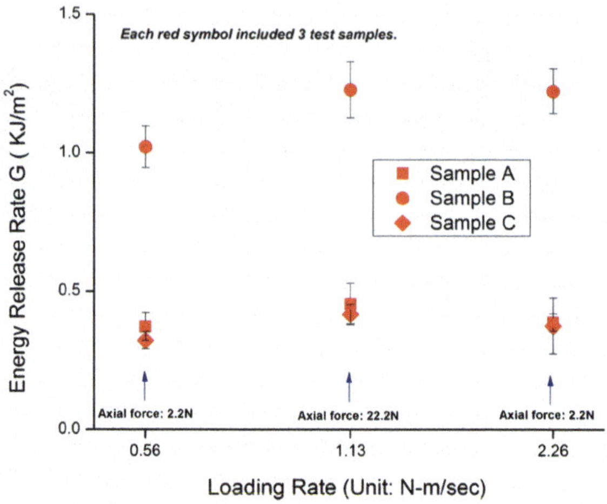

Fig. 8 Energy release rates of GFRP composite samples. Adapted from Tan et al. [19]

intensity factors were extracted from ABAQUS to find out the domain fracture modes for type A and type B CRFP composites (Fig. 7b–c). The results indicated type A was mainly in mode I, whereas for type B, it was mostly the mixture of mode I and mode III.

Energy release rates of GFRP composites at three different loading rates were shown in Fig. 8. Type B specimens had the highest energy release rates in all three loading rates. Those of type A and type C composite specimens were very close while type C showed slightly lower energy release rate than type A. Consistent results were obtained for the same type of composite for each loading rate. Among the three loading rates, the energy release rates at 1.13 N m/s showed higher values than those at 0.56 and 2.26 N m/s loading rates.

4.3 *Torque Drops*

For unidirectional CFRP composites, the wiener filter [23–25] was applied to process the high temporal resolution stress-time curves. Due to the extended stress tails, statistical analyses were performed to evaluate torque drops of B type. Each torque drop was separated based on the starting and ending points defined in Fig. 4b. The distributions of avalanche durations and sizes of B type torque drops showed that Gaussian models described distributions well (Fig. 9a, b).

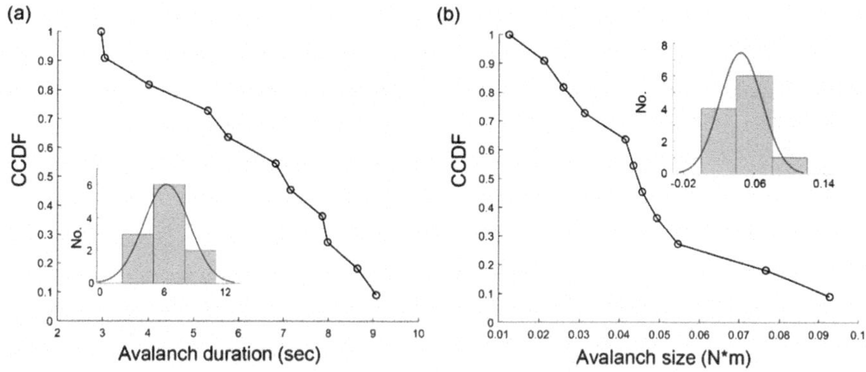

Fig. 9 Complementary cumulative distribution functions (CCDF) of type B specimens **a** torque drop durations **b** torque drop sizes. Adapted from Du et al. [20]

4.4 Fractographic Characterization

For both unidirectional CFRP composites and GFRP composites, different types of specimens exhibited distinctive failure characteristics. For type A CFRP specimens, a fractural surface occured across of the gauge section (Fig. 10a, b). By loading torsion, the principal tensile stresses are perpendicular to the spiral notch (Fig. 1a). Figure 10c exhibited that fractural areas were characterized by debonded fibers, cusps and river patterns [20]. Broken fibers and embedded fibers in matrices were observed, whereas empty holes were found within matrices (Fig. 10d). For GFRP specimens, similar failure modes included fiber failure and strand delamination (Fig. 10e, f). The fracture occurred along the spiral notch and the crack propagated parallel to the fiber orientation. Cusps and shear marks along the interfaces clearly demonstrated the fast cracks during the catastrophic drop. Type C specimens of GFRP exhibited similar behavior as type A specimens. This type of specimen also separated into two halves as type A (Fig. 10g), whereas a curvature was observed on the fractural surfaces (Fig. 10h).

However, the type B speicmens (Fig. 11) did not separate into two halves. The reason was the fiber orientations were parallel to the longitudinal axis, cracks cut along the fiber-matrix interfaces rather than within the fibers. MicroCT analysis exhibited that axial cracks propagated between plies (Fig. 11b), and around the unnotched cores (Fig. 11c). Similarly, cracks started from the spiral notch and grew around the unnothed cores (Fig. 11d-f). Surprisingly, the circular cracks propagated alternatively between fiber-matrix interfaces and polymeric matrices in a unidirectional CFRP composite. The cross-section of the GFRP composite (Fig. 11g-i) revealed that similar patterns of multiple cracks, including (1) cracks propagate on the fabric interfaces; (2) cracks between fiber strands; and (3) a circular crack initiated from spiral notch of cross-section plane (Fig. 11i).

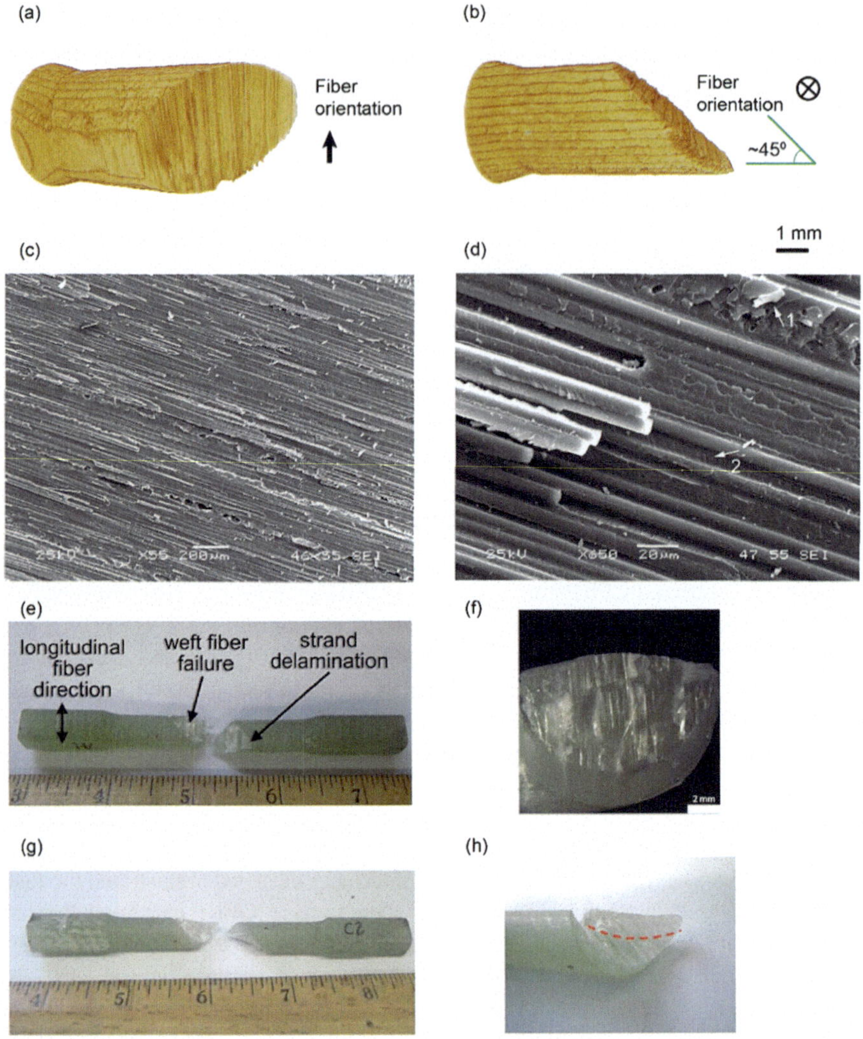

Fig. 10 CFRP composite type A specimen surfaces images after failure. **a** MicroCT image front view, **b** and side view. **c** SEM images at low magnification, and **d** high magnification. GFRP composite specimen images of **e** a failed type A specimen, **f** type A fracture surface, **g** a failed type C specimen, and **h** type C fracture surface. *Note* Figure **e** and **g** ruler length unit: inch. **a**–**d** adapted from Du et al. [20]; **e**–**h** adapted from Tan et al. [19]

5 Summary and Conclusions

Spiral notch torsion test (SNTT) has been applied to study the fracture of unidirectional CFRP and GFRP composites in this study. Three types of composite specimens were prepared with different fiber orientation alignments. Type A specimen has a

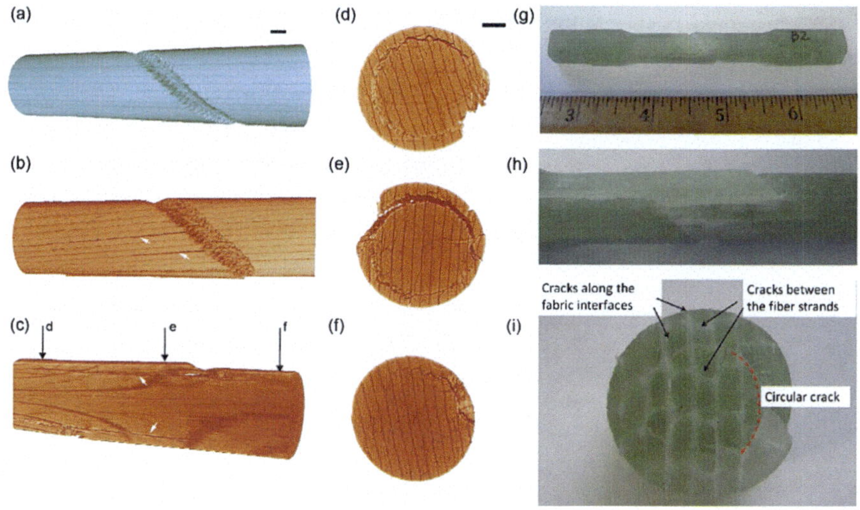

Fig. 11 Figures of CFRP composite type B specimens **a** before and **b** after loading. MicroCT analysis of a fractured type B specimen **c** along the axial direction **d**–**f** cross sections at different locations of the spirally notched specimen. White arrows point the longitudinal cracks. Scale bars are 500 μm. Figure **a**–**f** adapted from Du et al. [20]. A GFRP composite type B sample showing the **g** overview, **h** the back side, and **i** the cross-section. *Note* Figure **g** ruler length unit: inch. Figures **g**–**i** adapted from Tan et al. [19]

90° angle between the longitudinal axis of cylinder to the composite fiber orientation. This angle was 0° for type B, and 45° for type C, respectively. The SNTT tests were conducted for CFRP composite type A and type B, while three types, A, B and C, for GFRP composite. Due to the different fiber orientations in these cylindrical samples, the failure characteristics were different. For CFRP and GFRP composites, after SNTT testing, type A specimens were separated from the spiral V- shaped notch into two halves, while type B specimens failed without separation into two parts. Detailed observation of failed type B specimens showed that circular cracks propagated alternatively along interfaces and polymeric matrices. Type C specimens also separated into halves.

Finite element analysis has been conducted to disclose the displacement and crack-tip stress contours of the composite. The anisotropic linear elastic properties were used in the simulation. The results showed that *von* Mises stress contours of type A and type C specimens near the crack tip were asymmetric. However, type B specimens showed symmetric stress contours near the crack tip. The energy release rates were compared between CRFP and GRFP composites. Consistent results were obtained for both composites that type A specimens had lower energy rates, while type B specimens had higher values. Type C specimens of GRFP had very close values as type A. With the variation of loading rates during the test, the energy release rates changed slightly. Findings of unidirectional fiber-reinforced polymeric composites could further help in the study of advanced fiber-reinforced polymeric composites.

Acknowledgements This work was supported by Vermont NASA EPSCoR [grant NOs. NNX13AB35A, NNX14AN20A], the Wind Energy Program of Department of Energy and Oak Ridge National Laboratory under contract DE-AC05-00OR22725 with UT-Battelle, LLC., Vermont Technical College through Vermont additive manufacturing development fund [grant NO. 07120-OEA-AMP-VTC], University of Vermont clean energy fund [grant NO. 028613].

References

1. Yao SS, Jin FL, Rhee KY, Hui D, Park SJ (2017) Recent advances in carbon-fiber-reinforced thermoplastic composites: a review. Compos Part B 142:241–250
2. Wang X, Jiang M, Zhou Z, Gou J, Hui D (2017) 3D printing of polymer matrix composites: a review and prospective. Compos Part B 110:442–458
3. Mallick PK (2007) Fiber-reinforced composites: materials, manufacturing, and design. CRC Press, Boca Raton
4. Brondsted P, Lilholt H, Lystrup A (2005) Composite materials for wind power turbine blades. Annu Rev Mater Res 35:505–538
5. Rosen BW (1964) Tensile failure of fibrous composites. AIAA J 2(11):1985–1991
6. Hine PJ, Brew B, Duckett RA, Ward IM (1989) Failure mechanisms in continuous carbon-fiber reinforced PEEK composites. Compos Sci Technol 35(1):31–51
7. Vinod MS, Nayaka V, Shenoy R, Murali MS, Nafidi A (2001) Fractography of compression failed carbon fiber reinforced plastic composite laminates. J Mech Eng Res 2(1):1–9
8. Budiansky B, Fleck NA (1993) Compressive failure of fiber composites. J Mech Phys Solids 41(1):183–211
9. Iwamoto M, Ni QQ, Fujiwara T, Kurashiki K (1999) Intralaminar fracture mechanism in unidirectional CFRP composites: part I: Intralaminar toughness and AE characteristics. Eng Fract Mech 64(6):721–745
10. Whitney JM, Browning CE, Hoogsteden W (1982) A double cantilever beam test for characterizing mode I delamination of composite materials. J Reinf Plast Compos 1(4):297–313
11. Broughton WR, Kumosa M, Hull D (1990) Analysis of the iosipescu shear test as applied to unidirectional carbon-fiber reinforced composites. Compos Sci Technol 38(4):299–325
12. Walrath DE, Adams DF (1983) The iosipescu shear test as applied to composite materials. Exp Mech 23(1):105–110
13. Czabaj MW, Ratcliffe JG, Davidson BD (2014) Observation of intralaminar cracking in the edge crack torsion specimen. Eng Fract Mech 120:1–14
14. Li J, Lee SM, Lee EW, O'Brien TK (1997) Evaluation of the edge crack torsion (ECT) test for mode III interlaminar fracture toughness of laminated composites. J Compos Tech Res 19(3):174–183
15. Wang JA, Liu KC, McCabe DE, David SA (2000) Using torsional bar testing to determine fracture toughness. Eng Mater Struct 23:45–56
16. Wang JA, Liu KC, McCabe DE, David SA (2001) Using torsion bar testing to determine fracture toughness, K_{tc}. Fatigue Fract Engng Mater 23:45–56
17. Wang JA, Liu KC, McCabe DE (2003) An innovative technique for measuring fracture toughness of metallic and ceramic materials. Fatigue Fract Mech 33:757–770. [ASTM STP 1417]
18. Wang JA, Wright IG, Lance MJ, Liu KC (2006) A new approach for evaluating thin film interface fracture toughness. Mater Sci Eng A 426:332–345

19. Tan T, Wang JJA, Ren F, Lara-Curzio E, Agastra P, Mandell J, Bertelsen WD, LaFrance CM (2013) Investigating fracture behavior of polymer and polymeric composites using the spiral notch torsion test. Eng Fract Mech 101:109–128
20. Du F, Alghamdi S, Riabbans B, Tan T (2019) An experimental study on the fracture of a unidirectional carbon fiber-reinforced composite under quasistatic torsion. Compos B Eng 172:547–554
21. Composites D (2017) Materials data sheets. Allred and Associates, Elbridge
22. Kelly A (ed) (2012) Concise encyclopaedia of composite materials. Elsevier, Amsterdam
23. Antonaglia J, Wright WJ, Gu X, Byer RR, Hufnagel TC, LeBlanc M, Uhl JT, Dahmen KA (2014) Bulk metallic glasses deform via slip avalanches. Phys Rev Lett 112(15):155501
24. Dahmen KA, Ben-Zion Y, Uhl JT (2011) A simple analytic theory for the statistics of avalanches in sheared granular materials. Nat Phys 7(7):554
25. Dahmen KA, Ben-Zion Y, Uhl JT (2009) Micromechanical model for deformation in solids with universal predictions for stress-strain curves and slip avalanches. Phys Rev Lett 102(17):175501
26. Alghamdi S, Tan T, Hale-Sills C, Vilmont F, Xia T, Yang J, Huston D, Dewoolkar M (2017) Catastrophic failure of nacre under pure shear stresses of torsion. Sci Rep 7:13123
27. Alghamdi S, Du F, Yang J, Tan T (2018) The role of water in the initial sliding of nacreous tablets: findings from the torsional fracture of dry and hydrated nacre. J Mech Behav Biomed Mater 88:322–329
28. Alghamdi S, Du F, Yang J, Pinder G, Tan T (2020) Tensile and shear behavior of microscale growth layers between nacre in red abalone. J Mech Phys Solids 138:103928
29. Alghamdi S, Liu Z, Du F, Yang J, Dahmen KA, Tan T (2020) Sliding avalanches between nacreous tablets. Nano Lett. https://doi.org/10.1021/acs.nanolett.0c01148
30. Younes R, Hallal A, Fardoun F, Chehade FH (2012) Comparative review study on elastic properties modeling for unidirectional composite materials. In: N Hu (ed) Composites and their properties. InTech, London
31. Affdl JC, Kardos JL (1976) The Halpin-Tsai equations: a review. Polym Eng Sci 16(5):344–352
32. Chamis CC (1989) Mechanics of composite materials: past, present, and future. J Compos Technol Res ASTM 11:3–14
33. Huang ZM (2001) Simulation of the mechanical properties of fibrous composites by the bridging micromechanics model. Compos Part A 32:143–172
34. Agastra P, Mandell JF (2010) Testing and simulation of damage growth at ply drops in wind turbine blade laminates. In: Proceedings of society of the advancement of materials and process engineering (SAMPE), Seattle, WA
35. Samborsky DD, Agastra P, Mandell JF (2020) 3-D static elastic constants and strength properties of a glass/epoxy unidirectional laminate, internal report. Department of Chemical and Biological Engineering, Montana State University, Bozeman, MT. https://www.coe.montana.edu/composites/documents/3D%20Static%20Property%20Report.pdf, last accessed 2020/06/16

Comparative Numerical Study of Circular-Shaped Steel Tubes Subjected to Cyclic Horizontal Loading

Qusay Al-Kaseasbeh

Abstract Circular-shaped steel tubes have increasingly used in a range of structural and architectural engineering applications. This paper investigates the influence of biaxial cyclic horizontal loading patterns on the seismic behavior of the stepped CHS tubes. A numerical comparative study has been carried out in finite-element software ABAQUS with four uniaxial (one-cycle (1 *N*), and three-cycle (3 *N*), line 22.5°, and line 45°) and four biaxial (square, circular, octagon, and ellipse) cyclic horizontal loading scenarios with the existing of superstructure dead load. The results revealed that the shape of the load–displacement hysteresis loops under uniaxial loading patterns is totally different than biaxial loading patterns. Moreover, the H_m/H_y values under biaxial cyclic horizontal loading patterns are less than uniaxial horizontal cyclic loading patterns. Among all the selected loading scenarios, the square loading pattern shows the worst behavior while the ellipse loading pattern shows the best behavior in terms of bearing capacity.

Keywords Circular-shaped section · Biaxial loading patterns · Cyclic · Seismic performance

1 Introduction

Over the last few decades, structural hollow steel tubes as a cantilever pier have increasingly used in a range of structural and architectural engineering applications including offshore oil structures, building structures, and huge steel bridges [1–6]. The widespread use of structural hollow steel tubes is due to their appearance, ease of maintenance and construction, excellent mechanical properties, high deformation, and bearing capacity [1–7]. During major earthquakes, circular-shaped hollow sections (CHSs) are exposed to local and global buckling damage which led to a significant degradation in the bearing and deformation capacities [1–6]. In the light of these observations, extensive research has been carried out on the hysteretic behavior

Q. Al-Kaseasbeh (✉)
Mutah University, Mutah, Al-Karak 61710, Jordan
e-mail: qusay.alkaseasbeh@mutah.edu.jo

M. Abdel Wahab (ed.), *Proceedings of 1st International Conference on Structural Damage Modelling and Assessment*, Lecture Notes in Civil Engineering 110,
https://doi.org/10.1007/978-981-15-9121-1_13

of CHSs to attain a better understating of such structures performance under uni/bi-axial cyclic horizontal loads [2–5, 8]. As a result of all these studies, the bearing capacity and deformation capacity are influenced by the radius-to-thickness ratio and slenderness ratio parameters [2, 4, 9]. In general, the earthquake excitations are complicated and 3-dimensional not 2-dimensional components as early assumed [10–13]. As expected, the CHSs hysteretic behavior is significantly severe than the uniaxial cyclic horizontal loads [3, 4]. Accordingly, the effect of the loading pattern should be considered in the seismic design of CHS structures [14, 15].

Al-Kaseasbeh et al. [2, 4] studied the bearing and deformation capacities of CHSs under uni/biaxial cyclic horizontal loading in which the wall plate thickness is stepped in different ranges. This study investigates the influence of biaxial cyclic horizontal loading patterns on the seismic behavior of the proposed stepped CHS tubes. A numerical comparative study has been carried out in finite-element software ABAQUS with four uniaxial (one-cycle (1 N), and three-cycle (3 N), line 22.5°, and line 45°) and four biaxial (square, circular, octagon, and ellipse) horizontal displacement-controlled loading scenarios with the superstructure dead load as an axial load. The adopted biaxial cyclic horizontal loading scenarios depict the expected random patterns that might be experienced during the major earthquakes.

1.1 Finite-Element Numerical Model

The finite element model is built in ABAQUS/Standard software in which nonlinearity of the geometric and material are taken into account [2, 8]. To simulate the inelastic material performance, von Mises yield criterion with associated plastic flow rule in ABAQUS software was selected in the present study [16]. To reduce the simulation time, the upper portion of the column is meshed into a 90 mm B31 element size, while the lower portion is modeled with the S4R shell element. Finer meshing is adopted for the bottom half of the column (i.e., equal to the diameter of the CHS tube) with 26 shell elements and the remaining part has meshed with 14 shell elements. Along the perimeter, 40 shell elements are selected as shown in Fig. 1. The aforementioned finite model is validated in the literature [2, 4]. A cyclic horizontal loading was applied on the top of the column by controlling the horizontal displacement at the column's top. The selected four uniaxial (one-cycle (1 N), and three-cycle (3 N), line 22.5°, and line 45°) and four biaxial (square, circular, octagon, and ellipse) horizontal displacement-controlled loading patterns are quasi-statically imposed to the column's top with an axial load.

1.2 Geometric and Material Properties

All the analyzed columns are assumed to be made of carbon steel ASTM A36 [17] which their material and geometric properties listed in Table 1.

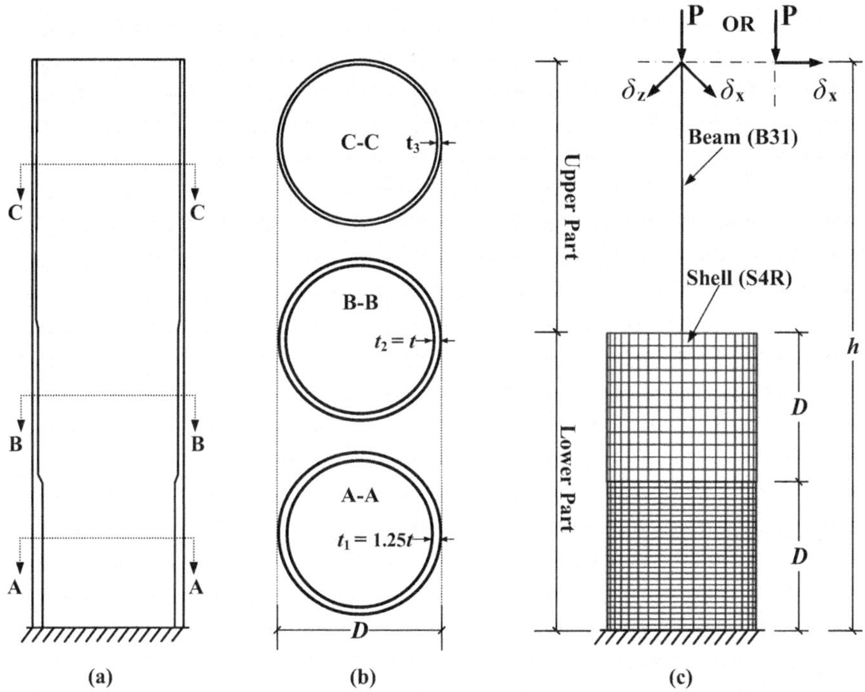

Fig. 1 Column Model: **a** Analyzed Column; **b** Cross Section; and **c** FE Meshing

Table 1 Geometric and Material Properties of the loaded columns

Parameter	Definition	Value
h (mm)	Column height	3403
D (mm)	Column Diameter	900
t (mm)	Plate thickness	9.0
R_t	Radius-to-thickness ratio	0.116
Λ	Slenderness ratio	0.26
H_o (KN)	Yield load	414.2
δ_o (mm)	Yield displacement	10.6
E (GPa)	Young's Modulus	206.0
σ_y (MPa)	Yield Stress	289.6
σ_u (MPa)	Ultimate Stress	495.0
P (KN)	Axial load	904.7
V	Poisson's ratio	0.3

2 Results and Discussion

The numerical results in the presented study are shown in terms of hysteresis loops, force trajectories, and damage deformation to make a significative comparison between the selected four uniaxial (one-cycle (1 N), and three-cycle(3 N), line 22.5°, and line 45°) and four biaxial (square, circular, octagon, and ellipse) horizontal displacement-controlled loading scenarios. Moreover, the presented findings in this study are expected to be useful for a better understanding of CHS tubes under such loading conditions.

2.1 *Hysteretic Loops*

Load–displacement hysteresis loops of the analyzed columns under all uni/biaxial horizontal cyclic loading patterns with constant axial load are shown in **Fig. 2**. As noted, the shape of the load–displacement hysteresis loops under uniaxial loading patterns (one-cycle (1 N), and three-cycle(3 N), line 22.5°, and line 45°) is totally different than under biaxial loading patterns (square, circular, octagon, and ellipse). Moreover, the normalized maximum bearing capacity (i.e., H_m/H_y) and corresponding maximum displacement (i.e., δ_m/δ_y) of the column are different under uni/bi-axial loading scenarios in the X-axis as listed in Table 2. In conclusion, all the uniaxial cyclic horizontal loading patterns show approximately the same H_m/H_y and δ_m/δ_y values. In case of biaxial cyclic horizontal loading patterns, circular and octagon loading scenarios show exactly same results (i.e., $H_m/H_y = 1.60$ and $\delta_m/\delta_y = 2.0$) but Square loading scenario shows the least $H_m/H_y = 1.47$ with the same $\delta_m/\delta_y = 2.0$. Among all the selected loading scenarios, the Ellipse loading pattern shows the best behavior with $H_m/H_y = 1.64$ and $\delta_m/\delta_y = 2.67$. The other point is that the H_m/H_y values under biaxial cyclic horizontal loading patterns are less than uniaxial horizontal cyclic loading patterns.

2.2 *Force Trajectories*

To facilitate a meaningful comparison between the column behavior under uniaxial and biaxial cyclic horizontal loading patterns, the normalized loads in both directions of the columns are plotted for biaxial loading patterns (i.e., $H_x/H_y\_H_z/H_y$). Moreover, the maximum bearing capacity of uniaxial loading patterns of the columns is superimposed as a circular pattern as shown in Fig. 3. As seen, the bearing capacity of the column under biaxial loading is less than in the uniaxial cyclic horizontal loading which emphasizes that biaxial cyclic horizontal loading should be considered in seismic design of CHS tubes.

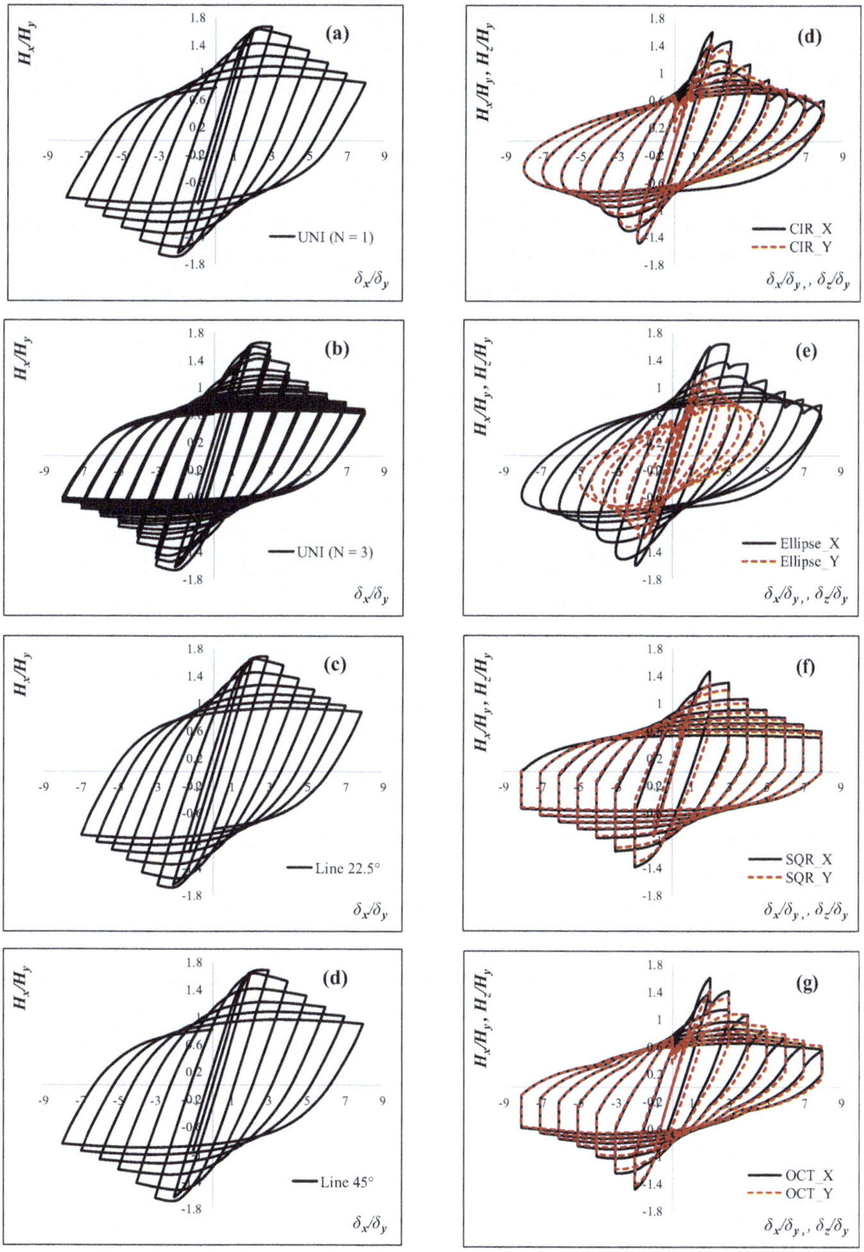

Fig. 2 Load–displacement hysteresis loops of the columns

Table 2 Normalized maximum bearing capacity of the columns

Load Pattern	H_m/H_y	δ_m/δ_y
	X-axis	X-axis
Uni (N = 1)	1.68	2.71
Uni (N = 3)	1.66	2.48
Line 45°	1.68	2.65
Line 22.5°	1.68	2.60
Circular	1.60	2.00
Octagon	1.60	2.00
Square	1.47	2.00
Ellipse	1.64	2.67

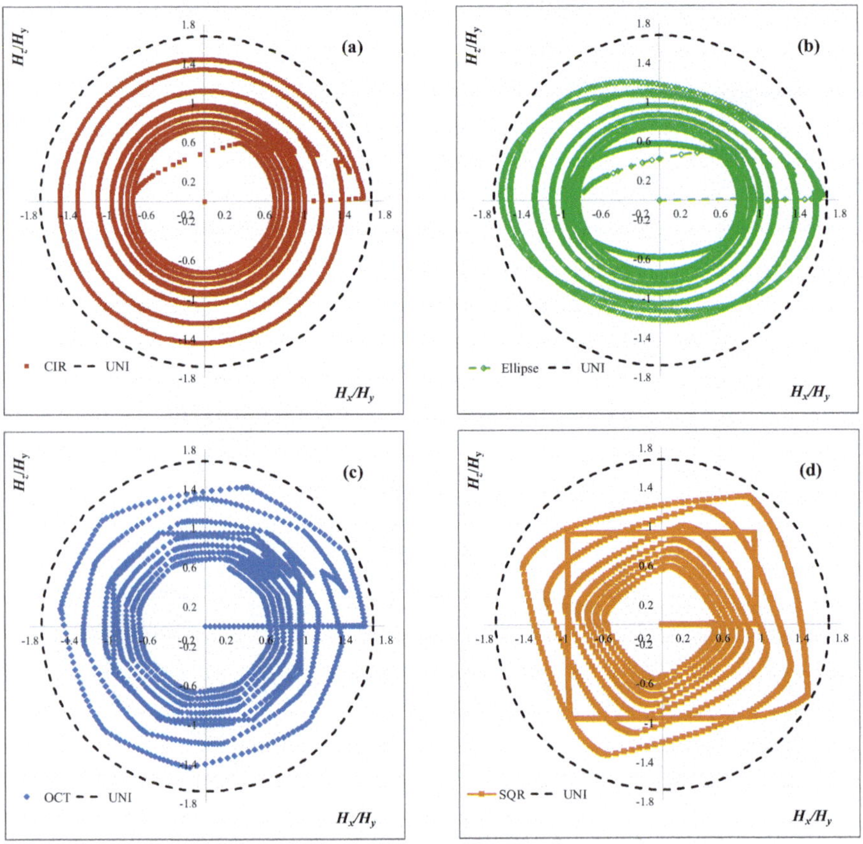

Fig. 3 Force trajectories of the columns

2.3 Damage Deformation of Columns

The deformation of the columns under uni/bi-axial cyclic horizontal loading is shown in Fig. 4. As noted, the damage deformation under biaxial cyclic horizontal loading patterns (see Fig. 4 a, b, c, and d) is severe than uniaxial cyclic horizontal loading patterns (see Fig. 4 e, f, g, and h). Moreover, the severest damage is reported under SQR (see Fig. 4 c) loading scenario which conforms with the results in Table 2 which shows the least maximum bearing capacity amongst all cyclic horizontal loading patterns.

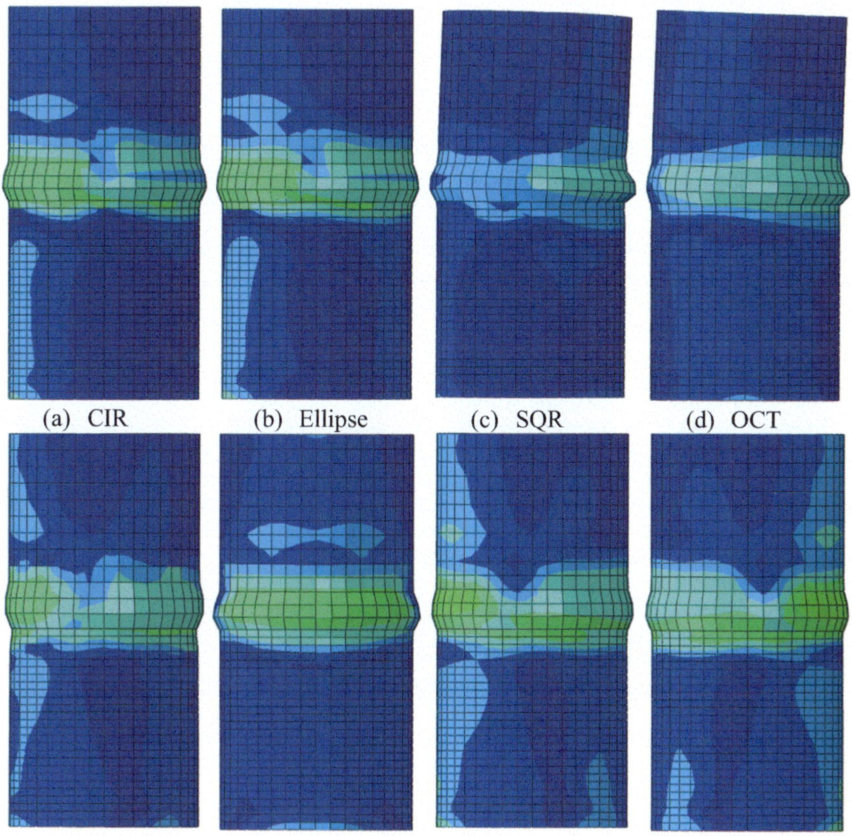

(a) CIR (b) Ellipse (c) SQR (d) OCT

(e) UNI (1N) (f) UNI (3N) (g) Line 22.5 (h) Line 45

Fig. 4 Damage deformation of the columns

3 Conclusions

This study investigates the influence of biaxial cyclic horizontal loading patterns on the seismic behavior of the stepped CHS tubes. The results obtained for this study indicate the shape of the load–displacement hysteresis loops under uniaxial loading patterns is entirely different than biaxial loading patterns. Moreover, the H_m/H_y values under biaxial cyclic horizontal loading patterns are less than uniaxial horizontal cyclic loading patterns. In specific, square loading scenario shows the least $H_m/H_y = 1.47$ with $\delta_m/\delta_y = 2.0$ and ellipse loading pattern shows the best behavior with $H_m/H_y = 1.64$ and $\delta_m/\delta_y = 2.67$. In conclusion, uniaxial cyclic horizontal loading patterns lead to over-estimated bearing and deformation capacities while biaxial cyclic horizontal loading patterns give an approximately good estimation for the capacity of CHS tubes.

References

1. Al-Kaseasbeh Q, Mamaghani IHP (2019) Performance of thin-walled steel tubular circular columns with graded thickness under bidirectional cyclic loading. Struct Congr 2019, Orlando, FL: American Society of Civil Engineers, pp 1–10. https://doi.org/10.1061/9780784482230.001.
2. Al-Kaseasbeh Q, Mamaghani IHP (2018) Buckling strength and ductility evaluation of thin-walled steel tubular columns with uniform and graded thickness under cyclic loading. J Bridg Eng 24:04018105. https://doi.org/10.1061/(ASCE)BE.1943-5592.0001324
3. Al-Kaseasbeh Q, Mamaghani IHP (2020) Thin-walled steel stiffened square box columns with uniform and graded thickness under bidirectional cyclic loading. Eng Struct 219:110919. https://doi.org/10.1016/j.engstruct.2020.110919
4. Al-Kaseasbeh Q, Mamaghani IHP (2019a) Thin-walled steel tubular circular columns with uniform and graded thickness under bidirectional cyclic loading. Thin-Walled Struct 145:106449. https://doi.org/10.1016/j.tws.2019.106449
5. Aoki T, Susantha KAS (2005) Seismic performance of rectangular-shaped steel piers under cyclic loading. J Struct Eng 131:240–249. https://doi.org/10.1061/(ASCE)0733-9445(2005)131:2(240)
6. Xu SH, Qin GC, Zhang ZX (2016) Experimental research on hysteretic characteristics of steel plates artificially corroded by neutral salt spray. Adv Mater Sci Eng. https://doi.org/10.1155/2016/7645763.
7. Ucak A, Tsopelas P (2014) Load pattern effects in circular steel columns under bidirectional lateral cyclic loading. J Struct Eng 141:1–11. https://doi.org/10.1061/(ASCE)ST.1943-541X.0001057
8. Al-Kaseasbeh Q, Mamaghani IHP (2019b) Buckling strength and ductility evaluation of thin-walled steel stiffened square box columns with uniform and graded thickness under cyclic loading. Eng Struct 186:498–507. https://doi.org/10.1016/j.engstruct.2019.02.026
9. Gao S, Usami T, Ge H (1998) Ductility evaluation of steel bridge piers with pipe sections. J Eng Mech 124:260. https://doi.org/10.1061/(ASCE)0733-9399(1998)124:3(260)
10. Dang J, Yuan H, Igarashi A, Aoki T (2017) Multiple-spring model for square-section steel bridge columns under bidirectional seismic load. J Struct Eng 143:04017005. https://doi.org/10.1061/(ASCE)ST.1943-541X.0001735
11. Watanabe E, Sugiura K, Oyawa WO (2000) Effects of multi-directional displacement patterns on the cyclic behaviour of rectangular hollow steel columns. Struct Eng Eng 17:51. https://doi.org/10.2208/jscej.2000.647_79

12. Anderson EL, Mahin SA (2004) An evaluation of bi-directional earthquake shaking on the provisions of the AASHTO guide specifications for seismic isolation design. In: Proceedings on 13th World Conference Eq. Eng
13. Okazaki TUA (2003) Elasto-plastic dynamic analysis of steel bridge piers subjected to bi-directional earthquakes. J Struct Eq Eng, JSCE 27:1–8
14. Dang J, Aoki T (2013) Bidirectional loading hybrid tests of square cross-sections of steel bridge piers. Earthq Eng Struct Dyn 42:1111–1130. https://doi.org/10.1002/eqe.2262
15. Goto, Jiang K, Obata M (2006) Stability and ductility of thin-walled circular steel columns under cyclic bidirectional loading. J Struct Eng 132:1621–31. https://doi.org/10.1061/(asce)0733-9445(2006)132:10(1621)
16. Hibbit, Karlsson S (2014) ABAQUS 2014 Documnetation 2014
17. ASTM (2014) ASTM A36/A36M—14 Standard Specification for Carbon Structural Steel. ASTM Int West Conshohocken, PA 2014:12–4. https://doi.org/10.1520/A0036.

On the Asymptotics of Solutions of the Wave Operator with Meromorphic Coefficients

Maria V. Korovina, Hovik A. Matevossian, and Ilya N. Smirnov

Abstract We study the problem of wave propagation in the medium whose velocity characteristics change under an external impact. We will study a three-dimensional case. The aim of our study is constructing the asymptotic of the solution for the wave operator with a variable time depended coefficient of the Laplacian at infinity. The case of meromorphic and holomorphic coefficients is considered.

Keywords Wave operator · Asymptotics · Meromorphic function

1 Introduction

In this paper the problem of wave propagation in the medium whose velocity characteristics change under an external impact in three-dimensional case is considered. We study the problem of constructing the asymptotics of solutions for a wave equation with a variable coefficient that depends on time at the Laplacian and is a meromorphic function in a neighborhood of infinity. For the first time, a physical interpretation of such a problem was considered in [1], in which the phenomenon of light self-focusing was studied This phenomenon is one of the effects of self-action of light and manifests itself in the concentration of the light-beam energy in the nonlinear medium with a refractive index that increases with increasing light intensity. It was also shown there that the ionizing, thermal, and separating effect of the beam of intensive radi-

M. V. Korovina · I. N. Smirnov (✉)
Lomonosov Moscow State University, Moscow 119991, Russia
e-mail: ismirnov@cs.msu.ru

M. V. Korovina
e-mail: betelgeuser@yandex.ru

H. A. Matevossian
Federal Research Center "Computer Science and Control", Russian Academy of Sciences, Moscow 119333, Russia
e-mail: hmatevossian@graduate.org

Moscow Aviation Institute, National Research University, Moscow 125993, Russia

M. Abdel Wahab (ed.), *Proceedings of 1st International Conference on Structural Damage Modelling and Assessment*, Lecture Notes in Civil Engineering 110,
https://doi.org/10.1007/978-981-15-9121-1_14

ation on the medium can be so strong that it leads to a drastic difference between the medium properties in the beam and outside it, which results in the waveguide propagation of the beam and eliminates its geometrical and diffraction divergence; this interesting phenomenon can be called the self-focusing of the electromagnetic beam. Later, the foundations of a mathematically rigorous theoretical description of this phenomenon were laid in [17]. However, still in 1875, J. Kerr discovered the Kerr effect, or the quadratic electro-optic effect—the phenomenon of a change in the value of refractive index of an optical material caused by an applied electric field and being proportional to the square of the electric field strength. The Kerr effect can be observed in all media, however, for some liquids, it is more pronounced than for other substances.

The main operator that describes these effects is the D'Alembert wave operator with a variable time-dependent coefficient of the Laplacian:

$$\left(\frac{d}{dt}\right)^2 u(x,t) - a^0(t)\Delta u(x,t) = 0. \tag{1}$$

It is assumed that the function $a^0(t) \geq 0$ has a pole at infinity (or is holomorphic in a neighborhood of infinity), which means that there exists an exterior of the circle $|t| > R$ such that the function $a^0(t)$ is expanded in it in the Laurent series

$$a^0(t) = t^k \sum_{j=0}^{\infty} \frac{a_j}{t^j}, \tag{2}$$

$k \in Z$, in this case we can always choose k so that $a_0 \neq 0$ holds. Indeed, if $k \leq 0$, then series (2) in a neighborhood of infinity is a Taylor series, if $k > 0$—is a Laurent series.

Below, we construct the asymptotic of solutions to Eq. (1) at $t \to \infty$.

Using the variable separation method $u(x,t) = Y(x)v(t)$, we obtain

$$\Delta Y(x) + \lambda Y(x) = 0, \tag{3}$$

$$\left(\frac{d}{dt}\right)^2 v(t) + a^0(t)\lambda v(t) = 0. \tag{4}$$

Equation (3) is the Sturm–Liouville problem for the Laplace operator. This problem is already well studied in [13]. In this case, problem (3) is reduced to solving a homogeneous Fredholm equation of the second order with a symmetric kernel. In the general case, the solution $Y(x)$ can be represented by using the Green's function

$$Y(x) = \lambda \int_D G(x,\hat{x})Y(\hat{x})dV_{\hat{x}},$$

where $G(x, \hat{x})$—Green function of the internal Dirichlet problem for the Laplace operator.

Let's proceed to the construction of the asymptotic of Eq. (4) at $t \to \infty$. Let us make the substitution of variable $t = \frac{1}{r}$.

Rewrite Eq. (4) in the form:

$$r^k \left(-r^2 \frac{d}{dt}\right)^2 v(r) + a(r)\lambda v(r) = 0, \tag{5}$$

where $a(r) = \sum_{j=0}^{\infty} a_j r^j$.

Infinity, as generally case, is an irregular singular point of Eq. (4). In case when it is a regular singular point, the problem of constructing the asymptotics of solutions is solved in [5], where it is shown that the asymptotics are conormal, namely, they have the form

$$\sum_j t^{s_j} \sum_{i=0}^{k} a_i^j \ln^i \frac{1}{t},$$

where a_i^j, s_j—some complex numbers.

The problem of constructing asymptotics for solutions of ordinary differential equations in a neighborhood of an irregular singular point was formulated in the final form by Poincare [14] and has still not been solved in general form.

For Eq. (5), the singular point is the point $r = 0$. It was shown in [3] that any homogeneous ordinary differential equation with holomorphic coefficients of order n can be represented in the form

$$\hat{H}u = H\left(r, -r^k \frac{d}{dr}\right)u = 0, \tag{6}$$

where $\hat{H}$—is the differential operator with holomorphic coefficients with symbol

$$H(r, p) = \sum_{i=0}^{n} a_i(r) p^i.$$

Here, $a_i(r)$ are holomorphic functions; $a_n(0) \neq 0$; a formula for calculating the minimum integer non-negative value of k is obtained in [3]. Depending on this value of k, we can divide the equations into three types; each of them corresponds to its own type of asymptotic behavior:

(i) If $k = 0$, then $r = 0$ is a non-singular point and the solutions are holomorphic functions;

(ii) If $k = 1$ then $r = 0$ is a regular singular point and, as mentioned above, the asymptotic behavior of the solution is conormal;

(iii) If $k > 1$, then $r = 0$ is irregular. In this case, equation (6) is an equation of non-Fuchsian type.

Let us divide the equations of non-Fuchsian type into two classes as follows:

- the first class the equations such that the polynomial $H_0(p) = H(0, p)$, which is called the principal symbol, has only simple roots. We call such equations the non-Fuchsian equations of the first type;
- The second class includes the remaining equations, i.e, the non-Fuchsian equations such that the principal symbol has not only simple, but also multiple roots. We call them the non-Fuchsian equations of the second type

Earlier, the case of non-Fuchsian equations of the first type for linear equations and their systems was considered, for example, in particular, in Sternberg [15]. In these papers, the asymptotic expansions of solutions to the non-Fuchsian equations of the first type were constructed; they were obtained in the form of products of the corresponding exponentials by divergent power series as follows:

$$\sum_{j=1}^{n} e^{\frac{\alpha_j}{r^k}+\sum_{i=1}^{k-1}\frac{a_{k-i}^j}{r^{k-i}}} r^{\sigma_j} \sum_{i=0}^{\infty} b_i^j r^i. \tag{7}$$

In the particular case, this asymptotics has the form

$$u = \sum_{j=1}^{n} e^{\frac{\alpha_j}{r}} r^{\sigma_j} \sum_{i=0}^{\infty} b_i^j r^i,$$

where $\alpha_i, i = 1, \ldots, n$ are the roots of polynomial $H_0(p)$, σ_i and b_i^j are some complex numbers.

In case when the asymptotic expansion

$$u \approx u_1 + u_2 + \cdots + u_n = e^{\frac{\alpha_1}{r}} r^{\sigma_1} \sum_{i=0}^{\infty} b_1^i r^i + e^{\frac{\alpha_2}{r}} r^{\sigma_2} \sum_{i=0}^{\infty} b_i^2 r^i + \cdots$$
$$+ e^{\frac{\alpha_n}{r}} r^{\sigma_n} \sum_{i=0}^{\infty} b_i^n r^i \tag{8}$$

has at least two terms corresponding to values α_1 and α_2 with different real parts (for definiteness, let us assume that $\Re e\alpha_1 > \Re e\alpha_2$), there is a significant difficulty in interpreting the obtained expansion. The fact is that all terms in the first element corresponding to the value α_1 (the dominant element) are of greater order at $r \to 0$ than any of the terms in the second (recessive) element. Therefore, to interpret expansion (8), it is necessary to summarize the series (divergent) corresponding to the dominant element.

Consideration of the recessive components of the expansion of the solution u of Eq. (6) is important, in particular, for constructing uniform asymptotics of the solutions in the complex case, where the point r moves on the complex plane and the dominant and recessive elements of the expansion can exchange their roles In other words, the plane is conditionally divided into sectors in which one of the elements is dominant, another is recessive, and when passing from one sector to another, their leadership changes the recessive element becomes the dominant one and vice versa. However, in the vicinity of the boundaries of these sectors, several elements are of equal order and none of them can be neglected. This phenomenon occurs, for example, when considering the Euler example, as well as when constructing the asymptotics of the solution for problem (1) and, in general, for all non-Fuchsian asymptotics (7). As a result, to study the asymptotic expansions of solutions to Eq. (1), it is necessary to introduce a regular method for summing up divergent series to construct uniform asymptotics of solutions with respect to the variable r, i.e., representing asymptotic expansions not only in certain sectors, but also in the entire vicinity of the considered point.

Method suitable for summing up such series based on the Laplace-Borel transform and the concept of a resurgent function was first introduced in [2] is called resurgent analysis. The main idea of which is that the formal Borel transform $\tilde{u}_1(p)$, $\tilde{u_2}(p)\ldots$ are the power series with respect to the dual variable p, p that converge in the neighborhood of the points $p = \alpha_j$. The inverse Borel transform gives a regular way to sum up these series. However, it is necessary to prove an infinite extendibility of functions $\tilde{u}_j(p)$, i.e., their extendibility along any path on the Riemann surface $\tilde{u}_j(p)$, not passing through some discrete set depending on the function $\tilde{u}_j(p)$. The proof of this fact, as a rule, was very complicated when applying resurgent analysis to constructing asymptotic forms of solutions to differential equations. For equations with degenerations, the proof of infinite extendibility was obtained in [6, 9]. This result allows applying the resurgent analysis methods to constructing uniform asymptotics for solutions of linear differential equations with holomorphic coefficients in the spaces of functions of exponential growth.

Owing to this result, in [7, 9], uniform asymptotics of solutions were constructed for the case where the roots of the principle symbol $H_0(p) = H(0, p)$ are simple. Thus, the question of constructing uniform asymptotics of solutions for this case was closed by directly applying the Laplace–Borel transform to the corresponding equation.

We also note papers [10–12], in which an asymptotic expansion of solutions of the basic boundary value problems for the elasticity system and the biharmonic (polyharmonic) equation in the exterior of a compact set and in a half-space, including in the form of a co-normal asymptotic, is obtained.

2 Definitions and Auxiliary Statements

Let's introduce some notions of the resurgent analysis.

Let $S_{R,\varepsilon}$ denote the sector $S_{R,\varepsilon} = \{r| - \varepsilon < \arg r < \varepsilon, \quad |r| < R\}$.

Definition 1 We will say that the function f is analytical on $S_{R,\varepsilon}$ and is of exponential growth no more than k, if there are nonnegative constants C and α such that in the sector $S_{R,\varepsilon}$ the following inequality is valid:

$$|f| < Ce^{a\frac{1}{|r|^k}}.$$

Let us denote by $E_k(S_{R,\varepsilon})$ the space of functions holomorphic in the domain $S_{R,\varepsilon}$ and of k-exponential growth in zero; by $E(C)$—the space of integer functions of exponential growth.

Definition 2 The k-Laplace–Borel transform of the function $f(r) \in E_k(S_{R,\varepsilon})$ is the mapping $B_k : E_k(S_{R,\varepsilon}) \longrightarrow E(\tilde{\Omega}_{R,\varepsilon})/E(C)$:

$$B_k f = \int_0^{r_0} e^{-p/r^k} f(r) \frac{dr}{r^{k+1}},$$

where r_0 denotes an arbitrary point of the sector.

The inverse k-Laplace–Borel transform is defined by the formula:

$$B_k^{-1} \tilde{f} = \frac{k}{2\pi i} \int_{\tilde{\gamma}} e^{p/r^k} \tilde{f}(p) dp.$$

The contour $\tilde{\gamma}$ is shown in Fig. 1.

It should be noted that for the k-Laplace–Borel transform, the following formulae are true:

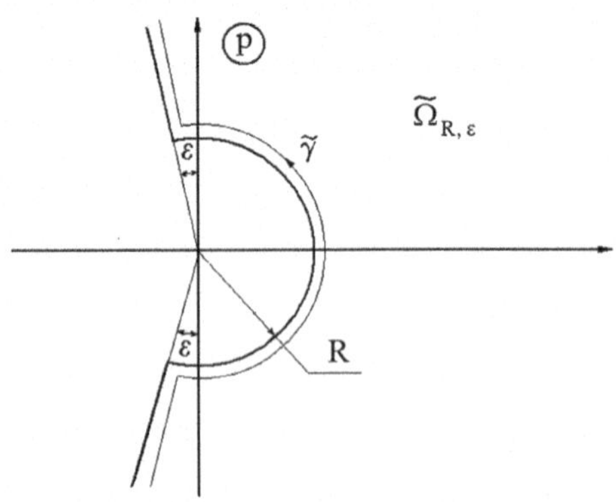

Fig. 1 Contour $\tilde{\gamma}$

$$B_k \circ \left(-\frac{1}{k} r^{k+1} \frac{\partial}{\partial r}\right) f(r) = p B_k f,$$

$$\frac{\partial}{\partial p} \circ B_k f = -B_k \left(\frac{1}{r^k} f(r)\right).$$

In [8] it is shown that if $k \in N$, the equality is satisfied:

$$B_k r^k B_k^{-1} \left[\tilde{f}\right] = -\frac{1}{k} \int\limits_{p_0}^{p} \tilde{f}(p') dp'.$$

Definition 3 The function $\tilde{f}$ is called infinitely extendable, if for any number R, there is a discrete set of points Z_R in C such that the function $\tilde{f}$ is analytically extended from the initial domain of definition along any path with a length smaller than R, which does not pass through Z_R.

Definition 4 The element f of the space $E_k(S_{R,\varepsilon})$ is called the k-resurgent function, if its k-Laplace–Borel transform $\tilde{f} = B_k f$ is infinitely extendable.

Theorem 1 *Let f be a resurgent function, then the solution of the equation*

$$H\left(-r^2 \frac{d}{dr}, r\right) u = f$$

is a resurgent function. If the polynomial $H_0(p)$ has simple roots at the points $p_1, \ldots, p_m$, then the asymptotic behavior of the solution of the homogeneous equation has the form

$$u(r) \approx \sum_{j=1}^{m} \exp\left(\frac{p_j}{r}\right) r^{\sigma_j} \sum_{i=0}^{\infty} b_i^j r^i, \tag{9}$$

where the sum is taken by the union of all the roots of the polynomial $H_0(p)$.

For equations with $k + 1$-order degeneracy, where $k \in N$, namely, for equations of the form

$$H\left(r, -\frac{1}{k} r^{k+1} \frac{d}{dr}\right) u = 0$$

in the case when the roots of the main symbol are simple, the asymptotic behavior have the form

$$u(r) \approx \sum_{j=1}^{m} \exp\left(\frac{p_j}{r^k} + \sum_{i=1}^{k-1} \frac{\alpha_{k-i}^j}{r^{k-i}}\right) r^{\sigma_j} \sum_{i=0}^{\infty} b_i^j r^i. \tag{10}$$

If $k+1=\frac{m}{n}, m \in N, k \in N, m > k$, the asymptotic behavior of the solution will be:

$$u(r) \approx \sum_j \exp\left(\frac{p_j}{r^{\frac{m}{k}-1}} + \sum_{i=1}^{m-k-1} \frac{\alpha^j_{m-k-i}}{r^{\frac{m-i}{k}-1}}\right) t^{\sigma_j} \sum_{i=0}^{\infty} b^j_i t^i. \tag{11}$$

A proof of this theorem can be found in [6, 7, 9].

3 Main Results

At first, consider the case of $k \le 0$, i.e. the construction of asymptotics of solutions of Eq. (5) for $r \to 0$, when the function $a(r)$ is holomorphic.

Lemma 1 *(i) Let $k=0$, then if $\lambda \neq 0$, then all the asymptotics of the solutions of Eq. (5) in the space of exponentially growing functions in a neighborhood of infinity have the form:*

$$v(r) \approx e^{\frac{c_1}{r}} r^{\sigma_1} \sum_{i=0}^{\infty} A^1_i r^i + e^{\frac{c_2}{r}} r^{\sigma_2} \sum_{i=0}^{\infty} A^2_i r^i,$$

where the numbers c_i, $i=1,2$ are the roots of the polynomial $p^2 + a_0\lambda$.

(ii) If $k=-1$, $\lambda \neq 0$, then the asymptotic behavior of the solutions has the form:

$$v(r) \approx e^{\frac{c_1}{\sqrt{r}}} r^{-\frac{1}{4}} \sum_{i=0}^{\infty} A^1_i r^{\frac{i}{2}} + e^{\frac{c_2}{\sqrt{r}}} r^{-\frac{1}{4}} \sum_{i=0}^{\infty} A^2_i r^{\frac{i}{2}}.$$

where the numbers c_i, $i=1,2$ are the roots of the polynomial $p^2+4a_0\lambda$, and $\sum_{i=0}^{\infty} A^j_i r^i$, $j=1,2$ are the asymptotic series.

(iii) If $k \le -2$ or $\lambda = 0$, then the asymptotic behavior of the solution is conormal.

Now we consider the case $k \in N$, i.e. consider the case when the coefficient $a^0(t)$ in Eq. (1) is a meromorphic function. We assume that $\lambda \neq 0$, otherwise the asymptotic behavior of the solution of Eq. (5) is co-normal. Since the equality is satisfied

$$r^k\left(-r^2\frac{d}{dr}\right)^2 = \left(r^{\frac{k}{2}+2}\frac{d}{dr}\right)\left(r^{\frac{k}{2}+2}\frac{d}{dr}\right) - \frac{k}{2} r^{k+3}\frac{d}{dr},$$

we can rewrite Eq. (5) in the form:

$$\left(-r^{2+\frac{k}{2}}\frac{d}{dr}\right)^2 v(r) - \frac{k}{2} r^{1+\frac{k}{2}}\left(r^{2+\frac{k}{2}}\frac{d}{dr}\right) + a(r)\lambda v(r) = 0. \tag{12}$$

From Theorem 1 it follows that the solution of Eq. (12) is a resurgent function and the following is true

Lemma 2 *Let $k \in N$ and $\lambda \neq 0$, then, provided k is even, that is, $k = 2n$, the asymptotics of the solution to Eq. (5) has the form*

$$u(r) \approx \exp\left(\frac{p_1}{r^{n+1}} + \sum_{i=1}^{n} \frac{\alpha_i^1}{r^{n+1-i}}\right) r^{\sigma_1} \sum_{i=0}^{\infty} b_i^1 r^i$$

$$+\exp\left(\frac{p_2}{r^{n+1}} + \sum_{i=1}^{n} \frac{\alpha_i^2}{r^{n+1-i}}\right) r^{\sigma_2} \sum_{i=0}^{\infty} b_i^2 r^i. \quad (13)$$

Let $k = 2n + 1$ is odd, then

$$u(r) \approx \exp\left(\frac{p_1}{r^{n+\frac{3}{2}}} + \sum_{i=1}^{2n+2} \frac{\alpha_i^1}{r^{n+\frac{3}{2}-\frac{i}{2}}}\right) r^{\sigma_1} \sum_{j=1}^{\infty} b_j^1 r^{\frac{j}{2}}$$

$$+\exp\left(\frac{p_2}{r^{n+\frac{3}{2}}} + \sum_{i=1}^{2n+2} \frac{\alpha_i^2}{r^{n+\frac{3}{2}-\frac{i}{2}}}\right) r^{\sigma_2} \sum_{j=1}^{\infty} b_j^2 r^{\frac{j}{2}}. \quad (14)$$

Here p_1, p_2 are the polynomial roots $\left(\frac{2}{2+k}\right)^2 a_0\lambda + p^2$; the corresponding numbers are denoted by α_i^j, σ_j, $j = 1, 2$, $\sum_{i=0}^{\infty} b_i^j r^i$, $j = 1, 2$ are the asymptotic series.

The question arises of finding the numbers $\alpha_i^j, \sigma_j, j = 1, 2$ included in the asymptotics (13), (14). We give an algorithm for calculating them. For definiteness, we assume that k is even, for odd k the algorithm for calculating the numbers α_i^1, $i = 1, \ldots, n$ does not differ from the one given below.

In order to find the numbers α_i^1, σ_1, $i = 1, \ldots, n$, in case when k is even, we need to shift the root p_1 to zero by using the replacement $v = e^{\frac{p_1}{r^{n+1}}} v_1$. We obtain the equation

$$\left(r^{2+n}\frac{d}{dr}\right)^2 v(r) + 2p_1(n+1)\left(r^{2+n}\frac{d}{dr}\right) v(r)$$

$$-nr^{1+n}\left(p_1(n+1) + \left(-r^{2+n}\frac{d}{dr}\right)\right) v(r) + (a(r) - a(0))\lambda v(r) = 0. \quad (15)$$

Since $a(r) - a(0) = \sum_{j=1}^{\infty} a_j r^j = \sum_{j=1}^{n+1} a_j r^j + \sum_{j=n+2}^{\infty} a_j r^j$, then Eq. (15) can be rewritten as

$$\left(-r^{2+n}\frac{d}{dr}\right)^2 v(r)+2p_1(n+1)\left(-r^{2+n}\frac{d}{dr}\right)v(r)-nr^{1+n}\left(-r^{2+n}\frac{d}{dr}\right)v(r)$$

$$+\lambda\sum_{j=1}^{n}a_jr^jv(r)+(\lambda a_{n+1}-n(n+1)p_1)r^{n+1}v(r)+\lambda r^{n+2}\sum_{j=0}^{\infty}a_{j+n+2}r^jv(r)=0. \tag{16}$$

In order to find the numbers α_i^1, $i=1,\dots,n$ we need to substitute $v=\exp\left(\sum_{i=1}^{n}\frac{\alpha_i^1}{n+1-i}\right)v_1$ and select the numbersα_i^1, $i=1,\dots,n$ so that the terms of the form $\lambda a_jr^jv(r)$, where $j\le n$. For example, to nullify the term $\lambda a_1rv(r)$, we need to replace $v=\exp^{\frac{\alpha_n^1}{r^n}}v_n$, then Eq. (16) can be rewritten in the form:

$$\left(-r^{2+n}\frac{d}{dr}\right)^2 v_n(r)+2p_1(n+1)\left(-r^{2+n}\frac{d}{dr}\right)v_n(r)$$

$$+\alpha_n^1nr\left(-r^{2+n}\frac{d}{dr}\right)v_n(r)-nr^{1+n}\left(-r^{2+n}\frac{d}{dr}\right)v(r)$$

$$+\lambda\sum_{j=2}^{n}a_jr^jv_n(r)+(\lambda a_{n+1}-n(n+1)p_1)r^{n+1}v_n(r)$$

$$\lambda r^{n+2}\sum_{j=0}^{\infty}a_{j+n+2}r^jv_n(r)+(\alpha_n^1n)^2r^2v_n(r)$$

$$+(2p_1(n+1)\alpha_n^1n+\lambda a_1)rv_n(r)=0$$

And set $\alpha_n^1=-\frac{\lambda a}{2p_1(n+1)n}$, then we get an equation that can be written in the form

$$\left(-r^{2+n}\frac{d}{dr}\right)^2 v_n(r)+2p_1(n+1)\left(-r^{2+n}\frac{d}{dr}\right)v_n(r)$$

$$+\alpha_n^1nr\left(-r^{2+n}\frac{d}{dr}\right)v_n(r)-nr^{1+n}\left(-r^{2+n}\frac{d}{dr}\right)v(r)$$

$$+\lambda\sum_{j=2}^{n+1}a'_jr^jv_n(r)+\lambda r^{n+2}\sum_{j=0}^{\infty}a'_{j+n+2}r^jv_n(r)-n(n+1)r^{n+1}v_n(r)=0.$$

Here, by a'_j, $j = 2, \ldots, n, \ldots$ denotes the corresponding numbers. And so on, making substitutions of the form $v_{i-1} = \exp^{\frac{\alpha_i^1}{r^i}} v_i, i = 2, \ldots, n$, all the terms entering in the sum $\lambda \sum_{j=2}^{n} a'_j r^j v_n(r)$.

To find σ_1, we need to make the substitution $v_{n+1} = r^{\sigma_1} v_1$, and select σ_1 so that the term with $n+1$ degree of r is canceled. All other terms are minor and do not affect the form of the asymptotics.

Lemmas 1 and 2 imply the following

Theorem 2 *If $k = 0$, $\lambda \neq 0$, then all the asymptotics of the solutions of Eq. (1) in the space of exponentially growing functions in a neighborhood of infinity by t have the form:*

$$u(x,t) \approx \left(e^{c_1 t} t^{\sigma_1} \sum_{i=0}^{\infty} A_i^1 t^{-i} + e^{c_2 t} t^{\sigma_2} \sum_{i=0}^{\infty} A_i^2 t^{-i} \right) Y(x),$$

where numbers c_i, $i = 1, 2$ are the roots of the polynomial $p^2 + a_0\lambda$.

If $k = -1$, $\lambda \neq 0$, then all the asymptotics of the solution has the form

$$u(x,t) \approx \left(e^{c_1 \sqrt{t}} \sum_{i=0}^{\infty} A_i^1 t^{-\frac{i}{2}} + e^{c_2 \sqrt{t}} \sum_{i=0}^{\infty} A_i^2 t^{-\frac{i}{2}} \right) Y(x).$$

Here, numbers c_i, $i = 1, 2$ are the roots of the polynomial $p^2 + 4a_0\lambda$, and $\sum_{i=0}^{\infty} A_i^1 r^i$ denotes asymptotic series. Let $k \in N$ and $\lambda \neq 0$, then, provided k is even, that is, $k = 2n$, the asymptotics of the solution to Eq. (1) has the form:

$$u(x,t) \approx \left(\exp\left(p_1 t^{n+1} + \sum_{i=1}^{n} \alpha_i^1 t^{n+1-i} \right) t^{\sigma_1} \sum_{i=0}^{\infty} b_i^1 t^i + \right.$$

$$\left. + \exp\left(p_2 t^{n+1} + \sum_{i=1}^{n} \alpha_i^2 t^{n+1-i} \right) t^{\sigma_2} \sum_{i=0}^{\infty} b_i^2 t^i \right) Y(x).$$

Let $k = 2n + 1$ be even, then

$$u(r) \approx \left(\exp\left(p_1 t^{n+\frac{3}{2}} + \sum_{i=1}^{2n+2} \alpha_i^1 t^{n+\frac{3}{2}-\frac{i}{2}} \right) t^{\sigma_1} \sum_{j=1}^{\infty} b_j^1 t^{-\frac{j}{2}} + \right.$$

$$\left. + \exp\left(p_2 t^{n+\frac{3}{2}} + \sum_{i=1}^{2n+2} \alpha_i^2 t^{n+\frac{3}{2}-\frac{i}{2}} \right) t^{\sigma_2} \sum_{j=1}^{\infty} b_j^2 t^{-\frac{j}{2}} \right) Y(x).$$

Here p_1, p_2 are the polynomial roots $\left(\frac{2}{2+k}\right)^2 a_0\lambda + p^2$; the corresponding numbers are denoted by α_i^j, σ_j, $j = 1, 2$, $\sum_{i=0}^{\infty} b_i^j r^i$, $j = 1, 2$ are the asymptotic series.

If $k \leq -2$ or $\lambda = 0$, then the asymptotic behavior of the solution of Eq. (1) can be represented as of the function $Y(X)$ by the corresponding co-normal asymptotic behavior.

Remark 1 Theorem 2 is also valid if $a^0(t)$ in equation (1) is a meromorphic function and has a pole of order k at zero.

Remark 2 If equation (5) has the form

$$\left(\frac{d}{dt}\right)^2 v + a^2 \lambda v = 0,$$

where $a \neq 0$ is a real constant, then all asymptotics of solutions to this equation in the space of the functions of exponential growth in the neighborhood of infinity with respect to variable t can be represented in the form

$$u(x,t) = (A_1 e^{c_1 t} + A_2 e^{c_2 t}) Y(x).$$

Here, $c_i, i = 1, 2$, are the roots of the polynomial $p^2 + a^2\lambda$; A_i are arbitrary constants.

References

1. Askaryan GA (1973) Effect of self-focusing. Usp Fiz Nauk 111(10):249–260 (in Russian)
2. Ecalle J (1984) Cinq applications des fonctions r?surgentes, Prepub. Math. d'Orsay, 84:62 , 110 p
3. Kats DS (2015) Computation of the asymptotics of solutions for equations with polynomial degeneration of the coefficients. Diff Equations 51(12):1589–1594
4. Kats DS (2016) Coefficients of series in asymptotic expansions of solutions of equations with degenerations. Int J Open Inf Technol 4(9):1–7 (In Russian)
5. Kondrat'ev VA (1967) Boundary value problems for elliptic equations in domains with conical or angular points. Trudy Moskov Mat Obshch [Trans Moscow Math Soc] 16:209–292
6. Korovina MV (2011) Existence of resurgent solutions for equations with higher-order degeneration. Diff Equations 47(3):346–54
7. Korovina MV (2011) Asymptotics solutions of equations with higher-order degeneracies. Doklady Math 83(2):182–184
8. Korovina MV (2012) Asymptotics of solutions of equations with higher degenerations. Diff Equations 48(5):717–729
9. Korovina MV, Shatalov VE (2010) Differential equations with degeneration and resurgent analysis. Diff Equations 46(9):1267–1286
10. Matevossian OA (2001) Solutions of exterior boundary-value problems for the elasticity system in weighted spaces. Sb Math 192(12):1763–1798
11. Matevossian HA (2003) On solutions of mixed boundary-value problems for the elasticity system in unbounded domains. Izvestiya Math 67(5):895–929
12. Matevossian HA (2020) On the mixed Neumann-Robin problem for the elasticity system in exterior domains. Russ J Math Phys 27(2):272–276
13. Petrovskii IG (1961) Lectures on partial differential equations. GIFML, Moscow (in Russian)
14. Poincare H (1886) Sur les integrales irregulieres des equations lineaires. Acta math 8:295–344

15. Sternberg W (1920) Uber die asymptotische Integration von Differencialgleichungen. Verlag von Julius Spriger, Berlin
16. Sternin B, Shatalov VE (1996) Borel-Laplace Transform and Asymptotic Theory. Introduction to Resurgent Analysis. CRC Press, London
17. Talanov VI (1964) On self-focusing of wave beams in nonlinear media. Pis'ma Zh Eksp Teor Fiz 2(5):218–222 (In Russian)

Durability Performance of Binary Blended Geopolymer Concrete

Addepalli Mallinadh Kashyap, T. Chandra Sekhar Rao, and N. V. Ramana Rao

Abstract The present study is the investigation of durability constraints of geopolymer concrete made up of binary blending of industrial waste exhaust materials viz., Class F fly ash, GGBS under alkali activation. Two grades of concrete M30 and M70 were developed for OPC and GPC concretes. The developed concretes were tested for durability under sulphate, chloride attacks and also tested for sorptivity and water absorption tendencies. The two grades of GPC and OPC concrete were exposed to 1, 3 and 5% concentrated solutions of H_2SO_4, HCl and NaCl for 28, 56 & 90 days. The alkaline liquid to binder ratio adopted in this study was 2.5. The molarity concentration of NaOH adopted for this study was 12. The properties like density, surface deterioration, compressive strength were tested. SEM image characteristics and XRD analysis were also performed on samples to study microstructural properties. From the test results, it was inferred that water absorption is less for higher grades compared to normal grade in both the two types of concretes. The resistance of GPC concrete against sulphate and chloride attack shows better performance than OPC concrete irrespective of the grade. Regression analysis was performed between the concentration of solutions and compressive strength for two types of concretes. The test results indicated that binary blended GPC concrete in both the grades for specified time period achieved superior strength properties than OPC concrete under aggressive media.

Keywords Binary blended · Durability · Strength properties · Aggressive media

---

A. M. Kashyap (✉)
Ph.D Scholar, Jawaharlal Nehru Technological University, Hyderabad, India
e-mail: mallinadhkashyap@gmail.com

T. C. S. Rao
Bapatla Engineering College, Bapatla, India

N. V. R. Rao
National Institute of Technology, Warangal, India

M. Abdel Wahab (ed.), *Proceedings of 1st International Conference on Structural Damage Modelling and Assessment*, Lecture Notes in Civil Engineering 110,
https://doi.org/10.1007/978-981-15-9121-1_15

1 Introduction

Durability enables any cement-based concrete as widely accepted material for the construction of any structure. The typical considerations of decline the thermostat, reduction in carbon footprint, reduction of greenhouse gases while cement production, utilization of the industrial waste products in concrete are highly being considered [1]. The philosophy of geopolymer and its effectiveness in the precast industry is well known since long back as implementing no or zero cement technology. Utilization of industrial waste exhausts efficiently to meet the slogan 'Waste to Wealth' by commercializing the GPC concrete to decrement the carbon footprint has also become good practice in many of the countries now-a-days. GPC materials are the resultant final products of the chemical reaction between the industrial waste exhausts or materials from geological existence with Sodium or Potassium based alkaline materials [2]. The durability properties of geopolymer mortars indicated to have superiority in aggressive media than OPC mortar [3]. Tests were performed on geopolymer concrete prepared with fly ash of Class C grade that yielded appreciable performance in both durability and strength [4–6]. Increased strength and durability were reported with GPC when blended with GGBS and RHA [7]. GPC is highly resistive to acid attacks than OPC [8, 9]. The deterioration of concrete was controlled by permeability in an aggressive environment since the deterioration process like carbonation and sulfate and chloride attacks are directed by the movement of fluid in concrete [10]. The durability of concrete is a very important aspect of any structure's performance in aggressive environments. It was observed that the sulphate attack produce significant degradation in the concrete structure [11]. The GPC mortar specimens with fly ash of Class C grade cast with changing alkaline activator percentage show varied deterioration extent under exposure to sulphuric acid [12]. It was revealed that GPC is highly resistive to sulphuric acid with very little loss in mass[13].GPC with fly ash of Class F grade specimens was found with better durable properties that attributed to its constant cross linked A-S polymeric structure with a weight loss of 1.96% under exposure to sulphuric acid [14, 15]. Fly ash with grade Class C exhibited better strength properties than fly ash with grade Class F at ambient temperature curing and found vice-versa at oven temperature curing [16]. As the literature reveals, it is obvious that GPC concrete shows better performance in strength and durability considerations than OPC concrete.

2 Materials and Methods

2.1 Ingredients

Fly ash of Class F grade obtained from VTPS, Vijayawada and GGBS from Vizag Steel Industry in India with 2.65, 2.86 the specific gravity values and 320 kg/m^2,

450 kg/m^2 specific surface area values were used respectively. Table 1 represents the chemical constituents in the source materials.

Commercial grade Sodium based alkaline solutions were used to activate the source materials. River sand and coarse aggregate with a specific gravity of 2.67 and 2.60 were used to prepare the specified GPC mixes.

2.2 Experimentation

All GPC specimens were cast with specified source materials by adopting Na_2SiO_3 to NaOH ratio as 2.50 by mass as the maximum strength properties were achieved under these conditions [17]. M30 grade concrete was prepared based on IS:456–2000 and M70 grade concrete was prepared based on ACI:211-4R(93). The sodium hydroxide molarity concentration was chosen as 12. The mix specifications of normal and high strength grade OPC as well as GPC concrete are furnished in Table 2. Keeping the fine and coarse aggregates in SSD condition, they were mixed with source materials in a pan mixer for 2 min and then alkali solution is mixed to achieve a homogeneous mix. 150 × 150 × 150 mm size specimens were cast with the prepared mix. After casting, the test specimens left to normal room temperature for ambient curing. For comparison purposes, OPC specimens are also cast.

The sorptivity tests were undertaken for cube specimens in compliance with ASTM C 1585–04. The absorbed water quantity in 30 min period of time is measured by the following equation:

$$S = I/\sqrt{T} \tag{1}$$

in which S represents the coefficient of sorptivity in mm and t represents pass by time in minutes. Consider

$$I = \Delta W/AD \tag{2}$$

Table 1 Chemical Constituents in Source materials

S. No.	Chemical distribution (%)	ClassF-FA (%)	GGBS (%)
1	SiO_2	66.80	39.18
2	Al_2O_3	24.50	10.18
3	Fe_2O_3	4	2.02
4	CaO	1.50	32.82
5	MgO	0.45	8.52
6	Na_2O	0.40	1.14
7	K_2O	0.22	0

#Taken from manufacturers manual

Table 2 Mix ingredients of OPC and GPC concrete

Mix	Source materials (50% + 50%) (Kg/m^3)	Cement (Kg/m^3)	W/C ratio	Na_2SiO_3 (Kg/m^3)	NaOH (Kg/m^3)	Fine aggregate (Kg/m^3)	Coarse aggregate (Kg/m^3)	
M30	–	382	0.48	–	–	798	20 mm—529.2Kg 12 mm—264.6Kg 6 mm ——88.2Kg	882
M70	–	410.70	0.37	–	–	770	20 mm–693.0Kg 12 mm–346.5Kg 6 m–115.5Kg	1155
M30	360	–	–	70.8	28.3	798	20 mm–529.2Kg 12 mm–264.6Kg 6 mm ——88.2Kg	882
M70	410.70	–	–	117.30	47	770	20 mm–693.0Kg 12 mm–346.5 Kg 6 mm–115.5 Kg	1155

ΔW = Change in weight = $W_2 - W_1$, in which W_1 represents oven-dried cube weight in grams and W_2 represents cube weight in grams after 30 min capillary suction, A represents specimen surface area by which water is penetrated and D is the water density.

For the water absorption test, the specimens were placed in submerged conditions for a period of 90 days. After this, the specimens were kept dried in an oven at 110 °C for 24 h until constant specimen mass is achieved and weighed again. W_1 is noted as specimen dry weight. Later, the specimen is immersed in hot water at 85 °C for a period of 3.5 h and weighed as (W_2).

Consider

$$\text{Percentage absorption of water} = \frac{W_2 - W_1}{W_1} \times 100 \tag{3}$$

where W_1 represents oven-dried cube weight (gms) and W_2 represents cube wet weight after 3.5 h (gms).

The performance of OPC and GPC specimens in Sulphuric acid (H_2SO_4), Hydrochloric acid (HCl), and Sodium chloride (NaCl) environments by immersing the cube specimens in 1, 3, and 5% solutions separately after 1 day of casting [18, 19]. The concentration and choice of solution were taken depending on the practical usage of concrete in mining, sewerage pipes, etc., The test specimens were kept in water by submerging them in solutions, with 4 times the specimen volume for a period of 90 days. For typical maintain of the solution concentration, these solutions were replaced every 7 days with fresh solutions of the same composition [20]. The solution effects on specimens were regularly monitored by visual inspection, measuring the change in weight, and testing the strengths. For testing the change in weight of specimens, they were primed for a period of 3 days in water prior to submerging them in specified solutions as the SSD weight is taken as initial weight. The specimens were then taken and weighed at different exposure conditions in similar stages and are taken as final weights.

3 Test Results and Discussions

3.1 Compressive Strength

In Figs. 1 and 2, the OPC and GPC compressive strength were presented. Test results showed that the use of ground granulated blast furnace slag formed additional C-A-S-H gel which complimented the compressive strength parameters of GPC. GPC achieved superior strength properties than OPC. With the increase in the curing period, compressive strength values were found to be increasing. GPC mix prepared by blending of fly ash of Class F grade with GGBS at levels of 50% content achieved greater compressive strength than at other levels of blends [21, 22].

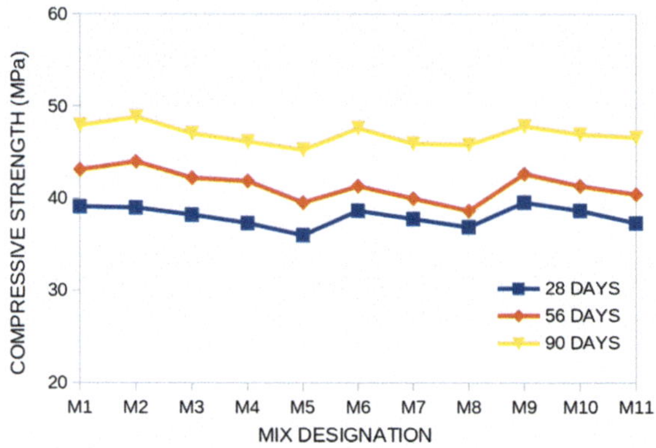

Fig. 1 M30 grade OPC and GPC concrete compressive strength under aggressive media. Note M1 represents M30 grade control OPC under ambient curing
M2 represents M30 grade control GPC under ambient curing
M3 to M5 represents M30 grade GPC -exposed to 1, 3 and 5% H_2SO_4
M6 to M8 represents M30 grade GPC -exposed to 1, 3 and 5% HCl
M9 to M11 represents M30 grade GPC -exposed to 1,3 and 5% NaCl

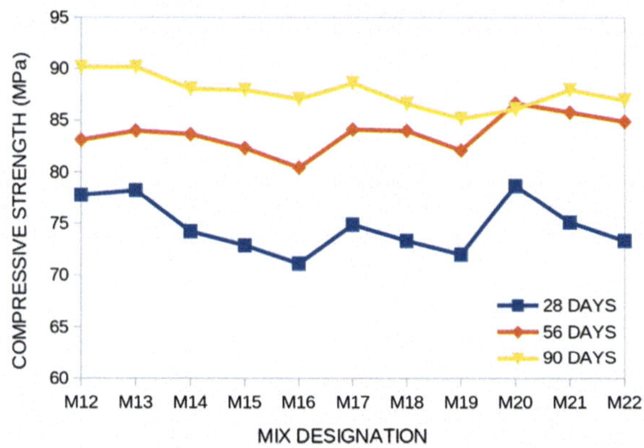

Fig. 2 M70 grade OPC and GPC concrete compressive strength under aggressive media. Note M12 represents M70 grade control OPC under ambient curing
M13 represents M70 grade control GPC under ambient curing
M14 to M16 represents M70 grade GPC -exposed to 1, 3 and 5% H_2SO_4
M17 to M19 represents M70 grade GPC -exposed to 1, 3 and 5% Hcl
M20 to M22 represents M70 grade GPC -exposed to 1, 3 and 5% NaCl

3.2 *SEM and XRD Analysis*

The SEM images of fly ash of grade Class F and GGBS along with the blended Class F-fly ash with GGBS in the percentage of (50% + 50%) for two grades were furnished in Figs. 3, 4, 5 and 6. The Fly ash of grade Class F particle's visual appearance is

Fig. 3 SEM image of Class F fly ash

Fig. 4 SEM image of GGBS

spherical in shape with surface smoothness in the range of size 1/10th of the maximum particle size of fly ash. The GGBS particle's visual appearance is a straight, rough-textured surface and also with flaky elongated shape and variations in size from 1 to 10 μ. From the images, the GPC prepared with the mentioned composition forms a condensed structure. By the interaction of fly ash with GGBS, C-S-H and C-A-S-H gels are formed. GGBS produced an additional content of calcium, which acts as a further binding agent, that affects the GPC hardening characteristics. The GPC with

Fig. 5 SEM image for M30 grade GPC (50 + 50%) FA & GGBS

Fig. 6 SEM image for M70 grade (50 + 50%) FA & GGBS

mentioned percentages attained enhanced strength because of the formation of extra C-A-S–H gel and a dense micro structure package.

In Fig. 7, XRD images showed an outstanding peak in fly ash is quartz (2-theta = 27^0), which is the identification of SiO_2 in fly ash. Mullite is the second outstanding peak in fly ash at varied range values of 2-theta. In GGBS, the outstanding peaks were observed for calcite and quartz at varied range values and CaO content identified is 33%. In Cement, the outstanding peaks were observed for Alite, Pentlandite, and Belite at varied ranges.

3.3 Sorptivity

The sorptivity test results are furnished in Fig. 8. From the figure, it is obvious that the pozzolan incorporation reduced the size of pores thus resulting in a paste with lower

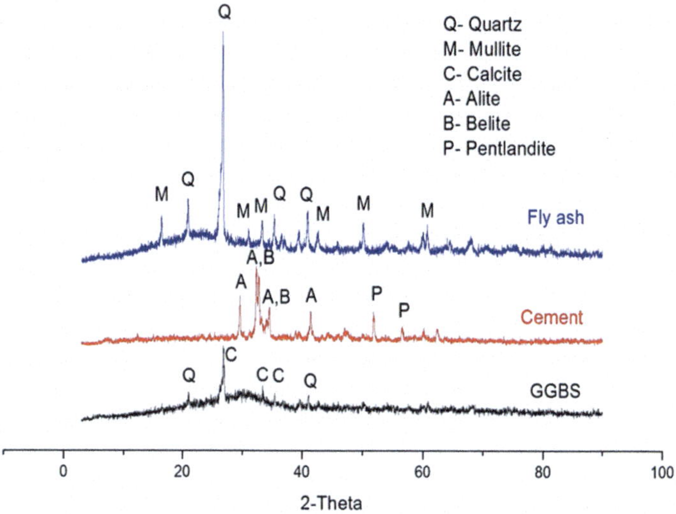

Fig. 7 XRD analysis of fly ash, cement and GGBS

Fig. 8 Sorptivity coefficient for two grades of OPC and GPC

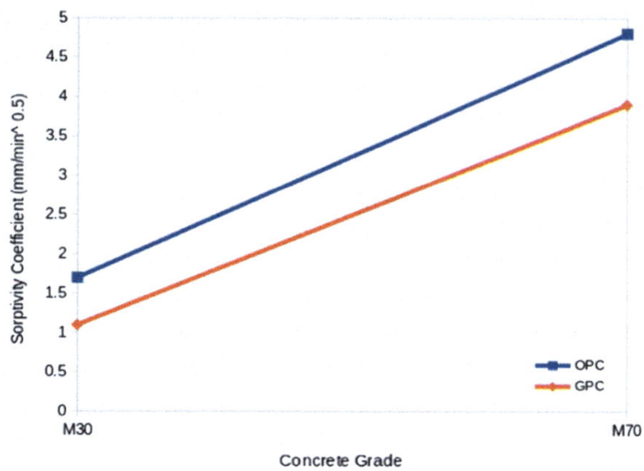

permeability and higher strength concrete [23]. To decrease the sulphate and chloride ingress into concrete, minimizing the sorptivity is important [24]. The geopolymer concrete sorptivity increases with the grade increase of concrete. The specimens with a low amount of sorptivity and low amount of water absorption result in increased compressive strength [25].

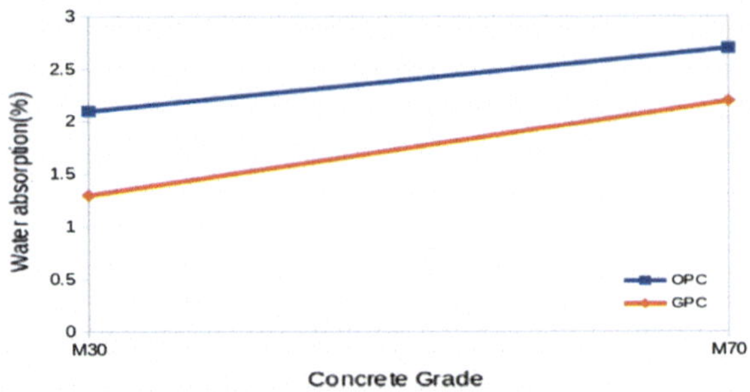

Fig. 9 Water absorption for OPC and GPC

Fig. 10 Visual appearance of GPC cubes exposed to 1, 3 and 5% of Sulphuric acid after 90 days

3.4 Water Absorption

Figure 9 indicates that concrete water absorption increases linearly with an increase in concrete grade which indicates the existence of voids due to incomplete polymerization. A low amount of water absorption was reported in GPC than OPC concrete in both grades. Higher silica content existence formed higher alumino-silicate content which provides better bonding between interstitial particles in which the silicate content engross the voids in between the source materials results in lower water absorption rate [26].

3.5 Resistivity to Sulphate Attack

Figure 10 is the representation of the images of GPC specimens exposed to a sulphuric acid solution for 90 days. The appearance of GPC & OPC was somewhat similar. The GPC cubes had no significant change in appearance. Efflorescence presence was identified on the specimen surface because of high quantities of calcium hydration compounds in GPC cubes [27].

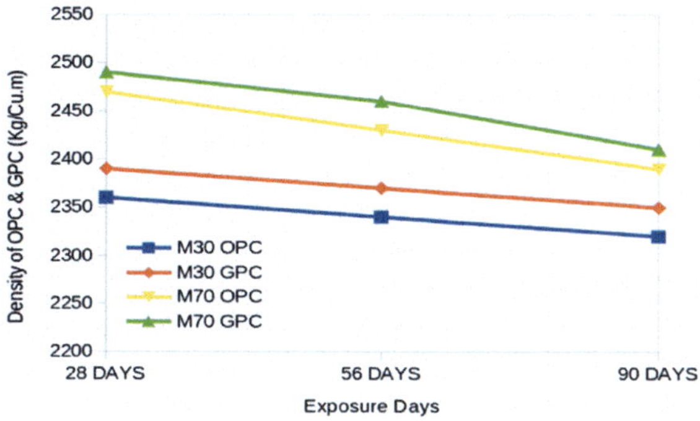

Fig. 11 GPC specimens density exposed to H_2SO_4 solution

Figure 11 is the representation of densities of both grades of OPC & GPC concretes under sulphuric acid exposure for 28, 56 and 90 days. As the period of exposure increased, the GPC specimen density gradually decreased. The decrease in density was observed in the range of 0.84–1.7% for M30 and 1.61–3.2% for M70 grades in OPC, whereas in GPC it was found to be 0.8–3.23% for M30 and 1.20–3.21% for M70 grades in GPC under sulphuric acid exposure. The decrease in density isdue to the presence of $Ca(OH)_2$ in concrete and in acid, induces tensile stress in nature, further results in the degradation of concrete [28]. Further, the silica content in source materials reacts to produce the C-S-H compound which fills the spaces in concrete. The source materials particle size allows a dense packing of the mixture resulting in the lower density of geopolymer concrete specimens. The fly ash & GGBS particle surface is not smooth and hence provides good bonding resulting in a lower amount of weight loss due to the attack of the H_2SO_4 solution.

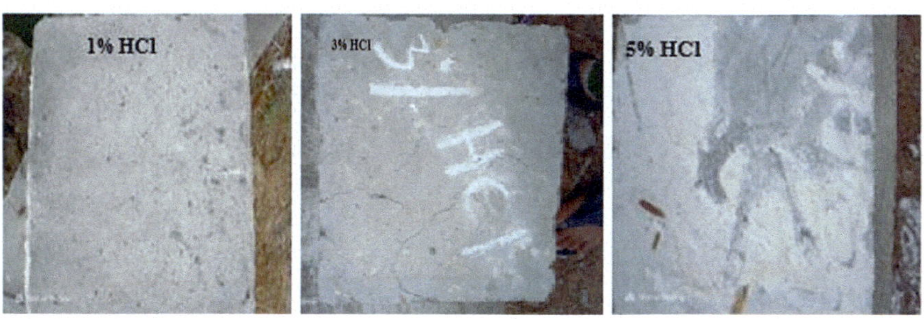

Fig. 12 Visual appearance of GPC cubes exposed to 1, 3 and 5% of Hydrogen Chloride after 90 days

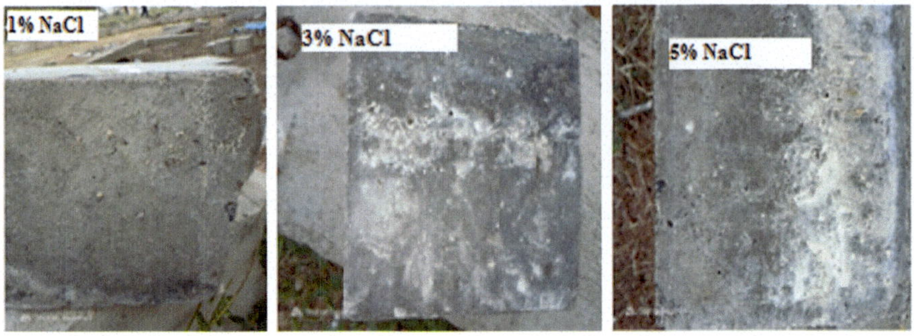

Fig. 13 Visual appearance of GPC cubes exposed to 1, 3 and 5% of Sodium Chloride after 90 days

Fig. 14 GPC specimens density exposed to HCl solution

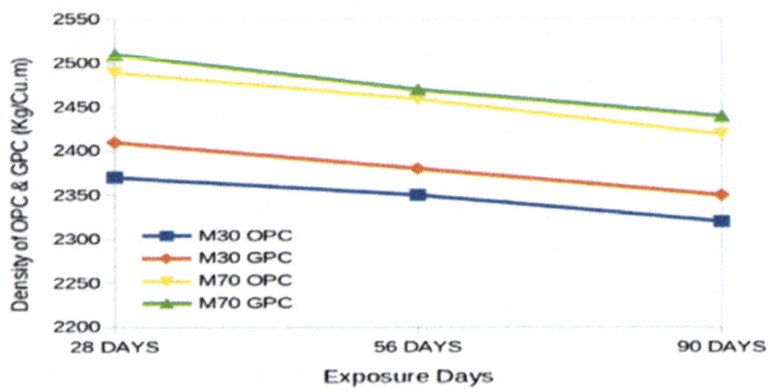

Fig. 15 GPC specimens density exposed to NaCl solution

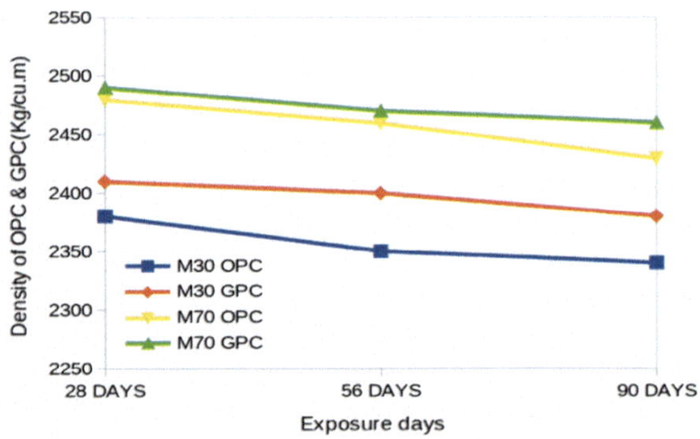

3.6 Resistivity to Chloride Attack

Figures 12 and 13 represents the visual appearance of GPC specimens after immersion in Hydrogen chloride and sodium chloride solution for 90 days. The appearance of GPC & OPC was somewhat similar. The GPC cubes had no significant change in appearance. Efflorescence presence was identified on the specimen surface because of high quantities of calcium hydration compounds in GPC cubes.

Figure 14 shows the density of two grades of OPC & GPC concrete exposed to Hydrochloric acid for 28, 56 & 90 days. As the exposure period increases, the GPC specimen density gradually decreased. The decrease in density was observed in the range of 0.84–2.10% for M30 and 1.2–2.81% for M70 grades in OPC, whereas in GPC it was found to be 1.24–2.48% for M30 and 1.68–2.78% for M70 grades in GPC under Hydrochloric acid solution exposure.

Figure 15 represents the density of two grades of OPC & GPC concrete exposed to Sodium Chloride solution for 28, 56 & 90 days. As the period of exposure increased, the GPC specimen density gradually decreased. The decrease in density was observed in the range of 1.26–1.68% for M30 and 0.8–2.01% for M70 grades in OPC, whereas in GPC it was found to be 0.41–1.24% for M30 and 0.8–1.20% for M70 grades in GPC when exposed to Sodium chloride solution. Geopolymer concrete exhibited superior resistivity from chloride attack. There was no significant degradation visible on the specimen's surface under the exposure of hydrochloric acid and sodium chloride solutions (Fig. 16).

From the regression analysis, an empirical equation has proposed for predicting the compressive strength of concrete for different concentrations of solutions of OPC and GPC.

$$y = ke^{-nx} \tag{4}$$

where k and n are the variables depending upon the percentage of concentration and exposure of the number of days of concrete.

From Fig. 16a, b, c, it is observed that the K values are varying from 39–49 MPa for M30 GPC and OPC grade concretes for the specified period of exposure and from Fig. 17a, b, c, it is observed that the K values are varying from 77–90 MPa for M70 GPC and OPC grade concretes for the specified period of exposure.

4 Conclusions

The present investigation represents the resistance on Sulphuric acid, Hydrochloric acid, and Sodium chloride of GPC and OPC concrete. The following conclusions were drawn.

1. The GPC concrete show more resistivity than OPC concrete in aggressive environments.
2. The SEM analysis depicted dense microstructure which leads to superior strength attainment for blended geopolymer concrete of different grades.
3. The geopolymer concretes have lower sorptivity and water absorption than OPC concretes.
4. The percentage decrease in density for two grades of concretes is 0.8–3.5% in Sulphuric acid exposure, between 0.8 and 3% in the case of Hydrochloric acid exposure and between 0.4 and 1.20% in case of sodium chloride exposure.

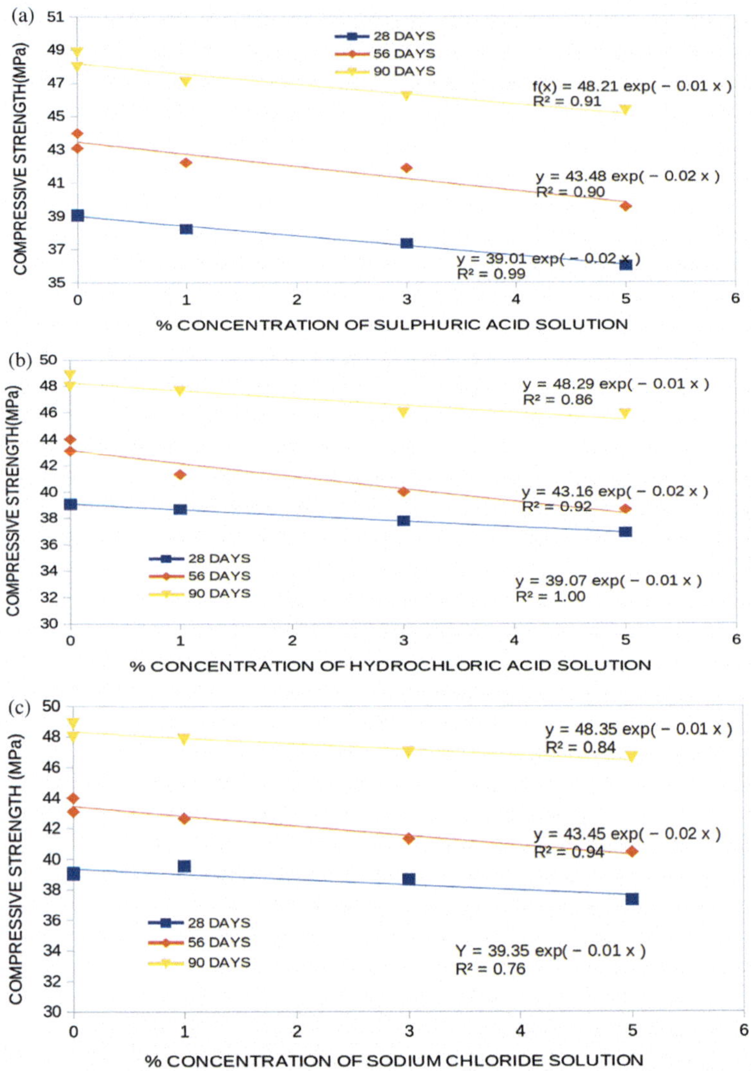

Fig. 16 a–c Regression analysis for percentage concentrations of solutions used and the compressive strength for M30 grade OPC and GPC concrete

5. The percentage decrease in compressive strength for two grades of concretes is 2 to 6% in Sulphuric acid exposure, between 3 and 7% in the case of Hydrochloric acid exposure, and between 2 and 6% in case of sodium chloride exposure.

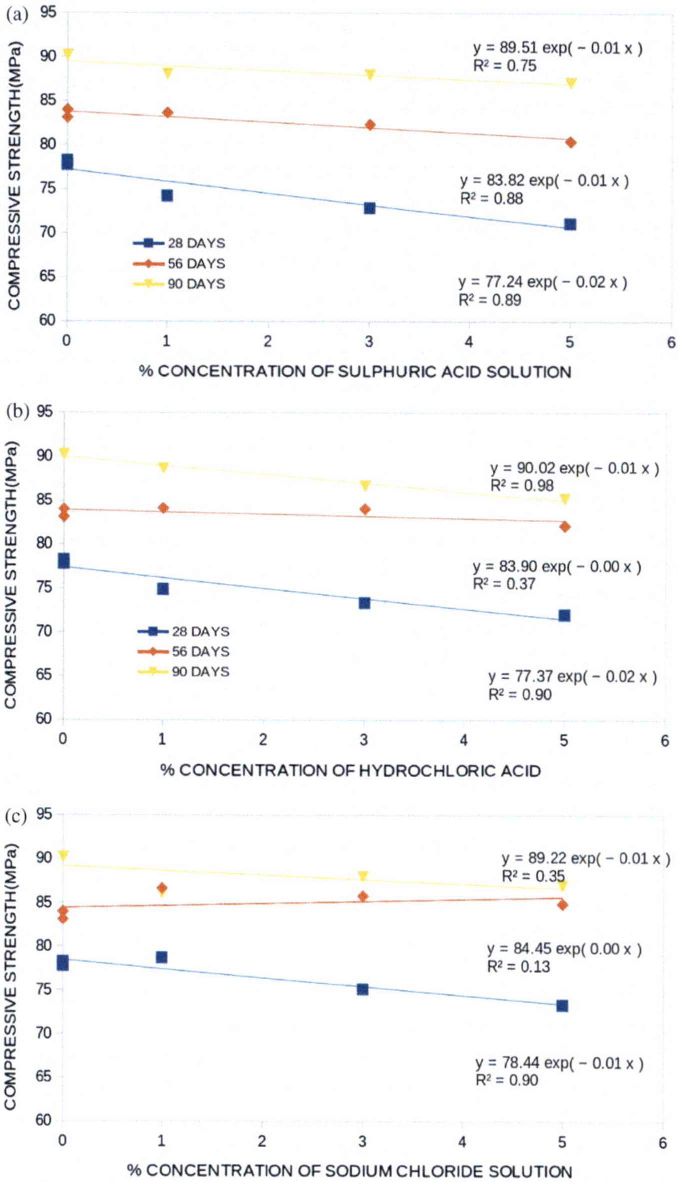

Fig. 17 a–c Regression analysis for percentage concentrations of solutions used and the compressive strength for M70 grade OPC and GPC concrete

References

1. Hardjito D, Wallah SE, Sumajouw DMJ, Rangan BV (2004) On the development of fly ash-based geopolymer concrete. ACI Mater J 101(6):467–472
2. Hardjito D, Vijaya Rangan B (2005) Development and properties of low-calcium fly ash-based geopolymer concrete, Research Report GC 1. Curtin University of Technology, Perth, Australia
3. Bingöl Ş, Bilim C, Atiş CD et al (2020) Durability properties of geopolymer mortars containing slag. Iran J Sci Technol Trans Civil Eng
4. Muthadhi A, Vanjinathan J, Durai D (2016) Experimental investigations on geopolymer concrete based on Class C Fly Ash. Indian J Sci Technol 9(5):1–5
5. Pattanapong T, Chindaprasirt P, Sata V (2015) Setting time, strength, and Bond of High-Calcium Fly ash Geopolymer Concrete. J Mater Civil Eng 27(7):1–7
6. Rattanasak U, Pankhet K, Chindaprasirt P, Sata V (2011) Effect of chemical admixtures on properties of High-Calcium Fly ash Geopolymer. Int J Minerals Metall Mater 18(3):364–369
7. Prasanna Venkatesan R, Pazhani KC (2016) Strength and durability properties of geopolymer concrete made with Ground Granulated Blast Furnace Slag and Black Rice Husk Ash. KSCE J Civil Eng 20(6):2384–2391
8. Srinivas T, Raju Vundi SP, Ramana Rao NV, Deepak Kumar S (2019) Resistance of acid attack on geopolymer concrete developed with partial replacement of coarse aggregate by recycled aggregate. Int J Rec Technol Eng 8(4)
9. Vidyasree B (2019) Durability studies on red mud and GGBS based geopolymer concrete. Int J Geol Geotech Eng 5(1)
10. Alexander MG, Magee BJ (1999) Durability performance of concrete containing condensed silica fume. Cem Concr Res 29(6):917–922
11. Bakharev T, Sanjayan JG, Yi BC (2003) Resistance of alkali-activated slag concrete to acid attack. Cem Concr Res 33(10):1607–1611
12. Sanni SH, Khadiranaikar RB (2012) Performance of geopolymer concrete under severe environmental conditions. Int J Civil Struct Eng 3(2): 396–407
13. Rajamane NP, Nataraja MC, Lakshmanan N, Dattatreya JK, Sabitha D (2012) Sulphuric Acid Resistant Eco-friendly concrete from Geopolymerisation of Blast Furnace Slag. Indian J Eng Mater Sci 19(5):357–367
14. Malviya M, Goliya HS (2014) Durability of fly ash-based geopolymer concrete using Alkaline Solutions (NaOH and Na_2SiO_3). Int J Emerg Trends Eng Dev 6(4):18–33
15. Bakharev T (2005) Durability of geopolymer materials in sodium and magnesium sulfate solutions. Cem Concr Res 35(6):1233–1246
16. Wardhono A (2019) Comparison study of class F and class C fly ashes as cement replacement material on strength development of non-cement mortar. IOP Conf Ser Mater Sci Eng 288 Jan 2019
17. Sanni SH, Khadiranaikar RB (2012) Performance of geopolymer concrete under severe environmental conditions. Int J Civil Struct Eng 3(2):396–407
18. Malviya M, Goliya HS (2014) Durability of fly ash-based geopolymer concrete using alkaline solutions (NaOH and Na_2 SiO_3). Int J Emerg Trends Eng Dev 6(4):18–33
19. Wallah SE, Rangan BV (2006) Low-calcium fly ash-based geopolymer concrete: long-term properties, Research Report GC 2. Curtin University of Technology, Perth, Australia, Faculty of Engineering
20. Kashyap AMN, Chandrasekhar Rao T, Ramana Rao NV (2018) Prediction of setting and strength characteristic of binary blended geopolymer matrix, i-Manager's J Struct Eng 6(4)
21. Mallikarjuna Rao G, Gunneswara Rao TD (2015) Final setting time and compressive strength of fly ash and GGBS-based geopolymer paste and mortar. Arab J Sci Eng 40(11):3067–3074
22. Chindaprasirt P, Jaturapitakkul C, Sinsiri T (2005) Effect of fly ash fineness on compressive strength and pore size of blended cement paste. Cem Concr Compos 27(4):425–428
23. McCarter WJ, Ezirim H, Emerson M (1992) Absorption of water and chloride into concrete. Mag Concr Res 44(158):31–37

24. Thokchom S, Ghosh P, Ghosh S (2009) Effect of water absorption, porosity and sorptivity on the durability of geopolymer mortars. J Eng Appl Sci 4(7):28–32
25. Ghosh K, Ghosh P (2012) Effect of %Na_2O and %SiO_2 on apparent porosity and sorptivity of fly ash-based geopolymer. IOSR J Eng 2(8):96–101
26. Ariffin MAM, Bhutta MAR, Hussin MW, Mohd Tahir M, Aziah N (2013) Sulfuric acid resistance of blended ash geopolymer concrete. Constr Build Mater 43:80–86
27. Aydin S, Yazici H, Yiğiter H, Baradan B (2007) Sulfuric acid resistance of high-volume fly ash concrete. Build Environ 42(2):717–721
28. Chindaprasirt P, Homwuttiwong S, Sirivivatnanon V (2004) Influence of fly ash fineness on strength, drying shrinkage and sulfate resistance of blended cement mortar. Cem Concr Res 34(7):1087–1092

Correlation Curves to Characterize Concrete Strength by Means of UPV Tests

Mariella Diaferio and Michele Vitti

Abstract In the design of interventions and/or in the evaluation of vulnerability of an existing reinforced concrete structure, a crucial role is played by the assessment of the concrete compressive strength. The estimation of such parameter may be obtained by destructive tests, which, even if may give an accurate estimation, usually can be performed on a limited number of points. Otherwise, non-destructive tests have been proposed which proceed by acquiring physical measurements which can be reported to the concrete compressive strength. In the present study, the results of an experimental campaign performed on an existing building in the area of Bari (Italy) are presented. The tests involved both destructive tests and Ultrasonic Pulse Velocity tests. Several possible test conditions are defined by combining the available data and varying the number of considered cores. The correlation curves have been evaluated for each one of the possible test conditions, and some considerations are discussed on the relationship between the parameters of such correlation models by varying the number of considered cores.

Keywords Ultrasonic pulse velocity tests · Concrete compressive strength · Correlation models

1 Introduction

The recent codes [1–4] have devoted great attention to the evaluation of performances of existing buildings, and several indications are given regarding the required tests and/or inspections. The characterization of the mechanical properties of concrete is a crucial issue in the process of knowledge of an existing r.c. building and may condition the subsequent interventions and/or may limit the use of the structure.

M. Diaferio (✉)
Politecnico Di Bari, DICATECh, 70125 Bari, Italy
e-mail: mariella.diaferio@poliba.it

M. Vitti
Structural Engineering, 70100 Bari, Italy
e-mail: ing.vitti@gmail.com

M. Abdel Wahab (ed.), *Proceedings of 1st International Conference on Structural Damage Modelling and Assessment*, Lecture Notes in Civil Engineering 110,
https://doi.org/10.1007/978-981-15-9121-1_16

Thus, many researchers investigated this topic and, in detail, the experimental procedures and the models which relate the measurements to the in situ compressive strength.

The problems connected to a reliable evaluation of the concrete compressive strength are mainly related to the nature of the material. In fact, many elements influence the concrete behavior: type of formwork, curing conditions, the age, the content and type of cement, the temperature, the humidity, the type and size of aggregates.

In light of the above, a possible way to obtain a more reliable estimation of the concrete compressive strength is to have a great amount of data on which performing analysis to verify the effective variability of the concrete strength and, thus, judging the uncertainties related to the adopted value.

However, this goal may be achieved only making use of non-destructive tests (NDT), as the destructive ones are usually restricted to a limited number of points due to economic and practical reasons. The NDTs do not directly measure the strength; thus, their reliability is strictly dependent on the curve which correlates the physical NDT measurement to the compressive strength.

To define such curve, the NDTs are performed in combination with destructive tests. In [1] and [2] the correlation model is proposed and the procedure for its calibration is indicated. In [2], destructive tests on at least 18 cores are requested for the application of NDTs.

The actual Italian Building Code [3–4] prescribes the use of the results of destructive tests to calibrate the law for the implementation of non-destructive tests.

In literature several studies [5–7] have been devoted to the analysis of the results obtained performing both destructive and non-destructive tests. In [8], the data scattering in the case of in situ concrete strength is investigated.

In this field several topics are still on debate: the number of cores for the calibration of the conversion models, the reliability of using single or combined experimental tests [9], the analytical elaboration of the measurements, etc. [10].

The present study deals with the assessment of the concrete compressive strength by means of Ultrasonic Pulse Velocity (UPV) tests.

In literature several correlation functions have been proposed to estimate the in situ strength by measuring the ultrasonic velocity, but their reliability appears strictly connected to the performed experimental campaign, and this circumstance limits the applicability of the UPV tests.

The challenge of the present study is to verify the existence of a possible relationship between the parameters of the correlation functions and the influence of the cores number on such relationship.

To this aim, the data acquired during an experimental campaign have been arranged in order to simulate several possible tests, and the calibration of the correlation models has been performed for each one of the obtained simulated tests.

The paper discusses the variability of the parameters of such models and a relationship between such parameters by varying the cores number considered for the calibration.

In the first part, a brief description of the destructive and UPV tests and of the laws which correlate the measurements to the on-site concrete strength is presented, while the second part is devoted to the description of the performed experimental test and to the discussion on the variability of the parameters of correlation models.

2 Destructive Tests

The direct determination of concrete strength proceeds through the extraction of samples which are, after, subjected to compression test till the failure.

The extraction, storage and testing procedures are specified by the following codes, UNI EN 12,504–1 [11], UNI EN 12,390–1[12], UNI EN 12,390–2 [13], UNI EN 12,390–3 [14], which have been adopted also for the experimental campaign here discussed.

For a reliable estimation of the on-site concrete strength, the results of the compression tests performed on the cores must be corrected to take into account the disturbance due to drilling process, the drilling direction, the ratio between core height and diameter, the presence of bars.

To this aim, many empirical relationships are available in literature. In the present study, the in situ compressive strength f_c has been evaluated as follows [15, 16]:

$$f_c = D_{l/d} C_{dia} C_a C_d f_{core} \tag{1}$$

where $D_{l/d}$ is the correction factor which takes into account the ratio between the core height l and its diameter d, and that is evaluated as $D_{l/d} = 2/(1.5 + d/l)$; C_{dia} is the correction factor related to the core diameter d, and it is assumed equal to 1.06, 1.00 and 0.98 for d, respectively, equal to 50 mm, 100 mm and 150 mm, C_a is the correction factor due to the presence of reinforcing bars which is assumed equal to 1 for no bars and varies from 1.03 for small diameter bars and 1.13 for large diameter bars. C_d is a correction factor which takes into account the disturbance due to the perforation, which is set equal to 1.06 in accordance with ACI 2003 [1], and f_{core} is the core strength.

3 Ultrasonic Pulse Velocity Test

The Ultrasonic Pulse Velocity (UPV) test proceeds by generating an ultrasonic pulse by means of a sending transducer and by recording the time lapse to arrive to the receiving transducer. Thus, the compressive strength is evaluated indirectly by means of the wave velocity through the sample.

The best way to perform the test is to locate the sending and the receiving transducers on the opposite side of the investigated element, in this manner the evaluated velocity may represent both the surfaces and the inner parts of the concrete; however, also other arrangements can be utilized in order to adapt the test to the in situ conditions.

The main advantages of UPV test are its adaptability to several on site settings, and the possibility of investigating a great number of points with a relative rapidity.

Nevertheless, many aspects may influence the determined velocity, for example the type, percentage and volume of aggregates [17], the water content, the temperature of the specimen, the curing conditions [18], the age of the specimen, the presence of cracks, etc.

Many Authors devoted their attention to the definition of a relationship between the calculated ultrasonic pulse velocity V and the compressive strength f_c [7, 19].

Several laws have been investigated, for example the exponential law [19], the polynomial law [20], etc. Here the power law is considered which is expressed as:

$$f_c = aV^b \tag{2}$$

where a and b are coefficients calibrated on the results of destructive tests. Following, the strength is expressed in MPa and the ultrasonic pulse velocity V in m/s.

The power law may be rewritten using the logarithmic notation:

$$Ln(f_c) = a + bLn(V) \tag{3}$$

where the parameters of the linear regression may be evaluated by means of the minimum weighted squares method, i.e. minimizing the mean square error ε:

$$\varepsilon^2 = \frac{1}{n}\sum_{i=1}^{n} f_{c,i}^2\left[Ln(f_{c,i}) - (a + bLn(V_i))\right]^2 \tag{4}$$

where n is the number of tests.

Following, some proposed correlation equations are listed:

$$f_c = 1.001 \times 10^{-28} V^{8.1272} \tag{5}$$

$$f_c = 1.2 \times 10^{-5} V^{1.7447} \tag{6}$$

$$f_c = 2.94 \times 10^{-16} V^{4.696} \tag{7}$$

where Eqs. (5, 6) and (7) have been proposed in [21, 22] and [23], respectively. The analysis of the Eqs. (5), (6) and (7) shows the high variability of the law parameters.

4 Case Study

A reinforced concrete building located near Bari (Italy) was investigated. A total number of $n = 16$ cores, with a ratio between the height and the diameter equal to 2, were extracted by drilling, and UPV tests were performed on each structural element chosen for core extraction (Fig. 1).

The core samples have been subjected to compression tests and the on-site compressive strengths have been evaluated in accordance with Eq. (1). The f_c average value is equal to 30.29 MPa, while the V average value is equal to 3824 m/s.

In Fig. 2 the experimental compressive strengths f_c, and the estimated values obtained by means of Eqs. (5), (6) and (7) for each investigated point are shown.

In order to evaluate the influence of the cores number, the experimental data have been arranged to simulate a huge number of possible experimental tests, more than 400 sets. In detail, the correlation models have been evaluated varying the number N

Fig. 1 The UPV test on the investigated building

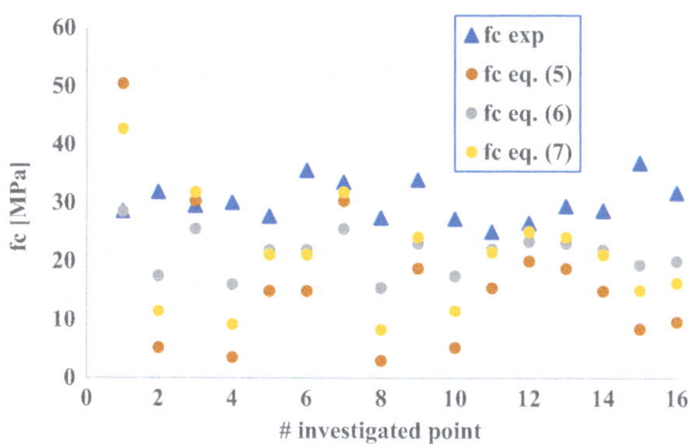

Fig. 2 Compressive strength *versus* # investigated point: in blue the value evaluated through compression test on the core, in orange, in grey and in yellow the values estimated by means of Eq. (5), (6) and (7), respectively

of the pair ($f_{c,i}$, V_i) considered for the calibration, i.e. minimizing the Eq. (4) for $n = N$ and varying$N$ from $N = 3$ to $N = 16$. In this way, it is investigated the influence of the cores number on the correlation equation. Obviously, excepted for the case in which all the available experimental data are considered (i.e. $N = 16$), for each value of N several combinations of data may be collected, as n–N data must be excluded from the calibration of correlation law; in this way, a huge number of tests are simulated and the parameters of Eq. (2) have been evaluated for each test.

To fix the ideas, the case $N = 5$ is now considered. In this case 118 different combinations of experimental data have been considered, and the parameters of the correlation model have been evaluated for each combination. The second step of the study was to compute the mean square error of the evaluated correlation model considering two groups of data: the first composed by the $N = 5$ data utilized for the calibration of the correlation law, called $RMSE_1$, and the second group obtained considering the other n–$N = 11$ remaining data, called $RMSE_2$:

$$RMSE_1 = \sqrt{\frac{\sum_{i=1}^{N}\left(f_{c,iexp} - \left(aV_i^b\right)\right)^2}{N}} \tag{8}$$

$$RMSE_2 = \sqrt{\frac{\sum_{i=1}^{n-N}\left(f_{c,iexp} - \left(aV_i^b\right)\right)^2}{n - N}} \tag{9}$$

The aim of this step is to verify the accuracy of each model.

The above described analysis gives the following results: the parameter a assumes values in the range [0.065; 8.986] and its mean value is equal to 2.717, while the parameter b ranges between 0.138 and 0.752 with a mean value equal to 0.378. It can be observed that the variability of the parameter a is greater than the one of the parameter b.

Moreover, the root mean square errors evaluated on the first group of five data (i.e. the ones considered for the model calibration) have a mean value equal to 2.57 MPa, which corresponds to 8.5% of thef_c mean value, and a standard deviation of0.68 MPa. Otherwise, the root mean square errors evaluated for the unconsidered data, i.e. the so called second group, have a mean value equal to 4.16 MPa and a standard deviation of 0.54 MPa.

In Fig. 3, the root mean square errors calculated for the 118 combinations considered for the case $N = 5$ for the two groups of data are plotted. The obtained RMSE values show that, even if the parameters of the correlation model are characterised by a great variability, the reliability of each model appears quite good.

In the light of above, the definition of a single conversion model appears quite difficult, thus, a relationship between the parameters a e b has been investigated; with this goal, in Fig. 4 the parameters b *vs.* parameters a are plotted for all the 118 tests.

The Fig. 4 shows that a logarithmic equation may be introduced to fit the obtained parameters, which is expressed as follows:

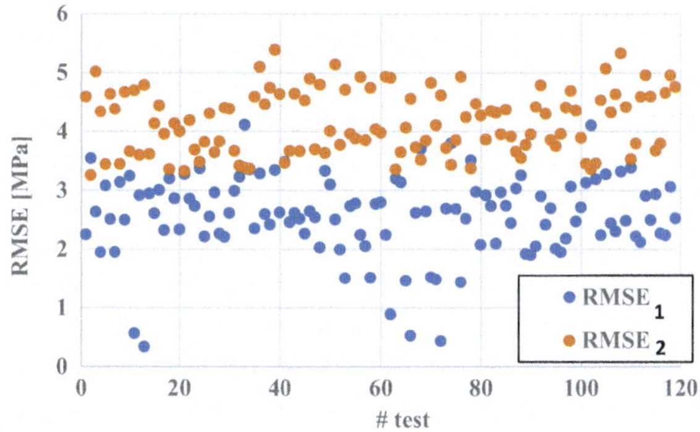

Fig. 3 The root mean square errors (RMSE) evaluated for all the defined tests for $N = 5$: in blue $RMSE_1$ the values related to the data groups composed by the five data considered for the correlation model, in orange $RMSE_2$ the values associated to the data group composed by the remaining eleven data

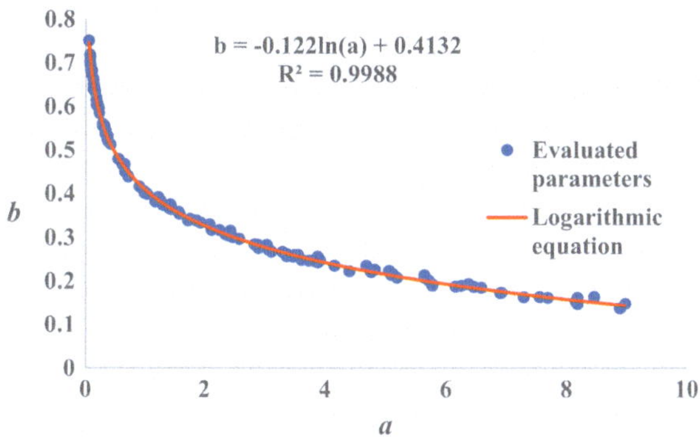

Fig. 4 Parameters b *vs.* parameters a for all the 118 tests considered for the case $N = 5$: in blue the values obtained through the linear regression analysis, in red the approximated logarithmic relationship

$$b = -0.122Ln(a) + 0.4132 \quad (10)$$

The determination coefficient is equal to 0.9988, showing a very good approximation.

The same procedure was applied calibrating the correlation models considering $N = 9$ cores. In this case, 34 combinations of data have been settled which can represent 34 possible experimental tests. Thus, the 34 correlation models have been determined, whose parameter a varies in the range [1.311;8.848], while parameter b varies in the range [0.142;0.383].

In Fig. 5, the RMSE values evaluated for the data group composed by the $N = 9$ cores considered for the calibration process, and the ones for the remaining n–N = 7 cores are plotted.

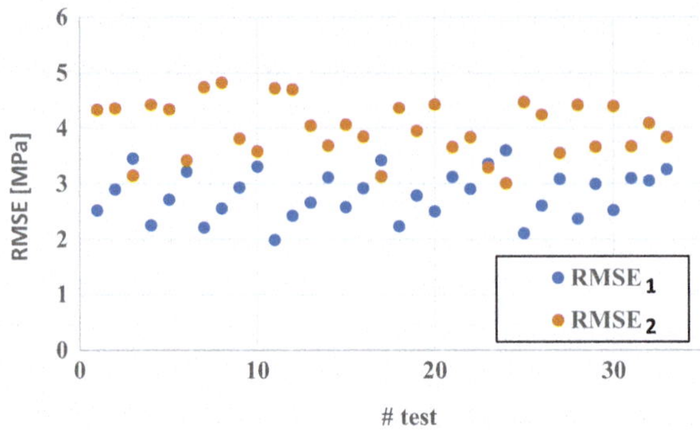

Fig. 5 The root mean square errors (RMSE) evaluated for all the defined tests for $N = 9$: in blue $RMSE_1$ the values related to the data groups composed by the nine data considered for the correlation model, in orange $RMSE_2$ the values associated to the data group composed by the remaining seven data

In this case, the $RMSE_1$ mean value is equal to 2.839 MPa, which corresponds to 9.37% of the f_c mean value, and a standard deviation of 0.42 MPa. Otherwise, the $RMSE_2$ mean value is equal to 3.97 MPa, which corresponds to 13.1% of the f_c mean value, and a standard deviation of 0.50 MPa.

In Fig. 6 the parameters *b* vs. parameters *a* are plotted.

The logarithmic equation that fits the points in Fig. 6 is expressed as follows:

$$b = -0.123Ln(a) + 0.416 \tag{11}$$

The determination coefficient is equal to 0.9974, showing a very good approximation.

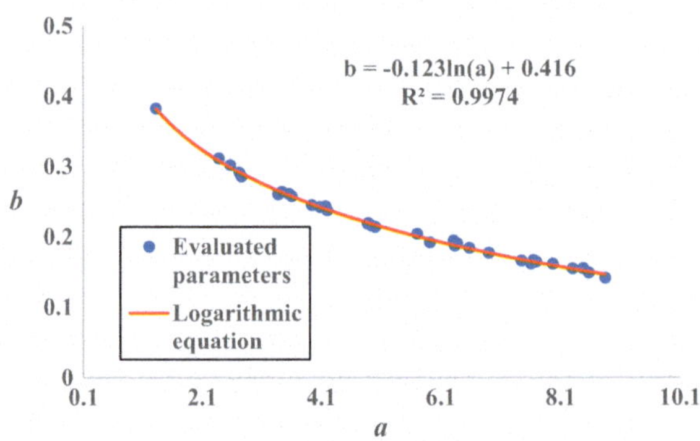

Fig. 6 Parameter *b versus* parameter *a* for all the 34 tests considered for the case $N = 9$: in blue the values obtained through the linear regression analysis, in red the approximated logarithmic relationship

The results confirm that, even if the parameters *a* and *b* of the conversion models may manifest a great variability, also in the case the same number *N* of cores are considered for the linear regression analysis, their relationship remains almost the same by varying the number N and it is characterized by a high value of the determination coefficient.

5 Conclusion

The study investigates the assessment of concrete compressive strength by means of Ultrasonic Pulse Velocity test. In detail, an experimental campaign was performed, and the data have been elaborated in order to simulate a huge number of possible experimental tests.

In this way, it can be examined the influence of the points chosen for the calibration of the correlation law. It can be verified that the parameters of the correlation laws manifest a high variability, even if the root mean square errors related to the estimated conversion laws are quite low.

On the contrary, the relationship between such parameters can be approximated by means of a logarithmic equation characterized by a high determination coefficient.

This behavior is also observed varying the number of cores utilized in the minimizing procedure, and it can be verified that the logarithmic equation remains almost the same.

Acknowledgements Ing. Giovanni Notarangelo is deeply acknowledged for his support. The project Politecnico di Bari FRA 2019—"Analisi del comportamento degli edifici esistenti e progettazione di interventi di risanamento e adeguamento funzionale" is acknowledged for the support given to the present research.

References

1. American Concrete Institute (ACI) (2003) Guide for obtaining cores and interpreting compressive strength results", ACI 214.4R-03, Detroit
2. EN 13791 (2007) Assessment of in situ compressive strength in structures and precast concrete. CEN, Brussels, p 28
3. D.M. 17 gennaio 2018. Aggiornamento delle «Norme tecniche per le costruzioni». Ministero delle Infrastrutture e dei Trasporti, G.U. n. 42 del 20 febbraio 2018. Supplemento Ordinario n. 8 2018 (in Italian)
4. Circolare 21gennaio 2019 n. 7 approvata dal Consiglio Superiore dei Lavori Pubblici. I*struzioni per l'applicazione dell'«Aggiornamento delle "Norme tecniche per le costruzioni"» di cui al decreto ministeriale 17 gennaio 2018*. Ministero delle Infrastrutture e dei Trasporti. (in Italian)
5. Qasrawi HY (2000) Concrete strength by combined nondestructive methods Simply and reliably predicted. Cem Concr Res 30:739–746
6. Nobile L (2015) Prediction of concrete compressive strength by combined non-destructive methods. Meccanica 50:411–417

7. Breysse D (2012) Nondestructive evaluation of concrete strength: an historical review and a new perspective by combining NDT methods. Constr Build Mater 33:139–163
8. Fiore A, Porco F, Uva G, Mezzina M (2013) On the dispersion of data collected by in situ diagnostic of the existing concrete. Constr Build Mater 47:208–217
9. Breysse D, Klysz G, Dérobert X, Sirieix C, Lataste JF (2008) How to combine several nondestructive techniques for a better assessment of concrete structures? Cem Concr Res 38:783–793
10. Malhotra VM, Carino NJ (2004) Handbook on nondestructive testing of concrete ASTM. 2nd ed. CRC Press
11. UNI EN 12390–1 (2012) Prova sul calcestruzzo indurito—Parte 1: Forma, dimensioni ed altri requisiti per provini e per casseforme
12. UNI EN 12390–2 (2019) Prove sul calcestruzzo indurito—Parte 2: Confezione e stagionatura dei provini per prove di resistenza.
13. UNI EN 12390–3 (2019) Prove sul calcestruzzo indurito—Parte 3: Resistenza alla compressione dei provini
14. UNI EN 12504–1 (2019) Prove sul calcestruzzo nelle strutture—Parte 1: Carote—Prelievo, esame e prova di compressione
15. Dolce M, Masi A, Ferrini M (2006) Estimation of the actual in-place concrete strength in assessing existing RC structures. In: Proceedings of the second international Fib Congress
16. Masi A, Vona M (2009) Estimation of the in-situ concrete strength: provisions of the European and Italian seismic codes and possible improvements. Eurocode 8 Perspectives from the Italian Standpoint Workshop, 67–7
17. Najim KB, Hall MR (2012) Mechanical and dynamic properties of selfcompacting crumb rubber modified concrete. Constr Build Mater 27:521–530
18. Samarin A, Meynink P (1981) Use of combined ultrasonic and rebound hammer method for determining strength of concrete structural member. Concr In 25–39
19. D'Ambrisi A, Cristofaro MT, De Stefano M (2008) Predictive models for evaluating concrete compressive strength in existing buildings. In: 14th World conf on earthquake eng. Beijing, China
20. Knaze P, Beno P (1985) The use of combined non-destructive testing methods to determine the compressive strength of concrete. Mater Struct 17(3):207–210
21. Pascale G, Di Leo A, Carli R (2000) Evaluation of actual compressive strength concrete by NDT. In: 15th world conference on non-destructive testing, Roma
22. Kheder GF (1999) A two stage procedure for assessment of in situ concrete strength using combined non-destructive testing. Mater Struct 32(6):410–417
23. Machado MD, Shehata LCD, Shehata IAEM (2009) Correlation curves to characterize concretes used in Rio de Janeiro by means of nondestructive tests. Ibracon Struct Mater J 2(2), 100–123

Damage in Mechanical and Materials Engineering

Numerical Investigation on the Effect of Wear Coefficient on Fretting Wear

S. Wang, D. G. Wang, G. Z. Xie, and Magd Abdel Wahab

Abstract Fretting happens when a relative slip occurs between two contact parts. Fretting wear is the main damage in gross slip regime. In most of current researches, the wear coefficient is commonly considered as a constant, which needs further experimental validation. In this paper, bi-linear decreasing wear coefficient numerical model is used to investigate the effect of the variation of wear coefficient on wear profile based on experimental data from literature. It can be concluded that the variation of wear coefficient has a significant effect on the wear profile. Moreover, contact pressure differences are also identified.

Keywords Fretting · Wear · Finite element method · Gross slip · Wear coefficient

1 Introduction and Background

Fretting wear is a dominant damage in gross slip regime, which can be detrimental to the contacting parts due to the degradation of the contact surface and fatigue crack initiation [1, 2]. Fretting can happens either in partial slip condition and gross sliding condition, which can be classified based on the relative slip amplitude.

A lot of research has been done to study the stress variation and loading conditions effect on the wear profile characteristics of the contact surface in gross sliding condition [3, 4]. Wang et al. [5, 6] carried out comparative analyses of torsional fretting, fretting fatigue and tension–torsion fretting fatigue behaviours of steel wires, and explored dynamic wear evolution and crack propagation behaviours of steel wires during fretting-fatigue. Urchegui [7] analysed fretting wear of thin steel wires using different operational variables like normal force, stroke and number of cycles.

S. Wang (✉) · M. Abdel Wahab
Department of Electrical Energy, Metals, Mechanical Constructions and Systems, Faculty of Engineering and Architecture, Ghent University, Ghent, Belgium
e-mail: Shengjie.wang@ugent.be

D. G. Wang · G. Z. Xie
Jiangsu Province Key Lab of Electromechanical Equipment, School of Mechatronic Engineering, China University of Mining and Technology, Xuzhou 221116, China

M. Abdel Wahab (ed.), *Proceedings of 1st International Conference on Structural Damage Modelling and Assessment*, Lecture Notes in Civil Engineering 110,
https://doi.org/10.1007/978-981-15-9121-1_17

Because the experimental method cannot obtain the slip and load information of the contact surface, some scholars have carried out the finite element analysis of fretting wear. Cruzado et al. [8, 9] carried out the finite element modelling of fretting wear scars of contacting steel wires with distinct crossing angles. Ding et al. [10] modelled the debris accumulated on the fretting interface as a layer structure with the mechanical properties being described by an anisotropic elastic–plastic material model to simulating the effects of debris on fretting wear. Fallahnezhad et al. [11] developed an adaptive finite element simulation to predict fretting wear in a head-neck taper junction of hip joint implant through a two dimensional (2D) model and based on the Archard wear equation. Bedolla et al. [12] developed a combined experimental and numerical simulation procedure based on a time- and space-resolved version of the Archard wear equation. Zeng et al. [13] proposed a finite element model to investigate the influence of stress relief groove on the fretting wear of the railway axle. Mohd Tobi et al. [14] presented a finite-element-based wear modelling methodology and a computational device for decoupling wear effects, the decoupling of wear effects facilitates the capture of plasticity accumulation on a particular wear-scarring profile after a specific number of cycles. Anders et al. [15] discussed how wear of the pad-to-rotor interface can be predicted using general purpose finite element analysis software, and developed a three-dimensional finite element model of the brake pad and the rotor to calculate the pressure distribution in the pad-to-rotor contact.

The commonly used experimental cylindrical-on-flat configuration is shown in Fig. 1 [16].

Wear coefficient is a parameter that have a dominant effect on fretting wear. Wear coefficient can be obtained by Archard's equation as:

$$k = \frac{V}{SP} \tag{1}$$

where V is total wear volume measured in the experiment, S is the local sliding distance, P is applied normal load and k is the Archard's wear coefficient.

Based on the wear profile characteristics, total number of cycles and applied displacements, the Archard's equation can be further written as:

$$k = \frac{Wbh_a}{4\delta \times N_t P} \tag{2}$$

Fig. 1 Experimental setup for fretting wear

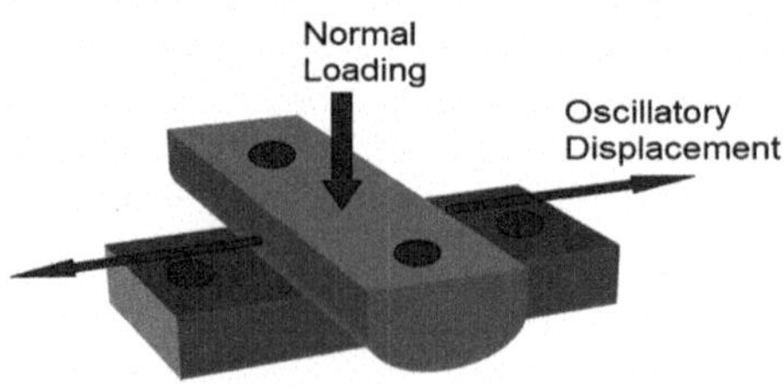

where W is the thickness of the specimen, b is the experimental wear width, h_a is the experimental averaged wear depth, δ is the applied displacement amplitude (half stroke) and N_t is the total number of cycles.

Archard's wear coefficient is an averaged value based on the experimental wear profile. In finite element (FE) model, the averaged wear coefficient is employed for the whole fretting process in most current research, which is not the real condition. In this paper, the bi-linear decreasing wear coefficient model is applied to the fretting wear process. And the wear profile from this model is compared with that obtained using the averaged wear coefficient model. Significant difference was found in the comparison. And differences in contact pressure as function of cycles were also found between the two models.

2 Finite Element Model and Wear Coefficient Model

Details of FE model is shown in Fig. 2. The bottom of the specimen is fixed, and normal load and oscillatory displacement are applied on the top centre of the cylinder. CPE4 elements are applied to the whole FE model. The size of the elements in Partition zone is 5 μm × 10 μm. And the partition of the specimen is set as the adaptive domain to simulate the wear profile updated in fretting process by subroutine UMESHMOTION. More details about this subroutine can be found in [17].

Material in the experiments was Super CMV steel. The Young's modulus is 200 GPa and Poisson's ratio is 0.3 [16]. The coefficient of friction after running in stage is 0.8 when the normal load is 185 N and applied displacement is 25 μm. More geometry and boundary condition details can be found in our previous work [18].

Fig. 2 Details of FE model

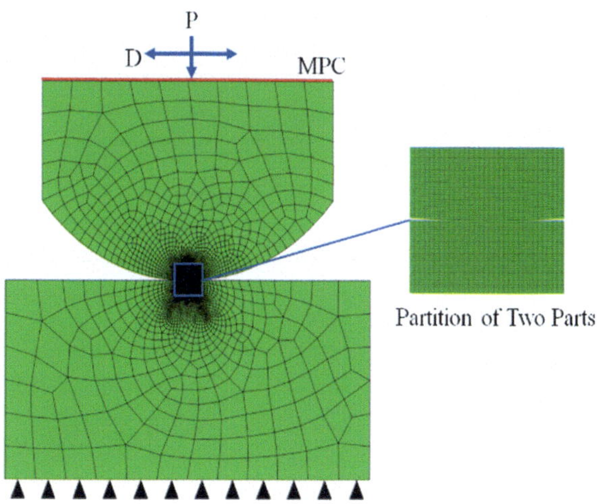

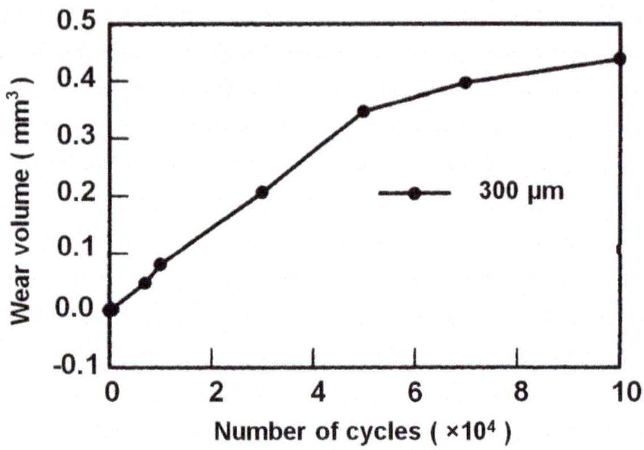

Fig. 3 Variation of wear coefficient in experiments

The variation of wear coefficient with cycles taken from experiments is shown in Fig. 3 [19]. The material used in the experiments was also steel and the setup was the same as that in [16]. The slope of the line in the figure is proportional to the wear coefficient based on Eq. (1). From Fig. 3, we can see that the slope of the spline is decreasing with cycles, which means that the wear coefficient decreases with the increasing of cycles. However, in most current FE models, the wear coefficient is regarded as a constant, which can be obtained from Eq. (2).

To make the simulations closer to the practical condition, bi-linear decreasing wear coefficient model is employed to the FE model, which is shown in Fig. 4. In most FE simulations, averaged wear coefficient model was applied, which is the yellow line in the figure. Based on experimental coefficient of friction variation with cycles under 185 N in [16], it can be concluded that the coefficient of friction tend to be a constant after 3000 cycle. Then the first 3000 cycles is regarded as the running in stage. We assume that in the first 3000 cycles, the rate of increase in the wear coefficient is high and after that, it decreases. But the wear volume in bi-linear decreasing model after 18,000 cycles is the same with that in averaged model, based on Eq. (1). The FE results of these two models are shown in next section.

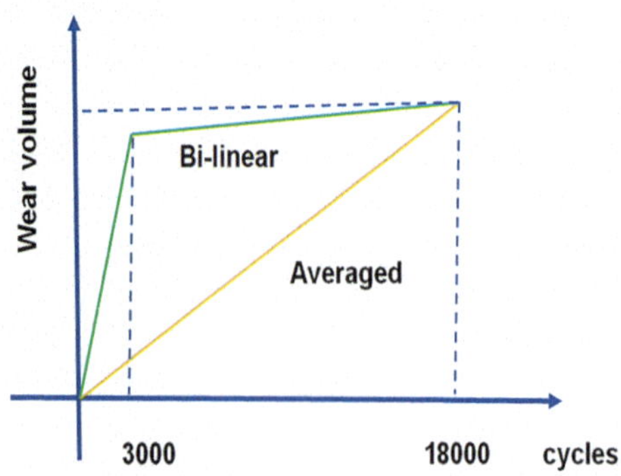

Fig. 4 Bi-linear model and averaged model

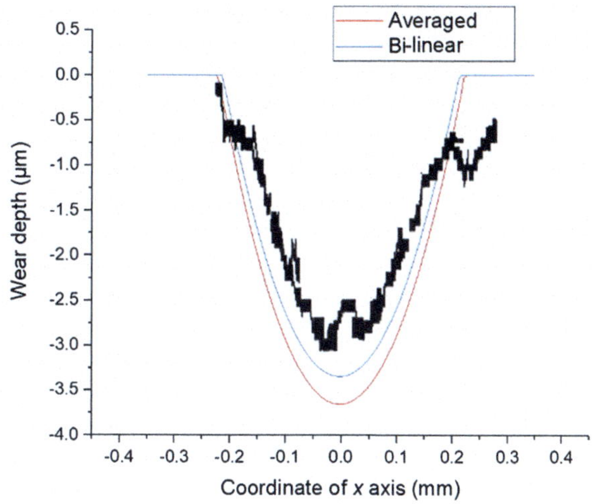

Fig. 5 Comparison of the simulation results with experimental data

3 Results and Discussion

Wear profiles after 18,000 cycles is shown in Fig. 5. The results of the averaged model and bi-linear decreasing model are compared with the experimental result [16]. It can be concluded that considering the decreasing of the wear coefficient in the FE model, the wear profile shows a better agreement with the experimental results.

The difference is mainly caused by the variation of the contact pressure in fretting process. In Archard's equation, all the parameters did not change with cycles. However, with the wear profile updating, contact pressure and shear stress are always changing, which is analysed in a lot of researches [20].

The contact pressure after 1000 cycles in both averaged model and bi-linear model are shown in Fig. 6. From the figure, we can see that contact pressure on the surface has a great difference when the wear coefficient is different for the first 1000 cycles. With lower wear coefficient, the maximum contact pressure is greater in averaged model than that in double-linear decreasing model, which causes the difference in the wear profiles after 18,000 cycles (Fig. 5).

4 Conclusions and Future Work

In this paper, the effect of variable wear coefficient on wear characteristics is analysed. The results shows that in bi-linear decreasing model the wear profile shows a better agreement with the experimental wear scar compared with that commonly used averaged wear model. The bi-linear decreasing model is designed based on the experimental variation of wear coefficient. Contact pressure on the surface also shows difference between two models. The peak value of contact pressure in case of

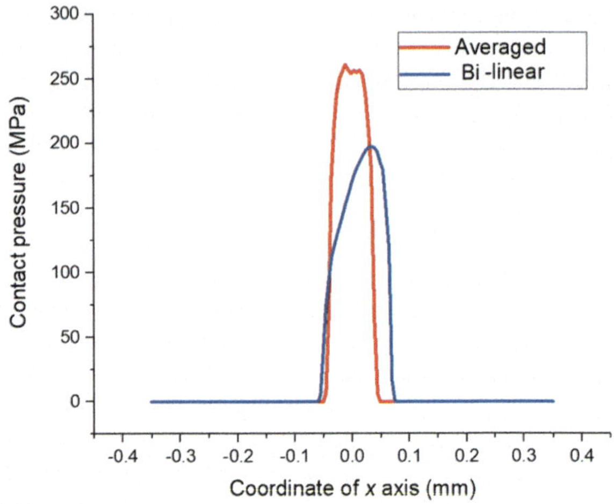

Fig. 6 Contact pressure difference after 1000 cycles

low wear coefficient model is greater after 1000 cycles, which means that the wear coefficient has a significant effect on the contact pressure.

In the next step, more loading cases and variable wear coefficient models can be further discussed to make the effect of the wear coefficient variation clearer. Moreover, experiments are needed to investigate the real wear coefficient variation in and after running in stage.

References

1. Li J, Lu Y (2013) Effects of displacement amplitude on fretting wear behaviors and mechanism of Inconel 600 alloy. Wear 304:223–230
2. Vingsbo O, Söderberg S (1988) On fretting maps. Wear 126:131–147
3. Yue T, Wahab MA (2017) Finite element analysis of fretting wear under variable coefficient of friction and different contact regimes. Tribol Int 107:274–282
4. Hurricks P (1970) The mechanism of fretting—a review. Wear 15:389–409
5. Wang DG, Li XW, Wang XR et al (2016) Dynamic wear evolution and crack propagation behaviors of steel wires during fretting-fatigue. Tribol Int 101:348–355
6. Wang XR, Wang DG, Li XW et al (2018) Comparative analyses of torsional fretting, longitudinal fretting and combined longitudinal and torsional fretting behaviors of steel wires. Eng Fail Anal 85:116–125
7. Urchegui MA, Hartelt M, Klaffke D et al (2007) Laboratory fretting tests with thin wire specimens. Tribotest 13:67–81
8. Cruzado A, Leen SB, Urchegui MA et al (2013) Finite element simulation of fretting wear and fatigue in thin steel wires. Int J Fatigue 55:7–21
9. Cruzado A, Leen SB, Urchegui MA et al (2014) Finite element modeling of fretting wear scars in the thin steel wires: application in crossed cylinder arrangements. Wear 318:98–105
10. Ding J, Mccoll IR, Leen SB et al (2007) A finite element based approach to simulating the effects of debris on fretting wear. Wear 263:481–491
11. Fallahnezhad K, Oskouei RH, Badnava H et al (2017) An adaptive finite element simulation of fretting wear damage at the head-neck taper junction of total hip replacement: The role of taper angle mismatch. J Mech Behav Biomed Mater 75:58–67

12. Bedolla PO, Vorlaufer G, Rechberger C et al (2018) Combined experimental and numerical simulation of abrasive wear and its application to a tillage machine component. Tribol Int 127:122–128
13. Zeng DF, Zhang YB, Lu LT et al (2019) Fretting wear and fatigue in press-fitted railway axle: a simulation study of the influence of stress relief groove. Int J Fatigue 118:225–236
14. Mohd Tobi AL, Sun W, Shipway PH (2017) Investigation on the plasticity accumulation of Ti–6Al–4V fretting wear by decoupling the effects of wear and surface profile in finite element modelling. Tribol Int 113:448–459
15. Anders S, Sören A (2009) Simulation of wear and contact pressure distribution at the pad-to-rotor interface in a disc brake using general purpose finite element analysis software. Wear 267:2243–2251
16. McColl I, Ding J, Leen S (2004) Finite element simulation and experimental validation of fretting wear. Wear 256:1114–1127
17. Abaqus V (2014) 6.14 Documentation. Dassault Syst Simulia Corporation 651:6.2
18. Wang S, Yue T, Abdel Wahab M (2020) Multiscale analysis of the effect of debris on fretting wear process using a semi-concurrent method. Comput Mater Continua 17–35
19. Chen G, Zhou Z (2001) Study on transition between fretting and reciprocating sliding wear. Wear 250:665–672
20. Paulin C, Fouvry S, Meunier C (2008) Finite element modelling of fretting wear surface evolution: Application to a Ti–6A1–4V contact. Wear 264:26–36

The Strength of Rigid and Flexible Adhesive Joints at Room Temperature and After Thermal Shocks

Anna Rudawska, Magd Abdel Wahab, Jakub Szabelski, Izabela Miturska, and Elżbieta Doluk

Abstract The aim of the article was to determine the strength of the adhesive joints of the carbon steel and aluminium alloy. Two types of two-component epoxy adhesives were used in the tests: rigid and flexible. Rigid epoxy adhesive contains epoxy resin based on Bisphenol A and triethylenetetramine (TETA) curing agent, in a ratio of 100 g resin to 10 g curing agent. Flexible epoxy adhesive contains epoxy resin based on Bisphenol A and polimaminoamide curing agent, in a ratio of 100 g resin to 100 g curing agent. The tested adhesive joints were subjected to conditioning at room temperature (23 °C) and humidity 35% (the first version) and subjected to thermal shocks (500 cycles: + 60 °C/−40 °C)—the second version. Strength tests of different variants of adhesive joints were performed on the Zwick/Roell Z150 testing machine in compliance with the DIN EN 1465 standard. Shear strength and elongation at break were determination. The obtained results have shown that in some cases, depending on the type of adherend, the adhesive joints exhibit greater elongation after a specified time of thermal shocks compared to non-thermal adhesive joints.

A. Rudawska (✉) · J. Szabelski · I. Miturska · E. Doluk
Faculty of Mechanical Engineering, Lublin University of Technology, Nadbystrzycka 36 Str., 20-618 Lublin, Poland
e-mail: a.rudawska@pollub.pl

J. Szabelski
e-mail: j.szabelski@pollub.pl

I. Miturska
e-mail: i.miturska@pollub.pl

E. Doluk
e-mail: e.doluk@pollub.pl

M. Abdel Wahab
Duy Tan University, Institute of Research and Development, 03 Quang Trung, Da Nang, Viet Nam
e-mail: magd.magdabdelwahab@duytan.edu.vn; magd.abdelwahab@ugent.be

Faculty of Engineering and Architecture, Technologiepark Zwijnaarde, Ghent University, Soete Laboratory, 903, B-9052 Zwijnaarde, Belgium

M. Abdel Wahab (ed.), *Proceedings of 1st International Conference on Structural Damage Modelling and Assessment*, Lecture Notes in Civil Engineering 110,
https://doi.org/10.1007/978-981-15-9121-1_18

Keywords Epoxy adhesive · Strength · Adhesive joints · Temperature · Thermal shocks

1 Introduction

Adhesive joints operate in a range of various environmental conditions [1–5]. An adhesive that bonds structural elements, like any polymer, is subject to ageing [6–10]. As a result, polymer degradation occurs, whose mechanism or speed is determined by the basic components (the type of polymers) of the adhesive [11–15].

Degradation is a process of structural changes causing deterioration of the original functional properties of the material that are known to stem from physical or chemical transformations in the polymer under long-term exposure to external factors, such as: heat, oxygen, ozone, visible light radiation, high-energy radiation, UV radiation and other chemicals including water and steam, as well as mechanical stress, particularly under cyclic dynamic loading (contributing to fatigue) [1, 12, 14, 16–20].

Thermosetting resins (reinforced or non-reinforced) exhibit substantially lower thermal sensitivity, i.e. the temperature-related loss of strength is less drastic compared to thermoplastics. Their characteristic properties, resulting from a strongly crosslinked structure that is unattainable in thermoplastic material, earmark thermosets as a reinforcement of glass, carbon, graphite, boron or aramid composites, among others. Although the resins exhibit greater resistance to adhesive failure, they are simultaneously more brittle and susceptible to fatigue, which becomes an important factor in applications that would involve contact loads [14, 16, 20–23].

Thermal degradation involves an array of concurrent reactions, whose relative relevance depends on the structure of resin and the type of curing agent. Strong intermolecular attractions in the polymer chain contribute to its rigidity and increase the resistance of polymers to high temperature [21].

Interestingly, in certain conditions, the mechanical strength and other material properties can improve in the initial stage of degradation. This occurs as a consequence of (for instance) heat-induced post-crosslinking of the polymer structure. It is only at later stages that other processes begin to exert a detrimental effect on the working properties of adhesives, e.g. excessive crosslinking or molecular weight reduction [22, 23].

This paper concerns with structural adhesive joints bonded with two-component structural epoxies that were subjected to thermal shocks.

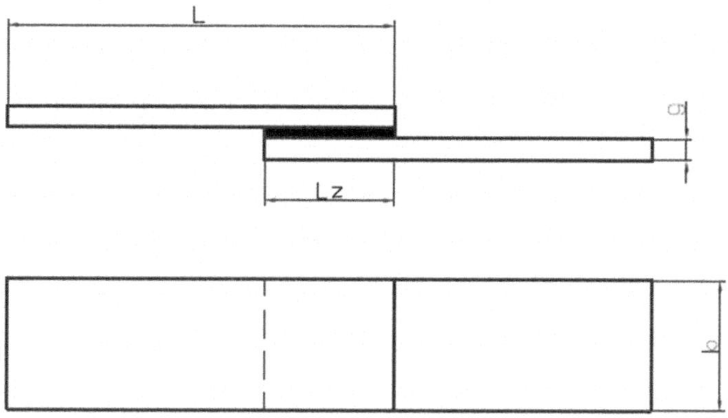

Fig. 1 The tested adhesive joints

2 Methodology

2.1 Adhesive Tapes

The single-lap adhesive joint specimens that were investigated in strength tests are shown in Fig. 1.

The adhesive joints were of the following dimensions: length L = 100 mm, overlap length lz = 20 mm, width b = 20 mm, adherend thickness g = 1.0 mm, adhesive layer thickness: g = 0.15 ± 0.02 mm.

2.2 Adhesive

Two two-component adhesive compositions were tested. The adhesive compositions used to produce the bond were obtained by combining the epoxy resin based on Bisphenol A, with a curing agent, which was either an amine (rigid adhesive—composition 1) or an amide curing agent (elastic adhesive—composition 2).

Composition 1 was an epoxy resin based on Bisphenol A (Epidian 57®, manufactured by Sarzyna Resins, Nowa Sarzyna, Poland) and triethylenetetramine (TETA) curing agent (Z-1®, manufactured by Sarzyna Resins, Nowa Sarzyna, Poland) in a stoichiometric ratio of 100:10 (designation: Epidian 53/Z-1/10:1). The description of epoxy resin and curing agent was presented in the [19, 20, 24, 25]. Composition 2 was a mixture of Epidian 57 epoxy resin and polymainoamide curing agent (PAC trade name, manufactured by Sarzyna Resins, Nowa Sarzyna, Poland) in a stoichiometric ratio of 100:100 (designation: Epidian 57/PAC:1:1).

Table 1 Properties of the analysed adhesive composition

Properties	Adhesive composition
	Epidian 57/Z-1/10:1
Bending strength, MPa	70–80
Compressive strength, MPa	65–75
Heat distortion temperature, °C	60–65

Some properties of the epoxy resin with triethylenetetramine (TETA) curing agent used in tests are given in Table 1. The other properties of epoxy resin and curing agents are included in [17, 19, 20, 24, 25].

2.3 Adherends

The adherends bonded using the tested adhesive tapes were:

1. C45 steel sheets, thickness g = 1 mm,
2. EN-AW 5754 aluminium alloy sheets, g = 1 mm,

Selected properties of the adherends are detailed in a previous work [26].

2.4 Adhesive Bonding Technology

Prior to adhesive joining, the surfaces to be bonded were subjected to degreasing using acetone, which was wiped onto the specimens in 3 repetitions and after the first two applications removed with dust-free swabs. Following the third deposition of acetone, the specimens were left for approx. 2 min so that the degreasing agent could evaporate. The temperature during the pre-treatment was 26±1 °C and humidity 40±1%.

Once the surfaces of adherends were prepared, the bonding process was carried out. The steel sheet and aluminium alloy sheet adherends were bonded with two two-component epoxy adhesives.

The compositions were prepared as follows. First, the specified proportions of resin and curing agent were weighed using a TP-2/1 balance (FAWAG SA Lublin, Poland) at a 0.1 g accuracy. Next, the two components of the adhesive were mechanically blended together using an anchor stirrer at the mixing stand in a polymer vessel. Mixing was carried out for 2 min at a speed of 480 rev./min. The operation was performed with due care in order to prevent trapping the air in cavities that would have formed in the case of imprecise and/or excessively dynamic mixing. Therefore, upon mixing, the composition was deaerated for 2 min under pressurized air. Finally, the adhesive was manually deposited onto one of the previously prepared surfaces and the adherends were set in place.

Following their assembly, the joints cured for 7 days in a procedure that was executed at a temperature of 23±1 °C and 35±1% and the pressure applied during the setting was equal to 0.20 MPa. In the next stage, the specimens were conditioned at a temperature of 23±1 °C and relative humidity of 28±1% for 24 h, which was followed by the visual inspection of dimensional and shape accuracy. Finally, the bonded assemblies were divided into two test groups.

The first group of joints (G1) was subjected to strength tests once the bond was fully cured, whereas the specimens from the second group (G2), prior to testing, were post-conditioned in a series of 500 thermal shock cycles. The thermal shock procedure was conducted in an STE 11 thermal shock chamber (a product of ESPEC, Klimatest, Poland). 10 adhesive joint specimens were prepared in each variant, *i.e.* 80 adhesively bonded joints in total: 2 adherend materials × 2 adhesive compositions × 2 test groups × 10 joint samples). The group variants are presented in more detail in Tables 2 and 3.

The strength testing conditions were as follows:

- Test temperature—23 ± 1 °C,
- Test speed—3 mm/min,
- DIN EN 1465 standard,
- Testing machine—Zwick/Roell Z150.

The strength properties of bonded specimens were determined in 80 tests. The presented results of individual quantities are mean values calculated from 7–10 results, after rejecting the outliers. The presented data account for the standard deviation, as a classic measure of variability of analysed quantities from the average.

Table 2 Description of test groups

Conditions	Test group variant	
	G1	G2
Curing	Temperature: 23±1 °C Relative air humidity: 35±1% Time: 7 days Load: 0.20 MPa	
Post-cure conditioning	Temperature: 23±1 °C Relative humidity: 28±1% Time: 24 h	
Thermal shock treatment	–	500 cycles 1 cycle: + 60 °C/15 min and −40 °C/15 min

Table 3 Test group variants designation

Adherend (sheets)	Adhesive tape	Test group	Designation of adhesive joints
Steel (C45)	Epidian 57/Z-1/10:1	G1	S/Epidian 57/Z-1/10:1/G1
		G2	S/Epidian 57/Z-1/10:1/G2
	Epidian 57/PAC/1:1	G1	S/Epidian 57/PAC/1:1/G1
		G2	S/Epidian 57/PAC/1:1/G2
Aluminium alloy (EN-AW 5754)	Epidian 57/Z-1/10:1	G1	A/Epidian 57/Z-1/10:1/G1
		G2	A/Epidian 57/Z-1/10:1/G2
	Epidian 57/PAC/1:1	G1	A/Epidian 57/PAC/1:1/G1
		G2	A/Epidian 57/PAC/1:1/G2

3 Test Results

3.1 Results of Steel Sheet Strength Tests

Figure 2 presents the comparison of shear strength performance of adhesive steel sheet joints that were and were not subjected to thermal shocks, and Fig. 3 summarises the results for elongation.

The conducted steel sheet adhesive joint strength tests show that a more flexible epoxy adhesive composition, Epidian 57/PAC/1:1, performs better than a more rigid

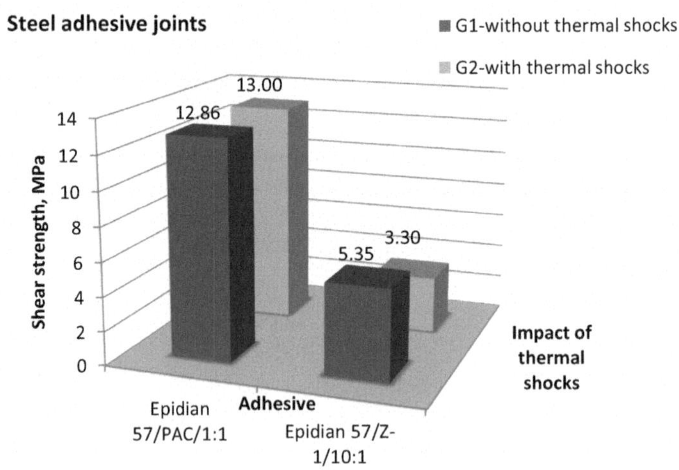

Fig. 2 The shear strength of C45 steel sheet adhesive joints subjected and not subjected to thermal shock testing

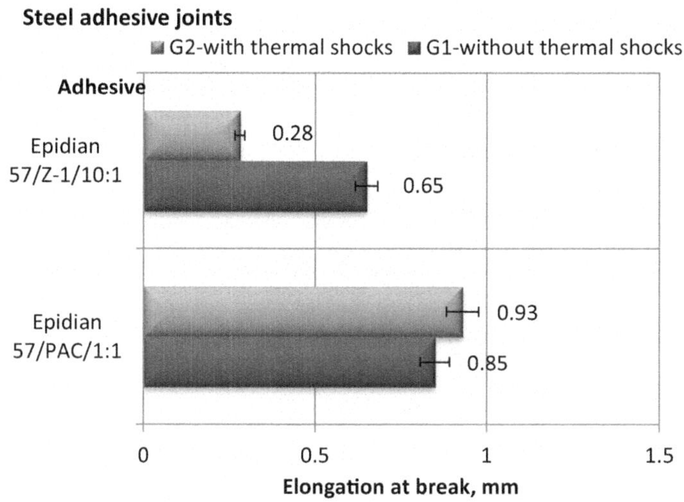

Fig. 3 The elongation at break of C45 steel sheet adhesive joints subjected and not subjected to thermal shock testing

Epidian 57/Z-1/10:1. The observation applies both to the assemblies that were post-treated and those that did not undergo thermal shocks. The difference in the strength characteristics of the joint variants tested in the study is approx. 60%; however, in the case of post-treated joints, where it amounted to 75%, the discrepancy was even more evident. Considering the effect of post-treatment on the strength of steel sheet adhesive joints bonded with Epidian 57/PAC/1:1 resin composition, it was found that it remained virtually unaffected by the 500 thermal shock cycles. However, Epidian 57/Z-1/10:1 adhesive was shown to respond negatively to the treatment—the reduction in the strength of these joints was about 40%. Similar relationships were observed in elongation of adhesive joints in the shear test: following thermal shock treatment, the flexible bond (Epidian 57/PAC/1:1) provided an increase in elongation of 9%, while the elongation of rigid epoxy composition (Epidian 57/Z-1/10:1) exhibited a substantial decrease of 57%.

The results from shear strength tests of adhesively bonded aluminium alloy sheet assemblies that were and were not subjected to thermal shocks are summarised in Fig. 4, whereas Fig. 5 reports the results from the elongation tests of the joint variants tested in the experiments.

The strength test results of aluminium adherends correlate with the steel sheet joint strength experimental data, and this concerns both untreated joints and those subjected to thermal shocks (Fig. 4). The assemblies bonded with a flexible adhesive composition, Epidian 57/PAC/1:1, have greater strength than with a more rigid adhesive, Epidian 57/Z-1/10:1. The difference in the performance of the tested epoxy adhesive compositions is 28%. However, it is even greater in the case of joints post-treated with thermal shocks, amounting to 66%.

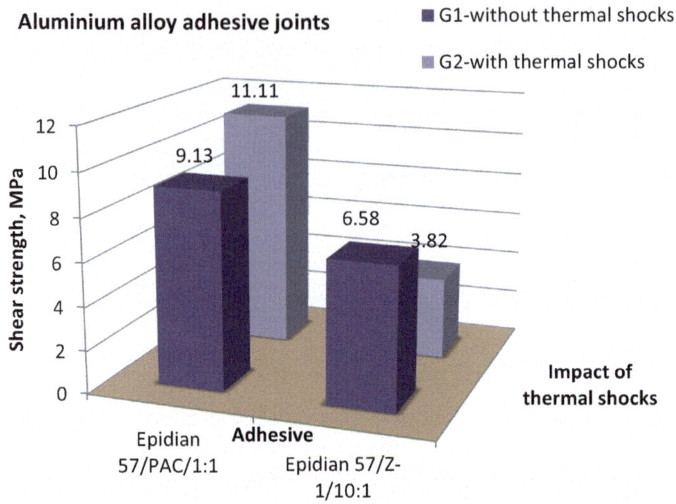

Fig. 4 The shear strength of aluminium alloy sheet adhesive joints subjected and not subjected to thermal shock testing

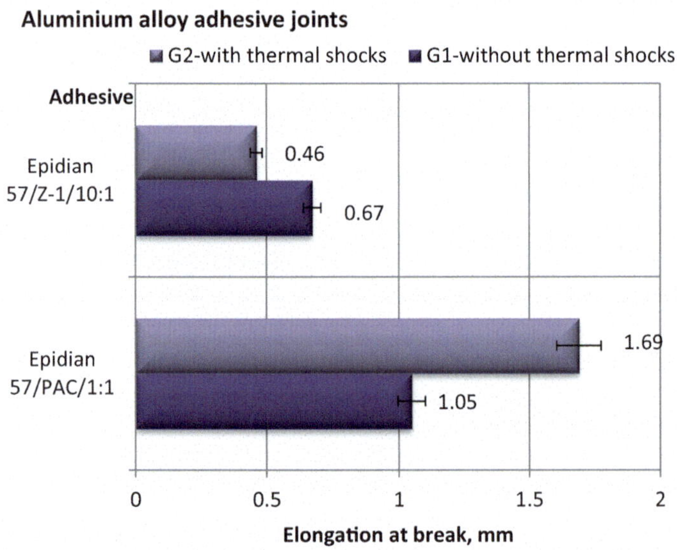

Fig. 5 The elongation of aluminium alloy sheet adhesive joints subjected and not subjected to thermal shock testing

Considering aluminium alloy sheet adherends, joints bonded using Epidian 57/PAC/1:1 adhesive and subjected to 500 thermal shock cycles (+60 °C/−40 °C) increased their strength by approx. 18%. However, the strength of the aluminium adhesive joints bonded with a rigid adhesive (Epidian 57/Z-1/10:1) and subjected to thermal shocks is in excess 40% lower.

Similar relationships were noted when analysing the elongation of shear-loaded adhesive joints. The flexible composition, Epidian 57/PAC/1:1, was found to promote elongation after thermal shock treatment: the increase in elongation was equal to 38%. The elongation of the bond decreased by about 30% in the joints made with a rigid epoxy (Epidian 57/Z-1/10:1).

3.2 *Comparison and Discussion of Results: Structural and Assembly Adhesive Joints*

This section collates the results obtained from the shear strength and the elongation tests of all the types of adherend and adhesive material investigated in the reported study. Figures 6 and 7 present the strength characteristics of the assembly (adhesive tape) [26] and structural (two-component epoxy adhesive) joints of steel and aluminium alloy sheets that were and were not subjected to thermal shock post-conditioning. The elongation data of the joints are resented in Figs. 8 and 9.

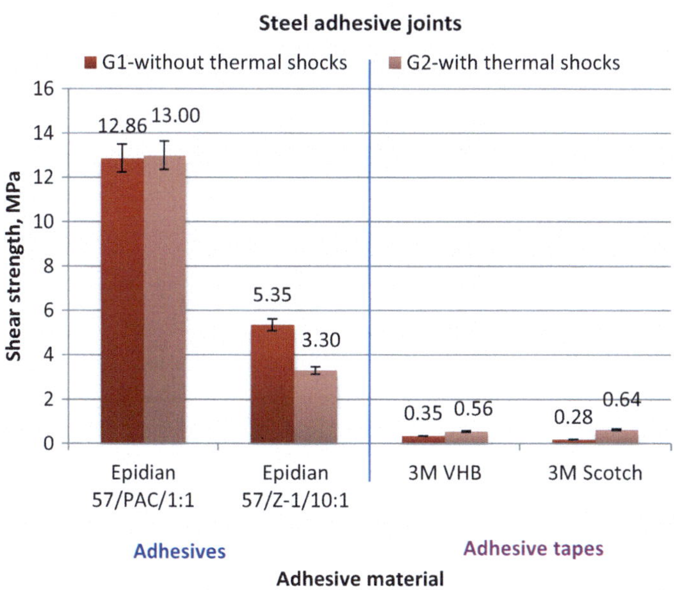

Fig. 6 The shear strength of C45 steel sheet joints subjected and not subjected to thermal shock testing

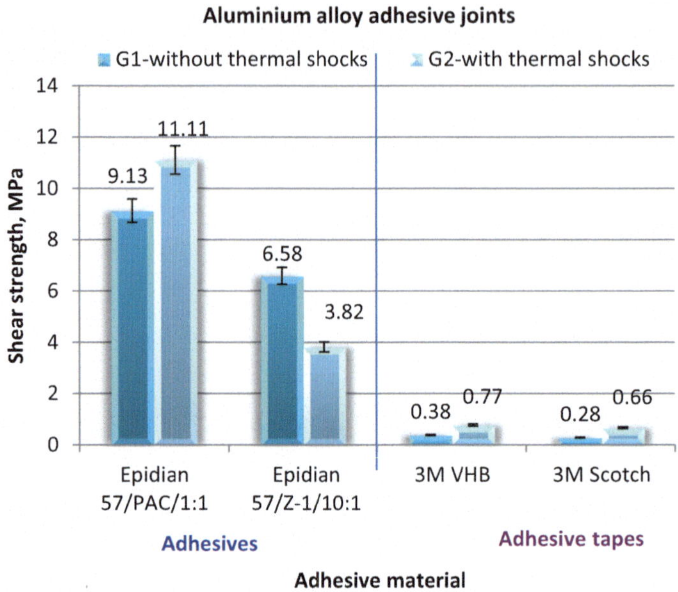

Fig. 7 The shear strength of aluminium alloy sheet adhesive joints subjected and not subjected to thermal shock testing

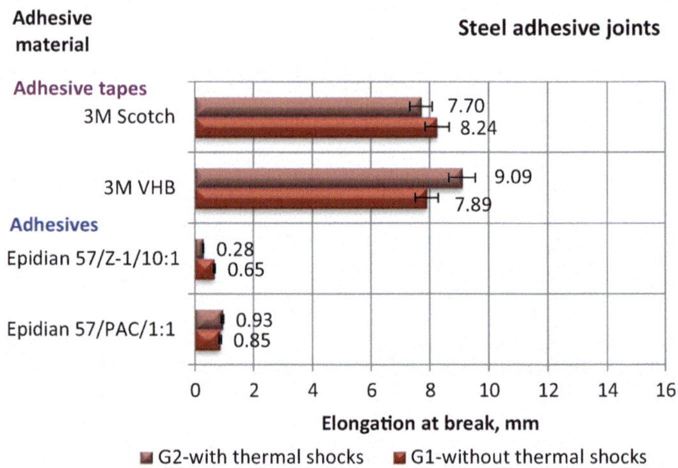

Fig. 8 The elongation of C45 steel sheet adhesive joints subjected and not subjected to thermal shock testing

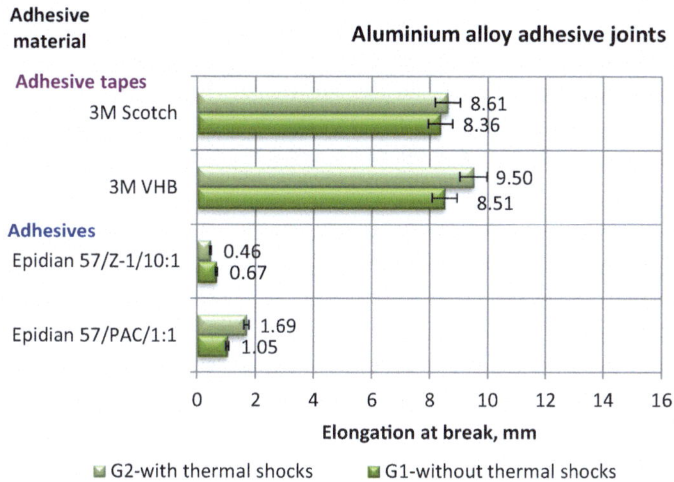

Fig. 9 The elongation of 5754 aluminium alloy sheet adhesive joints subjected and not subjected to thermal shock testing

The comparison of test results shown in Fig. 6 reveals that the strength of assembly joints bonded with adhesive tapes amounts to several per cent of the strength of adhesive joints formed using structural adhesives. It was, furthermore, shown that assembly joints responded positively to the varying operating conditions, simulated by means of the thermal shock post-treatment: their strength performance was considerably increased. What emerges from these observations is that assembly joints show good capacity for operating in the conditions involving thermal shocks; however, this conclusion is only valid for the applied testing conditions. In the case of structural steel joints, which were prepared with either flexible or rigid adhesives, no deterioration of the strength properties was observed after thermal shock treatment for the flexible bond variant, although in the rigid adhesive composition, there was a substantial drop in the strength, which amounted to approx. 40%.

A similar correlation emerged from the test results of aluminium alloy sheet joints—bonded using adhesive tapes (assembly joints) and epoxy structural adhesives (structural joints)—that were subjected and not subjected to thermal shock conditioning.

Although assembly adhesive joints had been expected to exhibit lower strength than structural adhesive joints, for practical reasons, it does seem important to account for the effect of thermal shocks on the performance of joints.

Comparatively analysed, the results from the strength tests of all adhesive joint and adherend variants (Figs. 8 and 9) indicate that the assembly joints (bonded with assembly tapes) develop a considerably higher elongation capacity (even twenty times). Even though their strength characteristics are largely equivalent to the structural joints', the tape-bonded assemblies (Figs. 6 and 7) ensure higher flexibility, yet it depends on the particular model of the assembly tape. They were, furthermore, found not to lose the flexibility under the effect of thermal shocks, which is an important implication regarding their future operating conditions and application.

4 Summary

The aim of the present research was to examine the effect of thermal shocks on bonded assemblies. The experimental data indicate that thermal shocks can positively or negatively affect the strength performance of structural adhesive joints, and that the decisive factor is the flexibility of the adhesive mix. The strength of a flexible adhesive was found not to deteriorate after thermal shocking or even to increase slightly by several per cent (although this is relative to the type of adherend). It can be, thus, assumed that the strength properties of adhesively bonded assemblies of steel sheets and aluminium alloy sheets exposed to thermal shocks are unlikely to deteriorate in the joints' lifetime. This slight improvement in performance may be associated with post-crosslinking of the adhesive resin, primarily under the impact of positive temperature. It could conceivably be hypothesised that exposure to elevated or high temperatures will exert a more pronounced effect on the increase in the strength of adhesive joints.

Concerning low-flexibility adhesives, it was found that the strength of the bond joining the tested adherends dropped as a result of thermal shocks. Perhaps it might be attributed to the use of a specific curing agent whose mechanical properties deteriorate under the detrimental effect of thermal conditioning (in positive and negative temperatures). The link between the strength of adhesive joints and thermal shocks that emerges from the results of the investigations presented in this work requires further scientific attention and will, thus, be the subject of future analyses and studies.

References

1. Brewis DM, Comyn J, Shalash RJA (1982) The effect of moisture and temperature on the properties of an epoxide-polyamide adhesive in relation to its performance in single lap joints. Int J Adhe Adhe 4:215–222
2. Hu P, Han X, Li WD, Li L, Shao Q (2013) Research on the static strength performance of adhesive single lap joints subjected to extreme temperature environment for automotive industry. Int J Adhes Adhes 41:119–126
3. Heshmati M, Haghani R, Al-Emrani M (2017) Durability of CFRP/steel joints under cyclic wet-dry and freeze-thaw conditions. Composite B 126:211–226
4. Cavalli A, Malavolti M, Morosini A, Salvini A (2014) Mechanical performance of full scale steel-timber epoxy joints after exposure to extreme environmental conditions. Int J Adhes Adhes 54:86–92

5. Rudawska A, Wahab MA, Müller M (2020) Effect of ageing process on mechanical properties of adhesive tubular butt joints in aqueous environment. Int J Adhes Adhes 96:1–11
6. Adams RD (2010) Adhesive bonding. science, technology and applications. Woodhead Publishing, London
7. Jiang X, Qiang X, Kolstein MH, Bijlaard FSK (2015) Experimental investigation on mechanical behavior of FRP-to-steel adhesively-bonded joint under combined loading—Part 2: After hygrothermal ageing. Compos Struct 125:687–697
8. Blackburn BP, Tatar J, Douglas EP, Hamilton HR (2015) Effect of hydrothermal conditioning on epoxy adhesives used in FRP composites. Constr Build Mater 96:679–689
9. Ameli A, Datla NV, Azari S, Papini M, Spelt JK (2012) Prediction of environmental degradation of closed adhesive joints using data from open-faced specimens. Compos Struct 94:779–786
10. Rudawska A (2019) The impact of seasoning conditions on mechanical properties of modified and unmodified epoxy adhesive compounds. Polymers 11:804. https://doi.org/10.3390/polym11050804
11. May CA (1988) Epoxy resins, chemistry and technology, 2nd edn. Marcel Dekker, New York, USA
12. Okba SH, Nasr E-SA, Helmy AII, Yousef IA-l (2017) Effect of thermal exposure on the mechanical properties of polymer adhesives. Constr Build Mater 135:490–504
13. Kolar V, Tichy M, Muller M, Valasek P, Rudawska A (2019) Research on influence of cyclic degradation process on changes of structural adhesive bonds mechanical properties. Agron Res 17(S1):1062–1070
14. Pertie EM (2006) Epoxy adhesive formulation. McGraw-Hill, New York
15. Hartshorn SR (1986) Structural adhesives. Chemistry and technology. Plenum Press, New York, London
16. Crocombe AD, Ashcroft IA, Wahab MMA (2008) Environmental degradation, Chapter 8. In: da Silva LFM, Öchsner A (eds) Modeling of adhesively bonded joints. Springer-Verlag, Berlin Heidelberg, pp 225–242
17. Rudawska A, Sikora JW, Müller M, Valasek P (2020) The effect of environmental ageing at lower and sub-zero temperatures on the adhesive joint strength. Int J Adhes Adhes 97:102487
18. Popineau S, Shanahan MER (2006) Simple model to estimate adhesion of structural bonding during humid ageing. Int J Adhes Adhes 26:363–370
19. Rudawska A, Brunella V (2020) The effect of ageing in water solution containing iron sulphate on the mechanical properties of epoxy adhesives. Polymers 12:2018. https://doi.org/10.3390/polym12010218
20. Rudawska A (2020) The effect of the salt water aging on the mechanical properties of epoxy adhesives compounds. Polymers 12:843. https://doi.org/10.3390/polym12040843
21. Czub P, Bończa-Tomaszewski Z, Penczek P, Pielichowski J (2002) Chemistry and technology of epoxy resins. WNT, Warsaw (in polish)
22. Tatar J, Hamilton HR (2016) Comparison of laboratory and field environmental conditioning on FRP-concrete bond durability. Constr Build Mater 122:525–536
23. Heshmati M, Haghani R, Al-Emrani M (2016) Effects of moisture on the long-term performance of adhesively bonded FRP/steel joints used in bridges. Composites B 92:447–462
24. Rudawska A, Worzakowska M, Bociąga E, Olewnik-Kruszkowska E (2019) Investigation of selected properties of adhesive compositions based on epoxy resins. Int J Adhes Adhes 92:23–36
25. Rudawska A (2020) The influence of curing conditions on the strength of adhesive joints. J Adhes 96:402–422
26. Rudawska A, Wahab MA. Resistance of adhesive tapes in adhesive joints strength at room temperature and thermal shocks (in press)

Service Life of the Cam Mechanisms

Monika Gromadova

Abstract Rolling contact fatigue is very closely connected to the service life of the cam mechanisms. The most common manifestation is pitting. This defect occurs on the contact surface and seems like a pit. This type of crack can lead to bigger damage of the functional parts of the mechanism. We can use two methods for service life estimation. The first method is the theoretical estimation. Its advantage is that it is quick and relatively easy. One disadvantage is that it can be very inaccurate and can be dependent on the input parameters accuracy. In addition, different theories can give different results. The second possibility is to estimate the service life by an experimental method. This method is based on the loading of the contact surfaces and observing the crack formations. We operate it on a special test rig. After this, we can find dependence between the loading force and the number of cycles leading to the surface damage. This information can be completed with data describing properties of the tested material e.g. hardness, purity, hardening parameters etc. We can estimate the service life of the real cam mechanism on basis of all this obtained information. This paper deals primarily with the surface damage and its influence on the service life of the cam mechanisms. Attention is paid to both methods of service life estimation. The test rig developed in our company is described here.

Keywords Service life · Theoretical method · Experimental evaluation · Pitting · Cracks

1 Rolling Contact Fatigue

The service life of the cam mechanism is closely connected to rolling contact fatigue (RCF). This type of surface damage occurs in functional surfaces that are exposed to repeated loading (high local pressure) during the movement of the functional surfaces of the machine components. It can be a relative rolling movement or a combination of rolling and sliding. This case is more common in practice [1]. The

---

M. Gromadova (✉)
VÚTS, a.s., Svarovska 619, Liberec, Czech Republic
e-mail: monika.gromadova@vuts.cz

M. Abdel Wahab (ed.), *Proceedings of 1st International Conference on Structural Damage Modelling and Assessment*, Lecture Notes in Civil Engineering 110,
https://doi.org/10.1007/978-981-15-9121-1_19

gradual accumulation of the cracks in the surface layer at the repeated contact stress is characteristic for RCF. This is typical loading type at cam mechanisms. RCF arises as a consequence of material imperfection (material has various limitations and imperfect structure) or working conditions e.g. lubrication, temperature). More information is in [2].

1.1 Types of the Rolling Contact Fatigue

There are more types of RCF. The most common (at cam mechanisms) is pitting (Fig. 1). We can observe micro- and macropitting by the size of the pits on the damaged surface [1].

Micropitting [3] is manifested by the presence of microscopic pits on the material surface in the contact zone. They are produced due to the repeated cyclic loading of contact at which occurs to rolling and sliding and they are formed by plastic deformations of the surface asperities. Micropitting can we more often observe at the higher surface hardness of the steel. The micropits are relatively small to the size of the contact zone, typically 10–20 μm deep and the size is usually smaller than 100 μm.

Macropitting [3] is presented by cracks formed at a certain depth under the surface, where the shear stress is maximal due to Hertzian pressure. It causes the

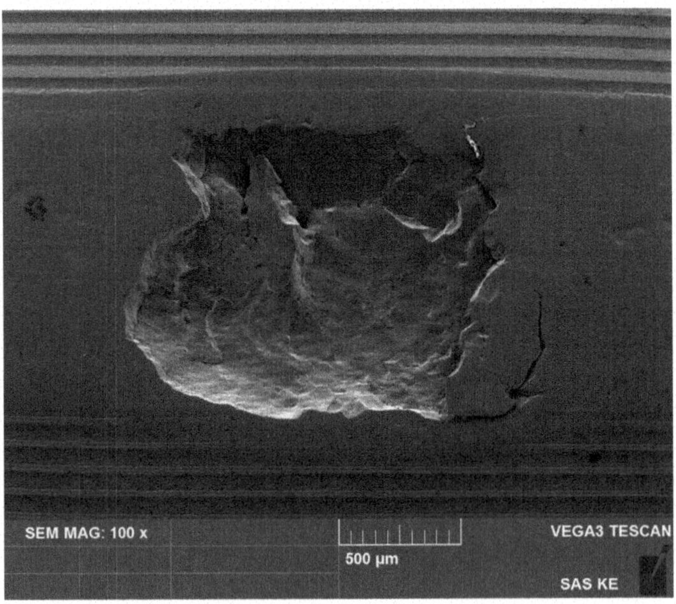

Fig. 1 Damaged surface (pitting)

formation of large pits (about 1 mm) on the material surface at the contact zone (see Fig. 1). These arise due to the spread of subsurface cracks what leads to flaking larger parts of the material from the surface.

We know other types of the damage mechanisms, e.g. galling, spalling, scoring, scuffing and their combinations. A description of all types of damage is in the literature, e.g. [4].

2 Methods of the Service Life Estimation

Basically, we can use two methods for service life estimation- theoretical and experimental methods, or their combination. Both of them have their own advantages and disadvantages. The next text will describe the principle of both approaches. More information is given in the literature, e.g. [4–6].

2.1 Theoretical Method

In most cases, the choice of the suitable material for cam production is based on the theoretical assessment. There is a huge amount of theories in the literature, e.g. [7]. We have chosen two theories for comparison. The first theory [5] deals with the service life estimation by means of the surface hardness. An advantage is that the calculation is simple. On the other side, the surface hardness does not describe all the mechanical properties of the examined material. The second theory [6] calculate with two material parameters. These parameters describe the mechanical properties of the material much better. A disadvantage is that the database of these parameters is limited. For materials outside the database is time consuming and very expensive to obtain the parameters [8].

The comparison of the both theories results shows that the service life determined by each theory can be very different [9]. An example of the service life estimation results is shown for the two above mentioned theories in Fig. 2. Therefore, we need to find a suitable experimental procedure, which can help verify the accuracy of the used theories. In addition, we need to understand all the processes, which can lead to surface damage.

2.2 Experimental Method

The experimental method lies in the testing of the material in required conditions. The results are evaluated in the next step. On their basis, we can estimate the service life of the contact surfaces.

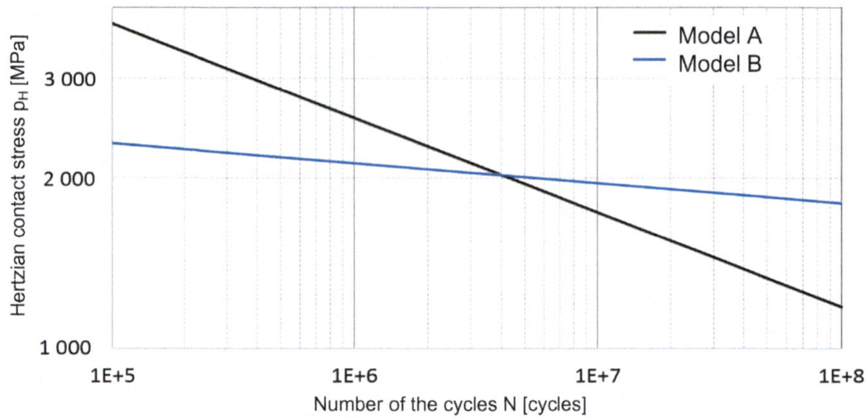

Fig. 2 S–N curves (steel 34CrNiMo6) for two theories [9]

A test rig developed in our company does testing of the rolling contact fatigue. Its kinematic diagram is shown in Fig. 3. A principle is that the specimen 5 is loaded between three discs 2, 3, 4. The loading force is exerted through a lever 6 and it is given by force N. All discs are rotating by angular velocity Ω_2 and specimen is rolling between them by angular velocity ω. Surface of the specimen is loaded in three places, so the test runs relatively quickly and the testing device is simpler than device for the testing of the real cam mechanisms.

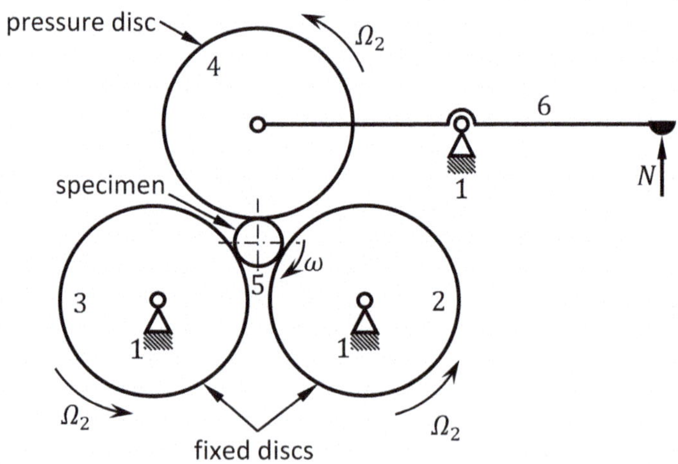

Fig. 3 Kinematic diagram of the test rig

2.3 *Testing Procedure [10]*

There are some pictures of damaged specimens in this paper. These examples allow us present the procedure of the RCF assessment.

The complete testing procedure can be divided into four phases. These are described below.

2.3.1 Phase 0: Theoretical Service Life

It is not necessary to do this phase, but it can help with the next phases. The aim of this phase is to find a dependence between the required service life and the loading force. According to this information, we can choose suitable material.

We can make a rough estimate of the surface service life based on the chosen theory. For more about this problem see in 2.1.

2.3.2 Phase 1: Testing Preparation

The first step is the preparation of the experiment. We need to select the suitable material. This choice is necessary to make according to the results from the calculations (FEM, analytic methods- see phase 0). Very important is to check the supplied material, its micropurity and chemical composition. In the next step, we have to determine the exact parameters of heat, respectively chemical heat treatment. The parameters of these technological processes are very important for required properties ensuring of the surface layers of the examined components. These are parameters of heating, carburizing etc.

We should take into account not only the material properties, but also the influence of the working conditions on the RCF forming. The most relevant are properties of the lubricant and its interaction with the contact surfaces. This interaction is important because it can influence on properties of the contact surface (chemical interaction, surface tension changes, etc.).

The shape of the contact surfaces plays a significant role too. The properties of the contact surfaces can have an effect on the rolling contact fatigue in two ways. The first one says that if there are cracks on the surface, it can lead to pitting. The second problem can be insufficient space for the oil film forming between the contact surfaces when they are too smooth.

In the next step, after verification of the relevant steel properties, the test specimens (Fig. 4) are produced.

Fig. 4 Specimen for rolling contact fatigue testing

2.3.3 Phase 2: Testing

The specimens can undergo the service life tests after determining the right test conditions, choosing the suitable material and production of specimens. More about these experimental tests in [4].

The test rig is shown in Figs. 5 and 6. The basis of the test is that the specimen (Fig. 4) is inserted between the three discs and is loaded by the force determined by the means of the theoretical estimation or from previous tests. This way of testing allows the shortening of the time testing to 1/3. This is a significant time saving because we need to test every specimen for 10^7 or 10^8 working cycles [10]. The tests are mostly run until detection a defect or until achieving the desired number of cycles, but there are more possibilities of the testing scheme.

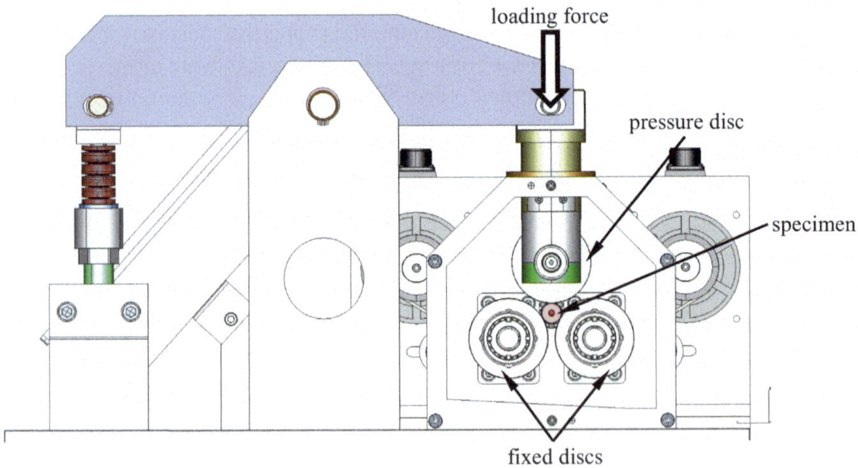

Fig. 5 Computer model of the test rig

Fig. 6 Real installation of the test rig

The test rig allows us to test individual cases, or combinations of them, as described below:

1. **The monitoring of changes in the material structure at the beginning of testing:** the surface of the test specimen has, after machining, its characteristic properties. This testing allows us to find out what happens in the surface layers in the first moments of contact. Theoretically, steel can be reinforced or softened (depending on the ratio yield strength Rm/tensile strength Re). In practice, this testing is performed by subjecting the specimen to a series of analyzes (destructive and/or non-destructive) after a given number of cycles.
2. **The determination of the cracking mechanism:** this test makes it possible to determine the mechanism of cracks forming under the surface. It takes several steps:

   a. the test specimen is subjected to pitting testing;
   b. the time we obtained in a. is divided in half; testing the specimen during this time, then subjecting it to analysis to determine the presence and nature of the cracks;
   c. if the cracks are present, re-divide the testing time in half and test another specimen, then analyze it;

d. repeat points b. and c. until the test layer without cracks is obtained;
The result of this test is a dependence of the time of the formation of cracks below the surface and their development until pitting. This information can be used, for example, to explain the mechanism of cracks forming, to analyze the effects on cracks forming, etc.

3. **Complete life curves**: this test allows obtaining a Wöhler diagram for different materials and this allows the prediction of the service life of the cam mechanism made of the given material.
All the tests described above can be performed in combination with the following parameters:

a. various surface roughness, textured surfaces
b. with sliding (various sliding values), non-sliding
c. lubricated surfaces (various types of lubricants), without lubrication.

2.3.4 Phase 3: Results Evaluating [8]

In the first step, the tested surface can be analysed by a stereomicroscope (or other suitable microscope) by low magnification after the service life testing. The task is to obtain information about surface damage from a macroscopic view.

There is the specimen CPM3V in Fig. 7, it is produced from steel CPM 3 V, loaded by a force of 6500 N. This value corresponds to Hertzian stress of 4217 MPa. Theoretical life was set on 0.08 mil. cycles (determined by the theory from [5]), in test reached number of 8.9 mil. cycles to damage.

There is the specimen 16MnCr5 in Fig. 8, it is produced from steel 16MnCr5 (carburised), loaded by force of 6000 N. This value corresponds to Hertzian stress of 2292 MPa. Theoretical life was set on 4.4 mil. cycles (determined by the theory from [5]), in test reached number of 61.7 mil. cycles to damage.

Fig. 7 Specimen CPM3V (damaged surface)

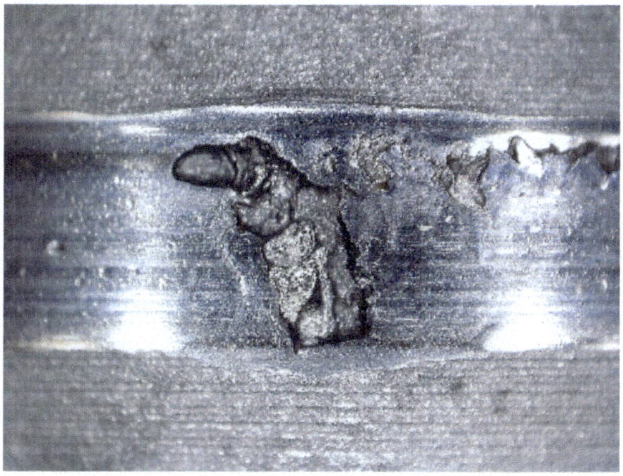

Fig. 8 Specimen 16MnCr5 (damaged surface)

When the specimens were looked through with the stereomicroscope, then can be analysed by the electron microscope. There is picture obtained on a Tescan electron microscope in Fig. 1 (shown pitting on specimen).

There are other possibilities of the specimen analysis on electron microscopes. These microscopes work with high magnification and this allows the detailed monitoring of the cracks. Some microscopes allows not only standard display, but also display in special (e.g. COMPO and TOPO) modes (see Figs. 9 and 10). Thanks to these modes, we are can get more information about the examined specimen. COMPO mode works on the basis of the differences in chemical composition and it

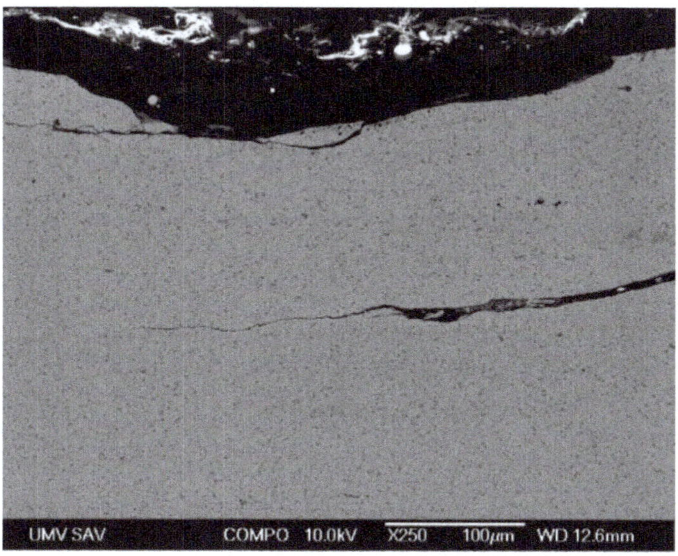

Fig. 9 Detail of the damaged surface (COMPO mode), specimen CPM3V

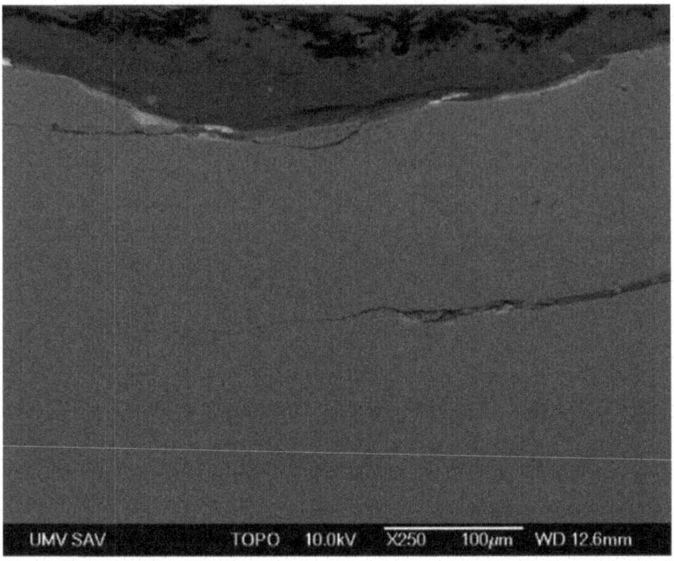

Fig. 10 Detail of the damaged surface (TOPO mode), specimen CPM3V

allows us to identify inhomogeneity. TOPO mode shows a topographic view at the examined surface.

The maximal depth of crack presence is marked in Fig. 11. From this picture, we obtained a value of 0.33 mm. For this case, the depth of 0.77 mm was for maximum shear stress. The value of the shear stress was determined by the analytical calculation according to [5]. This information says that the crack happened in depth smaller than the depth for maximum shear stress. It can be many reasons and it would need more research.

We can obtain information about the executed technological process from microscopic analyses (determining of uniformity and depth of the heat treatment) [8]. In

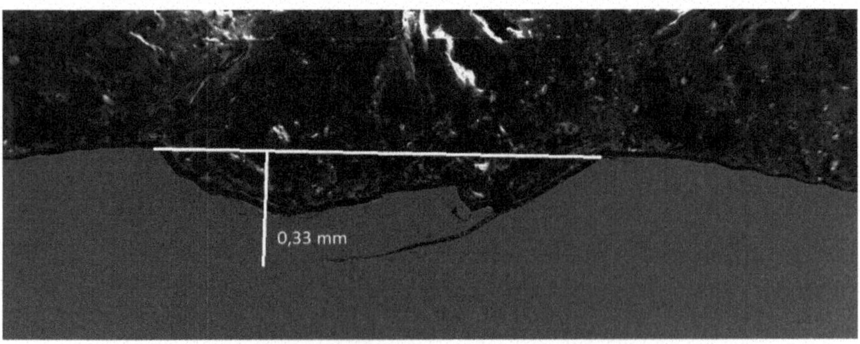

Fig. 11 Depth of damage, specimen CPM3V

Fig. 12 an image is shown with a marked depth of the carburised layer (obtained by the optical microscope).

In Figs. 13 and 14 a detail of the crack root is shown. We can see that the crack does not form on the grain boundaries in these images. There is not the increase presence of cavities, inclusions, impurities in the examined steel. We performed

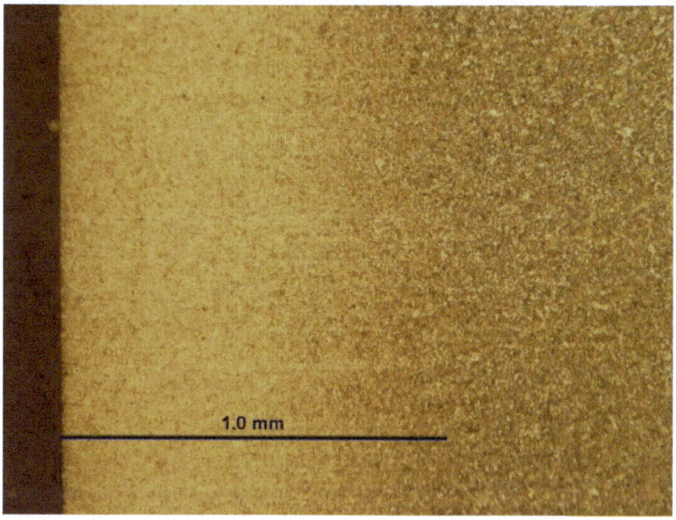

Fig. 12 Depth of the carburized layer (about 0.8 mm), specimen 16MnCr5

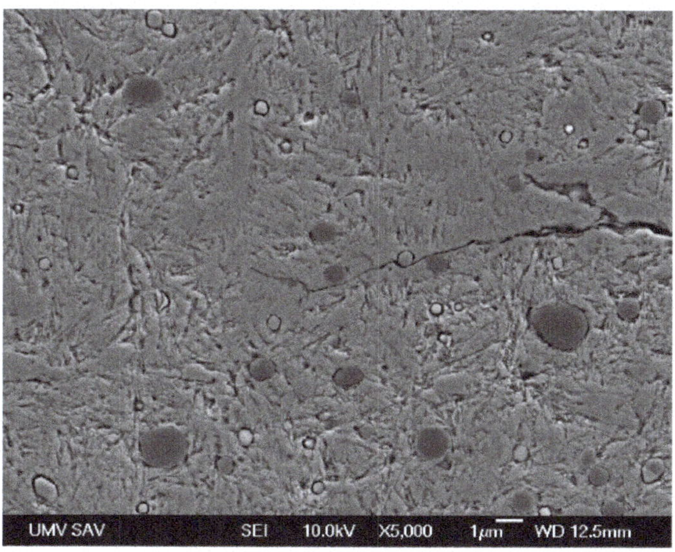

Fig. 13 Crack root, steel with globular carbides, specimen CPM3V

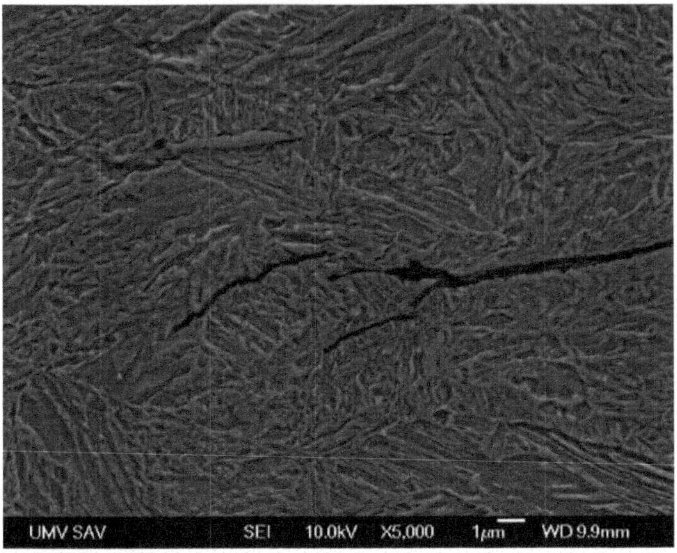

Fig. 14 Detail of the crack root, specimen 16mnCr5

their chemical analysis. The results is that the crack did not form from any cavity, impurity, or inclusion. The exhaustion of mechanical properties of the examined material leaded to formation of it.

The analyses give us information whether the structure is uniform, the micropurity of the examined material is convenient and allow us to determine the content of the inclusions and cavities. We can determine if the damage occurred in the steel and the cracks are formed on the basis of exhaustion of the mechanical properties of the examined material or on the basis of a non-uniform structure. It is useful to measure the microhardness.

Specimens should undergo continually (at service life testing) defectoscopic tests (detection of surface and subsurface defects).

This way we can have complete information about the examined steel [10]. Another example of the surface damage evaluation you can find in [11].

3 Conclusions

From results of the tests, we can say that the examined material will meet conditions for the service life respectively for loading force. After the evaluation of experimental testing, we will suggest an adjustment of the mathematical model and possibly technologic parameters of the heat treatment (mainly depth of the carburized layer and carbon concentration).

The most suitable solution of the material problematics of the rolling contact fatigue is to choose a material, which will be loaded in its elastic area. This is possible to achieve either a material with a high yield strength, with sufficient heat or chemical heat treatment surface layer or with a weld deposit [10].

Acknowledgements This paper was created within the work on the FV20235 project—Project supported by the Ministry of Industry and Trade of the Czech Republic.

References

1. Hejnova M.: Principle of the rolling contact fatigue formation. In: Fuis, V. (ed.) Engineering Mechanics 2017, pp. 374–377. Brno University of Technology, Brno (2017). ISBN 978-80-214-5497-2.
2. Knowledge system of the cam mechanisms (2020) https://ipti.vamaxis.cz/ipti/#/km/view//1. Last Accessed 12 Feb 2020.
3. Hejnová M, Ondrášek J (2017) Life estimation of the contact surfaces. In Beran J, Bílek M, Žabka P (eds) Advances in mechanism design II 2017, Springer, pp 43–49. ISBN 978-3-319-44086-6.
4. Hejnova M (2018) Materialova problematika vackovych mechanismu. VÚTS, Liberec (2018). ISBN 978-80-87184-79-0.
5. Koloc Z, Vaclavik M (1993) Cam mechanisms. Elsevier, Amsterdam. ISBN 0-444-98664-2
6. Morrison RA (1968) Load/life curves for gears and cam materials. Mach Des 40:102–108
7. Fatemi A, Yang L (1998) Cumulative fatigue damage and life prediction theories: a survey of the state of the art for homogeneous materials. Int J Fatigue 20(1):9–34
8. Hejnova M (2017) Assessment of the rolling contact fatigue. In Dede I, Itik M, Lovasz E, Kiper G (eds) Mechanisms, transmissions and applications 2017. Springer, Cham. ISBN 978-3-319-60701-6, pp 89–98
9. Ondrasek J (2018) Obecne kinematicke dvojice vackovych mechanismu. VÚTS, Liberec. ISBN 978-80-87184-77-6
10. Hejnova M (2017) Effective service life testing of the cam mechanisms. In Trebuňa F, Frankovský P, Kostka J (eds) Experimental stress analysis 2017. Technical University of Košice, Košice, pp. 93–94. ISBN 978-80-553-3166-9.
11. Rycerz P, Olver A, Kadiric A (2017) Propagation of surface initiated rolling contact fatigue cracks in bearing steel. Int J Fatigue 97:29–38

Review of Weld Quality Classification Standard and Post Weld Fatigue Life Improvement Methods for Welded Joints

Sachin Bhardwaj, R. M. Chandima Ratnayake, and Arvind Keprate

Abstract In typical fabricated structures, there are hundreds of joints, which acts as a potential location for fatigue cracking. Ageing and life extension (ALE) of such structures could be developed further by creating an optimal measure between the required quality level and fatigue performance defined in quality inspection code of welded joints. The quality system for welds is described in ISO 5817 standard, where acceptance criteria have been specified for different types of weld geometry imperfections. Challenge of establishing link between the fatigue performance and defined weld quality classes is not consistent in these inspection codes i.e. link between the design of weld having required fatigue strength and described fabrication quality as per acceptance criteria is missing. Furthermore, weld quality optimal assessment during the lifecycle of welded joints can sustainably improve their fatigue life by the adoption of good design practices. Fatigue loading is identified as one of the most important factors for degradation of complex welded structures, which are highly influenced by local and geometric features modifications at weld toe locations. In recent years, the International Institute of Welding (IIW) Commission XIII has recommended guidelines related to fatigue design and assessment of welded structures under which various fatigue life improvement techniques have become available lately. The focus of this manuscript is on recommendations specified by IIW on post weld fatigue life improvement of steel structures by High-frequency Mechanical Impact (HFMI) methods, a residual stress improvement technique. Residual stress improvement techniques contribute to fatigue strength increase by reducing harmful tensile residual stresses and by inducing beneficial compressive residual stresses at weld toe regions. These recommendations include major fatigue assessment based on nominal stress, structural stress, and effective notch stress methods. Major influencing factors namely thickness and size effects, influence of steel strength, loading

S. Bhardwaj (✉) · R. M. C. Ratnayake
Department of Mechanical and Structural Engineering and Material Science, University of Stavanger, Stavanger, Norway
e-mail: sachin.bhardwaj@uis.no

A. Keprate
Department of Mechanical, Electronics and Chemical Engineering, Oslo Metropolitan University, Oslo, Norway

M. Abdel Wahab (ed.), *Proceedings of 1st International Conference on Structural Damage Modelling and Assessment*, Lecture Notes in Civil Engineering 110,
https://doi.org/10.1007/978-981-15-9121-1_20

effects, stress ratio are reviewed in detail for fatigue design class improvement of HFMI treated welds.

Keywords HFMI · IIW · Residual stress · ISO 5817

1 Introduction

Welded structures are mostly vulnerable to adverse geometric and metallurgical effects induced by welding [1]. The geometries of most welded joints introduces sudden changes of sections leading to local stress concentrations during loading and offering sites for crack initiation [2]. Structural welded components exhibit poor fatigue properties in comparison to samples without any discontinuities due to the fact that majority of their resulting fatigue life are spent on propagating the crack [3]. Whereas fatigue crack initiation period can occupy majority of lives in unwelded components [3]. Increased service life of welded structures can be ensured by following a good design practice, use of high quality fabrication and post weld improvement techniques [4]. Operational parameters like appropriate welding process, weld penetration parameters, weld geometry, low stress concentration factors etc. [2] aligned with accepted weld quality inspection code, assists in improving fatigue strength of welded joints.

Weld class systems are part of standards like ISO 5817 [5] and Volvo STD 181-0004, [6] defining quality system for welds. However, inconsistencies have been found between weld quality systems and corresponding fatigue performance [7]. This led international institute of welding (IIW) to establish recommendations on weld quality level and corresponding fatigue strength [7] which are based on different assessment methods [8–10] namely nominal stress, hot spot stress and notch stress methods. Nominal and structural hot spot stress method does not take the effect of geometric parameters of the weld toe into consideration, whereas effective notch stress method and fracture mechanics considers the effect of toe radius. Weld toe geometry primary controlling parameters for fatigue properties are transition angle of weld toe, radius of weld toe and wall thickness, which have been used to derive stress raising notch effect of the toe by various researchers [11, 12].

In the last few decades there has been considerable work done in welded joints, post weld fatigue life improvement methods. These fatigue life improvement methods relies on principle of improving the stress fields around the weld toe [13]. These methods are mainly divided into two groups, namely modification of weld toe geometry and residual stress improvement methods around weld toe. In the first group, modification of weld toe geometry is done by grinding of weld toes or re-melting by TIG (tungsten inert gas) welding. Secondly, in residual stress improvement methods, shot/hammer/or needle peening and high-frequency mechanical impact (HFMI) techniques are well known established techniques [14]. In 2007, IIW commission XIII based on fatigue life of welded components and structures recommended guidelines on various techniques on post weld, fatigue improvement in welded joints [4]. HFMI treatment methods are included under IIW recommendations, based on numerous research studies results conducted between 2002 and 2012 [4].

In this manuscript, various weld quality level standards correlation with fatigue strength of welded joints is presented as recommendations to design engineers for adopting these standards as tool to measure weld performance during design and fabrication process. Fatigue life improvement techniques recommended by IIW namely HFMI, a residual stress improvement technique is discussed in detail. These recommendations also include fatigue assessment based upon nominal, structural and effective notch stress methods. Major influencing factors like thickness, influence of steel yield strength, fluctuating load effects, stress ratio is reviewed in detail for fatigue design class improvement of HFMI treated welds.

2 Classification of Weld Quality Level Standards

ISO 5817:2014 standard [5] defines quality levels for various imperfections in welded joints of steel, nickel etc. and their alloys. In this standard, defined quality levels are classified as B, C, or D. Level B corresponds to best level and level D towards lowest weld quality level. These levels correspond to different quality levels with each weld discontinuity and imperfection having defined acceptance limits. ISO 5817:2014 defines 40 different types of weld discontinuities and some imperfections corresponding to different quality levels as illustrated in Fig. 1. ISO 5817 contains 23 outside imperfections, 13 inside imperfections, two weld geometry imperfections and two types of multiple imperfections [5]. A non-consistent relationship has been

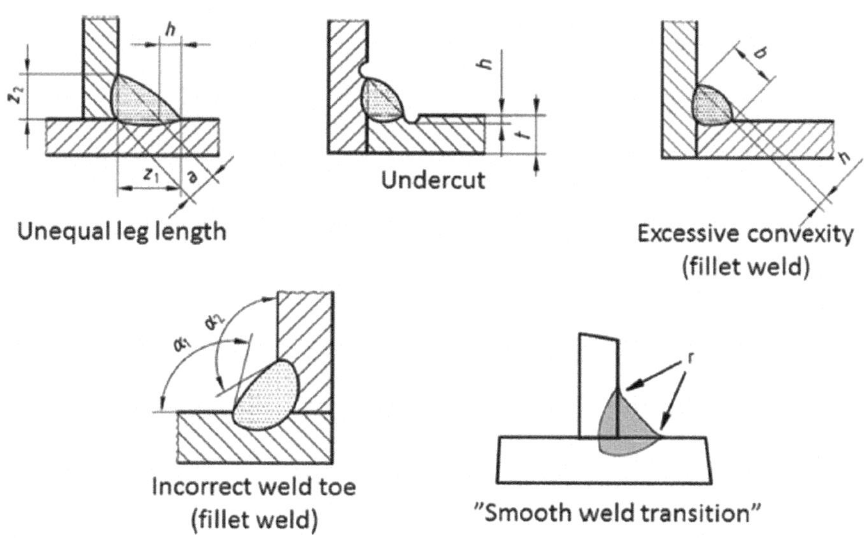

Fig. 1 Weld discontinuities mentioned in ISO 5817 [5]

determined between the quality level and required fatigue life, implying that weld having best quality level should have a longer fatigue life in comparison to a low quality level weld [15].Whereas, Volvo Group STD 181-0004 [16] has defined three different types of quality levels for fatigue. As-welded normal quality is classified as VD, high quality as VC and quality after post-weld treatment as VB. In Volvo Group STD 181-0004 [16], a change in quality level, leads to 25% increase in fatigue strength from previous class.

Welding introduces risk of fatigue failures by introduction of sharp notches and high stress raisers due to inhomogeneity created by welded geometries [3]. Improved weld quality is not achieved for welded component as fatigue strength remains same for welded steel as illustrated in Fig. 2a [17]. Weld defects, imperfections, irregularities are common in welded joints and various weld quality standards [5, 6] are available to determine acceptance criteria for them. Fatigue life is determined by weld defects size as shown in Fig. 2b and supported by Kitagawa diagram [17]. However, a weak correlation has been observed between weld quality levels and fatigue properties in ISO 5817 a weld quality assurance standard [5, 7, 16–18]. In ISO 5817, little influence on fatigue strength have been found for some weld imperfections as per their defined acceptance criteria. Whereas, imperfections that influence fatigue strength results in non-uniform changes in fatigue strength of their respective acceptance criteria between weld quality classes [7].

Current version of ISO 5817:2014, Annex C bridges the gap between required weld quality level and corresponding fatigue strength. As per IIW fatigue design recommendations [7, 9] notation, FAT 63 and FAT 90 class, has been related as imperfections limits for quality levels in Annex C of ISO 5817. Hobbacher and Kassner [19] had established this link between ISO 5817 and welds having consistent fatigue strength [7] in IIW recommendations [9]. IIW guidelines on weld quality in relationship to fatigue strength [7] presents a practical basis for ISO 5817, aligning quality levels consistent with fatigue strength. Such alignment can specify weld quality levels for required fatigue strength and vice versa [7].

Weld quality standards like ISO 5817 are so widely used for determining weld defects acceptance criteria that they are often interpreted as measure for structural

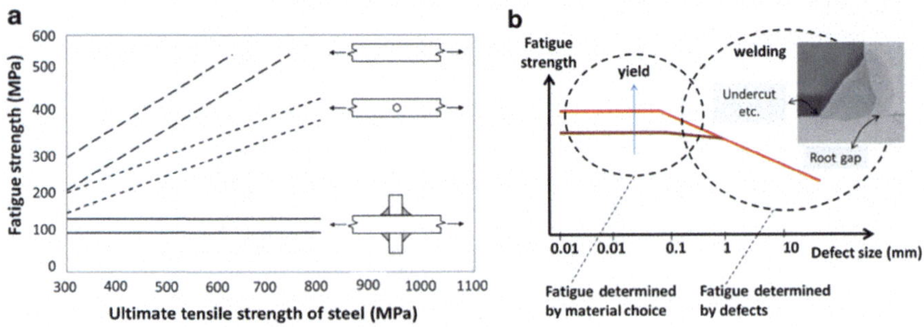

Fig. 2 **a** Fatigue strength vs tensile strength for, notched, unnotched and welded components. **b** Kitagawa diagram, fatigue strength versus defect size, with indicated weld positions [17]

performance, which is not their prime function. Design engineers often misinterpret these weld quality standards as a reference to determine high quality level welded joints. However design engineers should adopt methods like good design practices, high quality fabrications and various weld improvement techniques in their design that contributes to higher fatigue strength [2].

Adoption of best design practice can be achieved by minimization of fatigue loads e.g. by avoiding resonance and vibrations locations, using connections with low stress concentration factors and placing welds in area of low weld concentrations [2]. Conversely, high quality fabrication methods and using weld improvement techniques etc. results in increased fabrication cost. Hence, an effort should be made to increase fatigue strength in design, fabrication and inspection stage in line with IIW recommendations for fatigue improvement techniques [4, 7, 9, 20].

3 Fatigue Improvement Methods

The IIW commission has recommended techniques and procedures for improvement of welded steel structures fatigue strength [4]. As demonstrated in Fig. 3. an overview of various fatigue improvement techniques is presented [20]. These techniques are divided into, two broad categories. First category relies on the principle of reducing local stress concentration factor at weld toes, which improves the fatigue strength

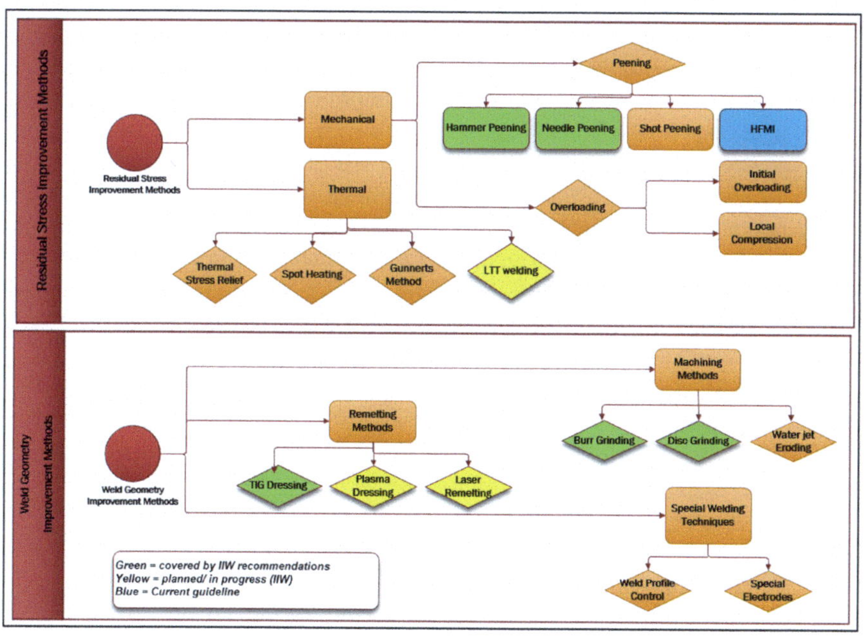

Fig. 3 Overview of Fatigue Improvement Techniques adapted from [21]

and ensures an uninterrupted transition between the profile of weld toe and base material. The second category ensures induction of beneficial compressive residual stresses and reduction of harmful tensile residual stresses in the weld toe region, eventually contributing to fatigue strength improvement [1, 5–7]. Fatigue improvement techniques may be applied during design, fabrication or post fabrication. These recommendations of IIW [4] on fatigue improvement of welded components and structures verifies all methods, e.g. component testing, nominal stress, structural hot spot stress in 2006 [8], notch stress method in 2008 [10] as well as fracture mechanics assessment procedures [3, 9, 10].

These guidelines cover four commonly applied post weld treatment methods. Reduction of stress concentration points by material removal mechanically are the basic principle on which, the first method (burr grinding or weld toe grinding) depends. It can be achieved by rise of the weld toe radius and fall of weld flank angle [22]. This method also helps in removal of any underlying defect close to weld toe. The second method (TIG re-melting) results in two main advantages, firstly reduction of stress concentration points around weld toe and secondly by improving hardness around weld toe zone region. Re-melting around weld toe results in effortless transition between the weld toe radius and the base material [23].

The first two techniques defined under the IIW recommendations namely burr grinding and TIG dressing benefits are claimed in FAT 90 class leading maximum improvement up to FAT 125 class as per IIW notation of SN curves. This corresponds to an increase in allowable stress range by a factor of 1.3 and 2.2 on life (for SN slope $m = 3$) [24]. For simplification purpose, an improvement of two (2)-fatigue class based on the IIW fatigue design notations [4].

For steels having yield > 355 MPa, improvement of welds by hammer or needle peening gives a upgrade by a factor of 1.5 in fatigue strength benefit, applied to the stress range as per IIW guideline [4]. In simplified terms, an increase of three (3) fatigue class [4]. Needle and hammer peening are grouped under the family of high frequency mechanical impact (HFMI) family. Reduction of stress concentration increase in hardness and introduction of beneficial compressive stresses at weld toes are major advantages of these methods. The affected region around weld toe becomes plastically deformed, which leads to large changes in the microstructure and corresponding local geometry. Simultaneously as mentioned before, residual stress state changes around welded toe region, leading to introduction of beneficial compressive residual stresses [23, 25].

Welding with special electrodes called low transformation temperature (LTT) electrodes, have emerged as other recent approach to improve welds fatigue strength. This technique has advantage of being more cost effective in contrast to HFMI techniques as it does not require any additional treatment after welding [26–29]. Special welding methods like friction stir welding, welded joints have shown excellent fatigue properties compared to conventional welds which shows fatigue strength comparable to base material [2]. In structural welding code of the American welding society (AWS), overall shape of weld can be controlled by achieving a concave profile on weld toe having a low stress concentration [30].

4 Fatigue Improvement by High Frequency Mechanical Impact Methods—HFMI

The innovation and the pioneer work of locally modifying the residual stress state using ultrasonic technology goes to the researchers based in the former Soviet Union [31, 32]. Since 2010, IIW Commission XIII uses the term HFMI as a universally accepted term to generalize several related techniques. Following are the names of different techniques having almost the same principle for which separate recommendations was published by the commission. Ultrasonic impact treatment, ultrasonic peening, ultrasonic peening treatment, high-frequency impact treatment, pneumatic impact treatment and ultrasonic needle peening [24, 33, 34] are generalized terms for these technologies [20]. These guideline are applicable to steel structures of plate thickness ranging from five to fifty mm (5–50 mm) and yield strength stretching from 235–960 MPa [20].

HFMI treatment is the more advanced term accepted for hammer or needle peening. Its principle relies on high frequency impacts of needles on areas around weld toes resulting in smooth indentations due to smaller spacing of needles. At the same time HFMI tools are found to be more comfortable and less noisy for the operator [35]. IIW guidelines clearly specifies that HFMI improved welds are recommended to improve the fatigue live of the welds, where probability of occurrence of failures are more from locations close to weld toe. Hence, occurrence of a failure originating from locations other than weld toe example weld root etc. must be taken into account before HFMI application, as illustrated in Fig. 4 a and b respectively [25]. As per IIW notation, fatigue strength (FAT value) at no of cycles to failure = 2×10^6 has been characterized as index point for defining improvement in fatigue for HFMI improved welds having a slope of 5 on SN curve [9].

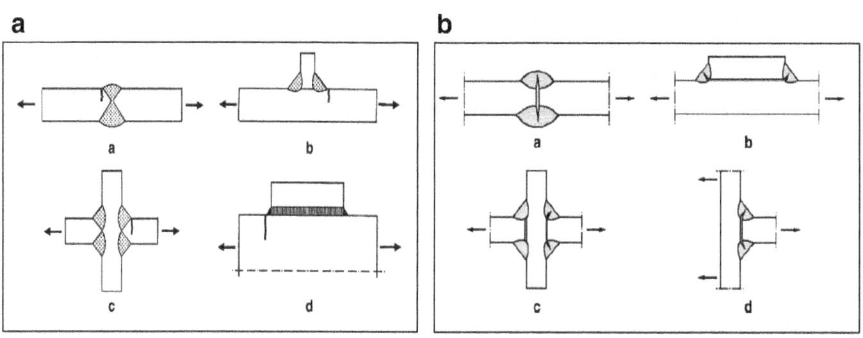

Fig. 4 **a** and **b** Weld joint configurations not suitable for post weld improvement [25]

4.1 Fatigue Improvement for HFMI Improved Welds Based on Nominal Stress

It has been validated and recommend by IIW guidelines [4], that fatigue lives of welds can be best observed in SN curve having slope greater than m = 3. Recent studies have established SN curve having slope m = 5, shows a best fit with available fatigue data for HFMI treated welds [36]. HFMI treated welds benefit can be claimed between FAT classes 50 and 90, as defined in IIW notation of SN curves in their recommendations. If failure dominates from root other than weld toe, than these recommendation tends to fail [20]. HFMI benefit of four-fatigue class can be claimed in IIW notation of SN curves for steel having yield strength ≤ 355 MPa [20].

As illustrated in Fig. 5 in black line, welded joint (without any post weld improvement) having yield less than 355 MPa are classified as FAT class 90 with a slope of 3, as per IIW recommendations on fatigue strength of welded joints. After HFMI treatment FAT class of 140 with a slope of five, is shown as red line in Fig. 5. This red line intersects with black line of as weld condition at about N = 72,000 cycles. This illustrates that welded structures with yield ≤ 355 MPa, fatigue strength improvement benefit cannot be claimed, if the fatigue life is less than N = 72,000 cycles. Similar cycle limit for steels having yield greater than 355 MPa is specified in the recommendations [22].

In nominal stress approach actual weld geometry is not considered hence its application of the assessment is in principle clear [22]. Due to high cost involved in conducting experiments in nominal stress approach and replication of its results to

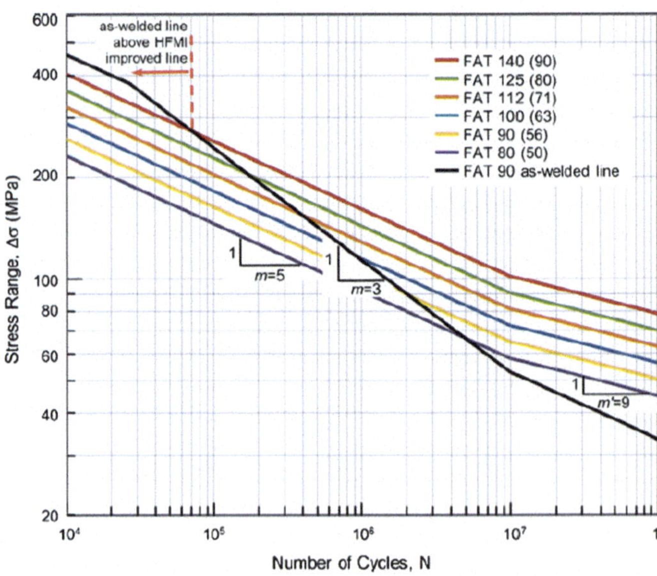

Fig. 5 Nominal stress SN curve for HFMI treated welds adapted from [9]

similar weld configurations may lead to overly conservative analysis. IIW recommendations for post-weld treatment methods [4] partitions a single division between steels having yield strength less or greater than 355 MPa. After a round robin exercise in 2003 supported with numerous studies and research results, IIW commission XIII a concluded relationship between yield and fatigue strength [36].

4.2 *Influence of Steel Strength Stress on HFMI Treated Welds*

Steels having different yield strengths showed a different trend in FAT class improvement when treated with HFMI [4, 20, 36]. Yildrim and Marquis [36] had proposed a distinction between yield strength range and corresponding improvement in fatigue class after assessment of published data. Steels having thickness ranging from 5–30 mm and yield strength ranging between 260 and 969 MPa were grouped under a yield strength correction method under IIW recommendations for HFMI treated welds [20]. It is recommended that there is an increase of one fatigue class for every 200 MPa increase in static yield strength as per IIW recommendations [20] as illustrated in Fig. 6.

For steels having yield strength greater than 950 MPa, the maximum increase in fatigue strength is up to 8 fatigue class [37]. As defined earlier all fatigue classes are defined at cycle to failure at $N = 2 \times 10^6$ cycles with slope of five in SN curve, for HFMI-treated welds as per IIW notation scheme. However, IIW recommends an increase of two-fatigue class for all improvement methods irrespective of its type as it classifies yield strength criteria into high and low strength steels only. Whereas in TIG dressing technique an increase of three fatigue class has been found for longitudinal stiffeners as an example and four for HFMI treatment welds [25]. Most importantly, it should not be missed that assumed SN slope for as welded joints is 3, $m = 4$ for TIG-dressed welds and $m = 5$ for HFMI-treated welds [14].

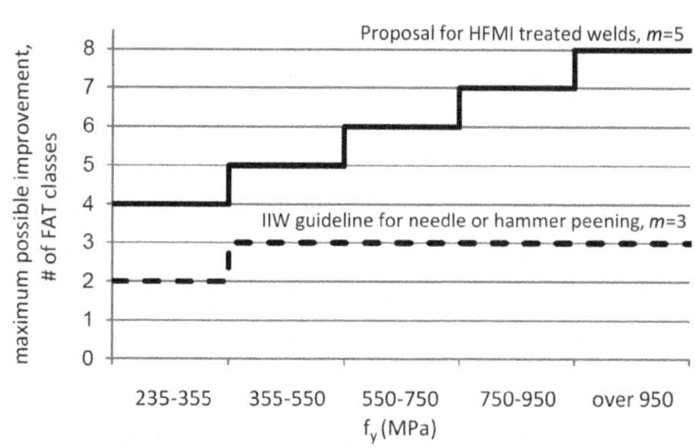

Fig. 6 Maximum improvement in fatigue class versus yield strength adapted from [21]

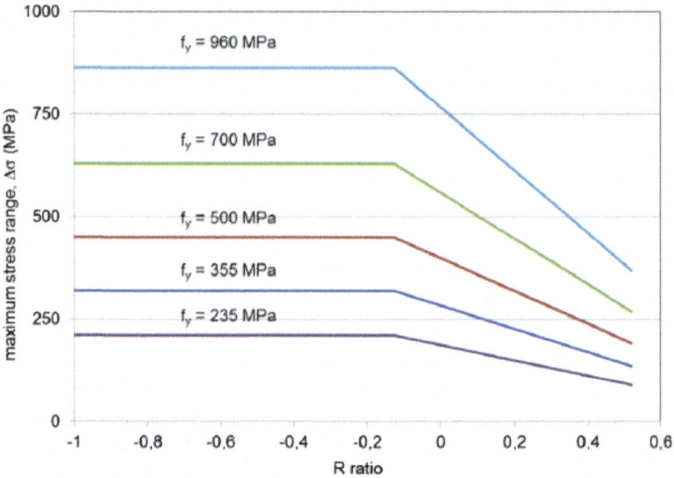

Fig. 7 Penalty of maximum constant amplitude stress range for HFMI treated welds [25]

4.3 Thickness and Size Effects on HFMI Treated Welds

As thickness of welded components are increased, its fatigue strength is reduced with increasing plate thickness of the load-carrying plate. Geometric features like weld fillet size, attachment plate size, main plate width etc. are known to make major influence in term defined as size effect [35]. Due to large stress concentrations created at weld toe critical locations, weld toe is largely affected by plate thickness and size of weld. Nominal stress and hot spot stress method uses the criteria of thickness reduction factor for thickness exceeding 25 mm [25]. Existing thickness reduction factor in IIW guideline [4] are based on Hobbacher [9], which is again referred for HFMI recommendations due to lack in extensive experimental data. Current guideline for HFMI improved welds, a thickness exponent of 0.2 is used as per IIW recommendations [20].

4.4 Loading Effects on HFMI Treated Welds

HFMI treated welds improvement is greatly benefited to induction of beneficial compressive residual stresses at weld toe regions. Areas where applied stress equals or nears yield value, amount of fatigue improvement decreases in regions of high mean stresses and R-ratios. Welded components operating at $R > 0.5$ (applied stress ratios) and applied stresses above 80% of yield value, are not suitable for improvement by HFMI technique [36]. IIW recommendations is valid for $R \leq 0.15$ as fatigue strength improvement benefit of hammer, penning is largely influenced by applied stress ratio [4]. Without testing, fatigue improvement cannot be claimed for R ratio greater than

Table 1 Penalty of FAT classes having high stress ratio for HFMI-treated welded joints [25]

Stress ratio	Minimum FAT class reduction
$R \leq 0.15$	No reduction due to stress ratio
$0.15 < R \leq 0.28$	One FAT class reduction
$0.28 < R \leq 0.4$	Two FAT classes reduction
$0.4 < R \leq 0.52$	Three FAT classes reduction
$0.52 < R$	No data available

0.5 [20]. For high stress ratios situations, IIW guidelines gives special limitation and FAT classes improvement for welded joints improved by needle or hammer peening [4]. Reduction in FAT classes are imposed as an penalty for high stress ratios for HFMI treated welds as illustrated in Table 1, referred form IIW guidelines [25].

Variable amplitude loading can drastically change the residual stress state around the weld toe in HFMI treated welds [20]. In practical situations variable loading are more prevalent, hence failure mode changes in HFMI treated welds in contrast to constant amplitude loading recommendations available from lab experiments [38]. IIW guidelines [20] follows the maximum amplitude stress ratio limitation criteria as illustrated in Fig. 7, for situations involving variable amplitude loading in HFMI treated welds. [25]. It should be noted that HFMI benefit cannot be claimed without fatigue testing, if the stress ratio exceeds the limit value for a given yield strength [20].

Available recommendations are based on constant amplitude loading results and limited data is available for variable amplitude testing which needs further studies for more accepted use of these recommendations in practical world. Accurate prediction of residual stresses in complex welded structures and its relaxation and redistribution during loading is a major challenge in defining improvement in FAT classes for post weld treatment methods of fatigue improvement.

4.5 HFMI Treatment Through Local Approaches

Effective notch stress method (ENS) and structural hot spot stress method (SHSS) are local approaches defined by IIW commission XIII of fatigue of welded components [9]. In SHSS, characteristic SN curves defined in the recommendations [8] were applied by Yildrim [39] on HFMI treated welds. Two types of fatigue class improvement class have been suggested for load and non-load carrying joints for HFMI treated welds. In SHSS approach, weld geometry is not considered in detail hence uncertainty in extrapolation often leads towards conservative results. Whereas in ENS method [10], only one type of SN curve have been proposed by Yildrim [39]. In case of HFMI treated welds, correct understanding of toe radius, microstructure of treated zone and residual stress needs to be understood for use ENS approach with actual radius of HFMI treated welds. Local approaches proposed in the guideline shave been found to be conservative and consistent with all type of weld geometries and existing fatigue data points [39].

4.6 HFMI Treatment of Ultra-High Strength Steels

As mentioned in Sect. 4.2, yield strength plays an important role in deciding improvement in fatigue strength or FAT class improvement as per IIW guidelines on HFMI treated welds. For steel having strength ranging between 960 and 1400 MPa, Berg and Stranghöner [40], investigated steel shaving strength greater than 1000 MPa. Their findings concluded an increase in fatigue strength of 15% for steels having yield between 1100 and 1300 MPa and 10% for steels having yield between 960 and 1100 MPa. Further studies and inclusion under IIW guidelines are needed to correlate HFMI improvement in ultra-high strength steels as various studies claims different degree of improvement.

5 Conclusion

Weld discontinuities are classified into a grouping scheme in ISO 5817 standard. This standard helps in measuring weld performance in terms of its quality. ISO 5817 is found to be inconsistent in terms of corresponding fatigue performance, as welds termed, as high quality in standard, should correspond to high fatigue strength joints. This led IIW fatigue design recommendations to include Annex C in ISO 5817, where consistency in quality levels in terms of fatigue properties of welded joints have been improved.

Under fatigue improvement methods, HFMI is the universal term under which various ultrasonic techniques have been grouped. Fatigue data analysis of HFMI treated welds are found to fit well with existing fatigue weld joints recommendations having a slope of five in comparison to classic characteristic SN curve slope of three as per IIW notation scheme. Stepwise grouping of material yield strength and corresponding improvement in fatigue class are included in IIW recommendations whereas penalty in fatigue class improvement due to varying loading does have a skeptical acceptance. Due to lack of experimental data for different weld geometries and varying load conditions, improvement in fatigue classes as per IIW recommendations should be validated beforehand application guidance.

Weld toe modification has been found one of the most critical weld geometry features, which has novice notion of acceptance by a smooth transition. In areas of low cycle fatigue, weld profile changes can be more effective way in contrast to areas of high fatigue loading, where residual stress modification can be more effective. As per IIW guidelines on post weld fatigue improvement, HFMI improved welds are found to be more effective when there are more chances of occurrence of failure from weld toe in contrast to weld root region. However, failures at weld toe have been found to move to other locations such as inter-bead toes and the weld root as per the loading condition. Hence, role of the weld root is more important in relation to weld improvement. Hence, designer engineers should establish appropriate weld quality levels on the drawings for fatigue improvement, considering interactions of weld imperfections and fatigue properties during design of cyclic loaded welded components.

During design stage, a designer should strike a balance between fatigue improvement methods and should design welds for purpose. Design engineers should consider possible methods for post-weld treatment like placing welds in areas of low stress concentration areas, reducing stress concentrations, improvement of weld with use of LTT electrodes and fatigue crack arrest (FCA) steels. For fatigue-loaded structures, quality guidelines should be more detailed in contrast to static loading, where role of local weld geometry profile and features can be overlooked.

Acknowledgements This work has been carried out as part of a Ph.D. research project, performed at the University of Stavanger, Norway. The research is fully funded by the Norwegian Ministry of Education.

References

1. (Ed) DR (1990) Design and analysis of fatigue resistant welded structures. Woodhead Publishing, Cambridge UK
2. Haagensen PJ (2011) 11 - Fatigue strength improvement methods. In: Macdonald KA (ed) Fracture and fatigue of welded joints and structures. Woodhead Publishing, pp 297–329
3. Maddox SJ (2011) 7 - Fatigue design rules for welded structures. In: Macdonald KA (ed) Fracture and fatigue of welded joints and structures. Woodhead Publishing, pp 168–207
4. Haagensen PMS (2013) IIW recommendations on methods for improving the fatigue lives of welded joints. In: Ltd WP (ed). International Institute of Welding, Cambridge
5. ISO (2014) ISO 5817:2014 in Welding—Fusion-welded joints in steel, nickel, titanium and their alloys (beam welding excluded)—Quality levels for imperfections. BSI
6. Group V (2013) Weld classes and requirements Life-optimized welded structures Steel, thickness ≥ 3 mm, in STD 181-0004. 2013, Volvo Group, Sweden
7. Jonsson B et al. (2016) IIW Guidelines on Weld Quality in Relationship to Fatigue Strength
8. Niemi E, Fricke W, Maddox S (2006) Structural hot-spot stress approach to fatigue analysis of welded components: designer's guide
9. Hobbacher AF (2009) The new IIW recommendations for fatigue assessment of welded joints and components—A comprehensive code recently updated. Int J Fatigue 31(1):50–58
10. Fricke W (2011) 5—Fatigue strength assessment of local stresses in welded joints. In: Macdonald KA (ed) Fracture and fatigue of welded joints and structures. Woodhead Publishing, pp 115–138
11. Lawrence FV, Ho ANJ, Mazumdar PK (1981) Predicting the fatigue resistance of welds. Ann Rev Mat Sci 11(1):401–425
12. Ida K, Uemura T (1996) Stress concentration factor formulae widely used In Japan. Fatigue Fract Eng Mater Struct 19(6):779–786
13. Kirkhope KJ et al (1999) Weld detail fatigue life improvement techniques. Part 1: review. Mar Struct 12(6):447–474
14. Yıldırım H (2017) Recent results on fatigue strength improvement of high-strength steel welded joints. Int J Fatigue

15. Karlsson N, Lenander PH (2005) Analysis of fatigue life in two weld class systems. In: Department of mechanical engineering 2005. Linköping University: SE-581 83 Linköping, Sweden
16. Volvo (2016) Fusion welding weld classes and requirementsLife-optimized welded structures Steel, thickness ≥ 3 mm. Volvo Group, Sweden
17. Barsoum Z et al (2012) Fatigue design of lightweight welded vehicle structures: Influence of material and production procedures. Proc Inst Mech Eng, Part B: J Eng Manuf 226:1736–1744
18. Stenberg T et al (2017) Quality control and assurance in fabrication of welded structures subjected to fatigue loading. Weld World 61(5):1003–1015
19. Hobbacher A, Kassner M (2012) On relation between fatigue properties of welded joints, quality criteria and groups in Iso 5817. Weld World 56(11):153–169
20. Marquis GB, IIW ZB (2016) Recommendations for the HFMI treatment for improving the fatigue strength of welded joints, ed. IIW. Springer
21. Haagensen PMS (2007) State of the art and guidelines for improved high strength steel welds. In: Proceeding of international symposium on integrated design and manufacturing of welded structures eskilstuna, Sweden
22. Baumgartner J, Yıldırım HC, Barsoum Z (2019) Fatigue strength assessment of TIG-dressed welded steel joints by local approaches. Int J Fatigue 126:72–78
23. Alkarawi H, Manai A, Al-Emrani M (2019) A Literature review on fatigue life extension of welded structures by peening and TIG dressing
24. Sonats. Available from: https://www.sonats-et.com
25. Marquis GB et al (2013) Fatigue strength improvement of steel structures by high-frequency mechanical impact: proposed fatigue assessment guidelines. Weld World 57(6):803–822
26. Alghamdi T, Liu S (2014) Low-transformation-temperature (LTT) welding consumables for residual stress management: consumables development and testing qualification. Weld J 93:243s–252s
27. Karlsson L (2009) Improving fatigue life with low temperature transformation (LTT) welding consumables. Svetsaren 64:27–31
28. Kromm A et al (2009) Determination of residual stresses in low transformation temperature (LTT -) weld metals using X-ray and high energy synchrotron radiation. Weld World 53(1):3–16
29. Ramjaun T et al (2014) Effect of interpass temperature on residual stresses in multipass welds produced using low transformation temperature filler alloy. Sci Technol Weld Joining 19:44
30. Society AW (2015) AWS D1.1/D1.1M:2015 Structural welding code—Steel. USA
31. Kudryavtsev YFTV, Mikheev PP, Statnikov EF, Burenko AG, Dobykina EK, Increasing the fatigue strength of welded joints in cyclic compression. International Institute of Welding in the World Paris, (1994). Document XIII-1596–94
32. Statnikov ESSU, Kulikov VF (1977) Ultrasonic impact tool for welds strengthening and reduction of residual stresses. Scientific Works: Metallurgy (1977). SEVMASH USSR (92):27–29
33. Pfeifer. Available from: https://www.pfeifer.de/
34. Pitec. Available from: https://www.pitec-gmbh.com/
35. Shams-Hakimi P, Yıldırım HC, Al-Emrani M (2017) The thickness effect of welded details improved by high-frequency mechanical impact treatment. Int J Fatigue 99:111–124
36. Yildirim HC, Marquis GB (2012) Fatigue strength improvement factors for high strength steel welded joints treated by high frequency mechanical impact. Int J Fatigue 44:168–176
37. Yıldırım HC et al (2016) Application studies for fatigue strength improvement of welded structures by high-frequency mechanical impact (HFMI) treatment. Eng Struct 106:422–435
38. Marquis G (2010) Failure modes and fatigue strength of improved HSS welds. Eng Fract Mech 77:2051–2062
39. Yildirim HC, Marquis GB, Barsoum Z (2013) Fatigue assessment of high frequency mechanical impact (HFMI)-improved fillet welds by local approaches. Int J Fatigue 52:57–67
40. Berg J, Stranghöner N (2016) Fatigue behaviour of high frequency hammer peened ultra high strength steels. Int J Fatigue 82:35–48

Analysis of VIN Errors in Information Systems, Causes, Consequences and Solutions

Roman Rak

Abstract Every vehicle manufactured after year 1986 is obligatorily marked with a VIN (Vehicle Identification Number), which is unique worldwide. This is the so-called individual vehicle identification, which is normally used as the primary information key in various computer registers, vehicle information systems—in the national vehicle register, in the police information systems (national and transnational databases of stolen vehicles, etc.), insurance companies, leasing companies, in eCALL technology (emergency call technology—implemented 2 years ago across all European countries for all new vehicles produced or imported to EU), etc. Any error in the VIN identifier means that the vehicle is then not found in the relevant information system and "looks" that everything is OK. But the reality is different. In national and international vehicle registers, the error rate ranges from 5 to 20%. Most errors in VIN arise in manual processes of rewriting, manual copying VIN from documents into information systems. These information systems are either in the competence of state, government or commercial entities. Especially sensitive to errors in VIN are all processes of forensic identification of vehicles, processes related to rescue service activities in the event of a vehicle accident, in the investigation of fraud and theft of vehicles by police and insurance units, The paper is based on five-year research and analysis of information systems during years 2014–2019. The paper deals with the anatomy of VIN errors in various processes and offers possibilities of their elimination and significant improvement of information systems containing VIN identifier. The errors in the VIN are categorized into several categories. For each category are listed its specificities.

Keywords Vehicle identification number · Information system · Vehicle register · Forensic · Error solution

---

R. Rak (✉)
Department of Forensic Science and Criminalistics, University of Finance and Administration, Estonska 500, Prague, Czech Republic
e-mail: roman.rak@irisident.cz

M. Abdel Wahab (ed.), *Proceedings of 1st International Conference on Structural Damage Modelling and Assessment*, Lecture Notes in Civil Engineering 110,
https://doi.org/10.1007/978-981-15-9121-1_21

1 Introduction

The development of the automotive industry is now closely linked to electromobility and digitization; electrification and electronization as such. In this direction, the vehicles themselves, transport and information technology infrastructure, automotive telematics are developing. The key processes are either ongoing or will be fully autonomous, without the primary human involvement. However, security, reliability of these processes, including data exchange and processing, will always remain a key aspect. These are the default attributes of all other processes or activities.

Each vehicle manufactured after 1986 must be identified with a unique VIN (*Vehicle Identification Number*) by the manufacturer [4]. This is so called individual vehicle identification, which is normally used as a primary data information key in various computer registers and vehicle information systems—in national vehicle registers, in police information systems (both national and international databases of stolen vehicles, etc.), in insurance company databases, leasing companies, in the eCALL technology [5, 7], etc. A VIN identifier error means that the vehicle cannot be found and it looks like everything is OK. The error rate of national and international registers is in the range of 5 to 20%. Most VIN errors occur in manual processes during transcribing or copying VINs from documents to information systems. This paper deals with the anatomy of VIN error occurrence in various processes and offers possibilities to eliminate them and improve the quality of the information systems that use VIN identifiers.

1.1 Research Background and Significance of the Subject

In this paper, we deal with so called individual errors [8], i.e., errors which are caused by persons who are transcribing the information from the vehicle documents to the information systems or directly from the vehicle to the documents or the information systems. These are subjective errors caused by fatigue, distraction, stress, using a PC keyboard or made deliberately [5]. The aim is to find the anatomy of the occurrence of these error types and to find the method of their general elimination.

This paper does not include objective errors, such as small letters of identifiers stamped on vehicles, corroded identifier surfaces, which make the characters illegible, etc. These errors are called "systematic" and they can be eliminated using systematic measures.

1.2 Analytic Approach and Examination Data Sources

Over 3.5 million real vehicle identifiers VIN were analyzed over 4 years [3]. This data came from computer databases of national central vehicle registers in the Czech

Republic and the Slovak Republic. Data from national central databases of selected insurance companies, leasing and other private companies were also used. A special software program—the Universal VIN Decoder (VINexpert [5]) was developed for this analysis. The VIN decoder was filled with VIN data structures of all vehicle models produced in 1986–2018.

Information about vehicles, including VIN, is very sensitive customer data [10]. No test data, reference samples, could be used to find real errors that were not yet known. Therefore, a specialized SW application was developed to incorporate the VIN structures of all models of the world's leading vehicle manufacturers for the production period 1986–2018. Combination options were created for the first 9 VIN positions. Their amount is about 2.5 million. The basic identification was done on the first 9 VIN positions in first step. Consideration was given to the fact whether specific manufacturers and their models use the so-called check digit. Subsequently, dozens of working hypotheses were created as to how a mistake could arise. The hypotheses were checked using specialized written SW, analyzed and some exceptions were compared manually. A more detailed description of the procedures used is far beyond the scope of this paper.

2 Individual Errors

When transferring the VIN from the vehicle or its documents into information systems, a few typical character confusions may occur, which are specific for these activities. Based on many years of experience and analyses of real vehicle registrations, the following typical confusions can be specified:

- Visual confusion
- Position confusion
- Kinetic confusion
- Phonetic confusion
- Logical thought confusion
- Computer keyboard setting confusion.

3 Visual Confusion

The visual confusion results from the similarity of characters used in VIN. It occurs in the human mind when reading the characters. This confusion is usually caused by the incorrect or currently unwell sight of the person who is reading the VIN, a lack of knowledge about the normal VIN structure (especially the WMI and VIS sections). A visually similar character is entered instead of the correct one. VIN illegibility may also be caused by improper lighting, the location of the VIN identifier in the vehicle (inaccessible areas, blind spots, covering VINs with certain vehicle components, lighting conditions), ageing and operational processes (corrosion, contamination),

Fig. 1 Possible confusions of "5" for the adjacent keys

inappropriate VIN font (stamping, engraving), etc. The character confusion is similar to the other general cases when someone is reading and transcribing the text.

In practice, the frequently confused characters are U-V, H-M, O-0, 5-S, etc. See Table 2 for more examples.

4 Position Confusion

The position confusion occurs when typing on a PC keyboard to enter data into an information system. A user presses an adjacent key on the keyboard instead of the correct one. These are the alphabetic as well as the numeric keys, which are usually entered using a small numeric keypad on the right side. The position confusions may be caused by a rush, distraction, insufficient experience in typing on the keyboard, small keys or large fingers (an inappropriately manufactured or selected keyboard in terms of ergonomics). With position confusion the adjacent characters on a keyboard are swapped. Position confusions are apparent from Figs. 1 and 2.

5 Kinetic Confusion

Kinetic confusion means confusion of the order of characters. The user wants to write two characters, e.g. "TM", but they reverse their order, which results in "MT". This error is common among older people or users who frequently work with PC for a long time. It is so called "overtaking keystrokes by individual fingers", which is caused by motor habits, which go beyond our reception capability. In the case

Fig. 2 Possible confusions of "J" for the adjacent keys

of kinetic confusion the adjacent characters in the VIN string are swapped. This is distinct from position confusion where the adjacent keys on a keyboard are swapped.

6 Fonetic Confusion

Phonetic (aural) confusion results from the same diction of inconsistently spelled letters. If a person pronounces "V" without accenting that it is actually "W", the user enters "V". Phonetic confusion is very much dependent on the user native language.

7 Logical Thought Confusion

This type of confusion results from unfamiliarity with the first three VIN characters (WMI). This confusion often occurs with vehicles manufactured in Germany. This applies specifically (among others) for the following brands (Table 1).

When specifying the WMI standard on a national level, lobbying of manufacturers played a significant role in reflecting the brand name in the WMI code. The first letter "**W**" comes from the former name of the state—**W**est Germany. The second and the third letter were created on the Germany level—these letters determined the brand name. According to the international ISO standard, however, the use of "O" letters was not allowed, therefore zero ("0") was used instead. The font used for stamping VINs does not allow to distinguish between "O" and "0" characters (if the user is not familiar with the standards). Due to this, many German vehicles have an incorrect "O" letter both in their documentation and information systems. This was caused by the fact that the information system designers are usually not familiar with the VIN standards and, especially in the past, the entry of illegal VIN characters in the information system was not checked. Similar to "O" and "0" characters, the letters "I" and "J" can also be confused ("I" letter should not be used). In other cases,

Table 1 Examples of frequent logical thought confusions

Brand	Incorrectly transcribed WMI	Correct WMI
Ford	WFO	WF0
Porsche	WPO	WP0
Opel	WOL	W0L
Bitter	WBI	WBJ
Heku	WHE	WHU
Holder	WHO	WH0
IAV Cars	WIC	WJC
Iveco Magirus	WIM	WJM
Kögel	WKO	WK0

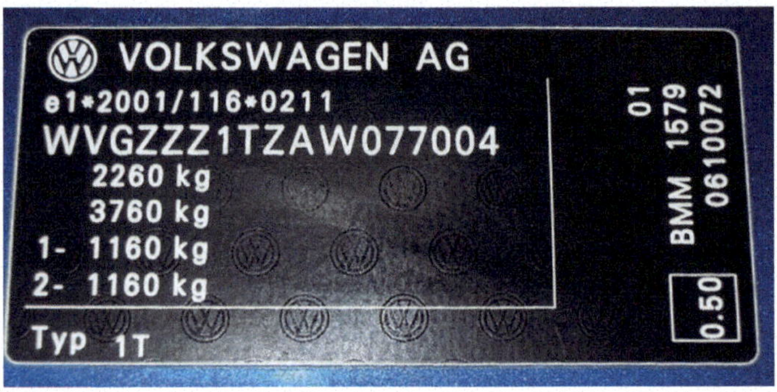

Fig. 3 Volkswagen WMI label = "WVG"

confusions are experienced due to a logical similarity to the brand, e.g., for the HEKU Company, the correct string is "WMI WHU", however, "WHE" would rather be expected (which is incorrect).

Similarly, a very common error occurs when WMI is recorded for the Volkswagen vehicles. The Volkswagen name evokes an idea that the WMI string should be **VWV** instead of the (correct) **WVW (or WVG)**, where the correct logic is **W** for **W**est Germany (which results from the ISO standards) and VW for Volkswagen (Fig. 3). Many incorrect strings WMI = VWV for the Volkswagen brand can be found in the registers; this is a quite common and typical error (Table 2).

8 Computer Keyboard Setting Confusion

The cause of the confusion is incorrect (forgotten) switching between national and English keyboard layouts (using Alt + Shift keys). This results in swapping **Y** and

Table 2 Basic examples of VIN confusion and practical examples

Type of confusion	Typical error situations	Incorrect VIN	Correct WMI
Visual confusion	U-V H-M E-F 0-O 5-S W-M 2 –Z 6-G G-C 6-C 3-B .... B-8 Y-V ...	JM2BJ14L201274161 JF1GCHLJ3WG067014 WAUZZZ4804N045331	JMZBJ14L201274161 JF1GCHLJ3W6067014 WAUZZZ4B04N045331
Position confusion	J-K J-U J-Z J-H J-M J-L .... J-N ([1]) ...	VF30A9HR8BS064344 JV1VW70821F709644 EF04XXGBB41K55965	VF30A9HR8BS064347 YV1VW70821F709644 WF04XXGBB41K55965
Kinetic confusion	TMB – TBM WFO – WOF WDB – WBD	SNJBAAN16U0314749 TBMGS46Y023304024 JMZBJ134221473662	SJNBAAN16U0314749 TMBGS46Y023304024 JMZBJ143221473662
Phonetic confusion	V-W	VVGZZZ5NZ8W000253 WAUZZZ4BZVN001678	WVGZZZ5NZ8W000253 WAUZZZ4BZWN001678
Logical thought confusion	WF0 – WFO WP0 – WFO W0L-WOL WVW-VWV	WFO4XXGBB41K55965 WOL0AHL3582003279 VWVZZZ1FZ9V028266	WF04XXGBB41K55965 W0L0AHL3582003279 WVWZZZ1FZ9V028266
Computer keyboard setting confusion	Y-Z Z-Y	VF1FLB1B1EZ535433 JMYBJ12P200249824	VF1FLB1B1EY535433 JMZBJ12P200249824

[1] See the keyboard layout pictures. Select any key and check the adjacent keys

Z characters in the VIN. This error is especially typical for programmers (using programmers setting of keyboard), persons who intensively use computers, extended applications, etc.

9 Conclusions

The VIN identifier correctness and quality is essential for efficient utilisation of information systems in the automotive area [12]. It is critical for vehicle search results, especially when it is necessary to obtain/disprove negative information on the vehicle [14] (vehicles reported as stolen, vehicles with negative indications—either scraped, obstructed by an enforcement procedure, indicated as doubled, i.e., there is another vehicle in another registration system with the same VIN, having modified odometer data [2], etc.). The required VIN identifier quality grows with the growing number of vehicles registered in information systems and with implementation of all new information systems (a general issue of continuous development of community computerisation), and the increasing number of various crimes and fraud related to vehicles. The demands are increasing with globalisation [1] (international exchange of information), European integration and cooperation, and with periodic stagnation periods in which the illegal grey or black economy is growing [10].

VINs have previously primarily been important for administrative and security reasons [3], i.e., for purposes of registration offices, police [2] and the other security forces, for forensic documentation [13]. With inevitable progress, VINs are now being implemented into rescue systems (eCALL[1] project). Here VIN quality determines the quality of information, organisational and technical readiness of rescue crews before they arrive to the scene of an accident, and solutions for the hazardous situations which the vehicle passengers who called for help are exposed to. The exchange of technical, administrative and ownership information occurs between the vehicle, the integrated rescue system, and the other co-operating systems. Based on the accident scene, its reachability, and the vehicle technical specifications, the rescue crews define the necessary means for the rescue operation within tens of seconds. In other words—the VIN quality within information systems determines whether the rescue crew obtains all the necessary and correct information from vehicle databases (like the central register of vehicles) in time.

Today, the VIN quality is also important in commercial processes. These are the information systems of insurance offices, leasing companies, vehicle producers, service shops, and assistance services. The communication intensity between various vehicle information systems also increases here. The aim is to exchange the information, improve services in general, and increase customer satisfaction as well as management comfort with growing profits.

In practice, VIN identifier quality in many information systems suffers from quality [12, 13], the error rate is relatively high, and the information system extraction is low. This article outlined the causes, mechanisms and ways that errors occur when VINs are manually transcribed into information systems.

In practice, it is necessary to carefully check the VIN quality at different levels. These are mostly single or repeated training courses, the analysis of errors occurred, and determining methods on how to remedy them in co-operation with the system key and methodology users, and not least with the management.

10 Future Work

In general, the data (VIN) quality in information systems (IS) can be ensured in two ways. The first way is various system and control measures (methods, SW, etc.) during manual VIN transcription into IS [7]. This means verification of entered VINs, like elimination of illegal characters (I, O, Q), incorrect VIN lengths (different from 17 characters), control of the logical VIN structure and digits (using VIN decoders). The second, more advanced way is to avoid manual entries and use automatic VIN readings from vehicles or documents for direct transfer to the information system (Fig. 4). This involves methods of using barcodes or 2D QR codes (that contain the VINs), reading VINs directly from vehicle control units with the use of the OBD[2]

[1]Emergency call.

[2]OBD—On Board Diagnostic.

Fig. 4 An example of a VIN with bar code in parallel for fast, comfortable, and especially error-free transfer of the VIN to information systems

interface. Another way is using reference databases, where the VIN correctness is guaranteed so it can be used as a reference for other information systems.

To eliminate errors in the VIN, the following steps can be implemented:

- Introduce trivial computer controls to forbidden characters in the VIN structure (the letters O, I, Q are forbidden to use) and the length of the text string that must contain exactly 17 positions. These measures must be implemented when writing a VIN to any information system;
- To convince global vehicle manufacturers to use a unique check digit mechanism defined by ISO 3779:2009 for VIN quality checking. It seems ideal to change the status of the recommendation to the obligation to introduce a check digit.
- Create, update special VIN decoders to check VIN quality when entering this item in all information systems.
- Train all users about the meaning of the VIN, its structure, and its completion policies [5].

References

1. Augustin P, Odler R (2013) The mission of the police in a democratic state in the context of globalization. In: Securitologia: czasopismo naukowe, pólrocznik, vol 18, no 2, pp 55–64. ISSN 1898-4509
2. Brunova M (2019) Forensics characteristics of vehicle theft. In: Porada V et al (ed) Criminalistics. Forensics science and cyber aspects. Plzen, Cenek, Czech Republic, pp 1015–1023. ISBN 978-80-7380-741-2
3. Felcan M (2008) Implementation of European Union legislation and regulations on road safety standards of the Slovak Republic In: Security of transport on the road, pp. 122–132. ISBN 978-80-232-0292-2
4. Kolitschova P, Kerbic J, Rak R (2018) Aspects of vehicle identification labels. Forensic Eng 29 (3):2–6 (CERM, Brno, Czech Republic). http://doi.org/10.13164/SI.2018.3.2 ISSN 1211-443X
5. Kopencova D (2020) Secondary education with security focus. INTED 2020 Proceedings. In: 14th international technology, education and technology and development conference. 2–4 Mar 2020, Valencia, Spain, pp 2477–2481. ISBN: 978-84-09-17939-8, ISSN: 2340-1079
6. Matouskova I, Moravcik L, Rak R, Tallo A (2015) eCall, intelligent transport system (legal, technical, informational and psychological aspects). Magnet Press Slovakia, Slovakia, Bratislava, 189–215 pp. ISBN 978-80-89169-31-3

7. Moravcik L, Jasksiewicz M (2018) Boosting car safety in the EU. In: Automotive safety 2018: proceedings of the XI international science-technical conference. Kielce University of Technology. ISBN 978-1-5386-4578-9, Web of Science 2018-07-09, 345 E 47th ST, New York, USA, ISBN 978-1-5386-4578-9, IDS Number: BK3OJ, Accession Number: WOS: 000435296000013
8. Moravcikova J (2016) Cross-border exchange of information about traffic offences through EUCARIS. In: Automotive safety 2016: proceedings of the X international science-technical conference. Kielce University of Technology, Poland, pp 227–234. ISBN 978-83-63792-70-1
9. Rak R, Kolitschova P, Kerbic P (2018) Forensic and technical aspects of vehicle identification labels. In: 11th international scientific and technical conference on automative safety, Casta Papernicka, Slovakia, 18–20 Apr, Web of Science Access Number: 000435296000046, 2018-07-09, 345 E 47th ST, New York, USA, IDS Number: BK3OJ, ISBN 978-1-5386-4578-9 (2018)
10. Roubal O (2017) Sociology of branding: "Just do it" in the "No Limits" World. Communication Today, FMK UCM Trnava: Fakulta masmediálnej komunikácie Univerzity sv. Cyrila a Metóda, year. 8, no 1, pp 40–52. ISSN 1338-130X
11. Roubal O (2019) The duality of hedonism in the ambivalent world of polarities. Eu J Sci Theol. 15(1):203–213. ISSN 1841-0464 (Technical University of Iasi, Iasi)
12. Sarkan B, Gnap J, Caban J (2017) Investigation of exhaust emissions of vehicles with the spark ignition engine within emission control. Proc Eng 187:775–782 (Transbaltica: transportation science and technology). https://doi.org/10.1016/j.proeng.2017.04.437
13. Sekyrova J, Kopencova D, Manas R (2010) Documentation methods of forensic crime scene, Czech Ministry of Interior, Prague, MV-39230-1/VO-2010, evidence number 9/2010
14. Skrucany T, Sarkan B, Figlus T, Synak F, Vrabel J (2017) Measuring of noise emitted by moving vehicles, dynamics of civil engineering and transport structures and wind engineering. MATEC web of conferences, vol 107, Article Number: UNSP 00072. https://doi.org/10.1051/matecconf/201710700072

Equilibrium and Limit States in Technical and Social Disciplines as Part of Risk Analysis and Management

Dagmar Kopencova, Miroslav Felcan, and Roman Rak

Abstract Basic terms such as 'phenomenon', 'event', 'state', and 'situation' are defined for risk analysis and security needs. Other important terms include 'equilibrium' and 'stability' (as well as related terms 'instability' and 'lability'). These terms can be found in both security and technical sciences as well as economics and social sciences. This paper generally defines and explains these key concepts across different disciplines and fields of human activity. Equilibrium states are known mainly from the field of physics—mechanics of solids. However, equilibrium states, stability and their changes are of great importance in the analysis of risks and safety, of various processes and objects in engineering practice (including automotive and other industry, commercial), but also in other areas of human activity. Steady state analyzes help to understand, define threats and risks and manage them effectively; improve the reliability, safety of technological and other components or units. The paper is mainly of a theoretical nature and originated in research within the security sciences. The paper redefines the concepts of equilibrium states, stability in the area of risk analysis and explains their importance for risk management. Impairing the stability of any object, subject or process can always become a dangerous risk that deserves our attention. The terms equilibrium states, stability belong to the basic concepts in the theory and practice of security sciences.

Keywords Equilibrium · Equilibrium position · Balance of forces · Stability introduction

D. Kopencova (✉) · R. Rak
Department of Security and Law, The College of Regional Development and Banking Institute—AMBIS, a.s., Lindnerova 1, Prague, Czech Republic
e-mail: dagmar.kopencova@seznam.cz

R. Rak
e-mail: roman.rak@irisident.cz

M. Felcan
Department of Administrative Law, Police Force Academy, Sklabinska 1, Bratislava, Slovak Republic
e-mail: felcan.m@gmail.com

M. Abdel Wahab (ed.), *Proceedings of 1st International Conference on Structural Damage Modelling and Assessment*, Lecture Notes in Civil Engineering 110,
https://doi.org/10.1007/978-981-15-9121-1_22

1 Introduction—Balance, Equilibrium State, Equilibrium Position, Stability

It is generally known that when objects, systems or processes, actions, or events [1, 2], are in equilibrium, equilibrium states, or equilibrium positions, and are capable of maintaining stability, they are, in a certain sense, safe over time, that is, there usually are no unexpected security incidents, actions, or events associated with them. In short, they behave as we expect (as they were designed, projected, manufactured, tested, etc. to do). Only after losing their equilibrium, equilibrium state, or equilibrium position, and consequently their stability, do security events or incidents occur, causing damage or harm [3, 4]. Therefore, equilibrium and stability are important characteristics of objects, systems, or processes that determine their behaviour in terms of security [5].

The concepts of equilibrium, equilibrium state, equilibrium position, balance of forces, and stability are not generally defined, but rather their definitions are usually targeted to certain occupational areas. Among the many synonyms of the word 'stability' are steadiness, fixity, equalisation, balance, constancy, constancy of properties, persistence, and consistency. In general, 'equilibrium' means 'being in line' or 'being mutually balanced'. The antonym of stability is 'lability' [6].

2 Balance, Equilibrium

Equilibrium is generally the state of a system when the actions in all directions are mutually aligned. The concept of equilibrium may have several meanings.

The concept of a balance of forces is encountered in various areas, such as the military, economics, business, sport, etc. By 'forces', we can mean things like the forces of nature, physics, military, politics, society, religion, hostile actions, criminality, etc.

By the 'balance of forces', we mean a balance that generally leads to stability. All the forces are balanced, that is, no force prevails over any other forces (Fig. 1).

3 Equilibrium Position

In physics (mechanics), the **equilibrium position** is defined as the position of a solid body at which the resultant of all the forces acting on the body is zero, and the resultant torque of all the forces is zero as well. The equilibrium position is a position resulting from the balance of forces.

And vice versa:

The **balance of forces** is the state when several forces act on the body, but their resultant is zero, and the resultant force moment resulting from the composition of all the force moments is also zero (Fig. 2).

We distinguish (in mechanics) three basic equilibrium positions:

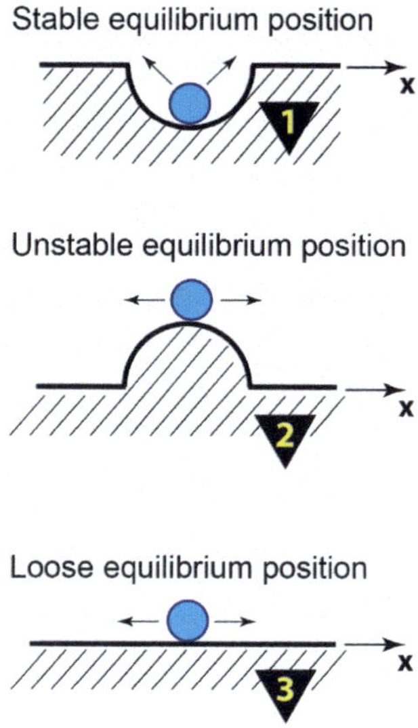

Fig. 1 Illustrative representation of equilibrium positions in solid mechanics

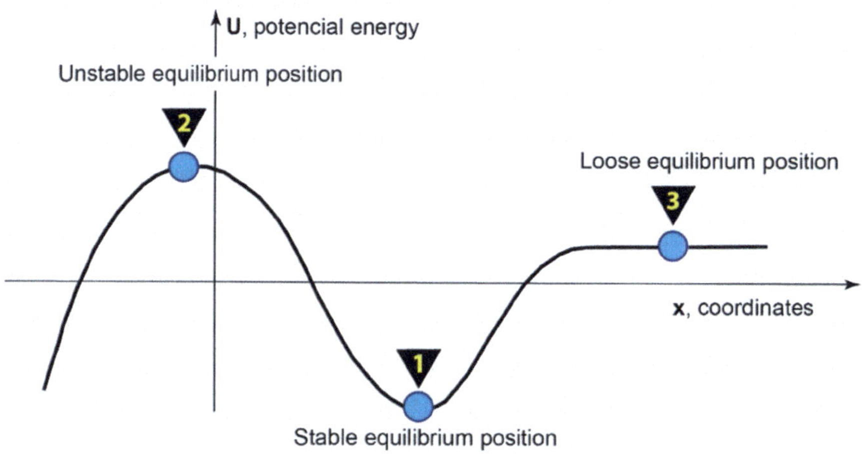

Fig. 2 The dependence of potential energy U on the x coordinate of the ball when it moves from the unstable equilibrium position to the stable and then to the free equilibrium positions

- **The stable equilibrium position** (also **the steady-state equilibrium position**) is a position wherein a body returns to its original position after it is deflected from this position, that is, the deflection gradually decreases. The potential energy of a body at steady-state equilibrium is the smallest, and it increases when deflected.
- An example is a ball located in a pit. When deflected, the ball will return to its starting position. The deflection increases the potential energy of the ball.

- **The unstable equilibrium position** (also **the labile equilibrium position**) is the position wherein a body does not return to its original position after deflection, but rather the deflection increases further. By deflecting from the unstable position, the body's potential energy decreases.
- An example is a ball located on top of a hill. When deflected from its position, the ball will always roll down and not return to its starting position on its own. The deflection decreases the potential energy of the ball.
- **The free equilibrium position** (also **the indifferent equilibrium position**) is the position wherein, when deflecting the body, neither the force resultant nor the resultant force moment acting on the body enacts change. When the body is deflected, the distance from the new position does not change (it neither increases nor decreases). When the body is deflected, its potential energy remains constant.

An example is a ball located on a horizontal plane. If we move the ball to another place, it will stay there and will neither move away from its original position nor return to it. The potential energy remains constant.

4 Equilibrium State

In thermodynamics, the equilibrium state is defined as the state of a thermodynamic system wherein there are no flows of extensive values (heat, mass, energy, etc.). Intensive values are often the same throughout the system in this case.

One of the thermodynamic postulates states that every system reaches its equilibrium. **Every system that is under constant external conditions from a certain point of time will spontaneously move to its equilibrium state after a certain period of time. It remains in this state as long as the external conditions remain the same**.

The aforementioned statement is general, in that it applies both to the natural, social, political [7, 8] and other equilibrium states that are characteristic of objects, systems, and processes. In the process of every development, evolution, and so on, the equilibrium state changes, stability falls, and consequently new equilibrium states are created regularly at different lengths of time. From the social, human, or security points of view, it is always a question of whether a new equilibrium state is desirable ('must the house necessarily burn, so we may build a new one?'). In some (numerous!) security cases, it is generally undesirable to lose the stability of any current object, system, or process [9].

Every state, every equilibrium state can be characterised by its characteristic, significant parameters and variables. Any change in a parameter that is significant for an equilibrium state (object, system, or process) can disrupt the equilibrium state and lead to the loss of stability [10].

The loss of stability of equilibrium states is a natural phenomenon that is part of the antagonistic world in terms of evolutionary and revolutionary development [11]. While the tendency to lose the stability of equilibrium states can be reduced to a large extent by preventive tools, in most cases it is not possible to prevent stability loss [12]. The loss of equilibrium state stability can be divided into two basic types according to the courses they follow [13]:

- Soft loss of stability,
- Hard loss of stability.

This issue is mainly addressed by the mathematical 'theory of disasters' developed by Russian mathematician Vladimir Igorevich Arnold (1937–2010), French mathematician and philosopher René Frédéric Thom (1923–2002), and especially Sir Erik Christopher Zeeman (1925–2016) of Great Britain.

5 Soft Loss of Stability

In the case of a soft loss of stability, an oscillating periodic mode becomes a steady mode of the system, which differs little from the equilibrium state in the beginning. The first symptoms of stability disruption may not be initially observable at all, as they arise slowly, gradually. However, a gradual change in parameters may ultimately cause a loss of system stability [13].

6 Hard Lost of Stability

Accidental or deliberate immediate and fundamental changes of parameters and their manifestations of disruption of a system to such an extent that stability is completely disrupted constitute hard losses of system stability. The system leaves its equilibrium state by jumping to another developmental mode. It may be another stable stationary mode, stable oscillation around an equilibrium state, or a more complicated uneven movement [13] (Fig. 3).

7 Stability

In physics (mechanics) **stability** is defined as the difference of the potential energy of a body between its unstable position and its stable equilibrium position, or in other words the amount of work that must be completed for the body to move from its stable equilibrium position to its unstable equilibrium position.

The stability of a body directly and proportionally depends on the weight of the body, inversely proportional to the height of its centre of gravity at its stable position and directly proportional to the height of the body's centre of gravity in its unstable position.

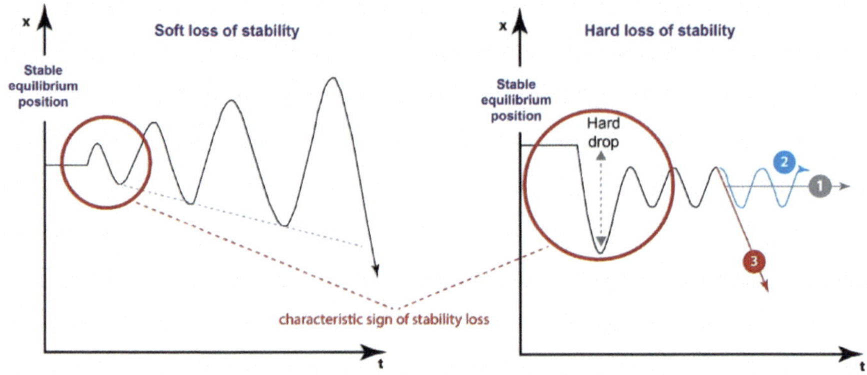

Fig. 3 Comparison of soft and hard losses of stability over time

In other physical disciplines (such as optics, chemistry, physics, electrical engineering, etc.), **stability** is defined as the **ability to maintain certain properties unchanged over time**; the humanities (such as linguistics, politics, etc.), are concerned with the **stability of links over time, links that keep a system whole**.

In a technical environment, stability is usually also defined as the ability to recover from failures and imbalances, that is, a given object, process, or system's ability to return to an equilibrium state.

Example:

Aircraft stability means the ability of an aircraft to maintain the flight mode established by the pilot. If the aircraft had no stability capability, in certain flight modes it could not be piloted, and a security incident would take place—an aircraft crash.

In practice, we talk about the stability of mechanical systems; chemical or thermodynamic processes; the geological stability of subsoil; currency stability; social, political, and economic stability; stability of our financial income; stability of health; etc. [1] (Fig. 4).

Example:

Stability can be explained by the example of a sailing ship exposed to a crosswind acting with the force Fv. The significant parameters for the loss of stability of the ship are the crosswind forces, the location of the ship' centre of gravity T (structural design, load fixing), and the ability of the captain and crew to handle the ship.

Under a certain maximum wind force Fv_MAX the mast tilts to the angle α. If the cargo the ship is carrying belowdecks is not sufficiently fixed, it will move with the tilting of the ship, meaning the ship's centre of gravity will change its location to T2. If this angle is $\alpha < \alpha MAX$, and the wind force decreases, the tilt of the sailboat mast will automatically return to its original, vertical, equilibrium position. However, if the wind force increases beyond its limit Fv_MAX, the ship will experience instability and will capsize. From this new equilibrium position, the ship will not right itself to its

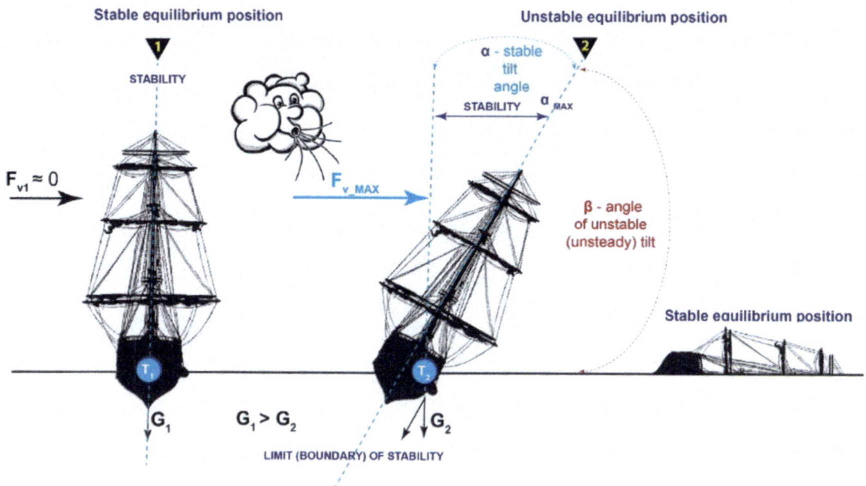

Fig. 4 A sailing ship's loss of stability as an example of the effect of significant external factors on equilibrium and equilibrium position

original vertical, stable position. The captain, to prevent his ship keeling over, must take the following measures: the heaviest cargo must be placed as deeply as possible in the underdecks; in case of a strong crosswind, the boat must be manoeuvred to eliminate the crosswind and turn it into tailwind by changing the position, or f, of the boat; or the sails must be scandalised (rolled up). Strong crosswinds are always a major change in the sailing environment (Fig. 5).

Every stability has a certain limit, boundary, or duration in the real world, a point where objects, systems, or processes are no longer stable and become unstable (unsteady), with all their negative security aspects and their impacts [12].

Stability can be a very relative term—some objects, systems, and processes do not change during the life of a single human, while other objects become unstable (unsteady) in a fraction of a second [14].

In the processes associated with threat, vulnerability, risk, and security analyses [15], we often analyse the properties, characteristics of objects, systems, and processes of interest. For this reason, it is necessary to pay attention to everything, in all contexts, that affects the stability/instability of the object of interest. Usually, our goal is to maintain the maximum stability to avoid potential security phenomena, events, disputes, conflicts, crises [13, 4] etc.

Stability (from the Latin *stabilis*—stable, permanent, fixed) can generally be characterised as a property or ability of an object, system, or process to automatically maintain its stable, essential characteristics, significant parameters, equilibrium state, equilibrium position, equilibrium, integrity, structure, functionality, complexity, or other significant features when faced with deviation from its equilibrium; that is, the ability to automatically return to an original equilibrium state, position, or equilibrium within some time.

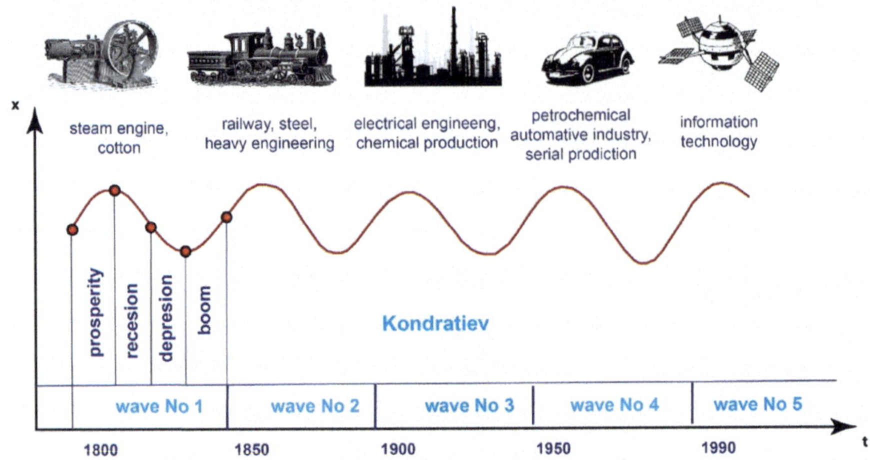

Fig. 5 The world economy develops in business cycles. Periodically, phases of prosperity, recession, depression, and conjuncture change. The Kondratiev cycle takes place, on average, over 50 years. There are more cycles of this kind in the economy, and the author of this overview in particular differs in what he believes explains the causes of prosperity—in the last 200 years, technological inventions prompted the population's demand for new, mass-produced goods. Periods of stability and lability alternate cyclically

8 Social Stability

Social stability is the equilibrium state of a social system when changes take the form of gradual adaptation to changes in conditions and the environment. Achieving such a state is at the heart of sociology. The various theories of sociology deal with the issue of social stability, for example in thoughts about the optimal relationships among the factors of social statics and social dynamics; applied sociology wishes to contribute to an increase in social stability by solving specific social issues or through engineering interventions in social relationships. The urgency of ensuring social stability often arises in a modern society that has replaced tradition, a highly effective stabilising factor, with the need for continuous innovation. If the social stability of a traditional society had been very close to social stagnation, modern stability often cannot be achieved at the expense of change, but rather through creating the right conditions for regulating changes in course. The issue of social stability [16], which at a theoretical level does not pose a grave issue, is in fact one of the most serious issues of practical politics. While totalitarian systems often ensure social stability by mortifying civil society, democratic societies largely achieve their social stability through excessive consumption available to the majority of the population.

9 Economic Stability

Economic stability is a state expressing the existence of assumptions under which, after every external disturbance of economic equilibrium, such equilibrium is restored either in its original form or as a new equilibrium. If any great deviation from equilibrium generates forces that return the system to its equilibrium state, we talk about a global stability; in the case of minor deviations, local stability.

The term economic stability means the steady GDP growth, price stability, and maintaining unemployment at the level of natural unemployment. This means that the product should grow evenly and at such a pace to ensure price stability and sufficient employment.

10 Financial Stability

Financial stability or **financial system stability** is a state of financial markets in an economy that prevents the creation of system risk, that is, the risks of providing necessary financial products and services that disturb the financial system so much that it may significantly affect economic growth and welfare. The origin of potential system risks in a financial system must be handled through macro-scale precautionary policies whose goal is to maintain financial stability.

The Czech National Bank defines financial stability as a situation where the financial system meets its functions without any serious failures or undesirable consequences for both the current and future development of the economy as a whole, and at the same time shows a high degree of resilience to shocks.

11 Instability as a Security Risk

Any fact, activity, event, or phenomenon that occurs in nature, society, the economy, technological processes, or elsewhere in a stable manner may, as a rule, become unstable (unsteady), risky, and uncontrollable with the real possibility of causing harm or losses to individuals, groups, institutions, companies, society, or the state. However, if one understands the causes of changes in objects, systems, or ongoing processes, if one is able to detect their symptoms in a timely manner, one may seek ways to eliminate or minimise threats (dangers), develop ways of reducing the risks of security events, avoid crisis phenomena, or eliminate negative phenomena before they affect any objects, systems, or processes. Detecting and continuously monitoring parameters that may become crisis factors for a given situation allows one to avoid negative impacts and to achieve a corresponding level of security [14].

12 Boundary States

The term 'marginal state' is a term behind a concept that may be applied to different entities in various fields of human activities (including security)—to man-made artefacts, to living or non-living 'products' of our planet's evolution, and so on. Generally, the cut-off state is the point in time when an entity (object, process, or system) is severely disrupted, causing the entity to lose its original functionality. Limiting states are applicable to humans; society; products; technical, natural, or social entities; morale [17] etc.

The knowledge of the existence of limiting states is also very important in security [12]. We work with different entities (objects, processes, systems) in risk analysis and countermeasure design processes. Usually, we use standard or transient states of these entities when there are no real threats; entities are in their steady, stable states over the long term. We often do not realise the limiting states of analysed entities until they appear and completely change the entity's properties, behaviour, and functionality. There is a situation when we cannot ensure security at the required level because the entity has changed fundamentally as a result of its marginal state and has ceased to fulfil its originally intended function or activity.

Limiting states may be divided into categories according to the type of entity in which they occur [18]:

- **Limiting states in technology** occur in technical objects, they are; therefore, the limiting states of technical objects. The reasons for removing a technical object from its function may be subdivided into technical (internal causes; corresponding technical limiting states) and technical-environmental (external causes, corresponding to technical and environmental states).
- **Limiting states of nature** (also ecological, environmental limiting states) occur in natural objects. Ecological limiting states may occur on various scales. They may be local, large-scale, or global. They may be caused by inappropriate human interventions into nature (e.g. amelioration, chemical substance application—fertilisation, preparations against undesirable flora and fauna, construction modifications—dams, motorways, toxic landfills), inappropriate effects of technical objects on nature (emissions, air pollution, soil and water pollution). The cause may also be nature itself—tornadoes, earthquakes, tsunamis, floods; space—collisions with cosmic bodies, etc.
- **Limiting states of humans (living beings)**. The object in this case is a person who may be characterised by their material body, intangible mental processes, and interactions with their surroundings, that is, people, artefacts created by them, nature, and the universe. Personality limiting states may be broken down by the causes that cause them, namely:
  - **Intrinsic**—this includes, for example, health limiting states associated with physical and mental illnesses; ethical limiting states that are related to human behaviour in relation to other people, their artefacts, and nature;
  - **Extrinsic**—this includes limiting states associated with the following factors:

Retirement;
Disagreements at work, in the family, within social or political organisations, sports clubs, other clubs, etc.;
Work overload—psychological limiting states, burnout syndrome.

- **Economic limiting states**. In the economic environment, it is possible to find a number of entities (objects, processes, systems) that reach their limiting states under specific conditions; to a greater or lesser extent threatening economic security (of individuals, institutions, firms, banks, insurance companies, leasing companies, individual industries, markets, communities, states). Some of the monitored system variables (parameters) of the economic state include indicators such as debt levels, loans, credits, mortgages, creditworthiness, ability to repay loans, inflation, unemployment, purchasing power, rates of exchange between local and world currencies, productivity, insolvency, secondary insolvency, personal, corporate, state bankruptcy, debt collection, etc.
- **Limiting social, professional (social, professional limiting states)**. The most widespread social limiting state is a **revolution**. This is a qualitative limiting state characterised by the fact that a certain social order cannot (or is not wished to) continue to fulfil its function. The stimulus for revolution usually does not come 'from above', from the ruling power groups trying to maintain a status quo. The source of revolutions is a long-term dissatisfaction felt by much of the population; thus the stimulus comes 'from below', from the 'oppressed' who no longer wish to continue in a given way. It is not true that in all the cases must there be a revolution. Dissatisfaction with the current state in democratic societies affects politics, elections to various representative bodies, strikes at various levels, public protests, etc., which are evolutionary processes, not revolutionary ones.
- **Morale limiting states**—this includes, in particular, the state of morale and associated limiting states of terrorism, violent religious expansions, etc.

13 Breakdown of Limiting States Per Their Properties

The number of limiting states of various objects may be quite vast, so we can group them, for example [18]:

- Per the nature of state variable changes
- **Qualitative limiting states**—a limiting state occurs if the value (quantity) of some of the state parameters becomes inadmissible for the object function.
- **Quantitative limiting states**—a limiting state occurs if the quality of some of the stated parameters change, that is, if they prevent the object from functioning.
- Per their origin time course
- **Instant limiting states**—the creation of a limiting state only depends on the instant values of the state parameters, which determine the limiting state origin.
- **Cumulative limiting states**—the origin of a limiting state depends on the accumulation of changes in the properties of an object's structure in the process of

progressively acting and influencing the object from its surroundings. Therefore, it does not depend on the instant values of the parameter of this acting and the influence on it, but on their time course.

- Per the potential sequence of limiting states
- **Excluding (disjunctive) limiting states**—two limiting states are referred to as 'excluding' if only one of them may arise, making the other non-achievable (i.e. it will not follow);
- **Causal (subsequent) limiting states**—two limiting states are referred to as causal if the origin of one of the limiting states creates conditions for the potential origin of another limiting state.
- Per the behaviour after object activation removal
- **Reversible limiting states**—after the removal of an object causing a limiting state, its consequences will subside.
- **Irreversible limiting states**—the consequences of reaching a limiting state remain after the object activations were removed.
- Per the character of consequences of reaching the limiting state
- **Disturbance limiting states**—achievement of a limiting state causes a disturbance whereby the object is unable to perform its function, but after the disturbance is removed the object is operational again with its full, original functionality.
- **Security limiting states**—related to object security. Achieving a limiting state will cause the destruction of a component that is part of a protective device against emergency conditions. The destruction of this 'protective component' protects the object from the emergence of other limiting states.
- **Emergency limiting states**—related to object emergency. Achieving the limiting state leads to object destruction.
- Per the possibility of their occurrence
- Per the size of the area featuring the limiting state
- Per the statistical concept of limiting states.
- **Deterministic limiting states**—limiting state characteristics (limiting surfaces) and reliability characteristics are unambiguously determined (the quantifier is defined by a single value).
- **Deterministic limiting states**—limiting state characteristics (limiting surfaces) and reliability characteristics are not unambiguously determined (the parameter is an interval number).
- Per the number of limiting parameters describing a limiting state
- **Single-parameter limiting states**—limiting condition only contains one limiting parameter (for example, a mortgage non-payment warning will come when a payment is 10 days overdue)
- **Multi-parameter limiting states**—the limiting condition contains more than one limiting parameter (e.g. water starts boiling at 100 °C and 1 atm pressure) etc.

14 Conlusions

The issue of limiting states in a systemic approach to security management in the general sense of the word has not been widely used in practice, at least for the time being. Limiting states are well known to engineers and designers, especially in the fields of civil engineering and other types of engineering, power supplies, transportation, etc. However, the knowledge of limiting states may, in fact, be generally applied to any object, providing we consider its specific characteristics. It does not have to be necessarily only products. The theory of limiting states is applicable to social [19], natural, and security sciences as well [7, 20]. Overcoming limiting states, in essence, the loss of stability (in both technical and human objects, processes, and systems), is a common cause of various emergencies, accidents, or, in a more global perspective,[1] diverse technical, social, or natural disasters.

References

1. Mamojka M, Müllerova J (2015) New methodology for crisis management RM/RA CRAMM and its legal frame. In: Production management and engineering sciences—scientific publication of the international conference on engineering science and production management, ESPM 2015. ISBN: 978-113802856-2. https://doi.org/10.1201/b19259-35
2. Müllerova J, Mamojka M (2017) Legal possibilities of the rescue forces during the emergency event. In: International multidisciplinary scientific geoconference surveying geology and mining ecology management, SGEM, vol 17, Issue 51, pp 605-612. ISSN: 13142704. https://doi.org/10.5593/sgem2017/51/s20.122
3. Boulding KE (1962) Conflict and defense: a general theory. Harper & Row, New York
4. Simak L (2015) Crisis management in public administration, University of Zilina. Faculty of security engineering, 259 pp. ISBN 978-80-554-1165-1
5. Simon L (2016) Public security activities, Bratislava, p 149. ISBN 978-80-8054-686-1
6. Hajdukova T (2018) Metodológia výskumu In: Aplikácia vedeckých metód na prípady z policajnej praxe. ISBN 978-80-8054-766-0, s. 8–35
7. Palkova M, Mullerova J, Endrizalova E (2018) Risk management system in Czech republic. In: International multidisciplinary scientific geoconference surveying geology and mining ecology management, SGEM, vol 18, Issue 5.2, pp 1049–1056. ISSN: 13142704. https://doi.org/10.5593/sgem2018/5.2/s20.135
8. Vlach F (2018) Evaluation of cooperation between educational institutions of the armed forces. In: New trends in police training III. International conference. Higher Police School and Secondary Police School of the Ministry of the Interior in Holesov, Holesov, pp 121–124. ISBN 978-80-7616-008-8
9. Blazek V, Dworzecki J, Buzalka J. a kol. Crisis screnarios in public administration. Akademia Policajneho zboru, Bratislava 304 pp. ISBN 978-80-8054-678-6
10. Matuszak Z, Nowak A, Żabińska I, Martínez de Osés FX (2019) Road safety on the example of the city of Bytom. The Archives of Automotive Engineering—Archiwum Motoryzacji 83(1):43–57. https://doi.org/10.14669/am.vol83.art3
11. Dougherty JE, Pfaltzgraff RL Jr (2001) Contending theories of international relations. Addison-Wesley, New York, London

[1] Augustin and Odler [15].

12. Robinson JA, Snyder RC (1965) Decision-making in international politics. In: Kelman HC (ed) International behavior: a social-psychological analysis, New York, pp 435–463
13. Simak L a kol (2006) Terminology dictionary of crisis management, Zilina, p 36. ISBN 80-88829-75-5
14. Makarova I, Shubenkova K, Buyvol P, Mavrin V, Giniyatullin I, Magdin K (2019) Safety features of the transport system in the transition to industry 4.0. The Archives of Automotive Engineering—Archiwum Motoryzacji 86(4): 79–99. https://doi.org/10.14669/AM.VOL86.ART6
15. Augustin P, Odler R (2013) The mission of the police in a democratic state in the context of globalization. Securitologia: Sci J Semiannual 18(2):55–64. ISSN 1898-4509
16. Hendrych D et al (2009) Law dictionary, 3rd issue. Beck, Praha. 459 pp. ISBN 978-80-7400-059-1
17. Jandoure Jan (2001) Sociological dictionary, 1st edn. Portál, Praha, p 128. ISBN 80-7178-535-0
18. Janicek P (2007) System conception of selected fields for technicians. Finding connections, vol 1. Brno, VUT, 682 pp. ISBN 978-80-724-555-6
19. Kopencova D (2020) Secondary education with security focus. INTED 2020 Proceedings. In: 14th international technology, education and technology and development conference. 2nd–4th March, Valencia, Spain, pp 2477–2481. ISBN: 978-84-09-17939-8, ISSN: 2340-1079
20. Odlerova M (2017) Information technology and operative-search activity. In: Act on police corps: application on practice, pp 196–217. Ales Cenek, s.r.o., Pilsen. ISBN 978-80-7380-682-8

The Scale and Shape Effects on the Characteristic Strength of a Rock Mass: Application to Mining Pillars

Youcef Cheikhaoui, Olivier Deck, Kamel Omraci, and Hamza Cheniti

Abstract This study suggests a methodology allowing the evaluation of the characteristic strength of a pillar in mining contexts taking into account the scale and the shape effects. The characteristic strength is understood in terms of probability of exceeding a certain value of strength in compression. An analytical formulation associating the approaches is developed, which takes into account the effects of scale with the concept of probability of occurrence of failure to assess the rupture risk, and the approach which considers shape effect by Back analysis. The first is Weibull approach (Weibull in A statistical theory of the strength of material (1939) [1]) that takes into account the scale effects in its formula and he is introduced the probabilistic aspect of the rupture based on a statistic scale effect (more the defects are bigger; more the probability of failure is greater). The second is Galvin et al.'s approach (Galvin in Int Soc Rock Mech News J 4(1) (1996) [2]) that takes into consideration shape and volume effects but with empirically determined coefficients. An application is proposed using data from Australian coal samples.

Keywords Strength · Pillar · Scale effect · Shape effect · Failure probability

1 Introduction

The influence of the scale effect on the assessment of the characteristic values of the rock mass mechanical properties is crucial for the mining structures, the management of risks associated with resource exploitation (mining, quarrying, drilling, etc.), and land (rock cuttings, tunnels, etc.) [3].

We will focus more specifically on the compressive strength of the pillars of underground structures (mines or quarries) exploited in fractured rock masses. The

---

Y. Cheikhaoui (✉) · K. Omraci · H. Cheniti
L3M Laboratory, National School of Mining and Metallurgy Annaba, Annaba, Algeria
e-mail: youcef.cheikhaoui@univ-lorraine.fr; youcef.cheikhaoui@ensmm-annaba.dz

Y. Cheikhaoui · O. Deck
Georessources Laboratory, University of Lorraine, CNRS, CREGU, Nancy, France

M. Abdel Wahab (ed.), *Proceedings of 1st International Conference on Structural Damage Modelling and Assessment*, Lecture Notes in Civil Engineering 110,
https://doi.org/10.1007/978-981-15-9121-1_23

study of stability in fractured rock masses requires an evaluation of their mechanical behaviour.

A common approach consists in apprehending these discontinuous environments by a continuous equivalent environment [4]. These approaches indirectly take into account a scale effect insofar as the description of the rock mass is likely to vary depending on the volume considered. The study of scale effect has been the subject of several researchers. Weibull [1] developed a probabilistic approach to scale effects based on the notion of weak links. Bazant (1990) [5] developed a theory adapted to quasi-brittle materials such as rock or concrete, but which does not take into account the stochastic distribution of heterogeneities. In rock mechanics, scale effect have been observed in laboratory compression tests. Zengchao et al. [6] have proposed to explain this scale effect from the fractal dimension of the discontinuities in volume.

The scale effect observed experimentally in solid mechanics can firstly be attributed to the presence of defects in the materials. As the volume gets larger, the probability of seeing a significant defect is higher. This results in a reduction in the average properties (strength, rigidity) with the increase in volume. It is therefore a stochastic scale effect. The approach was studied by Weibull [1], who used the concept of probability of occurrence of failure to assess the rupture risk. Weibull presents the failure criterion, which does not allow to deduce simply the resistance of a pillar which is practically related to the volume; but also to the shape (the w/h ratio which is Width/Height of the pillar). Hudson et al. [7] proved that the variation of the length-to-diameter ratio l/d of a specimen has a significant effect on the compressive strength of the rock, as well as on the shape of the stress-strain curve in the post-peak part. This is also what we found in the work of [2] where he developed an analytical formula which takes into account both of pillar geometry effect (through the ratio w/h), and volume.

This work aims to develop an approach to estimate the strength of a pillar via a combined model equations:

(1) Galvin (which allows consideration of shape characteristics and volume of the rock mass);
(2) Weibull allowing implicitly incorporate the concept of probability of occurrence of failure to assess the rupture risk.

In the present work the basic methodology of this combined approaches defined, and a validation of this approach via an application for estimating the compressive strength of the pillars in the coal case. Applying this approach to the Australian coal case study of Galvin et al. [2] is done to allow comparison of results. As well; this contribution is primarily validate the basic model of the proposed approach, and therefore; the presence of discontinuity effect on the compressive strength is not taken in consideration.

2 Methodology

The main proposition supposes that the strength of a pillar of volume is explained by the strength k (of the intact rock), by corrective functions of shape, volume and probability of survival (Eq. 1).

$$\text{Suggestion: } R_p = K.H(ps).F(f).G(v) \tag{1}$$

F (shape): It depends on the geometry of the pillar

G (volume): It is a function of the pillar volume

H (probability of survival): It is the probability of survival associated with an applied constraint such as the probability of survival (Ps) = 1 − probability of failure (Pf).

K: it is the strength of specimen.

By analogy between the power formulas of resistance given by Weibull [1] and Galvin et al. [2], we can rewrite a new resistance formula by the combination of the two approaches (Back Analysis—Probabilistic Aspect of Failure) as follows:

$$\text{Galvin: } R_{p(Ps0)} = R_0.(w/h)^b.V^a \tag{2}$$

where:

R_0: It is the resistance of a 1 m^3 specimen according to Galvin;

a and b: are parameters related to the material, they are determined by the Back-analysis method.

$$\text{Weibull: } R_{p(Ps)} = \sigma_0(-\ln(P_s))^{1/m}.(V_0)^{1/m}.V^{-1/m} \tag{3}$$

The comparison of Eqs. 2–3 allows to suggest the expressions using the Weibull parameters for the coefficients a and b, we will explain the method below. These coefficients a and b are estimated empirically in Galvin's law through tests on pillars (Back-Analysis); this will limit the use of this law because we have to repeat tests on pillars every time whenever the material changes.

Furthermore, if there is a set of N values of compressive strength measured experimentally on test pieces of the same material, of volume V_0 and of slenderness W_{ep}/H_{ep}, it is then possible to obtain the parameters of the Weibull's law m and σ_0.

Or:

m: is a material parameter characterizing the dispersion of the defects within the material;

σ_0: is the solicitation (constraint) associated with a probability of survival of 37%. Weibull defined from its law (Eq. 3), when the applied stress, so Ln (Ln (1/ps)) = 0 which implies ps = 0,37.

Equations 2 and 3 make it possible to give a physical meaning to the coefficient a and b which would be linked to the parameter m of Weibull's law.

$$\text{Such as: } a = -\frac{1}{m} \text{ and } b = \frac{\ln\left(V_0^{\frac{1}{m}}\right)}{\ln\left(W_{ep}/H_{ep}\right)}$$

These two coefficients, which characterize the pillar strength weakening functions:

$$G(volume) = V^a = V^{-\frac{1}{m}} \tag{4}$$

$$F(shape) := (w/h)^b = \left(\frac{w}{h}\right)^{\frac{\ln\left(V_0^{\frac{1}{m}}\right)}{\ln(Wep/Hep)}} \tag{5}$$

In addition, from Eqs. 2 and 3:

$$R_{0,(ps0)} = R_0 = \sigma_0(-\ln(Ps0))^{1/m} \tag{6}$$

As ps_0 represents the probability of survival of samples with an average strength of 1 m^3 proposed by Galvin. By extension, it is proposed that:

$$R_{0,(ps)} = \sigma_0(-\ln(Ps))^{1/m} \tag{7}$$

Suppose now that the Galvin parameters are known. Equation 6 is necessarily associated with a value of probability of survival Ps0 that can reasonably be taken equal to 0.5 if Rp is the mean value observed by Galvin or 0.95 if Rp is a value close to the 5% facile of distribution. The equalization of Eqs. 1 and 2, taking into account Eqs. 3, 4 and 5 then allow to obtain:

$$R_{p(Ps)} = \sigma_0.\ln(1/P_s)^{1/m}.(w/h)^{\frac{\ln\left(V_0^{\frac{1}{m}}\right)}{\ln(Wep/Hep)}}.V^{-1/m} \tag{8}$$

That is to say, in general form:

$$R_{p,P_s} = \sigma_0.H(P_s).G(volume).F(forme) \tag{9}$$

σ_0: is the stress (stress) associated with a 37% probability of rupture of the test pieces.

3 Validation of the Model to the Australian Coal Case

After obtaining the formula that takes into account at the same time the scale and the shape effects form (Eq. 9) a validation of this approach is required; the study is conducted by Galvin et al. [2] the Australian coal is used where:

Table 1 Coefficient of strength variation of different sizes for coal samples at a confinement of 0,2 Mpa [8]

Diameter (mm)	Number of samples	Coefficient of variation (%)
61	03	4,5
101	04	12,6
146	02	3,5

Table 2 Characteristic parameters of compression tests carried out on Australian coal samples (As reported by Galvin et al. [2])

Weibull parameters	Value	Geometric parameters	Value
m	11,56	W_{ep}/H_{ep}	0,5
σ_0 (Mpa)	9,52	$V_0(m^3)$	0,028

$R_0 = 9,1$ (Mpa) is the value of the representative strength of a sample of $V_0 = 0{,}028$ m^3(specimen with a diameter of 261 mm). Using the random generation of resistance values of an average $R_0 = 9,1$ (Mpa) , we set a coefficient of variation of 12% which means a standard deviation $\delta = 1$ (Mpa) on the basis of the study of Mudhurst and Brown [8] on coal samples. The following table presents the number of simple compression tests established on 61 mm, 101 mm and 146 mm samples with the coefficient of variation (Table 1).

We consider a set of 9 tests uni-axial compression to samples with V_0 (m^3) = 0,028 allowing the different parameters of Weibull's law to be calculated. The results are summarized in Table 2.

The strength formula of a coal pillar found using Weibull parameters is written as follows:

$$R_{p\,(Ps)} = 9.52\ln(1/P_s)^{0.086}(w/h)^{0.45}V^{-0.086} \tag{10}$$

The strength formula of a coal pillar given by Galvin is as following:

$$R_{p(Ps0)} = 7.2(w\ /h)^{0.59}V^{-0.067} \tag{11}$$

Such that:

$R_0 = 7,2$ (Mpa) represents the strength of a 1 m^3 of coal estimated by the Salamon formula (after Galvin et al. [2]), we can determine the value of ps_0 when estimating R_0 as shown below.

According to Weibull Eq. 3 and the Weibull parameters of the studied coal by Galvin presented in Table 2, we found that ps0 = 57%.

$$R_{0(Ps0)}(V = V_0) = \sigma_0(-\ln(P_s))^{1/m} \tag{12}$$

From the bibliography (Table 3), using Galvin's formula, suggests values a ϵ [−0,2; 0] and b ϵ [0,5; 1] for the coal case. We then observe a consistency between the values usually found using Weibull parameters.

Table 3 Power law coefficients defined by Eq. 2 from Salamon et al. [9]

Author	R0 [Mpa]	a	b	Comment	Rock type
Zern (1928)	–	0	0,5	–	–
Greenweld (1939)	19,3	−0,11	0,72	Large-scale in situ testing	'Coal'
Holland et Goddy (1957)	–	−0,166	0,83	Laboratory tests	'Coal'
Salamon and Munro [10]	7,2	−0,066	0,59	Back analysis	'Coal'
Wagner (1967)	11,0	0	0,5	Large-scale in situ testing	'Coal'
Bieniawski [11]	4,5	0	0,5	As reported by van Heerden (1975)	'Coal'
van Heerden (1975)	13,3	0	0,5	Large-scale in situ testing	'Coal'
Galvin et al. [2]	7,2	−0,067	0,59	As reported by Salamon and Munro (1966)	'Coal'
Cheikhaoui et al. (2019)	–	−0,086	0,45	Laboratory tests. As reported by Galvin and al. [2]	'Coal'

The table is modified

In this table the values of Salamon and Munro [9], are almost identical with those of [2] because the formula of Salamon and Munro by its nature, it makes it possible to take into account the shape and the scale effects. However, Galvin et al. [2] explicitly expressed the volume as well as the ratio of the form (w/h) of the pillar in their formula. In all cases a takes negative value. On the other hand, for b we have a positive value, which approaches 0.5–0.8. We also note that if $a = 0$ then the strength no longer depends on the volume in these formulas.

The curve in Fig. 1 shows a decrease in strength with an increase in volume. This phenomenon represents the scale effect, a result that was confirmed in the works of [12, 13].

Figure 2 shows that there is a shape effect where the strength increases with the increase in the w/h ratio of the pillar which is consistent with the results of [7, 14].

4 Conclusion

After a bibliographic synthesis, a new formula is established based on a relationship between two aspects (Back-Analysis 'Galvin et al. [2] formula' and the probabilistic aspect of Weibull [1]). This formula allows to take into account the scale effect, the shape effect and the probability of failure in estimating the pillar strength. An analysis of this formula indicates that the pillar strength decreases with increasing volume and it increases with increasing width (less slender pillar is more resistant than a slender pillar).

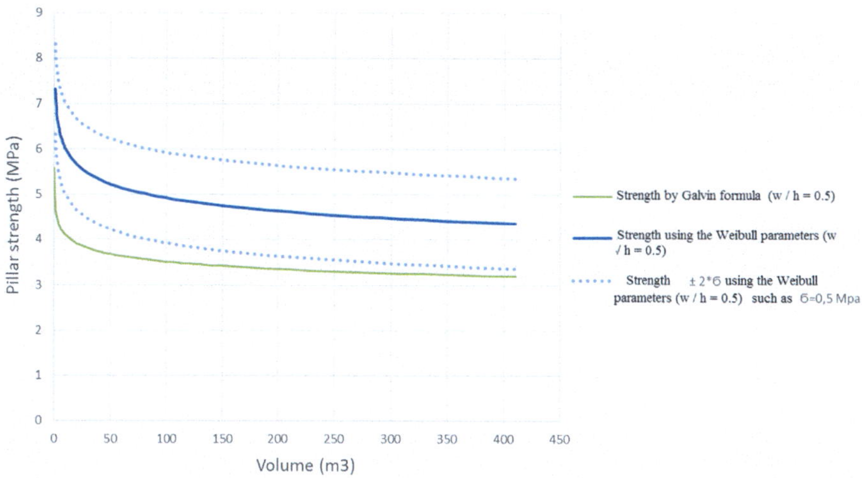

Fig. 1 The effect of volume on the strength of a Coal pillar with (w/h = 0.5) according to the formula of Galvin et al. [2] and the formula developed using Weibull parameters

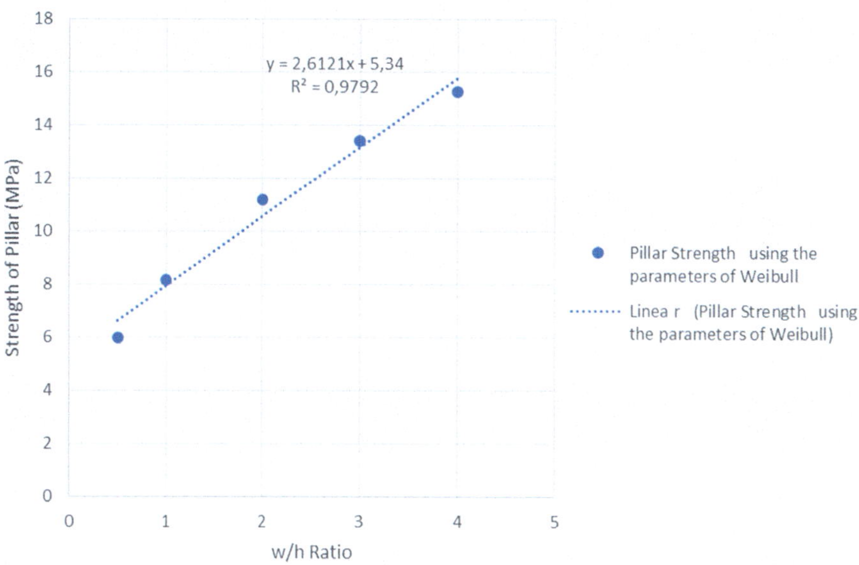

Fig. 2 The shape effect on the strength of a Coal pillar of a volume (10 m^3) according to the formula developed using Weibull parameters

Based on the Weibull parameters, and similarity to the formula Galvin; this approach allows the interpretation of the influence of volume and shape on the strength of pillar.

The confrontation of the results of this approach on samples of the Australian Coal with a priori probability of survival of 0,57, gave results close to those observed by Galvin and others. However, the influence of discontinuity in this basic formula is to be considered in our next contributions.

References

1. WeibullW (1939) A statistical theory of the strength of material. Ingeniorsvetenskaps - akademiens, Handlingar, NR 151, General-stabens Litografisca Anstalts Forlag, Stockholm
2. Galvin JM, Hebblewhite BK, Salamon MDG (1996) Australian coal pillar performance. Int Soc Rock Mech News J 4(l) Fall
3. Tahiri A (1992) Modélisation des massifs rocheux fissurés par la méthode des éléments distincts. Modélisation et simulation. Ecole nationale des ponts et chaussées - ENPC PARIS/MARNE LA VALLEE. Français. Pastel-00574092f
4. Hoek E, Brown ET (1980) Empirical strength criterion for rock masses. J Geotech Geoenviron Eng (ASCE) 106(GT9):1012–1035
5. Bazant ZP, Yu Q (2004) Size effect in fracture of concrete specimens and structures : new problems and progress. In: Proceedings of 5th international conference on fracture mechanics of concrete and concrete structures, vol 44, pp 7–15
6. Zengchao F, Yangsheng Z, Dong Z (2009) Investigating the scale effects in strength of fractured rock mass. Chaos, Solitons Fractals 41:2377–2386. https://doi.org/10.1016/j.chaos.2008.09.005
7. Hudson JA, Crouch SL, Fairhurst C (1972) Soft, stiff and servo-controlled testing machines: a review with reference to rock failure. Eng Geol 6:155–189. https://doi.org/10.1016/0013-7952(72)90001-4
8. Medhurst TP, Brown ET (1998) A study of the mechanical behaviour of coal for pillar design. Int J Rock Mechan Min Sci 35:1087–1105. https://doi.org/10.1016/S0148-9062(98)00168-5
9. Salamon MDG, Galvin JM, Hocking G, Anderson I (1996) Coal pillar strength from back-calculation. The University of New South Wales
10. Salamon MDG, Munro AH (1967) A study of the strength of coal pillars. J South African Inst Min Metall
11. Bieniawski ZT (1968) The effect of specimen size on compressive strength of coal. Int J Rock Mech Min Sci 5:325–335
12. Heuze FE (1980) Scale effects in the determination of rock mass strength and deformability. Rock Mech Felsmechanik Mecanique des Roches 12:167–192. https://doi.org/10.1007/BF01251024
13. Zhang Q, Zhu H, Zhang L, Ding X (2011) Study of scale effect on intact rock strength using particle flow modeling. Int J Rock Mech Min Sci 48:1320–1328. https://doi.org/10.1016/j.ijrmms.2011.09.016
14. Martin CD, Maybee WG (2000) The strength of hard-rock pillars. Int J Rock Mech Min Sci 37:1239–1246. https://doi.org/10.1016/S1365-1609(00)00032-0

Statistical Experimentation for Investigating Optimal Parameter Combination: Friction Stir Welding AA6082-T6 Alloy

Jan-Tore Jakobsen and R. M. Chandima Ratnayake

Abstract This manuscript proposes a statistical experimentation approach for investigating the combination of parameters that provides an optimal friction stir welded joint for the AA6082-T6 aluminum alloy. The statistical experimentation was carried out using a Mazak VCN-430a vertical milling machine. Plates with a uniform thickness of 3 mm were welded in a butt to butt joint configuration. Weldments were produced with FSW tools delivered by Stirweld. The shoulder diameter was 8.5 mm, with a threaded pin of 2.8 mm length. The statistical experimentation was performed using the L_9 orthogonal arrays suggested in Taguchi's engineering robust design approach, with ultimate tensile strength as the quality characteristic. Four factors with three levels were used: rotational speed 1200–1600 rpm, lateral travel speed 100–200 mm/min, dwell time 2–6 s and shoulder depth in the range 0.07–0.11 mm. Analysis of variance (ANOVA) was performed to estimate the error variance for the factor effects and variance of the predicted error. Subsequent joint performance strength testing was carried out using the existing resources. A verification experiment was performed, to verify the applicability of the suggested approach. The optimal parameters were found to be as follows: rotational speed of 1200 rpm, welding speed of 100 mm/min, dwell time of 4 s and shoulder depth of 0.07 mm. The optimal weld had an ultimate tensile strength of 222.7 MPa. X-ray photography in accordance with ISO 17636, radiographic examination, showed no defects. The results of initial and verification experiments are presented. The suggested approach is suitable for developing welding procedure specifications (WPS).

Keywords FSW · Robust design approach · Mechanical properties of friction stir welded AA6082-T6 alloy

---

J.-T. Jakobsen (✉) · R. M. Chandima Ratnayake
Department of Mechanical and Structural Engineering and Material Science, University of Stavanger, PO Box 8600 Forus, 4036 Stavanger, Norway

R. M. Chandima Ratnayake
e-mail: Chandima.ratnayake@uis.no

M. Abdel Wahab (ed.), *Proceedings of 1st International Conference on Structural Damage Modelling and Assessment*, Lecture Notes in Civil Engineering 110,
https://doi.org/10.1007/978-981-15-9121-1_24

1 Introduction

During the mid-1950s, friction welding (FW) of metals was originally proposed by the machinist A. I. Chudikov, and the friction welding approach was patented (No. 106270) during 1956 [1]. Although the possibility of utilizing friction heat for welding had previously been indicated in the literature, only A. I. Chudikov proved the possibility of making high-quality butt welds between two steels rods by friction-generated heat [1]. The Welding Institute (TWI) also adopted FW as a suitable fabrication technique during 1991 [2]. However, Swedish extruder, Sapa, has introduced the first commercial application of friction stir welding (FSW), to manufacture on-board-ship fish-freezing panels [3]. Since then, several other friction welding applications have been used to manufacture deck panels, floors, hulls, superstructures, and platforms [4]. Furthermore, FSW has also been used to manufacture prefabricated panels in workshop conditions and, later, the final assembly has been carried out in shipyards, using fusion welding processes [4]. However, there is no evidence of systematic optimal FSW parameter combination investigations being carried out in the industrial sector [5].

Friction welding has a higher potential in offshore and onshore applications, due to its advantages over traditional welding of Aluminium alloys [5]. It is a solid-state joining method, where the welding process is performed below the materials' melting point, and problems that arise from traditional welding, due to the physical properties of Aluminium alloys, such as high solidification shrinkage, high coefficient of thermal expansion and conductivity, sensitivity to oxide formation and high solubility of hydrogen in a liquid state can give defects that are not a problem with friction welding. Defects from the high heat input and melting of the base material give porosity, lack of fusion, hot cracking, residual stresses and softening in the heat affected zone (HAZ) [6]. FSW has been used to weld materials like Aluminum, copper, magnesium, titanium, and dissimilar material such as copper and Aluminum. Aluminum alloys like 2xxx and 7xxx, which previously were considered un-weldable, can now be welded by friction stir welding [7]. Hence, to enhance productivity, it is vital to investigate optimal FSW parameter combinations for different applications.

In order to obtain the optimal friction welded joint performance, it is very important to carry out statistical experimentation, to investigate the optimal parameter combination. This manuscript presents a statistical experimentation approach and experimental results that have been collected in investigating the optimal parameter combination that provides optimal friction stir welded joint performance for the AA6082-T6 aluminum alloy.

1.1 The FSW Process

The friction welding process for a butt to butt welding configuration is illustrated in [7, Fig. 1]. The plates need to be rigidly clamped together, so that they do not get apart

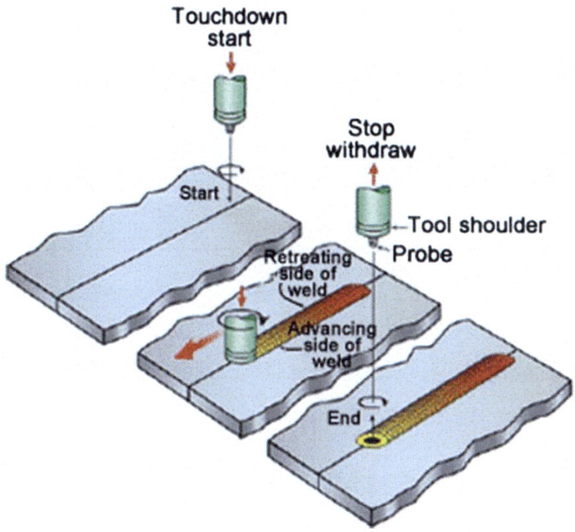

Fig. 1 FSW process for butt to butt configuration

during welding. The welding process starts as the special designed tool is plunged into the joint area between the plates. When the tooltip has penetrated the material and the tool shoulder touches the material, friction is generated by the rotating tooltip and shoulder. The material is softened due to friction and pressure generated by the high downforce. Further friction is maintained by keeping a constant downforce, and the shoulder generates friction during the transverse weld. The material gets into a plasticized state, and the tooltip will now move the material close to the pin with boundaries from the backing plate, shoulder and the material that is not softened by the frictional heat. The advancing side is the side where the flow is going in the same direction as the traverse welding direction, and the retracting side is when the flow is moving in the opposite direction to the welding directions. Friction stir welding can be seen as both a deformation and a thermal process, which generates very high strain at the welded joint [7].

1.2 Microscopic Weld Zones

The microstructure can be separated into four different zones, A-D, as shown in [8, Fig. 2]. In the unaffected zone "A", the material has experienced no change in the microstructure and mechanical properties. The area marked "B" is the heat-affected zone, known as "HAZ"; here, the material has been exposed to the thermal cycle generated from the weld, and the microstructure and/or the mechanical properties have been modified, but no plastic deformation has occurred. The thermo-mechanical zone, "TMAZ", marked "C", is where the material is plastically deformed, in addition to the thermal effect. For aluminum, this zone can have significant plastic strain and be without recrystallization.

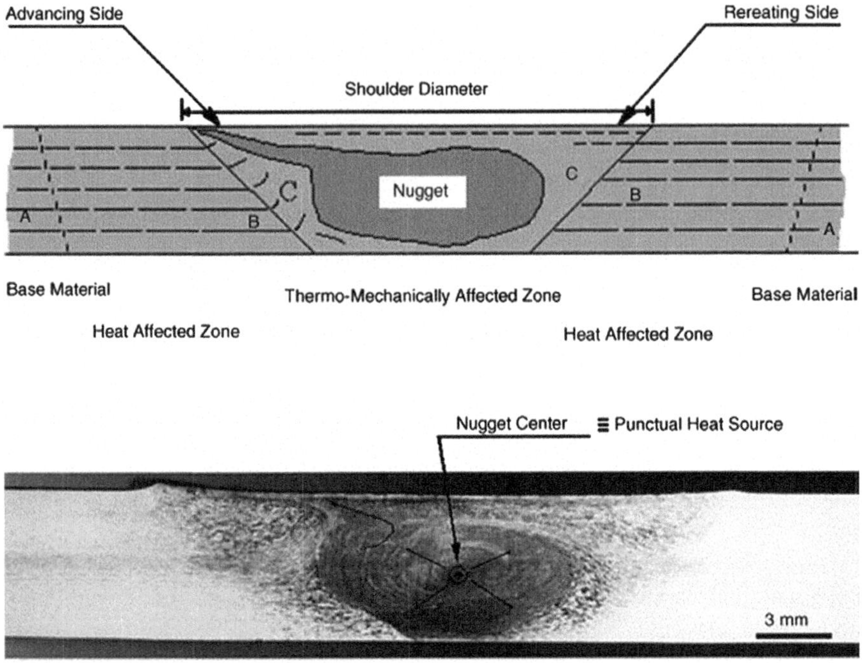

Fig. 2 The weld zones

It is possible to distinguish between the boundaries between the recrystallized zone and the deformed zone. Finally, the stir zone, D, called "nugget", is the zone where recrystallization occurs [7, p. 55], [8, 9].

1.3 Background and Industrial Challenge

Friction welding has been attractive for onshore and offshore applications, due to minimal distortion and good aesthetic appearance, leading to reduced need for post-welding remedial work, lowering the overall cost of the welding operation [4]. Also, it offers user-friendly and environmentally healthy operation, due to the lack of ultraviolet (UV) radiation and/or fumes. Furthermore, it enables the prevention of the flaws, such as hot cracking, porosity, etc., which inherently appear in conventional fusion welding [4]. The possibility for high-level automation of the friction welding process increases the availability of welding operators [4]. Moreover, it provides a high level of ability to weld in different places and/or orientations that are not influenced by gravity, due to the fact that there is no molten metal pool [4]. However, as usual in any welding process, the friction welding process also requires the design of an optimal parameter combination that provides optimal joint performance [10].

2 Methodology

2.1 Material Selection and Preparation of Specimens

The material used for this experiment is AA6082 aluminum alloy, EN-AW 6082-T6-AlSi1MgMn (Yellow). The 6000 series alloys are heat treatable, and the temper designation system for aluminum describes the T6 as a solution heat treated and artificially aged alloy. This alloy is often used in the offshore/marine environment because of its good corrosion-resistance characteristics. Magnesium and silicon are the main alloying elements, and the availability of a large amount of manganese enables the control of the grain structure, which in turn results in a stronger alloy [11]. The chemical composition is presented in [11, Table 1].

The mechanical properties of the AA6082-T6 alloy are presented below in [11, Table 2].

Table 1 Chemical composition

Element	wt%
Si	0.7–1.3
Fe	0.0–0.5
Cu	0.0–0.1
Mn	0.4–1.0
Mg	0.6–1.2
Zn	0.0–0.2
Ti	0.0–0.1
Cr	0.0–0.25
Al	Balance

Table 2 Mechanical properties

Property	Value
Proof stress	260 Min MPa
Tensile strength	310 Min MPa
Hardness Vickers	115 VH
Modulus of elasticity	70 GPa
Density	2700 kg/m^3
Melting point	555 °C
Thermal conductivity	180 W/m K

Fig. 3 Friction welding tool

2.2 *Preparation of Plates for the Welding*

A plate with dimensions 2000 × 1000 × 3 mm was cut into 300 × 100 × 3 mm coupons, using a Kimtech-Bosch waterjet cutter, to avoid any change in the mechanical properties of the material. The cut is along the rolling direction, such that the grain orientation was equal for all welds performed. Grinding with abrasive paper was performed to remove the oxide layer.

2.3 *Friction Welding Tool*

The welding tool used was delivered by Stirweld (see Fig. 3). The triangular flat threaded pin was machined to the desired dimension. The pin length chosen to weld the 3-mm plate was 2.8 mm, and the shoulder diameter was 8.5 mm. The scrolled shoulder is to improve the material stirring and, with this feature, a tool tilt angle will be unnecessary.

2.4 *Welding Rig and Clamping System*

The forces involved in the butt-welding configuration will try to lift and pull the workpieces apart, so it is necessary to have a proper clamping system. The backing plate is made from low-strength carbon steel (see Fig. 4). The backing plate is attached to the worktable and aligned using a measuring probe for precise alignment, so that repeatability can be easily reproduced for each experiment. The clamping used during the first test welds can be seen in Fig. 5. Flat steel bars with dimensions of 500 × 60 × 4 mm in different combinations with toe-clamps were tested to properly clamp the plates against the backing plate. This setup caused defects in every test weld performed. The measured curvature of the inserted plates had a height difference of ±0.1 mm in the weld direction. This uneven alignment of the plates made it difficult to keep a constant downforce and resulted in problems with excessive flash and wormholes.

Furthermore, problems due to a poor clamping system made the welded plates deform along the tool shoulder, as the material is softened in the TMAZ/HAZ during

Fig. 4 Backing plate fixed to the Mazak worktable

Fig. 5 Testing of different clamping systems

welding. The defects can be seen in Figs. 6 and 7. At the back side of the welded specimens are the edge defects mentioned above, which also cause lack of filling.

A more rigid clamping system was tested, as seen in Fig. 8. The steel bars were machined out from a 20-mm carbon steel plate; the surface pressing down against the weld specimens was flattened using a milling machine, to obtain uniform force distribution against the backing plate. The same defects as observed before could be detected using this setup. Several attempts using different toe clamps, applying the same torque on every bolt, and multiple toe clamps pushing down the steel bars, also failed. The measured height difference was slightly improved along the weld path. The most common defect that occurred from the test welds was a combination of wormholes, lack of filling and the previously mentioned edge defect.

In Fig. 9 the clamps are moved as near to the weld area as possible, as suggested by Anette O'Brien [12]. The gap between the clamps was reduced from 40 to 18 mm.

Fig. 6 Lack of filling and excessive flash

Fig. 7 Edge defect at the back side of the weld

Fig. 8 Clamping with 40-mm opening between clamps

Fig. 9 Clamping with 18-mm opening between clamps

This solved the problem that occurred with the edge defect at the back side of the welded specimens and the lack of filling. Figure 10 shows the interpolation of curvature for the weld path using the Mazak measuring probe. Two aluminum coupons ready to be welded can be seen in Fig. 11.

3 OA Matrix Experiment

The robust design approach gives a method for systematically finding solutions that make your design less sensitive to different causes of variation described here as noise factors. This method can be used for optimizing product design and for different manufacturing process designs. Phadke mentions that an OA matrix experiment can

Fig. 10 Measuring probe to find the curvature of the plate

Fig. 11 Specimens ready to be friction stir welded

be used to study the effects of control or noise factors, evaluate S/N ratios and find the optimal quality characteristics for an application [13].

This is a multicriteria optimization approach, using statistical experimentation suggested in the robust design approach (i.e. Taguchi approach). Hence, relationships between physics and factor effects have not been taken into consideration. In fact, the robust design approach has been used to overcome the difficulty of finding a relationship among physics and an optimal combination of the most influential parameters that provides the optimal parameter combinations. In this context, more experimental reiterations more the possibility to coverage into optimal parameter combinations.

To investigate all possible factor combinations for a full factorial experiment, we must complete a total of $3^4 = 81$ runs. With the same numbers of factors and levels using the Taguchi L_9 orthogonal array, only nine runs are necessary to find the optimal parameters [14]. Table 3 shows the experimental setup for the L_9 orthogonal array

Table 3 L_9 orthogonal array

Exp. No.	Levels of process parameters				Observations η (dB)
	A	B	C	D	
1	1	1	1	1	η_1
2	1	2	2	2	η_2
3	1	3	3	3	η_3
4	2	1	2	3	η_4
5	2	2	3	1	η_5
6	2	3	1	2	η_6
7	3	1	3	2	η_7
8	3	2	1	3	η_8
9	3	3	2	1	η_9

with four factors and three levels. All columns are equally orthogonal, meaning that, for every pair of columns, all combinations of factor levels occur an equal number of times.

The quality of the manufacturing process or the product is evaluated through a synthetic performance indicator called the signal to noise ratio (S/N). There are several criteria options, depending on the quality characteristic to be optimized. Equation (1)—The Smaller the Better criterion is applied to experiments when a minimization of the response is required for the output characteristic's data (e.g. surface defects count). Equation (2)—The Nominal the Better criterion uses the mean and standard deviations to find the S/N ratio (e.g. engineering design). Equation (3)—The Larger the Better criterion is applied to experiments when a maximization of the response is required for the output characteristic's data (e.g. finding the ultimate tensile strength).

$$\text{S/N ratio}\,(\eta_j) = -10\log\left(\frac{1}{n}\sum_{i=1}^{n} y_{ij}^2\right) \tag{1}$$

$$\text{S/N ratio}\,(\eta_j) = -10\log\left(\frac{\mu^2}{\sigma^2}\right) \tag{2}$$

$$\text{S/N ratio}\,(\eta_j) = -10\log\left(\frac{1}{n}\sum_{i=1}^{n} \frac{1}{y_{ij}^2}\right) \tag{3}$$

The signal to noise ratio (S/N) is for the calculations of the ratio for the jth experiment, where the overall mean is the average of the squares of the nine observations in the experiment (see Eq. 4), [13].

$$Overall\ \text{mean} = \frac{1}{9}\sum_{i=1}^{9}\eta_i \tag{4}$$

3.1 Running the OA Experiment

The quality characteristic for the experiment is ultimate tensile strength, and therefore the Larger the Better criterion is applied. A fishbone diagram was used to obtain an overview of all elements affecting the quality of the friction stir weld and other causes of noise. The diagram can be seen in Fig. 12. The factors and levels used for the experiment were chosen based on a dialogue with the tool supplier and articles in the literature [15–19]. The four factors used were rotation (rpm), welding speed (mm/min), dwell time (sec) and shoulder depth (mm). The factors with the appropriate levels are described in Table 4.

Table 5 shows the L_9 orthogonal array setup with factors and their respective levels.

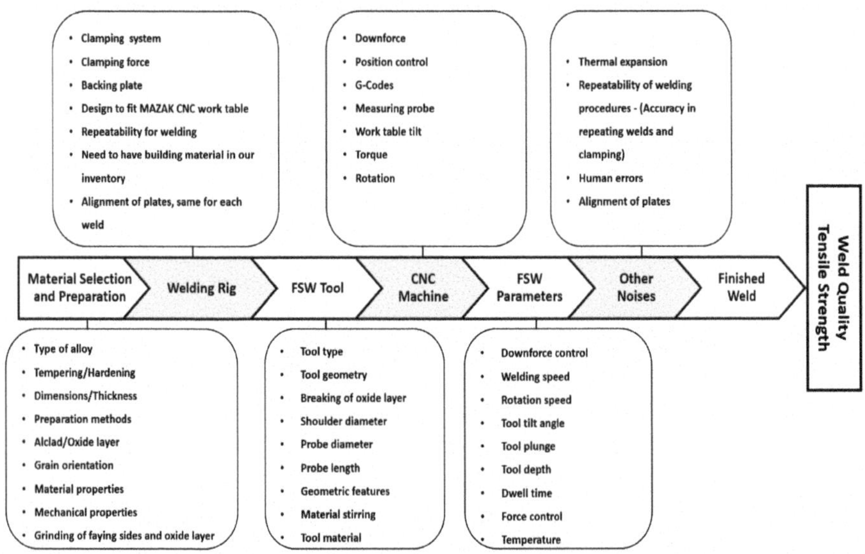

Fig. 12 Fishbone diagram

Table 4 Factors and levels

Factors	Levels			Units
	1	2	3	
Rotational speed	1200	1400	1600	[rpm]
Travel speed	100	150	200	[mm/min]
Dwell time	2	4	6	[s]
Shoulder depth	0.07	0.09	0.11	[mm]

Table 5 Experimental setup

Exp. No.	A Rotation speed	B Travel Speed	C Dwell time	D Shoulder depth
1	1200	100	2	0.07
2	1200	150	4	0.09
3	1200	200	6	0.11
4	1400	100	4	0.11
5	1400	150	6	0.07
6	1400	200	2	0.09
7	1600	100	6	0.09
8	1600	150	2	0.11
9	1600	200	4	0.07

3.2 G-Code Used for Welding

Friction stir welding was conducted using G-codes. Shown in Table 6 is the program used for this experiment, with a short description of each code. Line N03 is used to change the welding speed and the tool rotation parameters. Line N08 is the penetration of the tool tip and the selected shoulder depth of 2.91 mm (2.8-mm tool tip + 0.11 mm shoulder depth). Line 09 is the dwell time for the tool. Line N10 is the starting point, with the measured height difference caused by the curvature of the plates considered. Lines N11–N19 show the X and Z coordinates, where the Z coordinates are measured for every 20 mm in the weld direction, using Mazak's measuring probe. The curvature of the plate is then interpolated into the program.

3.3 Preparation of Test Samples for Tensile Strength Testing

The welded plates were cut into appropriate test specimens using a Kimtech-Bosch waterjet cutter, to avoid any change in the mechanical properties of the material. The cut-out test specimens can be seen in Fig. 13. Tensile test specimens I–III–IV were used for the tensile strength tests. The three samples in between the tensile

Table 6 ISO programming G-codes

Line	Code	Description
N01	G21 G90	Programming in mm/Absolute programming
N02	T14 M06	Select tool #14/Tool change
N03	M03 F150 S1200	Spindle CW/Feed rate in mm/min./Spindle speed in rpm
N04	G17	X–Y plane
N05	G0 X45 Y180.3 Z120	Rapid positioning/Linear motion to coordinates X/Y/Z
N06	X85	Linear motion in X direction
N07	Z5	Linear motion in Z direction
N08	G01 Z-2.91	Linear interpolation using feed speed
N09	G04 P2000	Dwell time, where P is the time given in milliseconds
N10	X105 Z-2.9211	Start of the welding session where Z-2.9211 is the interpolated value used to control the shoulder depth
N11–N19	The interpoled coordinates	Shoulder depth is changed for every 20 mm to follow the plate's curvature using linear interpolation
N20	X305 Z-3.0466	End point of the weld
N21	G4 P2000	Dwell time for the endpoint at 2 s' hold time
Z3	G0 Z300	Rapid positioning/Move to Z300 away from the weld
End		

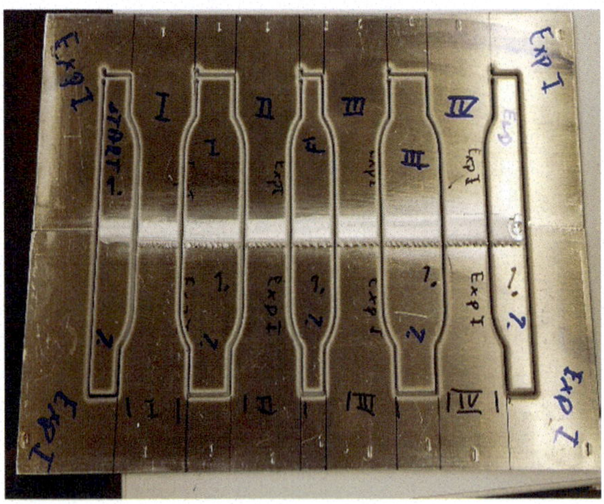

Fig. 13 Preparing test specimens

test specimens were used for microscopic examination, bending test and Vickers hardness testing.

3.4 Microscopic Examination Preparation

The sample used for microscopic examination was taken from the excess material between tensile specimens III and IV. A Struers laboratory abrasive saw with coolant was used to make samples for further examination. The cut-out piece was encased into mounting resin using epoxy. A Struers TigraForce-5 was used for the grinding and polishing process. SIC-paper with P-grade P320–P4000 was used for the coarser grinding and MD-MOL 9 to 1 μm. Barker's etchant solution was used for the anodizing process. From ISO/TR16060 [20, pp. 24–25], the solution is made from 940 ml water (H_2O) and 60 ml fluoroboric acid (HBF_4). The samples were inserted into the Struers LectroPol-5 unit and anodized using the following parameters: area 1 cm^2, temperature 25 °C, voltage 25-volt, flow rate 11, and time set to 120 s.

3.5 Microscopy LOM

The microstructure was revealed using an Olympus GX53 Inverted System Metallurgical Light Optical Microscope with a BX3M-CB/FM control box and several filters (see Fig. 14).

4 Results and Discussion

4.1 Tensile Strength Testing

Tensile specimens were in accordance with specifications in ISO 6892-1 [21].

Table 7 presents the results from the tensile strength tests for the matrix experiments, the mean value and the S/N ratio for each experiment.

To explain some of the results found during tensile strength testing, images of the weld in full, area of interest from the weld and the microstructure of the nugget are shown for experiments 1, 4 and 7. The tensile strength tests revealed significant variance within some of the experiments. The quality of the weld within an experiment also varied. However, the Taguchi model should be able to find the optimum combinations of settings, regardless of these variations.

The surface irregularities and excessive flash in the weld area for experiment 1 can be seen in Fig. 15. Furthermore, wormholes can be found during the microscopic

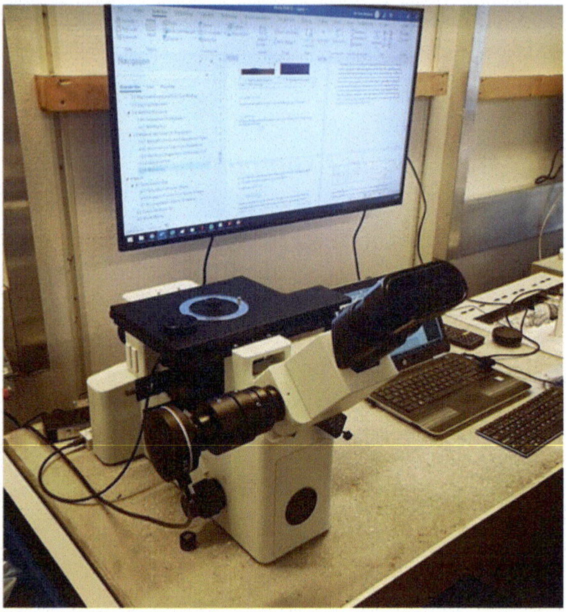

Fig. 14 Olympus GX53

Table 7 Tensile strength results

Exp. No.	Tensile strength results			Mean value	S/N ratio (dB)
	I	III	IV		
1	145.8	192.1	215.5	184.5	44.96
2	208.6	219.9	218.7	215.7	46.67
3	214.4	219.1	214.3	215.9	46.69
4	132.6	173.9	212.8	173.1	44.28
5	219.8	218.1	217.7	218.5	46.79
6	147.6	212.4	214.5	191.5	45.24
7	104.4	55.1	212.7	124.1	38.30
8	218.5	213.5	216.4	216.1	46.69
9	165.3	221.1	216.3	200.9	45.83

examination of the specimen. The back side of the weld has no sign of macroscopic defects and seems to have complete penetration.

The notable macroscopic defects for experiment 4 can be seen in Fig. 16. Lack of filling at the advancing side in the first 150 mm can be detected. Excessive flash increases along the retracting side throughout the weld. Furthermore, wormholes can be found during the microscopic examination of the specimen. The back side of the weld has no sign of macroscopic defects and seems to have complete penetration.

Fig. 15 Experiment 1

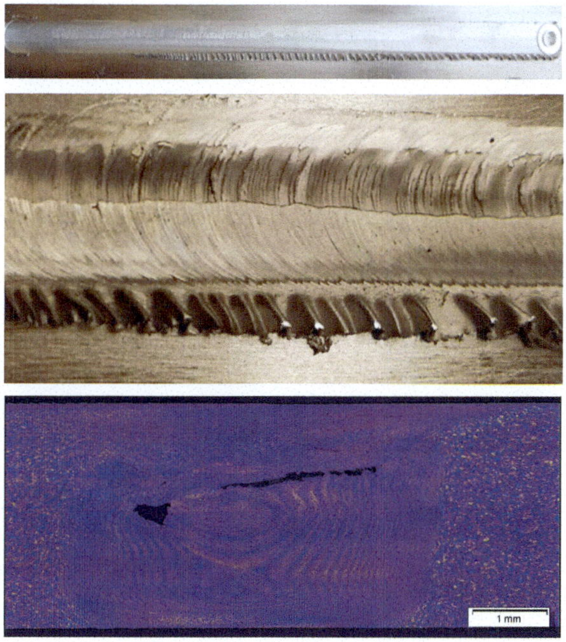

Fig. 16 Experiment 4

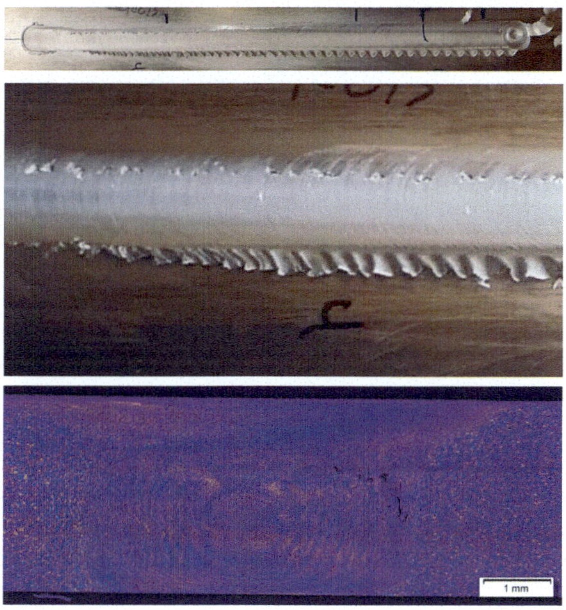

Fig. 17 Experiment 7

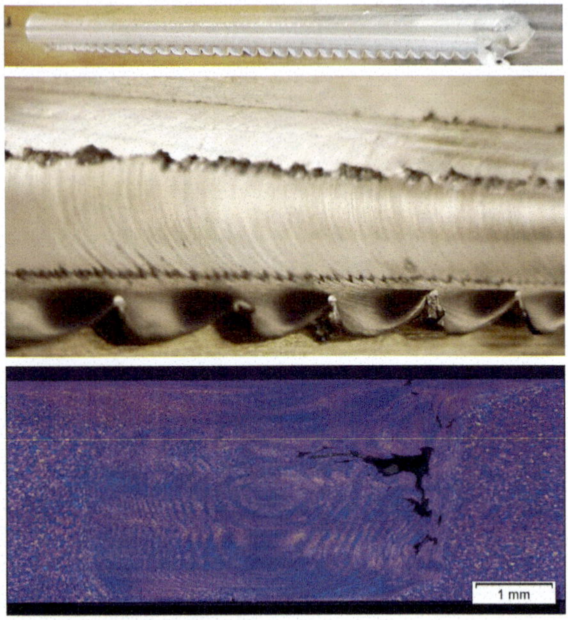

For experiment 7, the problem with the lack of filling in some parts of the welded area can be seen in Fig. 17. Excessive flash increases along the weld path. The surface at the end of the weld seems to be sound. Microscopic examination revealed wormholes and lack of filling.

4.2 Optimal Process Parameters

The effect of a factor level is defined as the deviation it causes from the overall mean [13, p. 45]. The average S/N is calculated for each factor and its corresponding levels. Because of the balancing properties from the orthogonal array, every factor and level has equal contributions of averages [13]. As seen in Fig. 18, the plot of the factor effect was calculated using the tensile test results from Table 7. The effect from factor and level A1 for the rotation speed parameter gave the highest η. For the welding speed parameter, the optimal effect on the process was factor and level B2. The dwell time had approximately the same values for the factor and levels C1 and C2, but C1 was slightly higher. For the shoulder depth parameter, D1 and D3 showed almost identical values, with D3 having the highest value. From this setup, there will be several possibilities for factors and levels for the confirmation experiment, but those with the highest values were chosen.

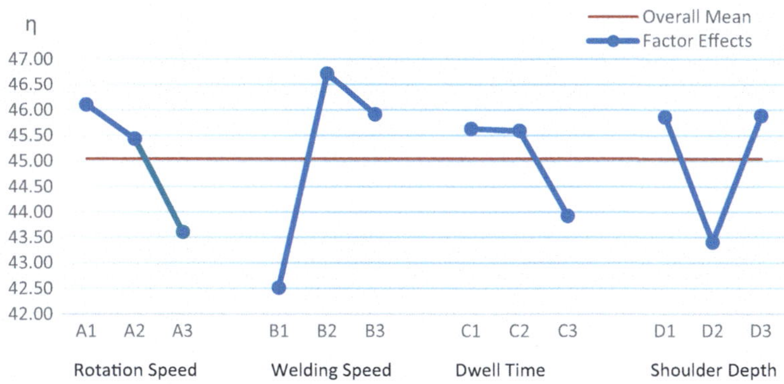

Fig. 18 Plot of factor effects for S/N ratios and mean value

Table 8 Finding the optimum factors

Factor	Level			Overall mean	Optimum level
	1	2	3		
A-Rotational speed	**46.106**	45.436	43.606	45.05	$\mathbf{A_1}$
B-Welding speed	42.513	**46.718**	45.917	45.05	$\mathbf{B_2}$
C-Dwell time	**45.632**	45.592	43.925	45.05	$\mathbf{C_1}$
D-Shoulder depth	45.860	43.403	**45.885**	45.05	$\mathbf{D_3}$

Table 8 shows the values and the parameters for finding the optimum factor levels. A confirmation experiment was performed, as the optimum combination of parameters did not exist in the original matrix setup (see Table 3).

4.3 *Verification of Experimental Results*

Having the highest values, factors and levels A1 and B2 were used to calculate the predicted η under optimum conditions. The η_{opt} was calculated to be 47.77 dB, meaning the predicted ultimate tensile strength is 244.62 MPa. The tensile strength test for specimen "I" had the highest value among all the experiments, and the same applies to for the average mean value for the verification experiment, with an average mean value of 220.7 MPa (see Table 9).

However, the predicted value was around 22 MPa higher than the achieved results, and, with the two-standard deviation confidence interval having a value of ±3.74 dB, this result is in between the limits. The reason for the wide confidence interval may be caused by the fact that the three tensile specimens used to calculate the mean average value for each experiment were taken from the same welded plate. The experiments revealed great quality differences within each weld, e.g. experiment 4 had a range

Table 9 Verification experiment

Result from verification experiment (MPa)			Calculations		
I	III	IV	Average mean	S/N ratio	η predicted
222.7	220.1	219.3	220.7	46.875	47.775

of tensile strength values from 55 MPa to as high as 212 MPa. The cost and time consumption did not allow for welding three plates within each experiment, so the average mean was calculated using specimens from one welded plate, as previously described in Fig. 13.

4.4 Appearance of the Optimum Weld

The full length of the weld and the back side of the weld can be seen in Figs. 19 and 20, respectively. After 50 mm of welding, some excessive flash developed. Figures 21 and 22 show enlarged views of zones from the welded plate.

Fig. 19 Front side of weld

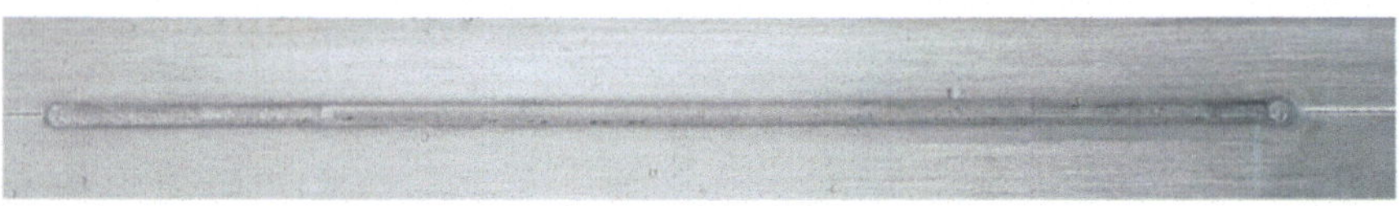

Fig. 20 Back side of weld

Fig. 21 First 50 mm of the weld

Fig. 22 Excessive flash developed

4.5 Analysis of Variance

A full description of the calculation shown in Table 10 can be found in Phadke's book, *Quality Engineering Using Robust Design* [13]. Variation in η for factor in percent gives the contribution to the process and quality characteristics. Welding speed is the main contributor with 51.70%. Second is the shoulder depth with 21.09% and third the rotation speed with 17.37%. The dwell time showed a minor contribution with only 9.84%. Interactions between parameters were not considered because of this project's time and budget restrictions.

Table 10 ANOVA table for η

Factor	Degree of freedom	Sum of squares	Mean square	F-ratio	Variation in η for factor %
A-Rotation speed	2	10.05	5.02	1.28	17.37
B-Welding speed	2	29.90	14.95	3.80	51.70
C-Dwell time	2	5.69	2.85	0.72	9.84
D-Shoulder depth	2	12.20	6.10	1.55	21.09
Error	0	0			
Total	8	57.84			100
Pooled error	4	15.74	3.94		

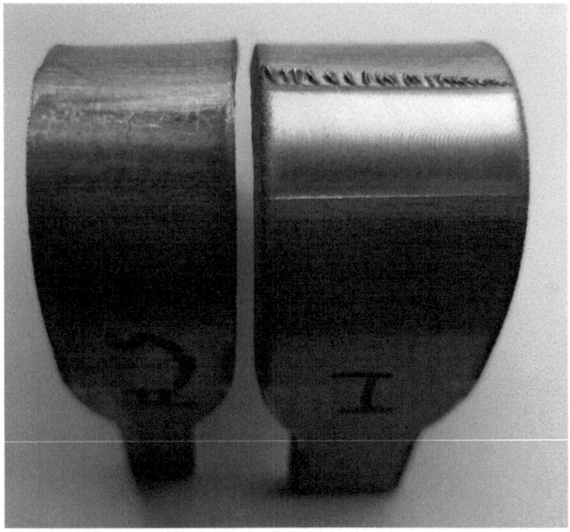

Fig. 23 Bending test

5 Microstructure and Mechanical Properties for Optimum Weld

5.1 Bending Test

A bending test was performed, using a 20-kN Zwick-Roell Z020 tensile machine. A flexure test, found in Zwick software, following standard DIN-EN-ISO 178, was used. The test speed was set to 20 mm/min and pre-load 0.1 MPa. Samples I–II as seen previously in Fig. 13 were used for the bending test. The test revealed no defects in the weld area, and the weld had great ductility (see Fig. 23).

5.2 Vickers Hardness Test

Vickers hardness testing was performed in accordance with ISO 6507-1 [22]. The test specimens used in the hardness test were taken from the optimal weld between tensile specimens III–IV (see Fig. 13). Parameters used were HV0.5 with a nominal value of the test force set at 4.9025 N for 10 s. Lowest hardness was detected in the TMAZ/HAZ zone on the retracting side, with a value of 58.2 HV (see Fig. 24). The heat affected zone seemed to continue throughout all specimens, as the reduction in hardness to around 75 HV was measured over 65 mm away from the TMAZ in both directions of the weld.

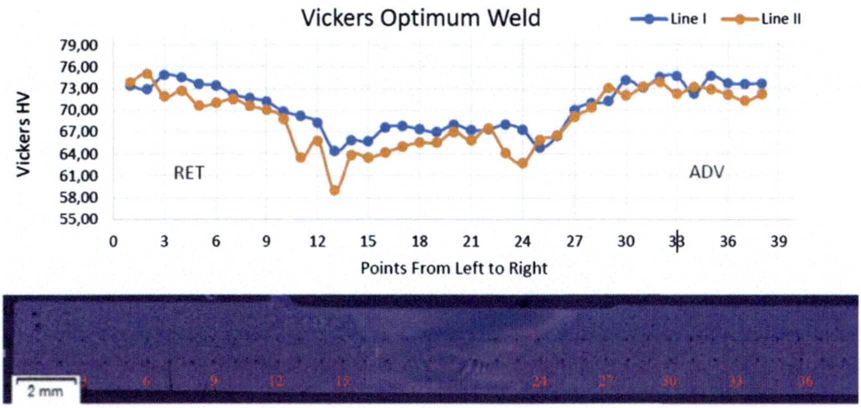

Fig. 24 Vickers hardness test

5.3 Tensile Specimen and Fracture Area

The tensile strength specimens from the optimum weld can be seen in Fig. 25. The tensile specimens all failed in the TMAZ/HAZ region on the advancing side, as seen in Fig. 26. The tensile test revealed values of 222.7 MPa, 220.1 MPa and 219.3 MPa, respectively, for specimens I, III and IV.

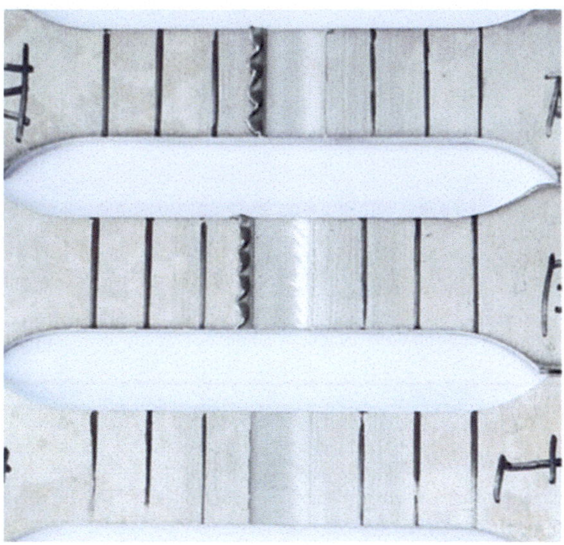

Fig. 25 Tensile specimens I, III and IV from below and above

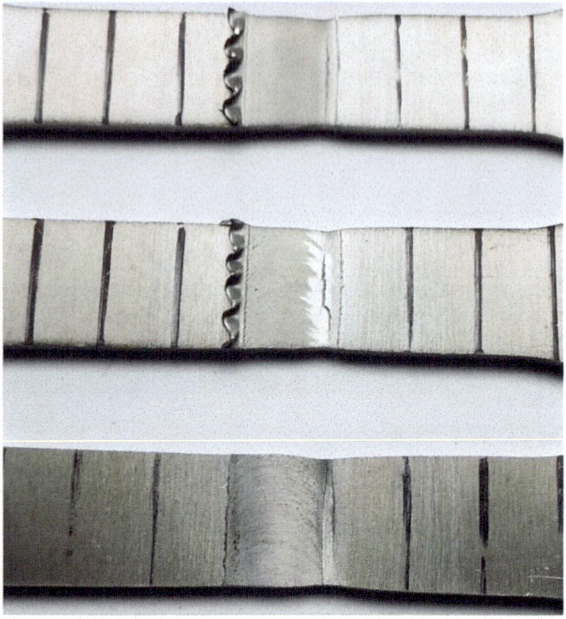

Fig. 26 Failure areas from the tensile strength testing

5.4 Microstructure from the Optimum Weld

The microstructure is revealed as seen in Fig. 27. The mapping was done using the Olympus software, and for Fig. 27 was the 5× lens used to map the image. HAZ was detected over the whole specimen; this was confirmed by the Vickers hardness test, where a drop in hardness was measured 65 mm from the nugget on both sides. Figure 28 is a zoom of the stir zone of the above image. Figure 29 shows a higher

Fig. 27 Mapping with 5x lens with the advancing side to the left

Fig. 28 Zoom of stir zone using the 5x lens

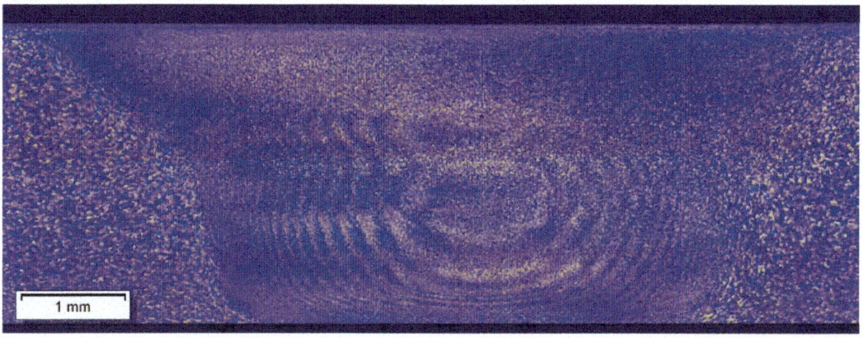

Fig. 29 Mapping with the 10x lens

magnification using the 10x lens and the mapping function. The weld nugget is well defined and "onion" rings are visible. TMAZ/HAZ are detected.

5.5 *X-Ray Confirmation of Weld Using Optimal Parameters*

IKM Inspections AS performed an X-ray in accordance with procedure/standard BPI-01/ISO 17636-1. No defects were detected. The radiographic film can be seen in Fig. 30.

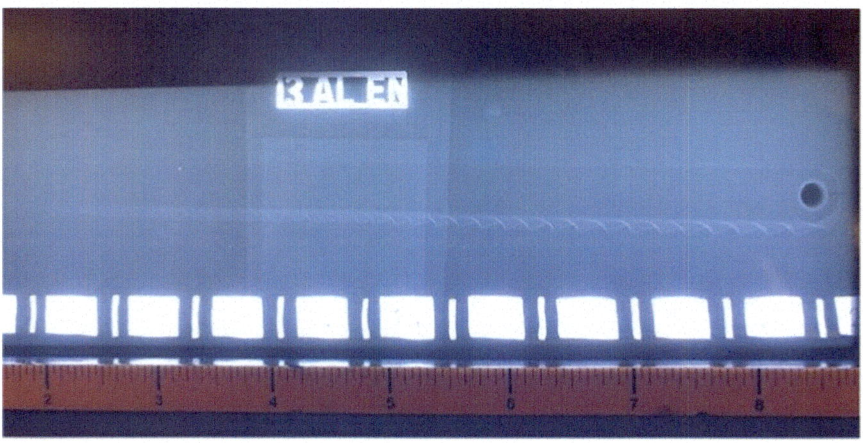

Fig. 30 X-Ray of optimum weld

6 Conclusion

The optimal combination of parameters was found from the selected levels, using the Taguchi robust design approach. Results from the experiment revealed the optimal parameters to be as follows: 1200 rpm rotational speed, 150 mm/min welding speed, 2 s' dwell time and a shoulder depth of 0.11 mm. However, a second experiment should be performed, using knowledge and experience gained from this experiment, to further improve the weld quality and tensile strength. Instead of dwell time, the Mazak thrust control feature can be used as a parameter. The severe nonlinearity in the effect of the shoulder depth could be due to interactions of factors, and a higher order experiment could be carried out, to further investigate the factors. The ANOVA revealed that welding speed had the greatest influence on the tensile strength of the parameters, with around 52% contribution to the process. Shoulder depth had the second most influence, with a contribution of around 21%, rotation around 17% and dwell time only around 10%.

The ultimate tensile strength from the confirmation experiment had the highest values among all experiments, with a value of 222.7 MPa. This is less than the predicted value of 244.62 MPa, but the value is within the two-standard deviation confidence interval. The matrix experiment did find the highest UTS.

The Vickers hardness test and microstructure revealed that the heat affected zone was as far out as 65 mm from the nugget on each side (this was as far out as it was possible to measure the Vickers hardness). The hardness was measured to around 75 HV, and the base material, unwelded, was measured to be 115 HV. For thin plate welding, an optimized welding rig with improved clamping is suggested. To understand the mechanism leading to the hardness drop throughout the welded specimens, a transmission electron microscope (TEM) should be used to investigate these mechanisms. Monitoring of temperature and downforce during welding would make it easier to understand the effect for each factor and the hardness drop experienced.

References

1. Vill VI (1962) Friction welding of metals. LCCCN 62-13420. Grafield Press, New York American Welding Society
2. Thomas WM et al (1991) Improvements relating to friction welding. European Patent Specifications 0615 48 B1
3. Miding OT, Kvale JS, Dahl O (1999) 'Industrialisation of the friction stir welding technology in panels production for the maritime sector', presented at the 1st International symposium on friction stir welding, Thousand Oaks, CA, USA (June 1999)
4. Martin J, Wei S (2016) Friction stir welding technology for marine applications. In: Mishra RS, Mahoney MW, Sato Y, Hovanski Y (eds) Friction stir welding and processing VIII. Springer International Publishing, Cham, pp 219–226
5. Ratnayake RMC, Brevik VA (2014) Underwater friction stud welding: evaluating optimum parameter settings for subsea intervention without a shroud. In: Volume 5: Materials technology; petroleum technology, San Francisco, California, USA (June, 2014), p V005T03A009. https://doi.org/10.1115/omae2014-23330

6. Texier D et al (2018) Fatigue performances of FSW and GMAW aluminum alloys welded joints: competition between microstructural and structural-contact-fretting crack initiation. Int J Fatigue 116:220–233. https://doi.org/10.1016/j.ijfatigue.2018.06.020
7. Threadgill PL, Leonard AJ, Shercliff HR, Withers PJ (2009) Friction stir welding of aluminium alloys. Int Mater Rev 54(2):49–93. https://doi.org/10.1179/174328009X411136
8. Kallee S, Nicholas D (2003) Joining materials. Friction Stir Welding at TWI. http://www.ansatt.hig.no/henningj/materialteknologi/Lettvektdesign/joining%20methods/joining-welding-friction%20stir%20weld.htm (accessed 05 June 2019)
9. Podržaj P, Jerman B, Klobčar D (2015) Welding defects at friction stir welding. Metalurgija-Sisak then Zagreb 54(2):387–389
10. Ratnayake RMC, Brevik VA (2014) Experimental investigation of underwater stud friction stir welding parameters. Mater Manuf Processes 29(10):1219–1225. https://doi.org/10.1080/10426914.2014.930891
11. Aluminium alloys - aluminium 6082 properties, fabrication and applications. AZoM.com (21 April 2005). https://www.azom.com/article.aspx?ArticleID=2813 (accessed 28 Apr 2019)
12. O'Brien A and American Welding Society (eds) Welding handbook. Vol. 3: Welding processes, part 2, 9th edn. American Welding Society, Miami, Fla
13. Phadke MS (1989) Quality engineering using robust design. Prentice Hall
14. Nourani M, Milani AS, Yannacopoulos S (2011) Taguchi optimization of process parameters in friction stir welding of 6061 aluminum alloy: a review and case study. ENG 03(02):144–155. https://doi.org/10.4236/eng.2011.32017
15. Jadhav GC, Dalu RS (2019) A study of effect process parameters in FSW on tensile strenght AA6061-T6. Int J Adv Res Eng Technol 10(2):482–490 (Feb 2019)
16. Naimuddin SK, Touseef M, Kampurath V, Ali Y (2016) Mechanical properties of friction stir welding joints of similar and disimilar aluminium alloys AA6061 & 6082. Int J Mech Eng Technol 7(4):256–266 (Aug 2016)
17. Adamowski J, Szkodo M (2007) Friction Stir Welds (FSW) of aluminium alloy AW6082-T6. J Achievements Mater Manuf Eng 20:403–406
18. Barlas Z, Ozsarac U (2012) Effects of FSW parameters on joint properties of AlMg3 alloy. Weld J 91(1):16.s (Jan, 2012)
19. Abd Elnabi MM, Elshalakany AB, Abdel-Mottaleb MM, Osman TA, El Mokadem A (2019) Influence of friction stir welding parameters on metallurgical and mechanical properties of dissimilar AA5454–AA7075 aluminum alloys. J Mater Res Techn 8(2):1684–1693 (Jan 2019). https://doi.org/10.1016/j.jmrt.2018.10.015
20. International Organization for Standardization (2003) ISO/TR 16060:2003(E), Destructive test on welds in metallic materials—etchants for macroscopic and microscopic examination
21. International Organization for Standardization (2016) NS-EN-ISO 6892-1, Metallic materials—tensile testing—Part 1: method of test at room temperature
22. International Organization for Standardization (2005) NS-EN-ISO 6507-1, Metallic materials—Vickers hardness test—Part 1: test method (ISO 6507-1:2005)

Experimental Investigation of Weld Joints Manufactured at Close Proximity in S420 Structural Steel

Magnus Larsson, Mattias Larsson, and R. M. Chandima Ratnayake

Abstract This manuscript presents the results from an experimental investigation performed to study the material and mechanical properties of two weld joints manufactured within close proximity. The experiment has been designed using seven welded S420G2 + M 500 × 300 × 15 mm structural steel plates, with varying distances between two parallel adjacent weld toes at 44, 12 and 1.3 mm. All these distances were less than the distance that has normally been recommended in standards, codes and specifications. A common requirement is four times the plate thickness, which in this experiment is 60 mm for a 15-mm-thick plate. The scope of the study was to analyze the effect the variation of distance had on the mechanical and material properties. A welding procedure specification was created and approved for the manufacturing of the weld joints. All subsequent welds were based on this welding procedure specification, in order to maintain conformance and uniformity. The non-destructive and destructive testing included: visual testing, radiographic examination, bending testing, Vickers hardness testing, Charpy V impact testing, tensile testing, as well as macro and microscopic examination. The results from the testing showed no adverse effects on the mechanical properties in the initial weld joint. This includes a complete heat-affected zone overlap in the specimen with 1.3 mm between the weld toes. These findings challenge the notion that the minimum distance requirements are based on metallurgical reasons. The outcome of this specific experimental setup can provide a baseline for performing further investigations.

Keywords Welding · Mechanical testing · Weld proximity · Offshore structural steel · Fine grained steel

M. Larsson (✉) · M. Larsson · R. M. C. Ratnayake
Department of Mechanical and Structural Engineering and Material Science, University of Stavanger, Stavanger, Norway
e-mail: magnus.larsson@dnvgl.com

M. Larsson
e-mail: mattias.larsson@kiwa.com

M. Abdel Wahab (ed.), *Proceedings of 1st International Conference on Structural Damage Modelling and Assessment*, Lecture Notes in Civil Engineering 110,
https://doi.org/10.1007/978-981-15-9121-1_25

1 Introduction

Welding guidelines from standards, codes and specifications require that a minimum distance is maintained between two weld joints in order to avoid the adverse effects of weld proximity. However, limited technical explanation has been given in the literature as to why such a minimum predetermined distance is required and what the consequences would be if that requirement were violated [1–4]. Although minimum distance requirements have been imposed by standards, codes and specifications, breaching these is sometimes unavoidable during the fabrication stage. Similar situations have also been discovered on existing structures. Hence, it is vital to perform investigations to gain an understanding about such circumstances, in order to minimize the life cycle cost related to frequent inspection.

It is important to differentiate between weld proximity and weld overlapping. Weld overlapping is the physical overlap of one weld on top of another, while weld proximity refers to two welds separated by a certain distance. Weld proximity issues arise when two initially approved welds conflict with the required minimum design distance. In this project, the interest has been to understand the implications of weld proximity. The requirements are stated in several international standards:

- BS 2633: "Class I arc welding of ferritic steel pipework for carrying fluids" states that the toes of adjacent butt welds shall, whenever possible, be no closer than four times the nominal thickness of the pipe [1].
- BS 2971: " Class II arc welding of carbon steel pipework for carrying of fluids" states that if design factors are such that the meeting of more than two welded seams cannot be avoided, then appropriate precautions shall be taken which shall be agreed between the contracting parties [2].
- BS 4515: "Specification for welding of steel pipelines on land and offshore" states that the proximity of weld toe-to-toe distance shall not be less than four times the pipe thickness [3].
- PD5500: "Specification for unfired fusion welded pressure vessels" states that where any part of a vessel is made in two or more courses, the longitudinal seams shall be completed before commencing the adjoining circumferential seam(s) and, where practicable, the longitudinal seams of the adjacent courses shall be staggered by four times the nominal thickness or 100 mm, whichever is the greater, measured from the toe of the welds [4].

The study performed in this manuscript has focused on increasing knowledge of the effects on the mechanical properties in a weld proximity scenario. In this context, there is a lack of clear guidance in welding standards, codes and specifications, which has led to uncertainties on how to address the specific scenarios. Hence, it is vital to identify the specific affected mechanical properties for optimizing the necessary precautions required in the assessment of an existing structure.

2 Industrial Challenge

Although the standards/codes and specifications indicate some guidance, it is not possible to follow them in all instances. There is limited information about how to deal with circumstances where the formal guidelines have been violated during fabrication or during the life cycle phase of a welded structure. In addition, there is no clear justification of why the given guideline-based distances (i.e. greater than four times the thickness of the plate) must be maintained during the manufacturing phase. Hence, it is vital to investigate the material and mechanical properties when two welded joints have been manufactured in close proximity.

3 Methodology

3.1 Process Description

The following experimental section describes the process of developing a welding procedure qualification record (WPQR) for S420G2 + M steel plates and subsequent mechanical testing. It aims to provide information on the weld fabrication process and the mechanical testing that has been carried out. This experimental process was divided into the four steps, as follows, that were completed during the study.

- Stage A—Material Selection and Cutting Process.
- Stage B—Welding Procedure Qualification Program.
- Stage C—Production Welding of Plates.
- Stage D—NDT, Specimen Preparation and Mechanical Testing.

Initially, an experimental investigation was performed in order to test and analyze parallel welds at the selected distances between the two welds. The choice of test setup was to produce two parallel butt welds on a S420G2 + M offshore structural steel plate. In order to maintain conformity of all welded plates, a welding procedure qualification record (WPQR) was initially established. This was achieved by welding an initial qualifying plate that was tested and approved in accordance with the requirements in ISO 15614-1:2017, level 2 [5]. All subsequent welding procedure specifications (WPS) to be used for production test plates were then based on the approved WPQR. The main objective of the welding procedure qualification program was to validate that the joining process intended for construction could produce joints with the intended mechanical properties. The welded joints were subjected to both destructive and non-destructive testing: visual testing, radiographic examination, Vickers hardness testing, Charpy V impact testing, tensile testing, fatigue testing, ultrasonic residual stress measurement, as well as macro and microscopic examination. The results from the fatigue test and ultrasonic residual stress measurement will be presented in a parallel article by the same authors.

3.2 *Stage A—Material Selection and Cutting Process*

The choice of steel was based on its regular occurrence in the offshore industry and its potential sensitivity to the following welding operations. The selected base material, S420G2 + M (MDS-Y30 Rev. 5), is classified according to NS-EN 10225:2009, EN 10020:2000, ISO/TR 20172:2009 and ISO/TR 15608:2017 as a fine grain steel in group 2.1 with steel number 1.8857 + M. The construction steel has a specified minimum yield point in room temperature at 420 MPa and a minimum average impact energy value of 60 J at −40 °C. The steel is of grade 2 and delivery condition thermomechanically rolled (M) [6–9].The mechanical test results and ladle analysis for the base material are shown in Table 1.

The dimensions of the plates were 15 × 300 × 500 mm. All cutting was done with waterjet to avoid inducing extra heat that could have an unwanted impact on the results, as the interest of the project was to analyze the adverse effect that a secondary weld could have on the initial weld metal (WM) and heat-affected zone (HAZ).

Table 1 Mechanical test results and chemical composition of the base material obtained from the inspection certificate

Base material S420G2+M (MDS-Y30 Rev. 5)					
Tensile test					
Yield point, ReH 420–540 (MPa)		Tensile strength, Rm 500–660 (MPa)		Elongation %≥ 19	
489		557		31	
Impact test					
Temperature		Impact energy ≥ 42 J		Average ≥ 60 J	
−40 °C		118	111	114	114
Chemical composition %					
C	Si	Mn	P	S	N
0.10	0.28	1.49	0.011	0.002	0.005
V	Ti	Nb	B	Mo	As
0.003	0.002	0.026	0.0001	0.002	0.002
Sb	Ni	Bi	Ca	Cr	Al
0.000	0.04	0.001	0.002	0.05	0.041
Sn	Pb	Cu	CEV	PCM	
0.005	0.001	0.02	0.36	0.19	

The rolling direction of the steel affects the microstructure and mechanical properties of the steel, thus affecting its anisotropy, which means that the strength can differ between directions. Therefore, the test plates have been produced consistently so that all welding passes have been done across the steel's rolling direction.

The filler material:

- The filler material used for root passes was NST's (Norsk Sveiseteknikk AS) NSSW SM-47A: a metal cored wire for low temperature pipe and steel applications down to −60 °C.
- The filler material used for hot, fill and cap passes was NST's SF-3AM, flux cored wire for low-alloyed steel, offshore applications, piping, etc (Tables 2, 3 and 4).

3.3 Stage B—Welding Procedure Qualification Program

The project started by developing a preliminary welding procedure specification (pWPS), as there was no available welding procedure for steel S420G2 + M. A welding procedure is necessary for planning and performing the welding process and for quality control purposes. This study adopted the method suggested for "Qualification based on a welding procedure test" in NS-EN ISO 15607:2007 [10]. The welding process started with the installation of strongbacks on the back side (see Fig. 1). According to the produced pWPS, the plate was mounted and welded in the welding position, PF (vertical up position).

Welding passes were placed according to Fig. 2. The root pass (1) was welded with method 138 MAG welding with metal cored electrode [11]. Fill and cap (2–6) were welded using method 136 MAG welding with flux cored electrode [11]. All fit-up, welding, logging and visual testing was done together with Kiwa TI in Stavanger.

The finished plate was transported to IKM Inspection for radiographic testing. After the approved radiographic test, the plate was transported to Quality Lab, where the remaining nondestructive testing (NDT) and mechanical testing were performed. All testing in the experimental study was performed according to ISO 15614–1:2017 Level 2 [5]. Figure 3 shows the extraction locations of the test specimens in the qualification test plate, PL1.

Table 3 shows the activities, procedures, acceptance criteria and verifying documents used during testing of the qualification test plate, PL1.

After approved test results, two different variants of welding procedure specifications (WPS) were produced. First, 'Primary WPS' was developed for weld A and, secondly, a 'Repair WPS' was developed for proximity welds (weld B). The 'Repair WPS' was based on the same WPQ as the 'Primary WPS' but was converted into a 'Repair WPS' in accordance with NORSOK M-101: 2011 Rev. 5 Sect. 6.11-Preheat and interpass temperature and Sect. 10.4-Repair welding procedure [26].

Table 2 Data of the filler material from NST's certified material test report

Root pass filler material—NSSW SM-47A					
Tensile test					
Yield point, ReH ≥ 460 (MPa)	Tensile strength, Rm 530–680 (MPa)				Elongation % ≥ 20
527	617				28
Impact test					
Temperature	Impact energy ≥ 32 J				Average ≥ 47 J
−60 °C	93	104		115	114
Chemical composition %					
C	Si	Mn	P	S	Cu
0.07	0.59	1.26	0.009	0.008	0.25
Ni	Cr	Mo	V	Nb	
1.02	0.02	0.01	0.01	0.01	
Filler and cap filler material—SF-3AM					
Tensile test					
Yield point, ReH ≥ 460 (MPa)		Tensile strength, Rm 530–680 (MPa)			Elongation % ≥ 20
556		614			26
Impact test					
Temperature °C		Impact tnergy ≥ 32 J			Average ≥ 47 J
−40		112	126	142	127
−60		82	80	87	83
Chemical composition %					

(continued)

Table 2 (continued)

Root pass filler material—NSSW SM-47A

C	Si	Mn	P	S	Cu
0.06	0.31	1.22	0.01	0.005	0.26
Ni	Cr	Mo	V	Nb	
1.02	0.02	0.01	0.01	0.01	

Table 3 Welding procedure testing: PL1-SW. Butt joint with full penetration with a single weld (SW)

Butt joint with full penetration with a single weld (SW)*			
Item #	Type of test	Extent of testing	Footnote
1	Discard 25 mm	–	a
2	Hardness test	1 specimen	b
2	Macroscopic examination	1 specimen	c
3	Transverse tensile test	2 specimens: T1, T2	d
4	Side bend test	4 specimens	e
5	Charpy V impact test, KV_8	4 sets	f
	Visual testing	100%	g
	Radiographic test	100%	h
	Surface crack detection (MT)	100%	i

*Tests and acceptance criteria are according to Table 2—For level 2 ISO 15614-1:2017 [5]
[a]According to the standard, 25 mm of material should be removed at the top and bottom of the plate
[b]Tested according to NS-EN ISO 9015-1:2011 and acceptance criteria ISO 15614-1:2017. Marco and hardness tests were carried out on the same specimen [5, 12]
[c]Tested according to NS-EN ISO 17639:2013 and acceptance criteria NS-EN ISO 5817:2014 (quality level B). Marco and hardness tests were carried out on the same specimen [13, 14]
[d]Tested according to NS-EN ISO 4136:2012, NS-EN ISO 6892–1:2016 Method A1 and acceptance criteria ISO 15614–1:2017, NS-EN 10225:2009 [5, 15, 16]
[e]Tested according to NS-EN ISO 5173:2010 and acceptance criteria ISO 15614-1:2017. For thicknesses ≥12 mm, four side bend specimens may be used instead of root and face bend tests; see ISO 15614-1:2017 Sect. 7.4.2 [5, 17]
[f]Tested according to NS-EN ISO 148-1:2016, ISO 9016:2012 and acceptance criteria ISO 15614-1:2017, NS-EN 10225:2009. One set in the welding material (W), one set in fusion line (FL), one set 2 mm next to the fusion line (FL2) and one set 5 mm next to the fusion line (FL5). All tests were conducted at the same temperature as the base material, −40 °C [18, 19]
[g]Tested according to NS-EN ISO 17637:2016 and acceptance criteria NS-EN ISO 5817:2014 B/C [14, 20]
[h]Tested according to ISO 17636-2:2013 and acceptance criteria ISO 10675-1:2016 [21, 22]
[i]Tested according to NS-EN ISO 17638:2016, NS-EN ISO 3452-1:2013 and acceptance criteria NS-EN ISO 23278:2015 Level B. Magnetic particle testing (MT) was performed to find surface cracks [23–25]

3.4 Stage C—Production Welding of Plates

In this section, the welding and cutting process for the production plates is described. In the previous stage B, only one plate was welded and tested. This was carried out to confirm that the welding process was correct and that the mechanical properties of the welds were in accordance with the requirements of ISO 15614-1:2017 Level 2 [5].

In Stage C, the production plates were welded according to the produced WPS. The production flow of test plates is shown in Fig. 4. The production flow should mimic a situation where a weld joint is prepared alongside an already existing joint.

Table 4 Production sample testing: PL3-DW, PL5-DW and PL7-DW

Butt joint with full penetration with a dual weld (DW).*			
Item #	Type of test	Extent of testing	Footnote
1	Discard 25 mm	–	a
2	Hardness test	1 specimen	b
2	Macroscopic examination	1 specimen	c
3	Transverse tensile test	2 specimens: (T1, T2)	d
4	Fatigue test	2 specimens (F1, F2)	e
5	Charpy V impact test, KV_8	4 sets	f
	Visual testing	100%	g
	Radiographic test	100%	h
	Surface crack detection (MT)	100%	i

*Tests and acceptance criteria are according to Table 2—For level 2 ISO 15614-1:2017 [5]
[a]According to the standard, 25 mm of material should be removed at the top and bottom of the plate
[b]Tested according to NS-EN ISO 9015:2011 and acceptance criteria ISO 15614-1:2017. Marco and hardness tests were carried out on the same specimen [5, 12]
[c]Tested according to NS-EN ISO 17639:2013 and acceptance criteria NS-EN ISO 5817:2014 (quality level B). Marco and hardness tests were carried out on the same specimen [13, 14]
[d]Tested according to NS-EN ISO 4136:2012, NS-EN ISO 6892-1:2016 Method A1 and acceptance criteria ISO 15614-1:2017, NS-EN 10225:2009 [5, 15, 16]
[e]This is a non-standard test piece that is not included in ISO 15614-1:2017. Fatigue specimens were checked for surface cracks before fatigue testing was performed. This was done with MPI
[f]Tested according to NS-EN ISO 148-1:2016, ISO 9016:2012 and acceptance criteria ISO 15614-1:2017, NS-EN 10225:2009. One set in the welding material (W), one set in fusion line (FL), one set 2 mm next to the fusion line (FL2) and one set 5 mm next to the fusion line (FL5). All tests were conducted at the same temperature as the base material, −40 °C [5, 6, 18, 19]
[g]Tested according to NS-EN ISO 17637:2016 and acceptance criteria NS-EN ISO 5817:2014 B/C [14, 20]
[h]Tested according to ISO 17636-2:2013 and acceptance criteria ISO 10675-1:2016 [21, 22]
[i]Tested according to NS-EN ISO 17638:2016, NS-EN ISO 3452-1:2013 and acceptance criteria NS-EN ISO 23278:2015 Level B. MT was performed to find surface cracks [23–25]

The root pass was made with method 138 MAG welding with metal cored electrode (energy per unit length/heat input around 1.29–1.61 kJ/mm). Hot pass, fill and cap were welded with method 136 MAG welding with flux cored electrode (energy per unit length/heat input around 1.21–1.99 kJ/mm) in the PF position [11].

In Steps 1 and 2, the plates were cut with a waterjet cutting process. In Step 3, strongbacks were mounted on the backside of the plate, and joint A was welded with the 'Primary WPS'. Preheat temperature was 20 °C and interpass temperature was 150 °C. Interpass temperature 150 °C was used due to it being the highest value measured during the qualification welding. The strongbacks were removed after the plates had air cooled to 20 °C. This removal temperature was not stated in any standards but was chosen in order to maintain consistency and manufacturing conformity between the plates.

Fig. 1 Qualification test plate, PL1-SW (single weld)

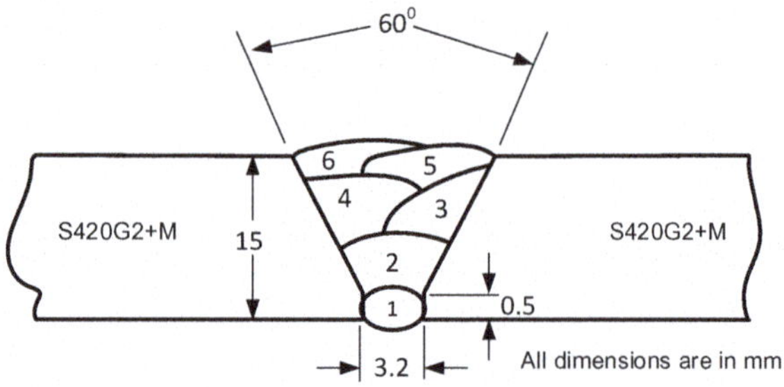

Fig. 2 Schematic diagram of the weld joint and the corresponding weld passes

In Step 4, weld joint B was cut using waterjet. In Step 5, strongbacks were installed, and joint B was welded with the 'Repair WPS', which was based on the same WPQR as the 'Primary WPS' but had an elevated minimum preheat temperature of 70 °C and interpass of 150 °C. The preheat temperature used during the second weld should be 50 °C higher than the previous weld when performing repair welds, in accordance with NORSOK M-101: 2011 Rev.5 [26]. Preheating was done with propane. The strongbacks were removed after the plates had air cooled to 20 °C, similar to weld joint A. Six production plates were prepared with three different spacings between the corners of the beveled edges. Two plates were produced for each selected distance between two welded joints (see Fig. 6).

The final distances between the weld toes (D1) were 44, 12 and 1.3 mm. The final groove angle (GA) ranged between 57 and 60 degrees (see Fig. 7).

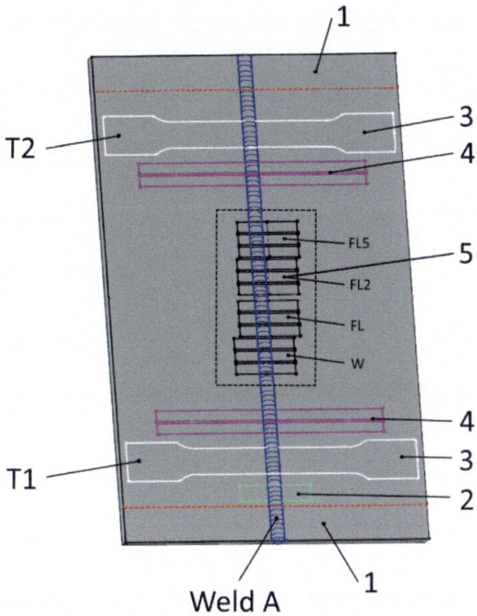

Fig. 3 Location of test specimens for the qualification test plate, PL1-SW. See Table 3 for explanation of number designation

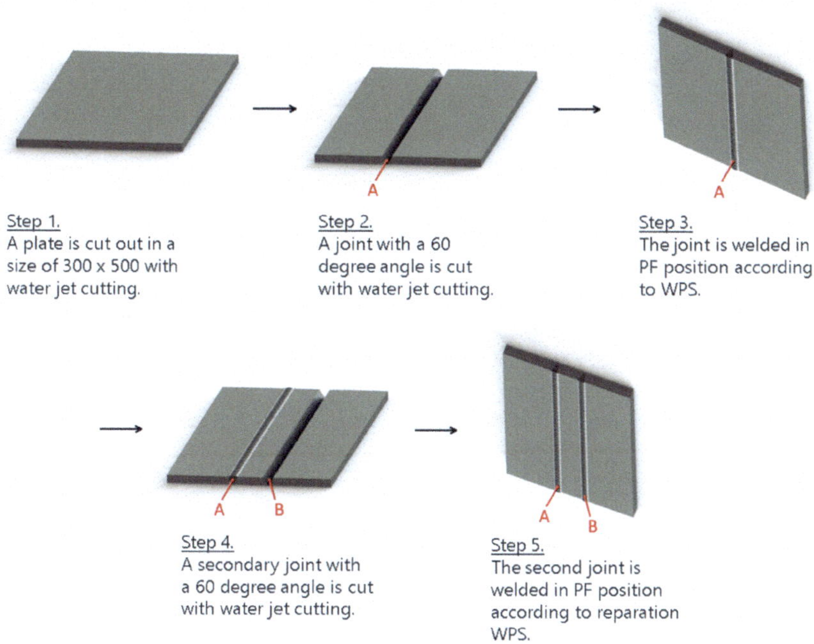

Fig. 4 Production flow of test plates for production test

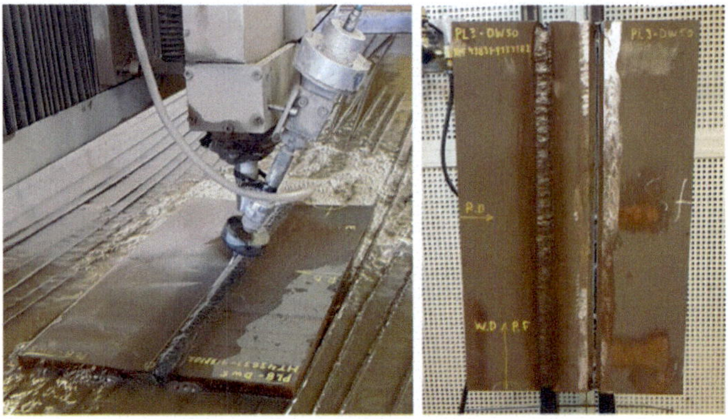

Fig. 5 Waterjet cutting of weld joint B and image of joint B before welding in PL3-DW (dual weld)

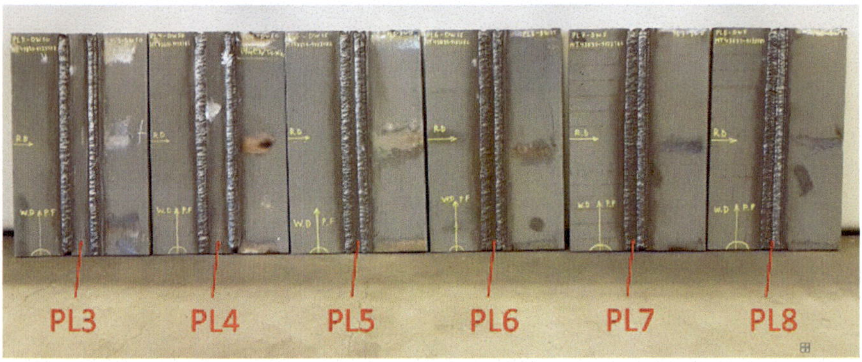

Fig. 6 All production plates ready to be forwarded for NDT and mechanical testing. In the figure, we can see that the plates have different distances between weld joints

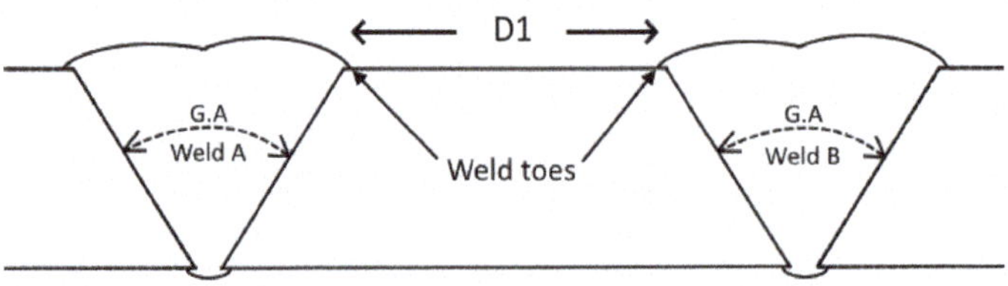

Fig. 7 Distance between the weld toes and groove angle

3.5 *Stage D—NDT, Specimen Preparation and Mechanical Testing*

The fabricated weld plates were tested using the same standard as the qualification plate, PL1, but the test scope in ISO 15614-1:2017 Level 2 had small changes [5].

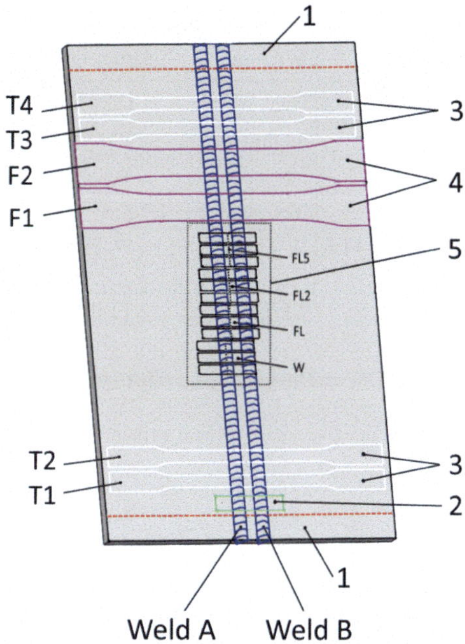

Fig. 8 Location of test specimens for the production test plates, PL3, PL5 and PL7. See Table 4 for explanation of number designation. shows the activities, procedures, acceptance criteria and verifying documents used during testing of the fabricated weld plates

Fatigue testing replaced the previous bending test. After visual and radiographic testing, the plates were transported to Quality Lab for mechanical testing. Plates PL3, PL5 and PL7 were mechanically tested at Quality Lab. Plates PL4, PL6 and PL8 were to be used for residual stress tests, presented in forthcoming articles. Figure 8 shows the extraction locations of the test specimens in the production test plates, PL3-DW, PL5-DW and PL7-DW.

4 Results and Discussion

The labeling of test specimens from the experimental investigation is shown in Table 5. All test specimens were assigned a specific serial number.

4.1 Microscopic Examination

The microscopic examinations were conducted using an Olympus GX53 optical microscope. The macrograph in Fig. 9 shows the weld joint of plate PL1-SW.

Table 5 Extent of mechanical testing and specimen identification

Plate ID	PL1-SW	PL3-DW	PL5-DW	PL7-DW
Weld configuration	Single Weld	Dual Weld	Dual Weld	Dual Weld
Weld toe distance	–	44 mm	12 mm	1.3 mm
Microscopic test	Yes	Yes	Yes	Yes
HV10 Hardness test	Yes	Yes	Yes	Yes
Charpy V impact toughness test, KV_8	W, FL, FL2, FL5	W, FL, FL2, FL5	W, FL, FL2, FL5	W, FL, FL2, FL5
Tensile specimen ID	T1-T2	T1-T4	T1-T4	T1-T4

PL = Plate, SW = Single Weld, DW = Dual Weld, T = Tensile, W = Weld, FL = Fusion Line, FL2 = Fusion Line + 2 mm, FL5 = Fusion Line + 5 mm, HV = Vickers Pyramid Number, KV_8 = Absorbed energy for a V-notch test piece using 8 mm striker

Note

The weld passes in the dual welded plates are divided into the initial weld (Weld A) and the secondary weld (Weld B)

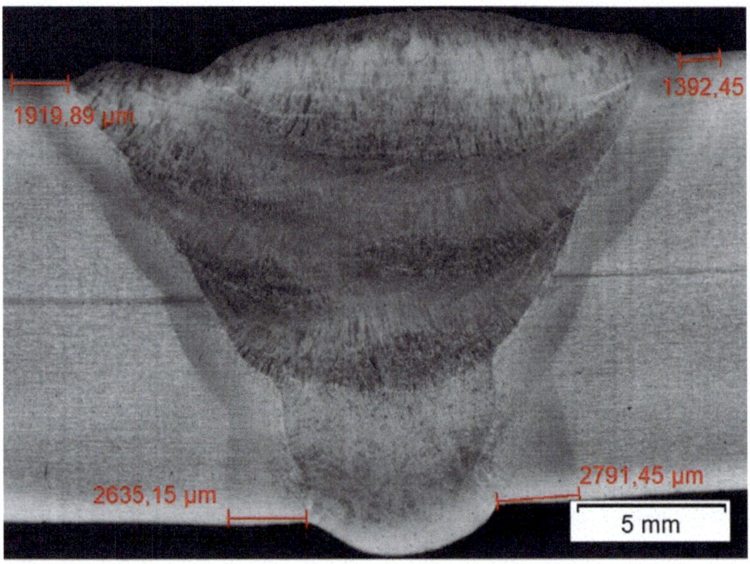

Fig. 9 Macrograph of PL1-SW, showing a multipass weld with corresponding HAZ. Optical microscope: Olympus GX53. Etching reagent: 2% nital. Magnification: 2.5X

The multipass weld consists of six weld passes in accordance with the schematic diagram in Fig. 2. Figure 10 shows the micrograph of the unaffected S420G2 + M thermomechanically treated base material at 500X magnification.

The banded ferrite-pearlite microstructure due to the thermomechanical treatment is clearly visible. Figure 11 shows the weld metal (WM) and the various regions of

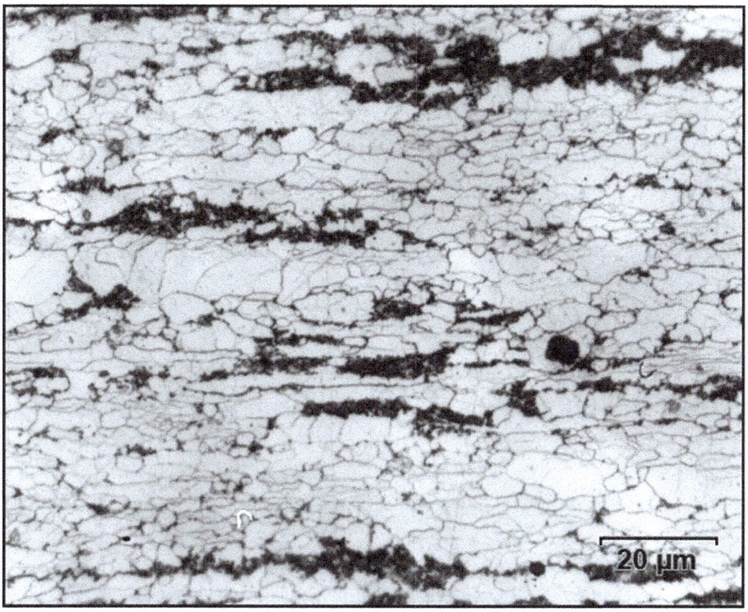

Fig. 10 Micrograph of the unaffected S420G2 + M base material. The steel has a banded microstructure due to the thermomechanical treatment. Etching reagent: 2% nital. Magnification: 500X

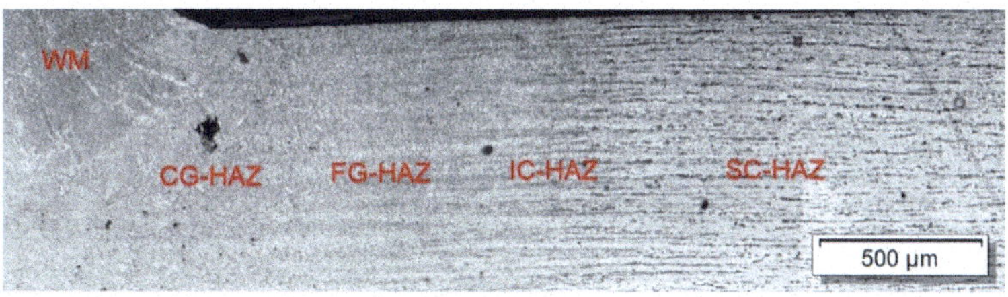

Fig. 11 HAZ outside the weld toe in PL1-SW. The WM, coarse-grained HAZ, fine-grained HAZ, tempered and partially austenitised HAZ and the unaffected base material are visible as a banded microstructure. Magnification: 30X

the HAZ, typical for a steel weld joint. This specific example shows the HAZ from the plate, PL1-SW, a single welded plate.

The visible regions are from the weld metal, coarse-grained HAZ (CG-HAZ), fine-grained HAZ (FG-HAZ), intercritical HAZ (IC-HAZ) and subcritical HAZ (SC-HAZ). Figure 12 shows the HAZ in plate PL5-DW, which had approximately 12 mm between the weld toes.

There was no sign of interaction between the two HAZs from the weld joints at this distance. The same can be stated about plate PL3-DW, which had approximately 44 mm between the weld toes. At these distances, the heat generated from the heat

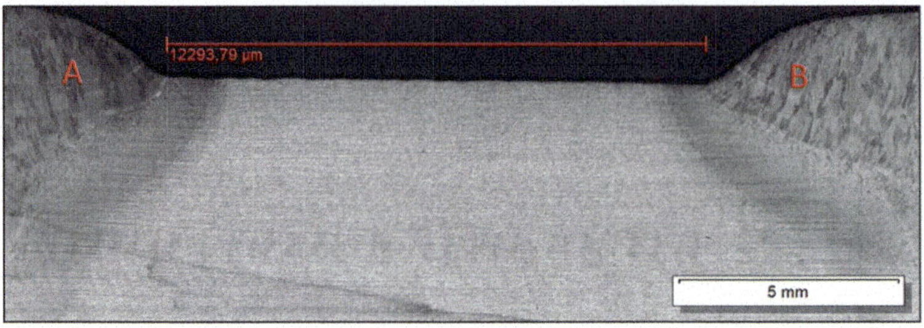

Fig. 12 Mutual HAZ of PL5-DW. There was no interaction between the two HAZ from weld A and weld B. Magnification: 5X

source did not have any visible impact on the microstructure. The macrograph from PL7-DW of the mutual HAZ between weld A and weld B is shown in Fig. 13.

The region had experienced a complete HAZ overlap. The distance between the weld toes was 1.3 mm. A higher magnification of the overlapping HAZ weld region is shown in Fig. 14.

The region appeared to have experienced a tempering and partially austenitizing effect from the welding procedure. No detrimental microconstituents were observed in the region that would affect the microstructural properties. Figure 15 shows the root of the weld joint in PL7-DW.

The root weld joints had no visible overlap of the HAZ due to the distance between the weld joints.

Vickers Hardness Test.

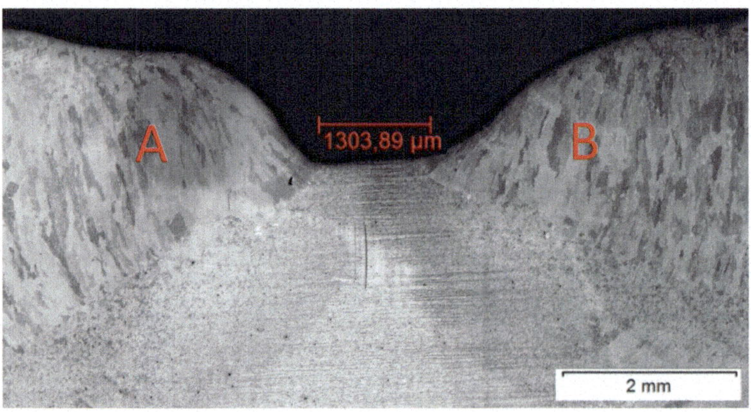

Fig. 13 Overlapping HAZ in specimen PL7-DW. Weld A is the initial weld and weld B is the secondary weld with overlapping HAZ. The distance between the weld toes is 1.3 mm. Magnification: 10X

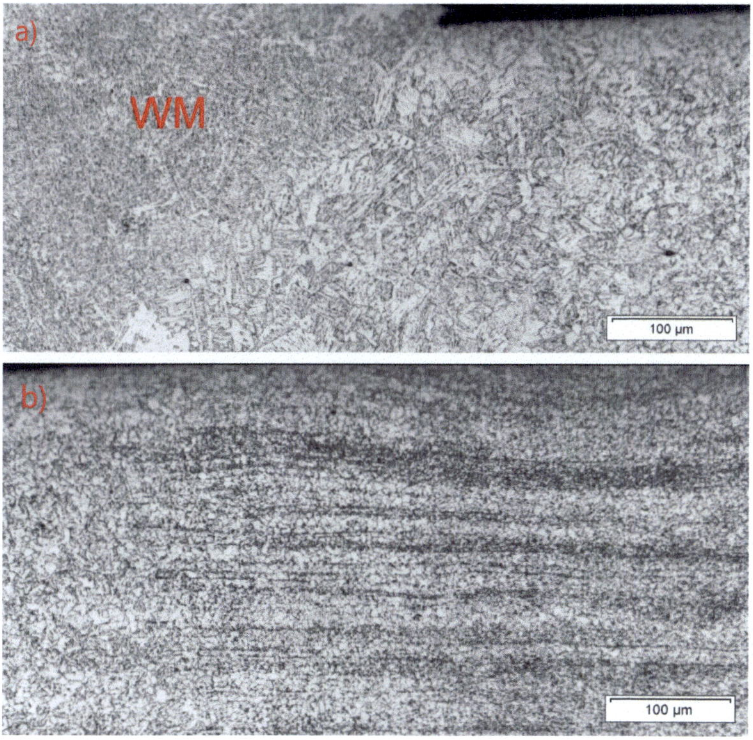

Fig. 14 A-b Overlapping HAZ in PL7-DW 1.3 mm between the weld toes. This micrograph is captured in the middle of weld A and B. Magnification: 150X

Fig. 15 Distance between root weld toes in PL7-DW. Magnification: 5X

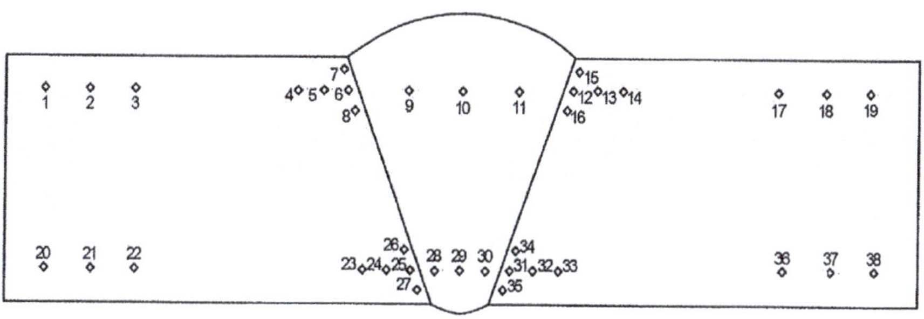

Fig. 16 Placement of hardness test indentations on the single welded plate

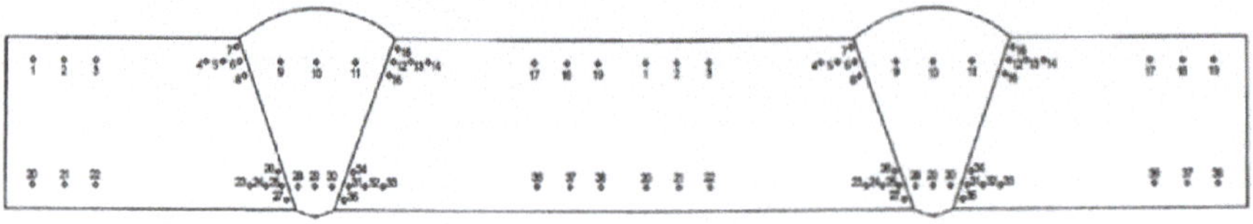

Fig. 17 Placement of indentations on the dual welded test specimen

Hardness tests had been carried out to investigate the possibility of martensitic formations. The indentations from the hardness test were placed as shown in Fig. 16 and Fig. 17, for single weld and dual weld, respectively.

These locations for the indentations in the base material (BM), HAZ and WM are standardized in ISO 9015-1:2011 [12]. The number of indentations in the cap had to be reduced in the sample with 1.3 mm distance between the weld toes, compared to 12 mm and 44 mm, due to space limitation. The hardness distribution in the sample with a single weld, PL1-SW, is shown in Fig. 18, obtained using Vickers hardness testing. Weld A corresponds to the initial weld.

The results showed a typical hardness distribution with the highest values in the coarse grain heat affected zone (CG-HAZ) and WM. The hardness values in the root were lower than in the cap. Figure 19 shows the hardness distribution of the sample PL3-DW with 44 mm between the weld toes. Weld A is the initial weld joint, while weld B is the secondary weld joint.

The distance between the two adjacent weld toes was too significant to cause any visible mechanical property changes. A similar result was observed in the hardness testing of sample PL5-DW with 12 mm between the weld toes, as shown in Fig. 20.

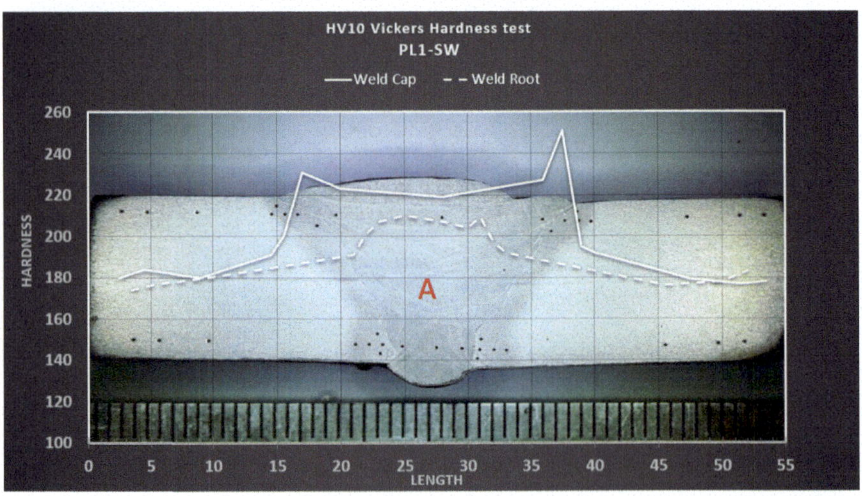

Fig. 18 PL1-SW HV10 vickers hardness test

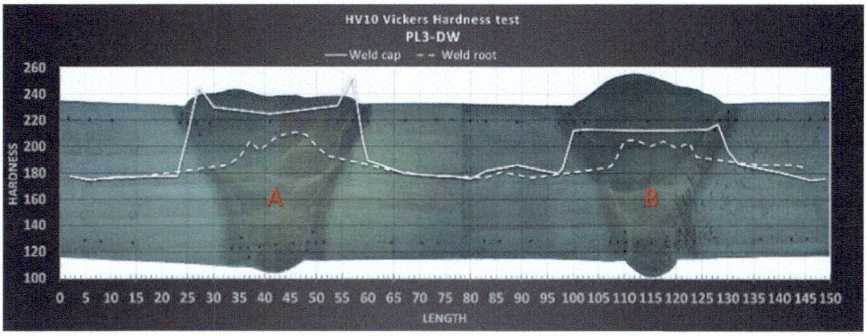

Fig. 19 PL3-DW HV10 vickers hardness test; 44 mm between the weld toes

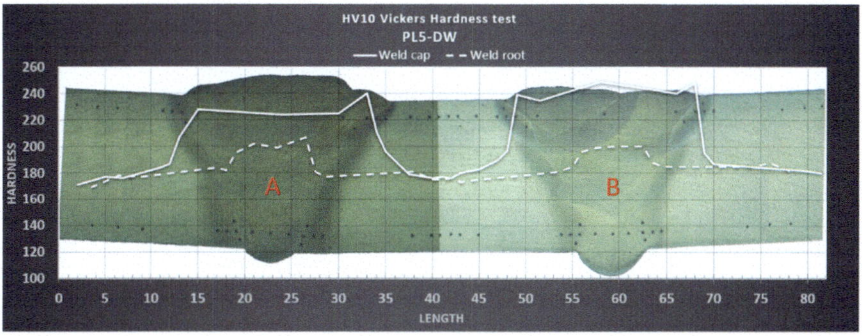

Fig. 20 PL5-DW HV10 vickers hardness test; 12 mm between the weld toes

The distance between the adjacent welds was not sufficient to have had any visible mechanical effect on the initial weld joint. The weld joint in PL7-DW is shown in Fig. 21 with 1.3 mm between the weld toes.

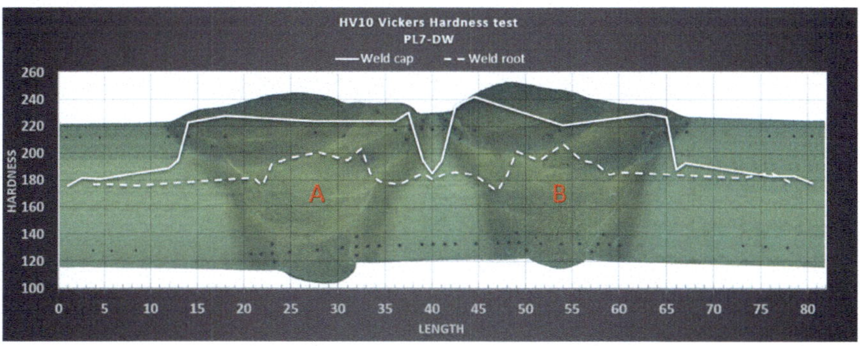

Fig. 21 PL7-DW HV10 vickers hardness test; 1.3 mm between the weld toes

The weld joint HAZ overlap did not seem to have any effect on the hardness properties. This seems to indicate that a reheating of the HAZ does not introduce any brittle zones or produce any excessive softening using this specific weld setup (WPS, base and welding material). Due to the inherent randomness of the hardness test results, it was not possible to derive whether the secondary weld B had an impact on the initial weld A.

4.2 *Charpy V Impact Toughness Test, KV_8*

The results from the Charpy V impact toughness test are presented in Table 6. Included in the table is a descriptive figure of the standardized testing locations on the specimen.

The test did not produce any abnormal results. All tests were well above the limit set in the requirements of MDS-S420G2 + M extracted from NS-EN 10225:2009 [6]. It was not possible to derive whether there were any changes in mechanical properties due to weld proximity.

4.3 *Transverse Tensile Tests*

The results from the tensile testing are presented below in Table 7.

All tensile tests failed in the BM outside the HAZ. It was not possible to detect any difference in results due to the effect of weld proximity. Four specimens were tested from each plate. A graphical presentation of the results is shown in Fig. 22.

The color coding from Fig. 22 represents the different samples and corresponds to the results in Table 7. The test results are separated by shifting to the right, in order to improve visualization.

The objective of this experiment was to obtain quantifiable data regarding the implications of having two adjacent weld joints in close proximity. The welding operation and the various weld toe distances did not seem to have any noticeable adverse impact on the material properties in the weld metal and HAZ, based on the results from the material and mechanical testing. This includes a complete HAZ overlap in the specimen with a distance of 1.3 mm between the two adjacent weld toes. The findings in this study can contribute in the assessment of weld proximity scenarios.

5 Summary and Conclusion

An experimental investigation was carried out in order to study degradation in material properties due to weld proximity. Material used was S420G2 + M and the

Table 6 Results from Charpy V impact toughness test, KV_8

Specimen dimensions:		10X10X55 mm			
Notch orientation:		T			
Test temperature:		−40 °C			
Plate	Indentation location	Single values [J] Min value 42			Average [J] Min. 60
		1	2	3	
PL1-SW	Weld	80	131	106	106
PL3-DW	Weld	121.3	120.1	112.2	117.9
PL5-DW	Weld	100.4	84*	91.4	91.9
PL7-DW	Weld	107.3	83.7*	81.3*	90.8
PL1-SW	FL	113	186	193	164
PL3-DW	FL	186.3	123.5	163	157.6
PL5-DW	FL	168.7	206.2	143.4	172.8
PL7-DW	FL	202.9	280.8	141*	208.2
PL1-SW	FL + 2	253	130	200	194
PL3-DW	FL + 2	394.6	340.5	404.3	379.8
PL5-DW	FL + 2	365.2	365.5	372.4	367.7
PL7-DW	FL + 2	244.7*	354.3*	395.9	331.6
PL1-SW	FL + 5	234	236	173	214
PL3-DW	FL + 5	196.2*	214.5	212	207.6
PL5-DW	FL + 5	360.9	319.1	326.9	335.6
PL7-DW	FL + 5	342.6*	305.9*	301.3*	316.6

(continued)

Table 6 (continued)

Specimen dimensions:			10X10X55 mm		
Notch orientation:			T		
Test temperature:			−40 °C		
Plate	Indentation location	Single values [J] Min value 42			Average [J] Min. 60
		1	2	3	

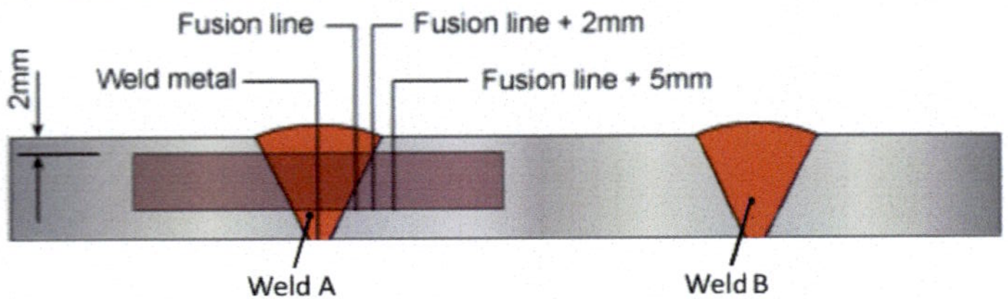

WM = Weld Metal, FL = Fusion Line, FL + 2 = Fusion Line + 2 mm, FL + 5 = Fusion Line + 5 mm,
*(Complete fracture of Charpy V Specimen)

Table 7 Table of tensile test results from the dual welded plates. Corresponding graph is presented in Fig. 22

Transverse Tensile test

	Specimen No.	Acceptance criteria [MPa]	R_{eH} [MPa]	R_m [MPa]	R_{eH}/R_m	A_5 [%]	Fracture
	PL3-DW-T1	500-660	440.54	528	0.83	-	BM
	PL3-DW-T2	500-660	458.94	535	0.86	23.03	BM
	PL3-DW-T3	500-660	-			24.45	BM
	PL3-DW-T4	500-660	444.78	525	0.85	-	BM
	PL5-DW-T1	500-660	448.19	526	0.85	-	BM
	PL5-DW-T2	500-660	-	537		28.06	BM
	PL5-DW-T3	500-660	-	530		28.50	BM
	PL5-DW-T4	500-660	442.26	517	0.86	-	BM
	PL7-DW-T1	500-660	461.07	531	0.87	-	BM
	PL7-DW-T2	500-660	-	534		28.13	BM
	PL7-DW-T3	500-660	-	533		29.11	BM
	PL7-DW-T4	500-660	455.23	525	0.87	-	BM

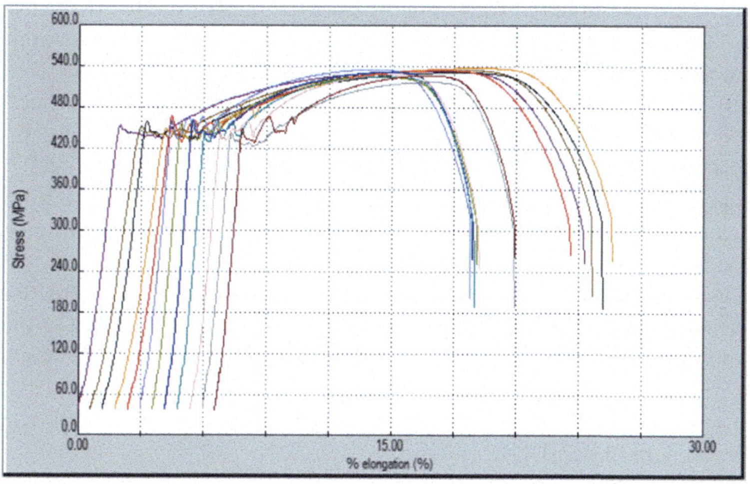

Fig. 22 Tensile test graph with corresponding color-coding identification shown in Table 7. The shift to the right is to separate the graphs for comparison

distances between adjacent welds were less than those normally recommended in standards, codes and specifications. The distances between the weld toes of the butt-welded joints were measured at 44, 12 and 1.3 mm. All mechanical testing was performed by the NS-EN ISO/IEC 17025-accredited Quality Lab and all testing was in accordance with ISO 15614-1:2017. It is important to note that the results are specific for this experimental setup. The results can be used as a baseline for further research when two welds are manufactured at a close proximity.

5.1 Based on the Experimental Work and Evaluation, the Following Can Be Concluded

The mechanical testing results indicated no degradation in material properties due to weld proximity. This statement was based on results from tensile tests, Charpy V impact toughness test and Vickers hardness tests.

No visible harmful phases were detected with the optical microscope in either the WM or the HAZ and the microconstituents were typical for this type of weld. The microstructural examination showed that there was only HAZ overlap in the weld joint with 1.3 mm between the weld toes. The weld joints with a distance of 12–44 mm between the weld toes had no overlap of HAZ.

The results from the Vickers hardness test did not show any reduction in material hardness properties due to the adjacent secondary weld. Due to the spread in results in the WM and HAZ, it was not possible to see whether the secondary weld had any softening effect on the initial weld. The welds with 12–44 mm distance from weld toe to weld toe did not show any adverse effect depending on the weld proximity and showed a similar toughness pattern as the welds at a distance of 1.3 mm.

The results from the Charpy V impact toughness test showed no reduction in toughness properties from the adjacent secondary weld. Due to the inherent spread of the test results, it was not possible to determine any distinct variation between the results at the different distances. However, all tests showed that the weld joint had toughness properties well above the requirements. The fracture appearance and lateral expansion were well within the limits.

All tensile test specimens failed in the BM outside the HAZ. This was as assumed, since the ultimate tensile strength (UTS) of the WM is higher than that of the BM. The results showed that the influence of the adjacent secondary weld did not have a negative impact on the tensile strength properties of the initial weld. The fracture also initiated a considerable distance away from the weld, showing no sign of reduction in tensile strength in the HAZ.

These test results indicate that weld proximity as close as 1.3 mm does not have any degrading impact on the material properties of this specific weld joint (WPS, base and welding material).

5.2 Based on the Evaluations and Conclusions, the Following is Recommended

The magnification of the optical microscope is limited, and further analysis with a scanning electron microscope (SEM) and a transmission electron microscope (TEM) is necessary, in order to ensure that no harmful phases are present either in the WM or the HAZ. Investigation with these microscopes is also needed, in order to confirm the exact nature of all phase transformations occurring in the overlapping HAZ.

Further research shall be carried out to assess residual stress distribution and to investigate the material and mechanical behavior under dynamic loading (i.e. fatigue performance).

Acknowledgements This work was part of a master thesis project at the University of Stavanger. Welding Engineer Arild Finnesand, KIWA TI in Stavanger, has been very helpful throughout the project. He has supported us with the welding of the steel plates as well as providing us with theoretical and technical knowledge. Metallurgist Petter Lunde, Qlab in Stavanger, helped us out with the machining and testing of all specimens.

References

1. Specification for Class I arc welding of ferritic steel pipework for carrying fluids, BS 2633, (1987)
2. Class II arc welding of carbon steel pipework for carrying of fluids, BS 2971, (1991)
3. Specification for welding of steel pipelines on land and offshore, BS 4515, (2009)
4. Specification for unfired fusion welded pressure vessels, PD5500, (2012)
5. Specification and qualification of welding procedures for metallic materials Welding procedure test Part 1: Arc and gas welding of steels and arc welding of nickel and nickel alloys, NS-EN ISO 15614-1, (2017)
6. Weldable structural steels for fixed offshore structures—technical delivery conditions, NS-EN 10225, (2009)
7. Definition and classification of grades of steel, NS-EN 10020, (2000)
8. Welding—Guidelines for a metallic materials grouping system, ISO/TR 15608, (2017)
9. Welding—grouping systems for materials—European materials, ISO/TR 20172, (2009)
10. Specification and qualification of welding procedures for metallic materials—general rules, NS-EN ISO 15607, (2004)
11. Welding and allied processes—nomenclature of processes and reference numbers, NS-EN ISO 4063 (2010)
12. Destructive tests on welds in metallic materials hardness testing Part 1: hardness test on arc welded joints, NS-EN ISO 9015-1, (2011)
13. Destructive tests on welds in metallic materials—Macroscopic and microscopic examination of welds, NS-EN ISO 17639, (2013)
14. Welding—fusion-welded joints in steel, nickel, titanium and their alloys (beam welding excluded)—Quality levels for imperfections, NS-EN ISO 5817:2014, (2014)
15. Destructive tests on welds in metallic materials—Transverse tensile test, NS-EN ISO 4136, (2012)
16. Metallic materials—tensile testing—part 1: method of test at room temperature, NS-EN ISO 6892-1, (2016)
17. Destructive tests on welds in metallic materials—bend tests, NS-EN ISO 5173, (2010)
18. Metallic materials—Charpy pendulum impact test—Part 1: Test method, NS-EN ISO 148–1, (2016)
19. Destructive tests on welds in metallic materials—impact tests—test specimen location, notch orientation and examination, NS-EN ISO 9016, (2012)
20. Non-destructive testing of welds—Visual testing of fusion-welded joints, NS-EN ISO 17637, (2016)
21. Non-destructive testing of welds—radiographic testing—Part 2: X- and gamma-ray techniques with digital detector, NS-EN ISO 17636-2, (2013)
22. Non-destructive testing of welds—acceptance levels for radiographic testing—Part 1: steel, nickel, titanium and their alloys, ISO 10675-1, (2016)

23. Non-destructive testing of welds—magnetic particle testing, ISO 17638, (2016)
24. Non-destructive testing—penetrant testing—Part 1: general principles, ISO 3452-1, (2013)
25. Non-destructive testing of welds—magnetic particle testing—acceptance levels, NS-EN ISO 23278, (2015)
26. NORSOK standard—structural steel fabrication, M-101, (2011)

Experimental Residual Stress Investigation of Weld Joints Fabricated at a Close Proximity in S420 Structural Steel

Magnus Larsson, Mattias Larsson, R. M. Chandima Ratnayake, and Xavier Ficquet

Abstract There is a lack of quantitative data regarding the determination of minimum distances between weld joints in offshore structures. Existing welding standards, codes and specifications provide recommendations for minimum distances between two adjacent weld joints. The issue arises when violations of the minimum recommendations are discovered in the field or need to be exceeded during construction. This paper is a continuation on a previous paper by the same authors, where the focus was to study the mechanical and material properties of welds in close proximity. The interest in this paper is to study the residual stresses that appear during the welding process, where metal undergoes heating and cooling cycles. This process induces significant levels of residual stresses, leading to increased fatigue failure potential during the operational phase of the structural elements. The measurements were obtained by using ultrasound measurement technique. The results were later verified using X-ray Diffraction and Incremental Centre-Hole Drilling. The ultrasound measurement technique detected a distinct change in residual stresses in the weld joints, with max peaks in the weld centre line. One interesting observation from the Incremental Centre-Hole Drilling measurement in one of the as-welded specimens weld joints was that the longitudinal residual stress was in tension while the transverse direction was in compression, which was also shown in the ultrasonic measurement. In the future, a contour plot measurement would be useful to confirm the results. Another interesting observation was the high compression stresses induced locally on the specimen's surface likely due to manual grinding with abrasive paper. The outcome of this specific experimental setup can provide a baseline for performing further investigations.

Keywords Welding · Residual stress · Weld proximity · Ultrasound · X-ray diffraction · Incremental centre-hole drilling

---

M. Larsson (✉) · M. Larsson · R. M. C. Ratnayake
Department of Mechanical and Structural Engineering and Material Science, University of Stavanger, Stavanger, Norway
e-mail: magnus.larsson@dnvgl.com

X. Ficquet
VEQTER Ltd, Bristol, UK

M. Abdel Wahab (ed.), *Proceedings of 1st International Conference on Structural Damage Modelling and Assessment*, Lecture Notes in Civil Engineering 110,
https://doi.org/10.1007/978-981-15-9121-1_26

1 Introduction

Welding guidelines from standards, codes and specifications require that a minimum distance is maintained between two weld joints in order to avoid the adverse effects of weld proximity. However, limited technical explanation has been given in the literature as to why such a minimum predetermined distance is required and what the consequences would be if that requirement were violated [1–4]. Although minimum distance requirements have been imposed by standards, codes and specifications, breaching these is sometimes unavoidable during the fabrication stage, and similar situation can be found on existing structures. Hence, it is vital to perform investigations to gain an understanding about such circumstances, in order to minimize the life cycle cost related to frequent inspection.

It is important to differentiate between weld proximity and weld overlapping. Weld overlapping is the physical overlap of one weld on top of another, while weld proximity refers to two welds separated by a certain distance. Weld proximity issues arise when two initially approved welds conflict with the required minimum design distance. In this project, the interest has been to understand the implications of weld proximity. The requirements are stated in several international standards:

- BS 2633: "Class I arc welding of ferritic steel pipework for carrying fluids" states that the toes of adjacent butt welds shall, whenever possible, be no closer than four times the nominal thickness of the pipe [1].
- BS 2971: "Class II arc welding of carbon steel pipework for carrying of fluids" (Sect. 10) states that if design factors are such that the meeting of more than two welded seams cannot be avoided, then appropriate precautions shall be taken which shall be agreed between the contracting parties [2].
- BS 4515: "Specification for welding of steel pipelines on land and offshore" (Sect. 11) states that the proximity of weld toe-to-toe distance shall not be less than four times the pipe thickness [3].
- PD5500: "Specification for unfired fusion welded pressure vessels" (Sect. 4.1.3), states that where any part of a vessel is made in two or more courses, the longitudinal seams shall be completed before commencing the adjoining circumferential seam(s) and, where practicable, the longitudinal seams of the adjacent courses shall be staggered by four times the nominal thickness or 100 mm, whichever is the greater, measured from the toe of the welds [4].

The scope of the study presented in this manuscript has focused on increasing the knowledge regarding the effects on the residual stress distribution characteristics in a weld proximity scenario. In this context, there is a lack of clear guidance in welding standards, codes and specifications, which has led to uncertainties on how to address the specific scenarios. Hence, it is vital to identify the specific affected mechanical characteristics for optimizing the necessary precautions needed in the assessment of an existing structure.

2 Industrial Challenge

Although the standards/codes and specifications indicate some guidance, it is not possible to follow them in all instances. There is limited information about how to deal with circumstances where the formal guidelines have been violated during fabrication or during the life cycle phase of a welded structure. In addition, there is no clear justification of why the given guideline-based distances (i.e. greater than four times the thickness of the plate) must be maintained during the fabrication phase. Hence, it is vital to investigate the residual stress characteristics when two welded joints have been fabricated in close proximity.

3 Description

3.1 X-ray Diffraction (XRD)

The x-ray diffraction residual stress measurement technique is a non-destructive, diffraction technique for use on polycrystalline materials. The x-rays used during the technique have wavelengths of the same order of magnitude as the typical inter-atomic/inter-planar distances in polycrystalline materials. Therefore, incident x-rays scattered from a polycrystalline material can constructively interfere, producing a diffracted beam; see Fig. 1. The angle $(\theta + \Delta\theta)$ at which the maximum intensities of the diffracted beam occur are measured and used to calculate the inter-planar spacing, d, of the diffraction planes, using Bragg's law. If residual stresses exist within a material, then the d spacing will be different from that of an unstressed

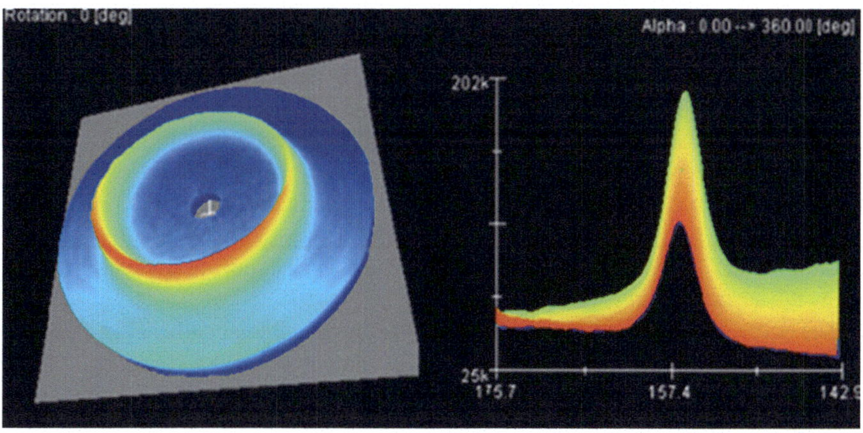

Fig. 1 Examples of the diffraction debye ring and peak for a nominal measurement

sample (i.e. d0). The difference is proportional to the magnitude of residual stresses present.

In principle, therefore, the material's inter-atomic planes are used as internal strain gauges for residual stress measurement. For this specimen, a Pulstec μ-X360n XRD machine analyzed the full Debye ring of diffracted x-rays emitted from the "211" atomic lattice plane of the ferritic steel phase, using "cos α" analysis [1–3].

Incremental centre-hole Drilling (ICHD)

The Incremental Centre-Hole Drilling technique is based on the measurement of the change in surface strain caused by the relief of stresses during the machining of a shallow hole in a component. The principle is that the removal of stressed material results in the surrounding material readjusting its stress state to re-attain stress equilibrium [4]. The measured surface strains allow for the back calculation of the previously existing residual stresses. The formulae and calculations derived for the back-calculation process were developed from a combination of experimental and Finite Element analyses for a flat plate.

The surface strains are usually measured using a special strain gauge rosette attached concentrically around the drilled hole. Incremental center-hole drilling will provide bi-axial results as a function of drilled depth.

The method does suffer from limited strain sensitivity and potential errors and uncertainties related to the dimensions of the hole (e.g. diameter, concentricity, profile, depth), surface roughness, surface flatness and specimen preparation. However, the ICHD technique is the most widely used technique, due to being cheap, quick and easily available, both in the lab and on-site.

3.2 *The Ultrasound Stress Measurement Technique (US)*

Ultrasound residual stress measurement is a non-destructive measurement technique that exploits the acousto-elastic property of common materials, in which the speed of sound through a material changes with stress. It is known [5, 6] that the speed of a longitudinal ultrasound wave through a material is most sensitive to stress changes in the direction of propagation of the wave. In essence, the speed of the longitudinal wave would decrease through tensile stress regions and increase through compressive stress regions [7]. By using Snell's law, it is possible to calculate a critically refracted longitudinal (L_{cr}) wave travelling parallel to the surface of a specimen reaching depths below the surface roughly equal to its wavelength [8, 9].

The US probe head used during this project is shown schematically in Fig. 2, which comprises 2-MHz ultrasound probes attached to a single acrylic wedge at specific angles, according to Snell's law. The transmitting probe (T) emits the longitudinal wave, which is then critically refracted at the specimen surface, travelling parallel to and just beneath the surface of the specimen before being detected by receivers 1 and 2 (R1 and R2). The time needed for the L_{cr} wave to travel between the transmitter and receivers is the Time-of-Flight (ToF) and is used to calculate the change in average

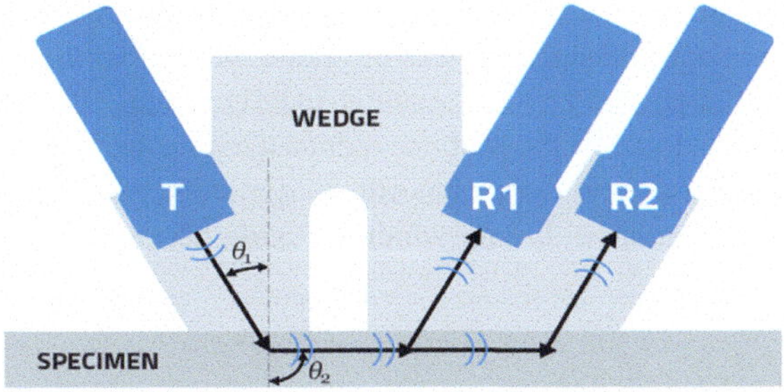

Fig. 2 A diagram showing the US residual stress measurement probe head

stress experienced in the specimen material between the ultrasound probes when referenced against a known stress field, i.e. ideally a stress-free region.

When measuring the ToF at different locations within a specimen, using the same gauge volume, it is possible to determine the relative change in stress experienced using the following formula:

$$\Delta\sigma = \frac{E(T - T_0)}{L_{11}.T_0} \tag{1}$$

where $\Delta\sigma$ is the change in stress between the two stress states, T is the Time of Flight in the unknown stress state, T_0 is the Time of Flight in the known stress state (usually the stress-free state but can be another known stress state) and L_{11} is the acousto-elastic coefficient. L_{11} is a constant property specific to the specimen material and is calibrated experimentally or using a different residual stress measurement technique [10].

For this project, the equipment was arranged to provide results from a gauge area 5 mm long, 4 mm wide and 2.8 mm deep below the material surface. Only the longitudinal ToF was measured using the US system, due to the gauge size. An ultrasonic couplant was applied over the measurement location to facilitate the transmission of the ultrasound. The transmitted and received signals were recorded by a fast sampling oscilloscope and processed off-line using in-house software. To convert the Time of Flight to a residual stress value, a known stress value needs to be defined. This can be done using another residual stress measurement technique. The results presented in this report were calibrated using an ICHD measurement carried out on an as-welded specimen.

3.3 Technique Accuracy

The uncertainty of the ultrasound residual stress measurement technique is determined by several parameters, with only some being accounted for in this analysis. The uncertainty is dependent upon the thickness of the ultrasound couplant, the gradients of residual stresses within the gauge volume, the material texture, the temperature and the resolution/accuracy of the oscilloscope recording equipment. It was assumed that the microstructure was the same throughout the material and, therefore, the L_{11} coefficient and *T0* were assumed to be constant. The US equipment applied a constant contact force on the probe to reduce the error from the variation in the couplant thickness. The oscilloscope recording equipment provides a state-of-the-art resolution of measurement of 0.4 ns, which is highly accurate.

4 Methodology

The following experimental section describes the fabrication process of specimens and the measurement test program.

4.1 Material Selection and Cutting Process

The choice of steel was based on its regular occurrence in the offshore industry and its potential sensitivity to the intended welding operations. The selected base material S420G2 + M (MDS-Y30 Rev. 5) is classified according to NS-EN 10225:2009, EN 10020:2000, ISO/TR 20172:2009 and ISO/TR 15608:2017 as a fine grain steel in group 2.1 with steel number 1.8857 + M. The construction steel has a specified minimum yield point in room temperature at 420 MPa and a minimum average impact energy value of 60 J at −40 °C. The steel is of grade 2 and delivery condition thermomechanically rolled (M) [11–14].

The dimensions of the plates were 15 × 300 × 500 mm. All cutting was done with waterjet, to avoid inducing extra heat that could have an unwanted impact on the results, as the interest of the project was to analyze the adverse effect a secondary weld could have on the initial weld metal (WM) and heat-affected zone (HAZ).

The rolling direction of the steel affects the microstructure and mechanical properties of the steel, thus affecting its anisotropy, which means that the strength can differ between directions. Therefore, the test plates have been produced consistently so that all welding passes have been done across the steels rolling direction.

The filler metal:

- The filler metal used for root passes was NST's NSSW SM-47A: a metal cored wire for low temperature pipe and steel applications down to −60 °C.
- The filler metal used for hot, fill and cap passes was NST's SF-3AM, flux cored wire for low-alloyed steel, offshore applications, piping, etc.

4.2 Welding of Production Plates

Six plates with adjacent welds were fabricated for the study. Welding was done according to a produced welding procedure specification (WPS), to maintain conformity. After welding, the plates were visually inspected, according to NS-EN ISO 17,637:2016 and NS-EN ISO 5817:2014 B/C, by KIWA TI [19, 20]. Before further mechanical testing, a radiographic examination was carried out, to verify that the welds were of good quality. This was performed at IKM Inspection AS. Figure 3 presents the plates PL3-PL8 that were used in the study. Weld A to the left was first welded, followed by the nearby weld B to the right.

Production test plates PL3, PL5 and PL7 were used for mechanical testing. The destructive testing included: Vickers hardness testing, Charpy V impact testing, tensile testing, fatigue testing, as well as macro and microscopic examination. Testing was carried out according to guidelines in ISO 15614–1:2017 and ASTM E466-15 [15, 16]. All mechanical testing was performed at the accredited Quality Lab. The results from mechanical and fatigue testing will be reported in other papers.

Plates PL4, PL6 and PL8 were used for residual stress measurements at Veqter in Bristol, England. The residual stress testing performed on these plates is the focus of this manuscript.

4.3 Design of Residual Stress Specimens in "Ready-To-Test" Conditions" and Specimen Preparation

Preparation of residual measurement specimens was carried out at the UiS laboratory. Plates PL4, PL6 and PL8 were first cut into rectangular sections. Two different types of specimens were produced from each plate. Figure 4 shows the location of the specimens and the appearance before machining.

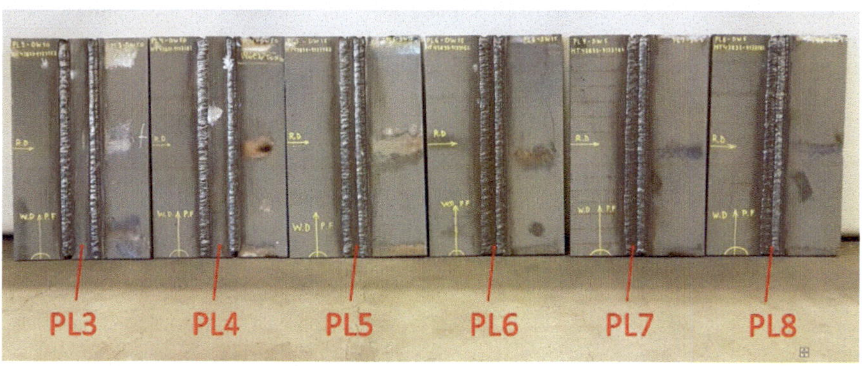

Fig. 3 Plates used for ultrasonic measurement were PL4, PL6 and PL8. The left weld on each plate was named A and the right weld, B

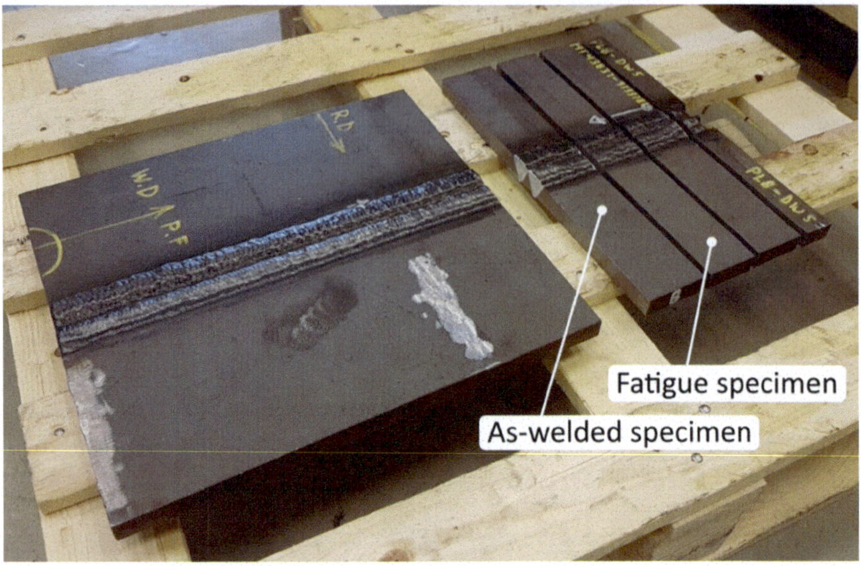

Fig. 4 The specimens before they were machined, showing the location, in the S420G2 + M steel plate, from which the specimens were taken. The marking R.D. on the plate means "Rolling Direction", W.D. Welding Direction and P.F means Vertical up position

The fabrication method for the two different specimens will be described below.

Fabrication of as-welded specimen.

The width of the as-welded specimen was 50 mm, with length around 310 ± 2 mm (see Figs. 4 and 5).

The unmachined as-welded specimen only had the weld caps grinded flush. The grinding was done with a flap disc. Finished 'as-welded' specimens are shown in Fig. 6.

The white markings are the welds and show the grinded cap side. There were three different distances between weld toes: 45, 9 and 1 mm^3. The figure also shows which side was cap side and which root side before grinding.

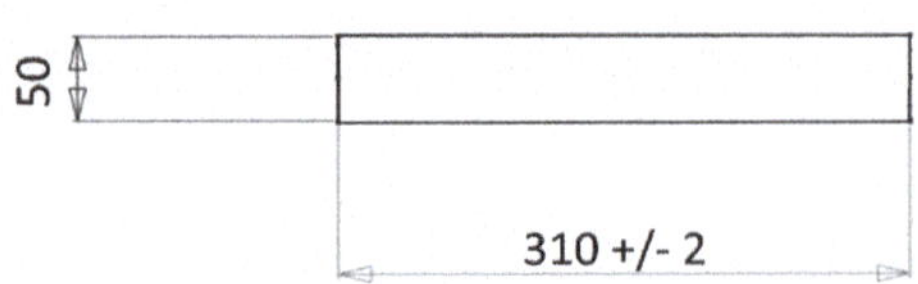

Fig. 5 Drawing of the 'as-welded specimen'. After cutting, joint preparation and welding, the width of the plates changed from 300 mm to 310 ± 2 mm

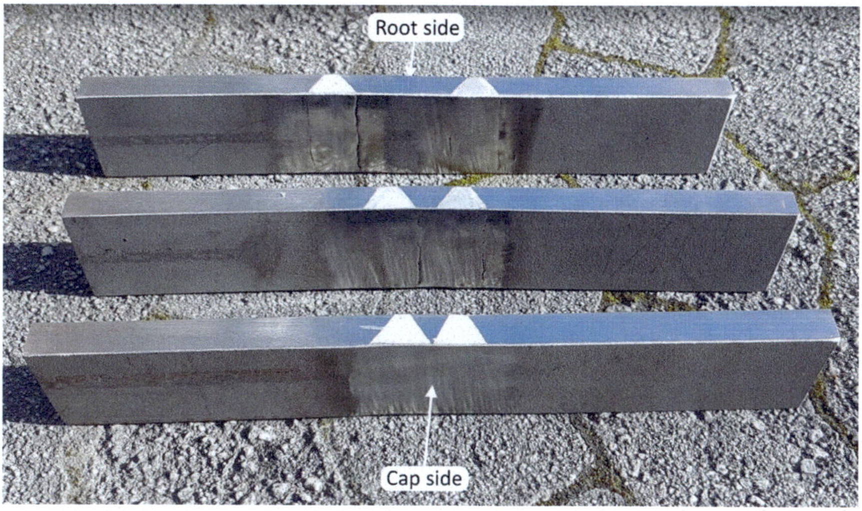

Fig. 6 Finished 'as-welded' specimens. Grinding was done with a flap disk. The width of the samples was 50 mm

Fabrication of Fatigue Specimen.

Since specimen preparation can strongly influence the fatigue data, the test pieces were prepared according to guidelines in ASTM E466-15, Appendix X1-Example of machining procedure [16]. The dimensions of the fatigue specimen are shown in Fig. 7.

A machining procedure from ASTM E466-15 Appendix X1 [16]. was used to minimize the variability of machining and heat treatment upon fatigue life.

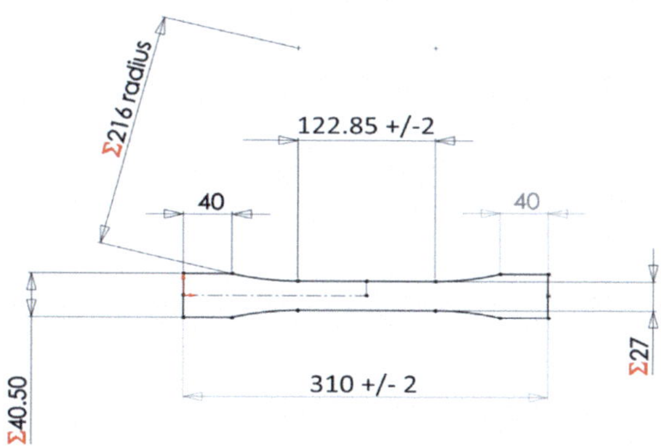

Fig. 7 After cutting, joint preparation and welding, the length of the plates changed from 300 mm to 310 ± 2 mm. This was corrected by extending the initial center distance from the base material fatigue specimen drawing from 112.85 mm to 122.85 ± 2 mm

1. The sides of the fatigue test specimens were milled according to the drawing above (Fig. 7). Machining was done gradually, where the second to last step was 0.4 mm and the last step was 0.2 mm. The thickness of the specimen after milling was 9.15 mm.
2. The next 0.1–0-2 mm on the cap and root sides of the specimens was removed with cylindrical grinding (surface grinding machine). The thickness of the specimen after surface grinding was 8.95 mm. It was not possible to use the surface grinding machine on the sides of the specimens, so this was done later with abrasive paper.
3. The final grinding was done manually with abrasive paper. Sanding was done in the following steps: Hermes WS Flex Waterproof P180 (grain size 82 μm), Struers FEPA P # 500 (grain size 30 μm), Hermes WS Flex Waterproof P1000 (grain size 18 μm), Struers FEPA P # 1200 (grain size 15 μm) and Silicon Carbide 1200/4000 (grain size 5 μm). The thickness of the specimen after final manual grinding was around 8.60 mm. This means that the manual grinding process removed approximately 200 μm on each side of the specimen. The manual grinding was done along the length of the specimen.
4. The welds were etched out with nital to become visible (see Fig. 8).

Figure 8 shows the machined, grinded and etched fatigue specimens, "Ready-To-Test".

Finished residual stress sample specimens.

The yellow dashed line in Fig. 9 below shows the finished sample pieces and placement, which are ready for testing.

Two different type of specimens were produced from each of plates PL4, PL6 and PL8, with a total of six specimens:

- 3 x 'As-welded specimen'.
- 3 x 'Fatigue specimen'.

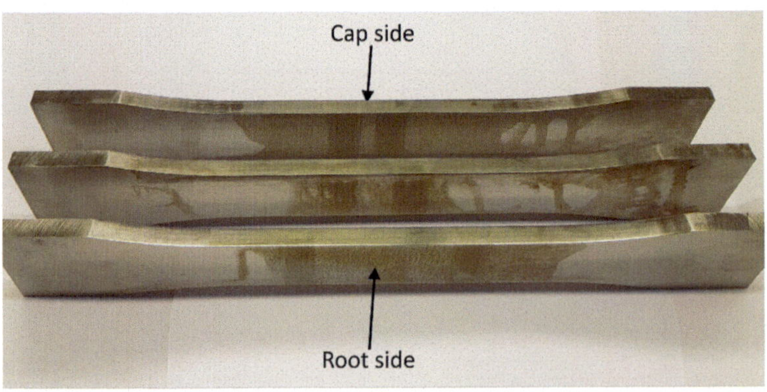

Fig. 8 Etched fatigue specimens. In the figure we see the "root side" of specimens

Fig. 9 Two specimens per plate were sent to Veqter for residual measurements. Here are shown, from the left, plate PL4 −45 mm between weld toes, PL6 −9 mm between weld toes and PL8 − 1 mm between weld toes

In Fig. 10, the picture below to the left shows how the finished machined specimens looked before testing, while the picture to the right shows the number of specimens. ***Test program—'Fatigue specimens'***:

- Test method: Ultrasonic Residual Stress Measurement.
- Test machine: VURSA.
- Number of specimens tested: 3 (test performed on all fatigue test specimens).
- Test description: Measurement of the average longitudinal and transverse residual stresses. The US measurement technique measures the average residual stress down to 2 mm. Both weld cap and weld root were measured.

The direction of measurement in ultrasonic test is shown in Fig. 11.

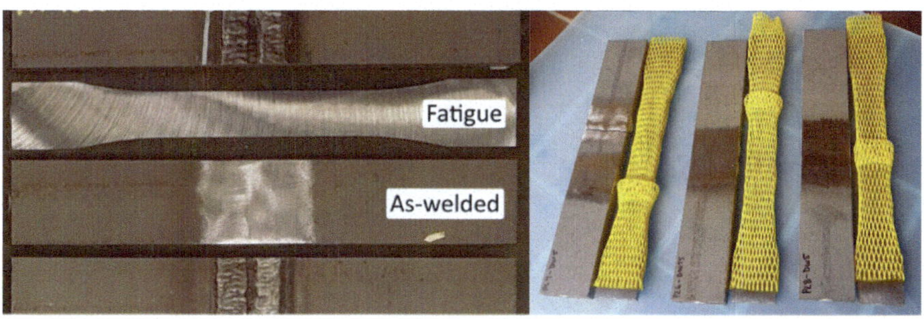

Fig. 10 Left: the appearance of finished specimens; right the six specimens produced

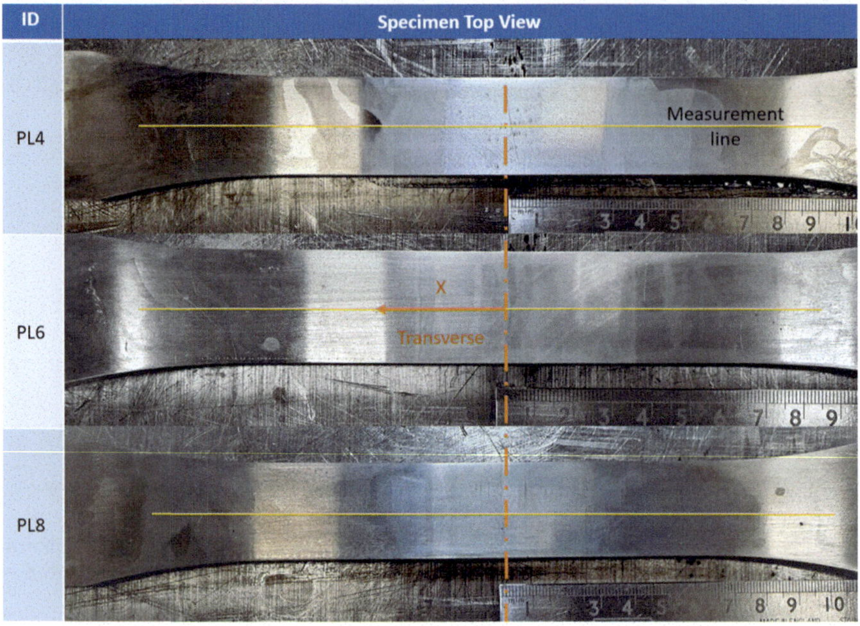

Fig. 11 ASTM E466-15 test sample ultrasound measuring direction

The measurements were performed along the centreline of the specimen. The probe head used during the ultrasound residual stress measurements is shown in Fig. 12.

Fig. 12 Ultrasonic residual stress measurement

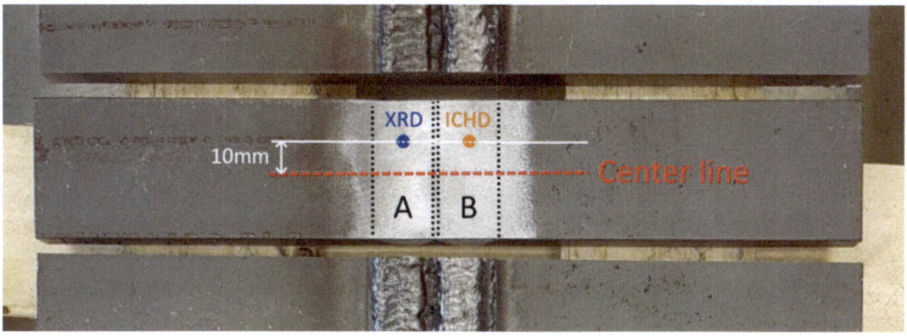

Fig. 13 Specimen test locations in the as-welded specimen, PL8. The test locations are performed 10 mm offset from weld centerline in y-direction. The figure also presents welds A and B

- Test method: XRD
- Test machine: VURSA
- Number of specimens tested: 1 (test performed on fatigue specimen PL8, 1 mm between the weld toes)
- Test description: Both longitudinal and transverse residual stresses were measured at a depth of 50 um. Place of measurement was in the x-direction along the specimen, in the weld centerline, 5–50 mm parallel to weld centerline on the weld cap face, see Fig. 12. The root face side was not measured.

Test program—as-welded specimens

Test method: ICHD.

- Test machine: VURSA.
- Number of specimens tested: 1 (test performed on PL8 As-welded, 1 mm between weld toes).
- Test description: ICHD was performed in weld B at a 10-mm offset from weld centerline in y-direction (see Fig. 13). The technique measured and obtained the stress profile at 1 mm depth.

Test method: XRD.

- Test machine: VURSA.
- Number of specimens tested: 1 (test performed on PL8 as-welded, 1 mm between weld toes).
- Test description: The test was performed in weld A at a 10-mm offset from weld centerline in y-direction (see Fig. 13). The test looked at bi-axial residual stresses at 50 μm and 150 μm from weld cap face side.
- The technique used a 2 mm hole diameter to measured and obtain the stress profile at 1 mm depth.

Figure 13 shows the as-welded test specimen.

In the figure, we can see the red centerline, weld A, weld B and the various tests performed on the test part.

5 Results and Discussion

The extent of testing of the different specimens is shown in Table 1 below.

Three different techniques were used for the analysis: ultrasound stress measurement, incremental center-hole drilling and X-ray diffraction.

5.1 *Fatigue Specimen*

Ultrasound stress measurement technique.

The US measurement technique measures the average residual stress down to 2 mm. The times of flight (ToF) measured using US on the three CNC-machined fatigue specimens on the top (weld cap) and bottom (root weld) surfaces across the welds are shown in Fig. 14. The measured ToF was shifted to show about 0 ns in the base material, in order to observe the change in time of flight relative to the parent material. The increases in ToF relate to increases in tensile stresses, and the decreases in ToF relate to increases in compressive stresses. Clear changes occur at the weld location and some drop in ToF was generally found on either side of the weld. Each of the maximum peaks was found to be center to the weld centerline at the bottom and top sides of the specimen. In the transverse direction, the ToF looks like a near mirror to the longitudinal direction, with the exception that each negative drop was balanced with a sharp increase in ToF.

The ToF seen in Fig. 14 was then converted, to show the corresponding residual stress, as seen in Fig. 15. In order to convert the ToF graph to show residual stress, it was necessary to determine two reference values: one for a maximum compressive/tensile peak and one for a minimum value. The stress distribution in the graph was then based on these two reference values. The values were appreciated to be

Table 1 Extent of testing of residual stress experiment

Test Specimen ID	Weld Adjacency	US	ICHD	XRD
PL4—Fatigue	45 mm	X	–	–
PL4—As-welded	45 mm	–	–	-
PL6—Fatigue	9 mm	X		
PL6—As-welded	9 mm	–	–	–
PL8—Fatigue	1 mm	X	–	X
PL8—As-welded	1 mm	–	X	X

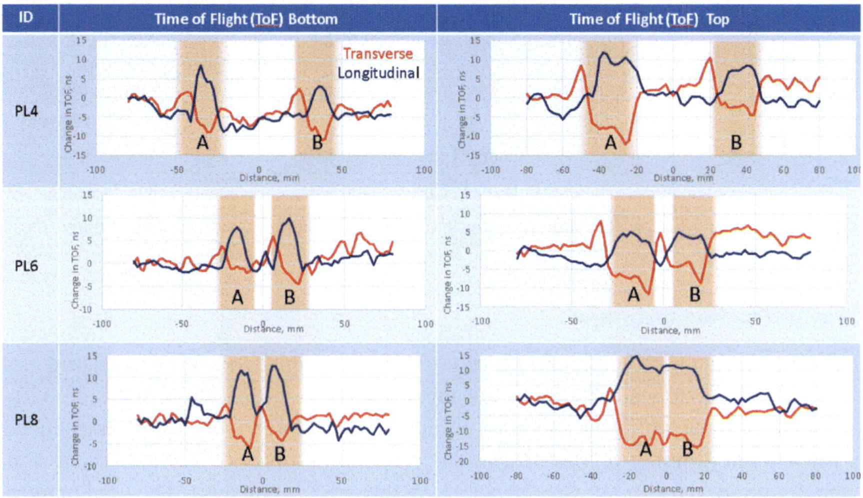

Fig. 14 Time of flight results using the ultrasound residual stress measurements. Longitudinal stresses are along the weld direction while transverse stresses are perpendicular to the weld direction. Bottom and top refer to the root side and weld cap side respectively

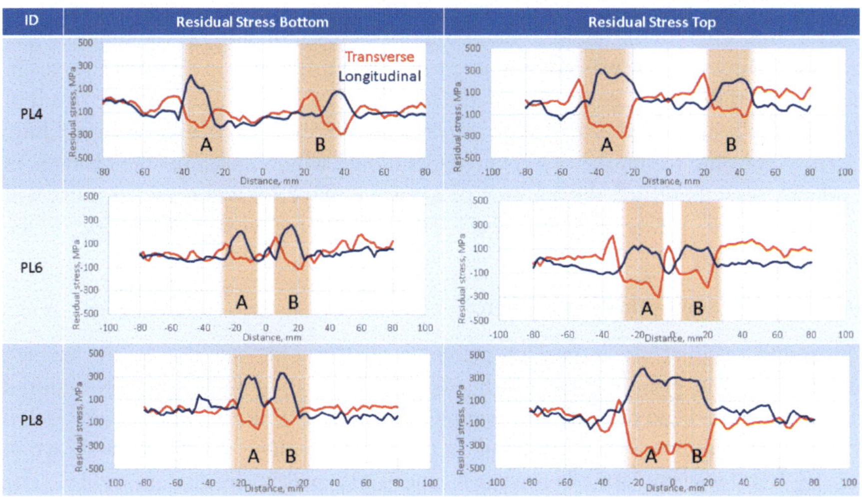

Fig. 15 Residual stress test results using the ultrasound residual stress measurements. Bottom and top refer to the root side and weld cap side respectively

0 MPa in the base material and 80% of the yield point at maximum tensile residual stress. Ultrasonic measurement can be severely affected by changes in texture which occur naturally in welds. Often, several additional calibration values need to be used.

Table 2 X-ray diffraction results from fatigue specimens

Testing location	XRD PL8 Top Transverse [Compressive]	XRD PL8 Top Longitudinal [Compressive]
Weld Center Line	336 MPa	440 MPa
5 mm from WCL	400 MPa	392 MPa
60 mm from WCL	300 MPa	346 MPa

X-ray diffraction technique.

The XRD carried out on the surface of the grinded fatigue specimen measured at a depth of 50 μm showed that the outer layers were in compression, between 320–450 MPa, as seen in Table 2. These compressive stresses were likely introduced by the manual grinding process during the preparation process, where 200 μm was manually removed with abrasive paper along the specimen length. The final manual grinding is intended to decrease the residual stresses induced by the machining and the machine grinding process [17].

The measurements were taken in the weld centerline, 5–50 mm^2 parallel to the weld centerline at 50 μm. The locations were picked to assess the difference in residual stresses in the weld metal compared to the heat-affected zone and base metal. The residual stress in the surface was similar in all tests, in both the longitudinal and transverse directions. It was not possible to see any difference in residual stresses in the vicinity of the weld metal. The results presented in Table 2 are likely the result of the grinding process and provided no information on residual stresses from welding.

The ultrasonic technique is a relative technique, which measures the changes in residual stress; therefore, the addition of a constant layer of compressive stress will not influence the US test results. The ICHD measurements were not made on the fatigue specimen.

Residual stress as-welded specimen.

Ultrasound residual stress measurement was performed, but, due to the uneven surface from the grinding, the results were inconclusive. The operation required a different probe head that was not available at the time of the experiment. XRD and ICHD measurements were carried out on the PL8-as-welded specimen, as shown in Fig. 13. XRD was performed on weld A, and ICHD was performed on weld B. The tests were not performed in the centreline of the specimen. This may have influenced the longitudinal residual stress. Figure 16 shows the results from the XRD and ICHD residual stress tests graphically. The y-axis shows the stresses, and the x-axis shows depth from surface, from 0 to 1 mm.

The XRD showed high local longitudinal and transverse compressive residual stresses in the top surface at 50–150 μm. As mentioned earlier, this was performed in weld A at a 10 mm offset from the specimen's centerline in y-direction (see Fig. 13). The residual stresses at 50 μm were near yield point, while the results from

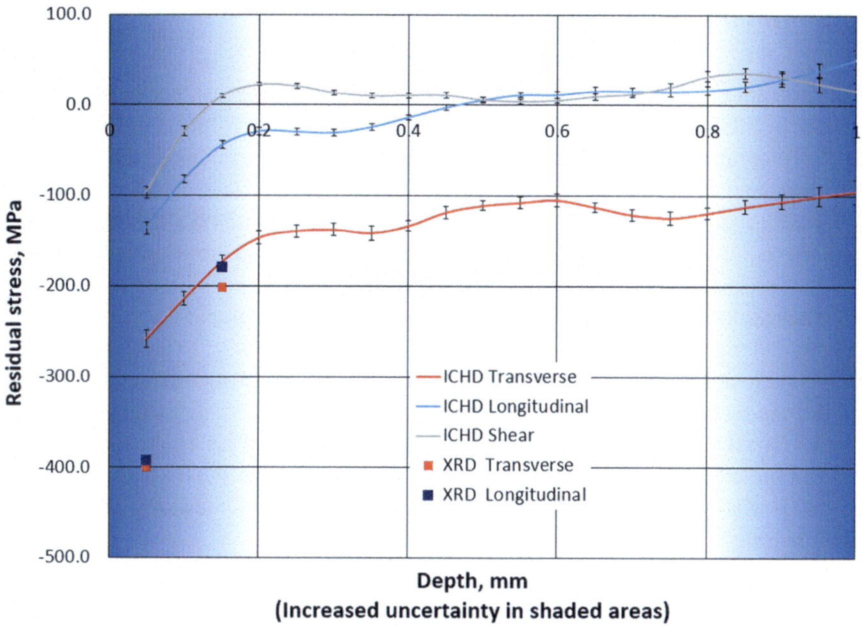

Fig. 16 XRD and ICHD measurements on PL8—as-welded test specimen top face. The XRD was measured at 50–150 μm from the top surface. The results from the ICHD show the stress distribution profile down to 1 mm depth (x-axis)

150 μm showed values approximately 200 MPa compressive residual stresses, in both longitudinal and transverse directions.

The results from the ICHD were performed in weld B at a 10-mm offset from the weld centerline in the y-direction. The technique measured and obtained the stress profile at 1 mm depth. It showed the same characteristics as the XRD in transverse direction at a depth of 150 μm, but the longitudinal direction differed somewhat. Similar to the XRD, it showed compressive residual stresses in both longitudinal and transverse directions. From 200 μm to 1000 μm, the compressive transverse residual stresses stayed approximately constant at 100 MPa. The longitudinal and shear residual stress was approximately 0 MPa and did not show any tendency for high tensile residual stresses. However, these results show that the longitudinal stress was higher than the transverse stress, which was also observed in the ultrasonic measurements. It is important to note that the results were conducted at an offset from the weld centerline, which may have influenced the longitudinal residual stresses.

6 Summary and Conclusion

Experimental work was carried out to examine the residual stress distribution of two adjacent welds at varying distances apart: 45, 9 and 1 mm[3] from weld toe to weld toe. Two variants of specimens were examined in this experiment: a set of *fatigue specimens* and an *as-welded specimen.*

The "*Fatigue specimens*" were prepared in accordance with *ASTM E466-15*. The standard requires the specimen to be subjected to a grinding operation before being tested, in order to remove as much of the residual stresses as possible to not contaminate the results. The final grinding, 200 μm, was performed manually with abrasive paper and along the specimen length. The fatigue specimens were examined by using ultrasound residual stress measurement technique (US) and x-ray diffraction (XRD).

The "*As-welded specimen*" was examined by using XRD and incremental-hole drilling (ICHD). The measurements were carried out on PL8 (plate with 1 mm between the weld toes). XRD was performed on weld A, and ICHD was performed on weld B. XRD was performed at a depth of 50–150 μm from the surface and ICHD was performed at a depth of 1000 μm from the surface.

6.1 Based on the Experimental Work and Evaluation the Following Can Be Concluded:

Fatigue specimens.

- Examined by using US measurement technique and XRD.
- The results from the US test showed that the longitudinal residual stress was in tension and the transverse residual stress was in compression.
- The effect weld proximity had on the residual stress profile was not noticeable in the PL4 and PL6 specimens.
- In specimen PL8, with 1 mm between the weld toes at the cap side, there was a noticeable increase in both tensile and compressive residual stress distribution.
- The US test showed a clear transition in residual stresses from the parent metal to the weld metal. The US was affected by changes in texture in the material which occur in the weld.
- The results from the XRD showed that there were local high compressive residual stresses 50 μm down the surface of the specimen, likely due to the preparation method. XRD was performed at the weld centerline, at a 5-mm and 50-mm off-set away from the welds. It was not possible to derive whether there was any influence on the residual stresses from the weld itself. The longitudinal and transverse residual stresses had similar magnitudes, and both were in compression.

As-welded specimens.

- Examined by using XRD and ICHD.
- The XRD and ICHD results were similar to the *fatigue specimens* showing high local longitudinal and transverse compressive residual stresses near the yield point in the outer surface.
- The ICHD results agreed with the results from the XRD showing high local compressive residual stresses approximately 200 μm into the outer surface of the specimen.

Further work.

Further research will be carried out to take additional residual stress measurements, and more measurements will be taken in order to calibrate the time of flight measured by the ultrasonic technique to provide the calibrated residual stresses. The ICHD measurements were not made on the fatigue specimens. This and other measurement technique, such as contour plot, will be carried out and presented in a future paper.

References

1. Sasaki T (1997) Influence of image processing conditions of debye scherrer ring images in x-ray stress measurement using an imaging plate. Int Centre Diffr Data
2. Hiratsuka K et al (2003) Development of measuring system for stress by means of image plate for laboratory x-ray equipment. Int Centre Diffr Data 46
3. Sasaki T et al (2009) X-ray multiaxial stress analysis using two Debye rings. Int Centre Diffr Data 46
4. Grant PV, Lord JD, Whitehead PS (2002) The measurement of residual stresses by the incremental hole drilling technique. Measurement good practice guide, no. 53, National Physical Laboratory, UK
5. Egle DM, Bray DE (1976) Measurement of acoustoelastic and third-order elastic constants for rail steel. J Acoust Soc Am, Am 60:741–744
6. Egle DM, Bray DE (1975) Non-destructive measurements of longitudinal rail stresses. Report No. FRA-OR&D-76-270, PB 272061, NTIS, Springfield, VA
7. Bray DE, Stanley RK (1997) Ultrasonic techniques for stress measurement and material studies, Chapter 9, non-destructive evaluation: a tool in design. CRC Press, Manufacturing and service
8. Belahcene F, Lu J (2002) Determination of residual stress using critically refracted longitudinal waves and immersion mode. J Strain Anal Eng Des 37(1):13–20. https://doi.org/10.1243/0309324021514790
9. Bray DE, Stanley RK (1996) Nondestructive evaluation: a tool in design, manufacturing and service. CRC press
10. Romac R, Cave D, McIntyre D, Ficquet X (2016) Characterisation of the effect of corrosion on the residual stresses in girth weld pipe using ultrasonic calibrated with strain-relieving measurement techniques,3rd- 7th ICRS
11. Weldable structural steels for fixed offshore structures—Technical delivery conditions, NS-EN 10225, 2009
12. Definition and classification of grades of steel, NS-EN 10020, 2000
13. Welding—guidelines for a metallic materials grouping system, ISO/TR 15608, 2017
14. Welding—Grouping systems for materials—European materials, ISO/TR 20172, 2009

15. Specification and qualification of welding procedures for metallic materials Welding procedure test Part 1: Arc and gas welding of steels and arc welding of nickel and nickel alloys, NS-EN ISO 15614-1, 2017
16. Standard practice for conducting force controlled constant fatigue tests of metallic materials, ASTM E466-15, 2015
17. Totten GE, Howes MAH, Inoue T (2002) Handbook of residual stress and deformation of steel. Materials Park, Ohio: ASM International, p vii, p 499

Investigation of Fatigue Strength Behaviour in Dual Weld S420 Steel Joints Fabricated at a Close Proximity

Magnus Larsson, Mattias Larsson, and R. M. Chandima Ratnayake

Abstract There is insufficient information available in existing welding standards, codes and specifications regarding the implications of having two or several adjacently fabricated weld joints in close proximity. This has introduced fatigue life performance uncertainties, due to a lack of understanding of the inherent behaviour of weld joints in the presence of adjacent welds. Hence, it is vital to investigate the fatigue life performance and obtain quantitative data for future assessments. This experiment has been designed using six welded, and one unwelded, S420G2+M 500 × 300 × 15 mm structural steel plate. Two of these plates were used for fatigue testing, having 44 mm and 12 mm between the adjacent weld toes. These distances were less than normally recommended in standards codes and specifications. The preparation of the samples was conducted in accordance with ASTM E466-15 for homogenous materials subjected to high-cycle fatigue. The experiment was not able to detect any detrimental effect by having welds in close proximity using this specific preparation method and weld setup. The weld metal and corresponding heat-affected zones had a higher fatigue strength than that of the surrounding base material. All tested specimens failed in the base metal. The material and welding method did not seem to introduce any degrading effect on the weld joint. Important to note is that the fabrication method required a surface roughness of 0.2 μm and had likely introduced an outer layer of compressive residual stresses at yield point level 50 μm into the surface and bottom surfaces of the specimens. This might have improved the resistance to crack formation in the surface. The residual stress was obtained by using ultrasonic stress measurement technique and x-ray diffraction presented in a parallel study by the same authors.

Keywords Welding · Fatigue · Offshore steel · Weld proximity

M. Larsson (✉) · M. Larsson · R. M. C. Ratnayake
Department of Mechanical and Structural Engineering and Material Science, University of Stavanger, Stavanger, Norway
e-mail: magnus.larsson@dnvgl.com

M. Abdel Wahab (ed.), *Proceedings of 1st International Conference on Structural Damage Modelling and Assessment*, Lecture Notes in Civil Engineering 110,
https://doi.org/10.1007/978-981-15-9121-1_27

1 Introduction

Welding guidelines from standards, codes and specifications require that a minimum distance is maintained between two weld joints in order to avoid the adverse effects of weld proximity. However, limited technical explanation has been given in the literature as to why such a minimum predetermined distance is required and what the consequences would be if that requirement were violated [1–4]. Although minimum distance requirements have been imposed by standards, codes and specifications, breaching these is sometimes unavoidable during the fabrication stage, and similar situations have been discovered on existing structures. Hence, it is vital to perform investigations to gain an understanding about such circumstances, in order to minimize the life cycle cost related to frequent inspections.

It is important to differentiate between weld proximity and weld overlapping. Weld overlapping is the physical overlap of one weld on top of another, while weld proximity refers to two welds separated by a certain distance. Weld proximity issues arise when two initially approved welds conflict with the required minimum design distance. In this project, the interest has been to understand the implications of weld proximity. The requirements are stated in several international standards:

- BS 2633: "Class I arc welding of ferritic steel pipework for carrying fluids" states that the toes of adjacent butt welds shall, whenever possible, be no closer than four times the nominal thickness of the pipe [1].
- BS 2971: " Class II arc welding of carbon steel pipework for carrying of fluids" (Sect. 10) states that if design factors are such that the meeting of more than two welded seams cannot be avoided, then appropriate precautions shall be taken which shall be agreed between the contracting parties [2].
- BS 4515: "Specification for welding of steel pipelines on land and offshore" (Sect. 11) states that the proximity of weld toe-to-toe distance shall not be less than four times the pipe thickness [3].
- PD5500: "Specification for unfired fusion welded pressure vessels" (Sect. 4.1.3) states that where any part of a vessel is made in two or more courses, the longitudinal seams shall be completed before commencing the adjoining circumferential seam(s) and, where practicable, the longitudinal seams of the adjacent courses shall be staggered by four times the nominal thickness or 100 mm, whichever is the greater, measured from the toe of the welds [4].

The study performed in this manuscript has focused on increasing the knowledge regarding the effects on the fatigue strength characteristics in a weld proximity scenario. In this context, there is a lack of clear guidance in welding standards, codes and specifications, which has led to uncertainties on how to address the specific scenarios. Hence, it is important to identify the affected mechanical properties in order to be able to assess the measures required to obtain safe structures.

2 Industrial Challenge

Although the standards, codes and specifications indicate some guidance, it is not possible to follow them in all situations. There is limited information about how to deal with circumstances where the formal guidelines have been violated during fabrication or during the life cycle phase of a welded structure. In addition, there is no clear justification of why the given guideline-based distances (i.e. greater than four times the thickness of the plate) must be maintained during the fabrication phase. Hence, it is vital to investigate the fatigue strength characteristics when two welded joints have been fabricated in close proximity.

3 Methodology

3.1 Experimental Test Background

The choice of test setup was two adjacent butt welds on a S420G2 + M offshore steel plate. The tests performed looked at changes in fatigue capacity and failure criteria due to variation in distance between the welds.

Fatigue test results are significantly influenced by factors such as material properties, history of the parent material, specimen preparation process, testing machine and testing procedures. The condition of the test specimen, the method of specimen preparation, the inspection program and testing procedure are therefore of the utmost importance. The study was therefore conducted in accordance with fatigue test standards, ASTM E466-15 and ASTM E468-11, which have clear procedures and guidelines for the above [5, 6]. The following experimental section describes the fabrication process and the process of fatigue testing that was carried out.

3.2 Material Selection and Waterjet Cutting

The choice of steel was based on its regular occurrence in the offshore industry and its potential sensitivity to the welding process. The steel type (MDS-Y30 Rev. 5) is classified, according to NS-EN 10,225:2009, EN 10,020:2000, ISO/TR 20,172:2009 and ISO/TR 15,608:2017, as a fine grain steel in group 2.1 with steel number 1.8857 + M. The steel is of grade 2, and delivery condition is thermomechanically rolled (M) [7–10]. Table 1 below shows the specification and properties for the base material used.

The dimension of the plates was 15 × 300x500 mm. These plates came cut from a larger plate of size 15 × 2500x12000 mm. All cutting was done with waterjets, to avoid inducing extra heat that could have an unwanted impact on the results, as the

Table 1 Specification and properties of base material S420G2 + M, Ilsenburger Grobblech - Inspection certificate

Form of product	Offshore steel plate			
Grade Designation	S420G2 + M/Q-Y30			
Heat Number	HT-43831–9,133,182			
Steel making	Basic oxygen process, vacuum degassed			
Last Mechanical and Thermal Treatment	Thermomechanically rolled and air cooled			
Chemical Composition	See NS-EN 10,225:2009 [7]			
Yield Point, ReH[a]	TR = 489 MPa, AC = 420–540 MPa			
Tensile Strength, Rm[a]	TR = 557 MPa, AC = 500–660 MPa			
Impact Energy[b]	Single values (≥42 J)			Average (≥60 J)
	118	111	114	114
Elongation[c]	TR = 31%, AC ≥ 19%			

AC-Acceptance Criteria, TR-Test Result.
[a]Tested in room temperature.
[b]Charpy V-Notch Impact test. Tested in -40 °C.
[c]A5: $L_0 = 5.65\sqrt{S_0}$

Table 2 Specification and properties of filler metal NSSW SM-47A

Trade Designation	NSSW SM-47A				
Material Classification	EN ISO 17,632-A-T 46 6 1Ni M M 1 H5				
Manuf. No	7U341AW996				
Yield Point, AW[a]	TR = 527 MPa, AC = Min 460 MPa				
Tensile Strength, AW[a]	TR = 617 MPa, AC = 530–690 MPa				
Impact Energy, AW[b]	Single values (≥32 J)			Average (≥47 J)	
	93	104	115	104	
Elongation[a,c]	TR = 28%, AC ≥ 20%				
Hydrogen Content of Deposited Metal	HDM (ml/100 g)			Ave	Spec
	1.0, 1.7, 1.0			1.2	5 Max

AC-Acceptance Criteria, TR-Test Result, AW = Condition, As-Welded
[a]Tested in room temperature
[b]Charpy V-Notch Impact test. Tested in -60 °C
[c]A5: $L_0 = 5.65\sqrt{S_0}$

interest of the project was to analyse the adverse effect a secondary weld could have on the initial weld metal (WM) and heat-affected zone (HAZ).

The rolling direction of the steel affects the microstructure and mechanical properties of the steel, thus affecting its anisotropy, which means that the strength can differ between directions. Therefore, the test plates have been produced consistently, so that all welding passes have been done across the steel's rolling direction.

The filler metal used for root passes was NST's (Norsk Sveiseteknikk AS) NSSW SM-47A: a metal cored wire for low temperature pipe and steel applications down to -60 °C (see Table 2).

The filler metal used for hot, fill and cap passes was NST's SF-3AM, flux cored wire for low-alloyed steel, offshore applications, piping, etc. (see Table 3).

Table 3 Specification and properties of filler metal NSSW SF-3AM

Trade Designation	NSSW SF-3AM			
Material Classification	EN ISO 17,632-A-T 46 4 Z P M 2 H5 EN ISO 17,632-A-T 46 6 Z P M 2 H5			
Manuf. No	7S041MP960			
Yield Point, AW[a]	TR = 556 MPa, AC = Min 460 MPa			
Tensile Strength, AW[a]	TR = 614 MPa, AC = 530–690 MPa			
Impact Energy, AW	Single values (≥32 J)			Average (≥47 J)
	82[b]	80[b]	87[b]	83[b]
	112[c]	126[c]	142[c]	127[c]
Elongation[a,d]	TR = 26%, AC ≥ 20%			
Hydrogen Content of Deposited Metal	HDM (ml/100 g)		Ave	Spec
	1.4, 1.2, 1.3		1.3	5 Max

AC-Acceptance Criteria, TR-Test Result, AW = Condition, As-Welded
[a]Tested in room temperature
[b]Charpy V-Notch Impact test. Tested in -60 °C
[c]Charpy V-Notch Impact test. Tested in -40 °C
[d]A5: $L_0 = 5.65\sqrt{S_0}$

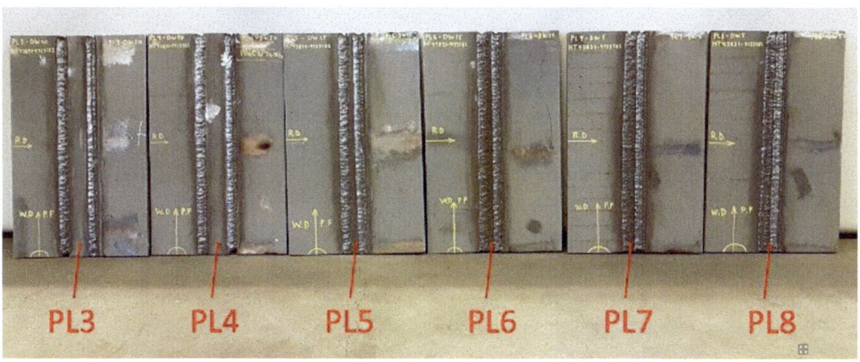

Fig. 1 All production plates, ready to be forwarded for NDT and mechanical testing. In the figure we can see that the plates have different distances between the weld joints

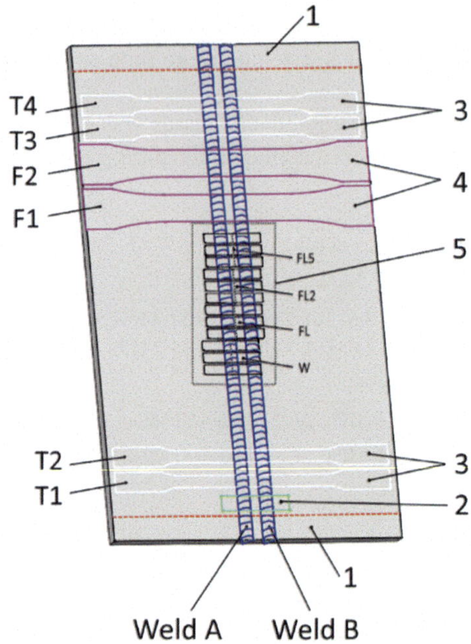

Fig. 2 Location of test specimen for the production test plates PL3, PL5 and PL7. 1-Material should be removed 25 mm at the top and the bottom of the plate, 2- Hardness/Macroscopic test, 3-Tensile test, 4-Fatigue test and 5- Charpy V impact test

3.3 Welding of Qualification and Production Plates

Table 4 below gives an overview and test description of the eight plates used in the whole study.

Plate PL1 was used to produce a Welding Procedure Qualification Record (WPQR). From this WPQR was later a WPS created. The WPS was used in the following welding operation in order to maintain conformity. Six plates (PL3-PL8) with adjacent welds were fabricated according to the produced WPS. After welding a visual inspected was performed, according to NS-EN ISO 17,637:2016 and NS-EN ISO 5817:2014 B/C, by KIWA TI [11, 12]. Before further mechanical testing, a radiographic examination was carried out on plate PL3-PL8 to verify that the welds were of good quality. Figure 1 shows plate PL3-PL8 used in the study.

Production test plates PL3, PL5 and PL7 were tested according to the test scope in ISO 15,614–1:2017 Level 2, although with small changes: fatigue testing replaced the bending test [11]. The destructive testing included: Vickers hardness testing, Charpy V impact testing, tensile testing, fatigue testing, as well as macro and microscopic examination. All mechanical testing except fatigue testing was carried out at the accredited Quality Lab. Figure 2 below shows the various specimens produced from plates PL3, PL5 and PL7. The results from mechanical testing will be reported in another paper.

Fig. 3 PL2 was an unwelded plate on which only fatigue testing was performed. Fatigue specimens from unwelded base material were machined according to the same procedure as welded plates PL3 and PL5. The aim of the test was to see the fatigue capacity of unwelded S420G2-M steel

This manuscript will focus on fatigue specimens, F1 (fatigue) and F2, in the figure above. Also, one unwelded plate, PL2, was used for fatigue testing (see Fig. 3).

Five pieces of fatigue specimens were produced from the unwelded plate, PL2. The plates, PL4, PL6 and PL8, were used for residual stress measurements at Veqter in Bristol, England. The yellow dashed line in Fig. 4 below shows the finished specimens and their location. The specimens were prepared for testing.

The fatigue specimens in, F1 and F2, were taken from the same location as the specimens for the residual stress measurements (see Fig. 4). This test was performed to gain an understanding of the residual stresses that arise in and around the welds. A more thorough analysis of the residual stress in the plate will be presented in a parallel study. The results from the x-ray diffraction measurement is included in the results and discussion chapter.

Fig. 4 Two specimens per plate were sent to Veqter for residual measurements. Here are shown, from the left, plates PL4 - 45 mm between weld toes, PL6 - 9 mm between weld toes and PL8 - 1 mm between weld toes. The marking R.D. on the plate means "Rolling Direction", W.D. Welding Direction and P.F means Vertical up position

Table 4 The studied eight plates, PL1-PL8

Plate ID	Weld Adjacency	Test Description
PL1-SW	–	Used for qualification of WPQ
PL2-DW	–	Used for fatigue testing of unw. BM
PL3-DW	44 mm	Used for mechanical testing
PL4-DW	45 mm	Used for residual stress measurements
PL5-DW	12 mm	Used for mechanical testing
PL6-DW	9 mm	Used for residual stress measurements
PL7-DW	1.3 mm	Used for mechanical testing
PL8-DW	1 mm	Used for residual stress measurements

BM = Base material, SW = Single weld, DW-Dual welds,

3.4 Fabrication of Fatigue Specimens

The preparation was performed in accordance with the requirements in ASTM E466-15 [6]. The standard covers the procedural steps in order to design a specimen in the fatigue stress range where the strains are predominately elastic. It is limited to axially loaded specimens subjected to constant amplitude loading at ambient temperature.

The high cycle fatigue test is designed for measuring the effects of variations in the material, geometry, surface condition and residual stress of metallic specimens [5].

Careful consideration of the procedural steps was therefore vital, in order to verify that the results were viable and reproducible. To achieve this, a tight control of variables was necessary, such as cleanliness, surface residual stress, surface finish, etc.[6].

Two different types of specimens were produced in accordance with the following guideline in ASTM 466–15 [6].

- 5 specimens from an unwelded base material plate
- 4 specimens from dual welded plates

The design of the specimen dimensions should be such that the eventual failure occurs in the reduced area in the test section. It was therefore vital, when reducing the area of the test section, that the radius introduced from the machining did not cause any detrimental stress concentrations [6].

The specimen's dimension was designed as a rectangular cross section with tangentially blended fillets between uniform test sections at the ends. According to ASTM E466-15 [6], the radius of the specimen is recommended to be eight times the specimen's test section width, to minimize the stress concentration. The specimen test section width should be 2–6 times the thickness, whereas the resulting area should lie in between 19.4–645 mm^2. The test section length should be 2–3 times the test section width. The width of the grip should be 1.5 times the test section width [6].

The length of unwelded specimens was 300 mm, resulting in a grip length of 40 mm on each side. The dual welded specimens became around 10 mm longer after cutting, joint preparation and welding. In order to fit both welds into the test specimen, the initial centre distance was extended from the base material fatigue specimens. The grip length remained 40 mm. The finished design of the two variants of specimens is shown Figs. 5 and 6.

Fig. 5 Dimensions of the unwelded fatigue specimens

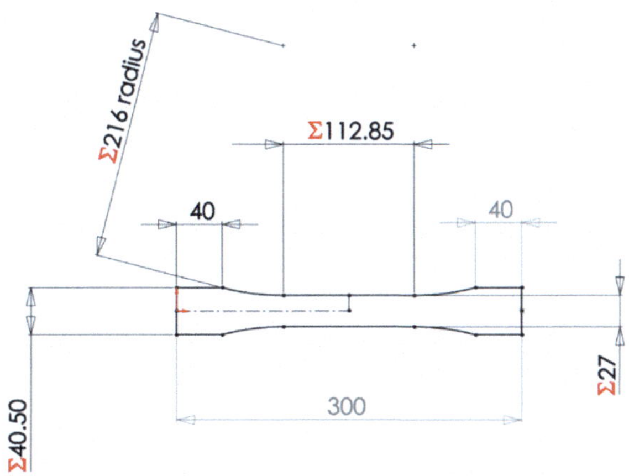

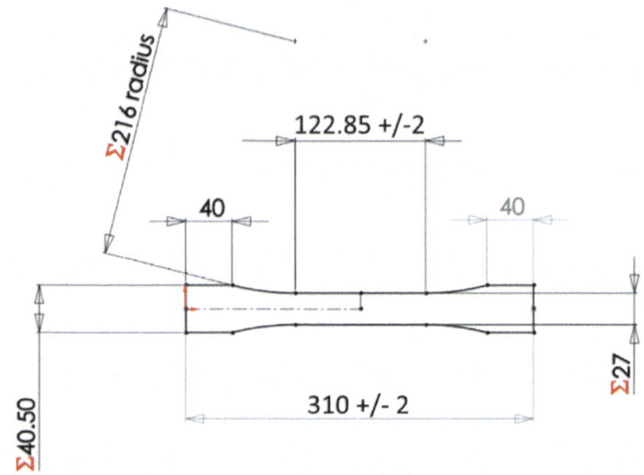

Fig. 6 Dual Welded Specimens. After cutting, joint preparation and welding, the length of the plates changed from 300 mm to 310 ± 2 mm. This was corrected by extending the initial centre distance from the base material fatigue specimen drawing from 112.85 mm to 122.85 ± 2 mm

The following procedural steps were taken to machine and prepare the specimens. The procedure are based on ASTM E466-15, Appendix X1 [6], but with some small modifications. Our procedure was the following:

1. Plates PL2, PL3 and PL5 were first cut into rectangular sections.
2. The rectangular section was then machined according to the drawings, see Figs. 5 and 6. Machining was done gradually where the second-last cut depth was 0.4 mm and last was 0.2 mm. The thickness of the specimen after milling was 9.15 mm. The dual welded specimens only had the weld cap flushed in order to keep the weld toe distances the same and to not introduce unnecessary residual stresses.
3. The next 0.1–0.2 mm on the cap and root side of the specimens were grinded with a plane surface grinding machine. All specimens were surface grinded at Castolin Trio OilTec Services, Stavanger. The thickness of the specimens after surface grinding was 8.95 mm. It was not possible to use the plane surface grinding machine on the sides of the specimens.
4. The final grinding was done manually with abrasive paper, instead of a polishing machine. Sanding was done in the following steps: Hermes WS Flex Waterproof P180 (grain size 82 µm), Struers FEPA P # 500 (grain size 30 µm), Hermes WS Flex Waterproof P1000 (grain size 18 µm), Struers FEPA P # 1200 (grain size 15 µm) and Silicon Carbide 1200/4000 (grain size 5 µm). The thickness of the specimen after final manual grinding was around 8.60 mm. This means that the manual grinding process removed approximately 200 µm on each side of the specimen. The manual grinding was done along the length of the specimen.
5. A requirement after grinding was that all slip marks shall be along the test direction of the test specimen. To verify this, a visual check was made with a 20 × magnifying glass, where no transverse grinding marks were accepted.

6. A roughness check was carried out. The requirement for maximum surface roughness is 0.20 µm in the longitudinal direction.
7. After surface treatment and control, specimens were lubricated with grease while waiting for testing. Specimens were stored in towels and "Ready-For-Test". The grease was removed before testing.

3.5 *Fatigue Test*

The fatigue testing machine was installed in the workshop of the Department of Mechanical and Structural Engineering and Materials Science at the University of Stavanger (see Fig. 7). The machine specifications are shown in Table 5.

Type of Test: Axial.

Test Frequency: 13 Hz.

Failure Criterion:Complete fracture.

Run-out: 5 million cycles.

Number of Tests:9

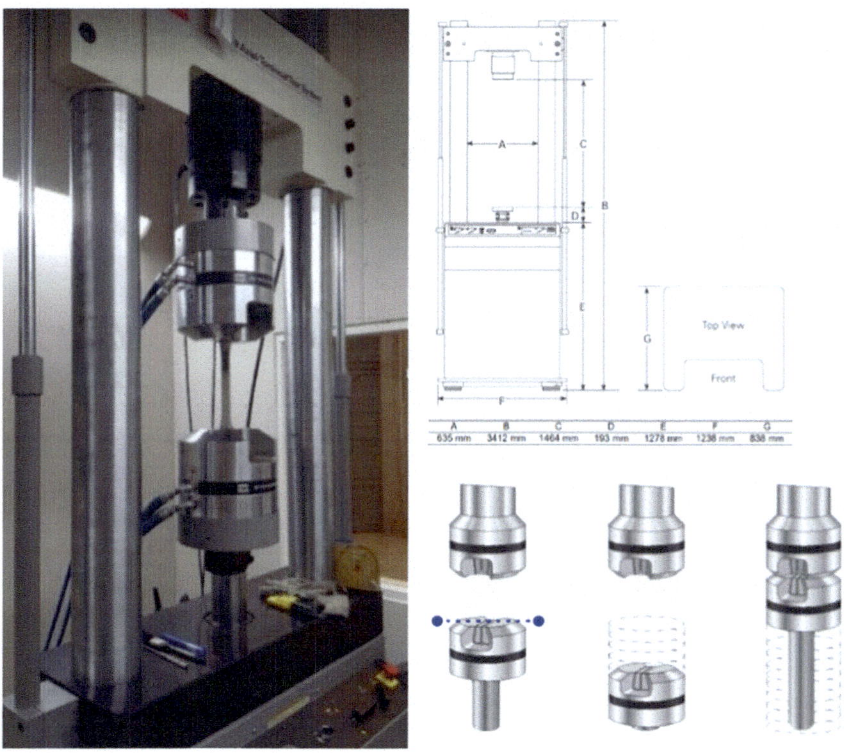

Fig. 7 MTS 809 Axial/torsional test system and dimensions. The lower grip is the actuator and the upper grip is static

Table 5 Specification of fatigue testing machine MTS 809 Axial–Torsional Test System

MTS 809 Axial–Torsional Test System	
Axial Capacity	Dynamic: 250 kN Static: 333kN
Torsional Capacity	2200 Nm
Axial Displacement	±75 mm
Controller Hardware	FlexTest 40
Wedge Grip Model	647.25 Axial–Torsional. Series 647 Hydraulic Wedge Grip, fixed against rotation about both axes
Axial–Torsional Load Frame, model	319.25
Hydraulic Actuator, model no	319.25
Dynamic forces Monitoring Procedure	MTS Force Transducer model # 662.10B-08

Stress Ratio, R: 0.1

Laboratory Temperature:Average, 23 °C; range, ± 2 °C.

Two different fatigue test series were performed as part of the experiment. The purpose of the first test was to determine a rough SN-curve for the base material. In the second test, the welded production plates were tested and analysed. Table 6 shows an overview of the different specimens that were tested and the checks that were performed on each test.A roughness test was carried out on pre-prepared specimens before testing, and a visual inspection of the specimen was performed before and after the test. Fatigue test data was logged before, during and after testing.

Table 6 Specimen appendix

Test Series 1 Specimens from unwelded base material plate			
Specimen Series	Roughness Test	Visual Examination	Fatigue Test Log
PL2-UW-BM1	Yes	Yes	Yes
PL2-UW-BM2	Yes	Yes	Yes
PL2-UW-BM3	Yes	Yes	Yes
PL2-UW-BM4	Yes	Yes	Yes
PL2-UW-BM5	Yes	Yes	Yes
Test Series 2 Specimens from welded production plates			
Specimen Series	Roughness Test	Visual Examination	Fatigue Test Log
PL3-DW-F1	Yes	Yes	Yes
PL3-DW-F2	No	Yes	Yes
PL5-DW-F1	Yes	Yes	Yes
PL5-DW-F2	Yes	Yes	Yes
UW = UnWelded, DW = DualWelded, PL = Plate, BM = Base Material, F = Fatigue			

4 Results and Discussion

The objective of the fatigue tests performed in this article was to assess whether weld proximity had any impact on the fatigue strength. Two fatigue test series were performed, as shown in Table 6.

The first series of fatigue specimens was fabricated from an unwelded S420G2 + M steel plate. The second test series consisted of welded fatigue specimens extracted from dual welded S420G2 + M steel plates; PL3 specimens had 44 mm between the weld toes, and PL5 specimen had 12 mm between the weld toes (see Table 4). The preparation was performed in accordance with ASTM E466-15 [6]. The criterion of failure was defined as when complete separation occurred. The endurance limit was determined to be at 5 million cycles based on unwelded steel of similar mechanical properties [12]. All specimens were visually examined before and after testing.

4.1 Unwelded Fatigue Specimen

Table 7 shows the fatigue test results from five unwelded base material test specimens. Specimen PL2-UW-BM2 was tested at a stress range of 350 MPa and R = 0.1. The test was aborted after 2,760,746 cycles. No visible crack formation was detected. The stress range was increased to 375 MPa for specimen PL2-UW-BM4 and was aborted after it was subjected to 6,770,000 cycles without showing any sign of crack formation. The endurance limit after this initial test was set at 5 million cycles at a load

Table 7 Test results from Plate PL2 Unwelded base material fatigue specimens

Fatigue test R = 0.1
Plate PL2 Unwelded base material fatigue specimens

Specimen ID	PL2-UW BM1	PL2-UW BM2	PL2-UW BM3	PL2-UW BM4	PL2-UW BM5
Run Sequence	1st run	1st run	1st run	1st run	1st run
Stress Range [MPa]	430	350	400	375	400
Area [mm^2]	237.5	234.8	240.14	233.9	228.2
Max Stress [MPa]	477.8	388.9	444.4	416.7	444.4
Min Stress [MPa]	47.8	38.9	44.4	41.7	44.4
Mean Stress [MPa]	262.8	213.9	244.4	229.2	244.4
Machine Stress [MPa]	262.8 ± 215	213.9 ± 175	244.4 ± 200	229.2 ± 187.5	244.4 ± 200
Cycles	15 643	2 760 746	394 701	6 770 000	426 615
Displacement Range [mm]	0.93	0.63	0.82	0.70	0.77
Run out/Fracture	Fracture	Run out	Fracture	Run out	Fracture

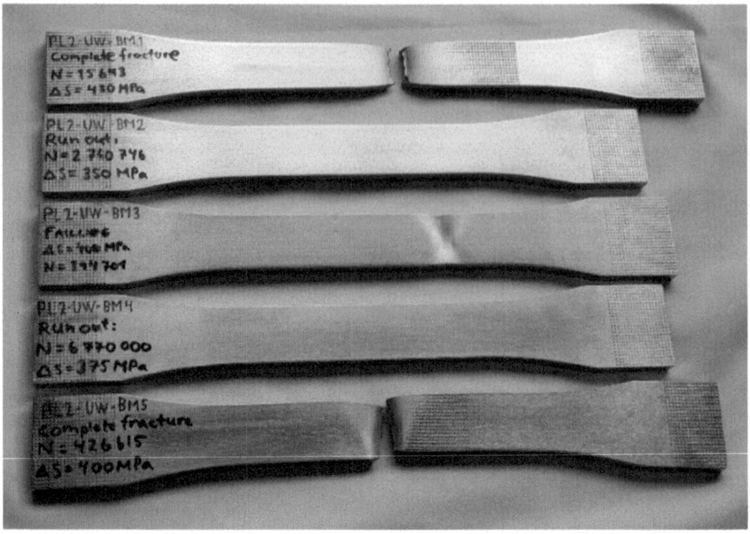

Fig. 8 Fractured unwelded base material fatigue specimen

cycle test frequency of 13 Hz. The maximum stress in this test was approximately 417 MPa. According to the inspection certificate, the yield point of the base material was 489 MPa. This means that PL2-UW-BM4 was subjected to a dynamic load with peak tensile stress at 85% of the yield point. The following specimens, all of which fractured, were tested at greater stress ranges: PL2-UW-BM1, which was tested at stress range 430 MPa, had a max load of 478 MPa at R = 0.1 equaling 98% of the yield point. This resulted in a failure after 15,643 cycles. Specimens PL2-UW-BM3 and PL-UW-BM5 were tested at a stress range of 400 MPa. The maximum stress was 444 MPa at R = 0.1, equaling 91% of the yield point. At this stress range, the fractures occurred after 394,701 and 426,615 cycles. Figure 8 shows the fractured unwelded fatigue specimens.

4.2 *Dual Welded Fatigue Specimens*

Four welded specimens were tested in total, as seen in Table 6. The results from the dual welded specimens are shown in Table 8. All fractures occurred in the base material of the welded specimens. The results show that the welded joints had a higher fatigue strength than the base material. The susceptibility to crack initiation in the base material, instead of the weld metal, is higher in weld joints with the weld cap grinded flush, as stipulated by Maddox [13]. This statement is based on a test specimen with a single weld. The results from the fatigue test showed that this was also the case for 12 mm and 44 mm between two parallel butt welds. Figure 9

Table 8 Test results from Plates PL3 and PL5 Unwelded base material fatigue specimens

Fatigue test R = 0.1
Plate PL3 and PL5 welded fatigue specimens

Specimen ID	PL3-DW F1		PL3-DW F2	PL5-DW F1		PL5-DW F2
Weld Toe Distance	44 mm		44 mm	12 mm		12 mm
Run Sequence	1st run	2nd run	1st run	1st run	2nd run	1st run
Stress Range [MPa]	375	400	400	375	400	400
Area [mm^2]	232.0	232.0	243.1	235.25	235.25	244.07
Max Stress [MPa]	416.7	444.4	444.4	416.7	444.4	444.4
Min Stress [MPa]	41.7	44.4	44.4	41.7	44.4	44.4
Mean Stress [MPa]	229.1	244.42	244.42	229.1	244.42	244.42
Machine Stress [MPa]	229.1 ± 187.5	244.42 ± 200.0	244.42 ± 200.0	229.1 ± 187.5	244.42 ± 200.0	244.42 ± 200.0
Cycles	5 000 000	439 404	322 187	4 953 349	483 574	306 418
Displacement Range [mm]	0.69	–	0.80	0.70	0.76	0.80
Run out/Fracture	Run out	Fracture	Fracture	Run out	Fracture	Fracture

Specimen ID	Description
PL3-DW-F1	1st run: Ran out at 375 MPa after 5,000,000 cycles 2nd run: Continued second run at 400 MPa and fractured after 439,404 cycles. Fracture occurred in the base material at a previously detected surface defect
PL3-DW-F2	1st run: Fractured at 400 MPa after 322,187 cycles. Fracture occurred in the base material on the side face. Visual inspection prior to testing had not detected any defect in the fracture area
PL5-DW-F1	1st run: Run out at stress range 375 MPa after 4,953,349 cycles 2nd run: Continued second run at 400 MPa and fractured after 483,574 cycles Fracture at surface defect in base material. An additional crack initiated in the base material at another surface defect

(continued)

Table 8 (continued)

Fatigue test R = 0.1
Plate PL3 and PL5 welded fatigue specimens

Specimen ID	PL3-DW F1		PL3-DW F2	PL5-DW F1		PL5-DW F2
Weld Toe Distance	44 mm		44 mm	12 mm		12 mm
Run Sequence	1st run	2nd run	1st run	1st run	2nd run	1st run
PL5-DW-F2	1st run: Fractured at 400 MPa after 306,418 cycles Fracture initiated in the base material. Visual inspection prior to testing had not detected any defect in the fracture area					

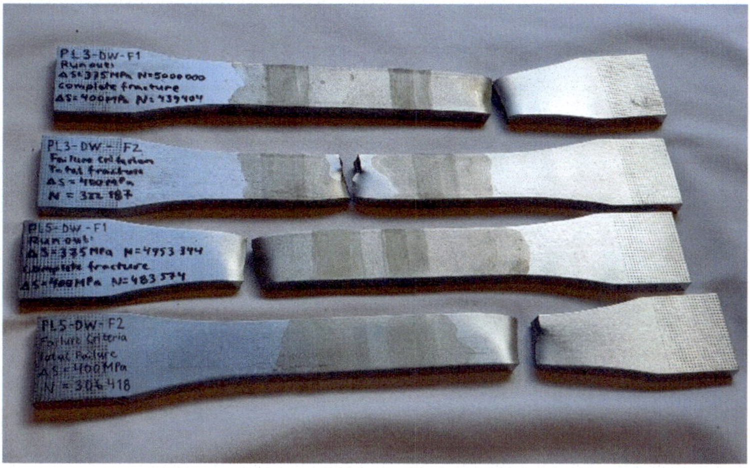

Fig. 9 Dual welded fatigue specimens, all fractures and cracks were in the base material

shows the fractured welded fatigue specimens. Figure 10 shows a surface crack in PL5-DW-F2.

A surface crack of 6 mm was detected in specimen PL5-DW-F2 at 300,000 cycles, and the fracture occurred after 306,4118 cycles. Figure 10 below shows the examination of specimen PL5-DW-F1 that was performed before and after the conducted

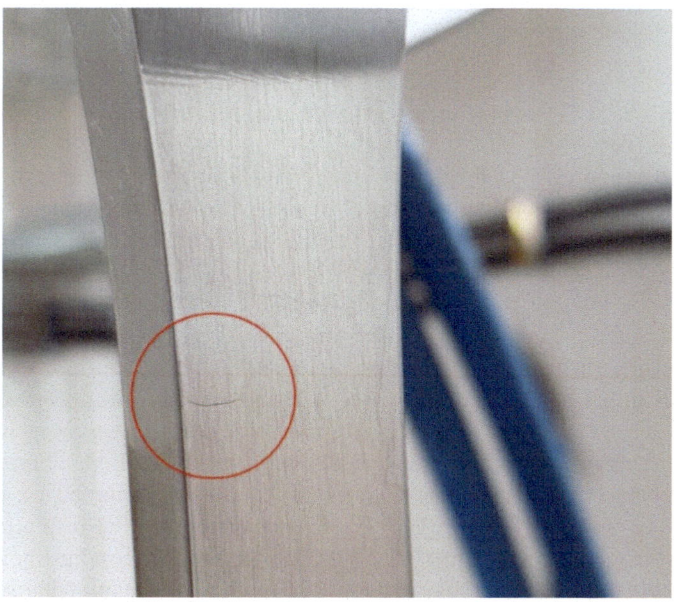

Fig. 10 PL5-DW-F2

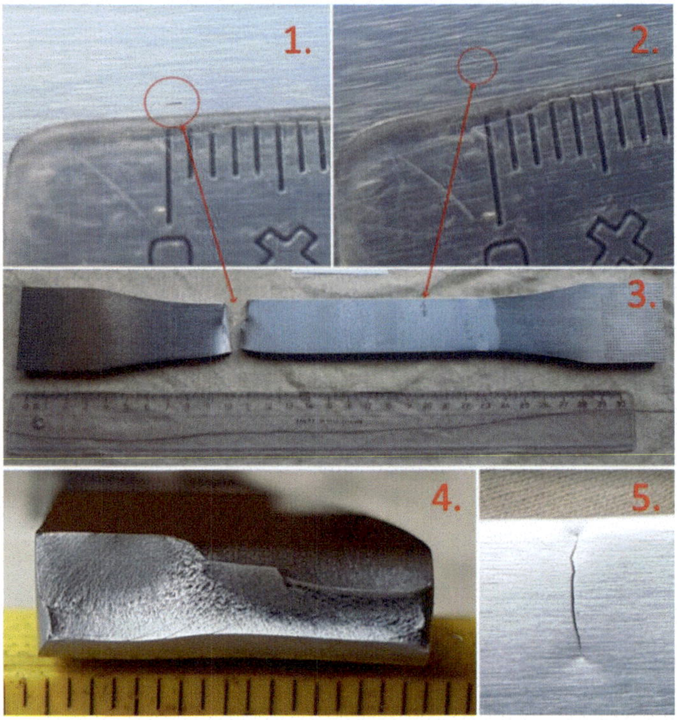

Fig. 11 The images are for specimen PL5-DW-F1. 1. Detected defect leading to fracture, 2. Detected defect leading to crack, 3. Fractured specimen, 4. Fractured specimen's surface, 5. Initiated crack from defect in picture number 2

test. Figure 1, 2, 11 is from the pretest examination, while Figs. 3, 4, 5, 11 is posttest examination.

The top two figures show the findings from visual inspection before testing. Small defects were detected in the base material surface of the specimens. The two lower figures show the fractured surface and the initiated secondary crack. Both are assumed to have originated from the defects shown above.

Figure 12 shows the combined results from the welded and unwelded fatigue specimens in a stress range vs. total cycles graph. Both the results from the fractured and run-out specimens are shown in the diagram.

The crack initiation occurred in the base material during fatigue testing of the dual welded fatigue specimens. The total cycles required for fracture to occur were also very similar between unwelded and welded specimens. In addition, all preparations and prior documentation of the state of the metal were also the same and in accordance with ASTM E466-15 [6]. This made it plausible to combine the two results in a single graph, to estimate the fatigue life of the S420G2 + M steel. When adjusting the stress ranges based on the averaged registered displacements, we obtain the graph shown in Fig. 13.

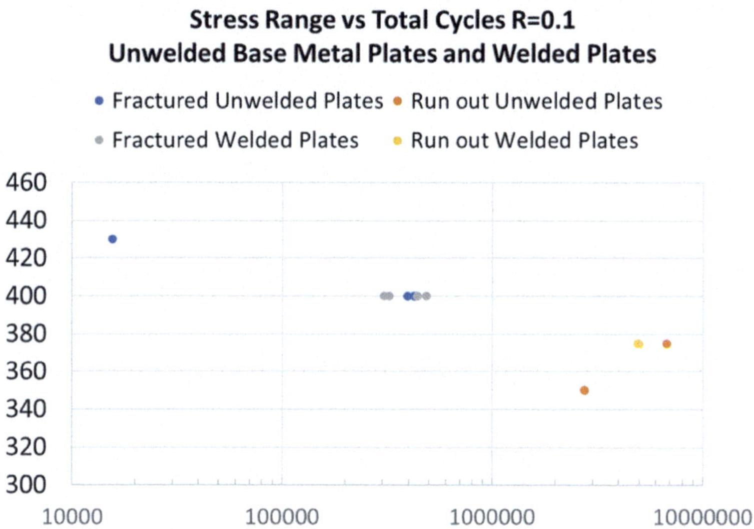

Fig. 12 Stress range (vertical axis) in MPa vs. total cycles graph (horizontal axis) of production plate specimens and unwelded base material specimens. R = 0.1

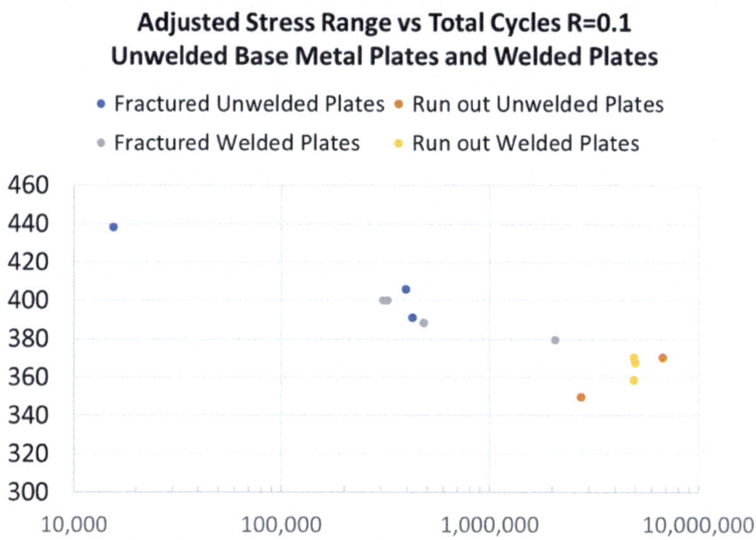

Fig. 13 Adjusted stress range vs. total cycles. R = 0.1

It was possible to derive a trend line using linear regression. By adjusting the stress ranges, we obtained the approximated stress range vs. total cycles curve. The lowest value at fracture was 380 MPa, while the highest value at run-out was 370 MPa. The endurance limit was determined to be 5 million cycles, based on unwelded steel

Table 9 X-ray diffraction results from fatigue specimens

	XRD PL8 weld cap side, transverse stress	XRD PL8 weld cap side, longitudinal stress
Weld Centre Line	−336 MPa	−440 MPa
5 mm from WCL	−400 MPa	−392 MPa
60 mm from WCL	−300 MPa	−346 MPa

with similar mechanical properties [12]. The number of cycles needed to fracture the specimen was very similar to that of unwelded base material. This may be due to the careful preparation process that all specimens underwent and because the lower strength of the base material. The preparation procedure might have affected all fatigue test results due to introducing compressive residual stresses that suppressed crack initiation. Residual stresses in the weld metal and adjacent areas are shown in Table 9.

The results were obtained using x-ray diffraction, measuring at a depth of 50 μm. Both longitudinal and transverse residual stresses were measured along the specimen length; in the weld centerline, 5 mm and 50 mm offset the weld centerline. The results show that an outer layer of high local compressive residual stresses has been introduced from the grinding process. The residual stresses in the surface of the specimen may have increased the fatigue life of both weld metal and base material.

5 Summary and Conclusion

Experimental work was carried out to assess whether weld proximity had any impact on fatigue strength. Two series of fatigue tests were conducted in this experiment and the material used was S420G2 + M. The first series of specimens was fabricated from an unwelded plate and the second series set was extracted from dual welded plates. The purpose of the unwelded specimens was to be used as reference for the welded specimens. The dual welded fatigue test specimens were fabricated with the intention to study the microstructures fatigue strength of a prior weld after an adjacent secondary weld is applied. Geometrical and angular weld stress concentrations were removed by machining and grinding of the top, bottom and side surfaces. All welds were subjected to radiographic testing, in order to detect any weld discontinuity.

5.1 Based on the Experimental Work and Evaluation the Following Can Be Concluded

- The preparation method introduced compressive residual stresses in the surface and removed the notch effect present in the weld toe.

- The test results indicated that having two parallel butt welds, in the absence of stress concentrations and welds discontinuities, does not negatively impact the fatigue performance of the weld joint at distances of 12 mm and 44 mm.
- No detrimental impact was observed in the specimens due to weld proximity having 12 mm between the weld toes, when using this specific preparation method.
- Both welded and unwelded specimens showed a similar fatigue resistance behaviour.

5.2 Based on the Evaluations and Conclusions, the Following is Recommended

- The sample size in this experiment was limited. The sample size was sufficient to obtain a basic understanding of the fatigue behaviour, but more extensive testing is needed in order to ascertain the results.
- In the experiment, the HAZ was measured to be between 1.4 - 2.8 mm from the fusion line; none of the dual welded specimens in this study had an overlapping HAZ. Additional testing should be performed, to analyse the situation where there is overlapping HAZ.

The findings in this report can be useful for understanding the implications the fabrication process has for a welded specimen and used as a baseline for future research.

References

1. Specification for Class I arc welding of ferritic steel pipework for carrying fluids, BS 2633, 1987.
2. Class II arc welding of carbon steel pipework for carrying of fluids, BS 2971, 1991.
3. Specification for welding of steel pipelines on land and offshore, BS 4515, 2009.
4. Specification for Unfired fusion welded pressure vessels, PD5500, 2012.
5. Standard Practice for Presentation of Constant Amplitude Fatigue Test Results for Metallic Materials, E468 – 11, 2011.
6. Standard Practice for Conducting Force Controlled Constant Fatigue Tests of Metallic Materials, ASTM E466–15, 2015.
7. Weldable structural steels for fixed offshore structures - Technical delivery conditions, NS-EN 10225, 2009.
8. Definition and classification of grades of steel, NS-EN 10020, 2000.
9. Welding — Guidelines for a metallic materials grouping system, ISO/TR 15608, 2017.
10. Welding — Grouping systems for materials — European materials, ISO/TR 20172, 2009.
11. Specification and qualification of welding procedures for metallic materials Welding procedure test Part 1: Arc and gas welding of steels and arc welding of nickel and nickel alloys, NS-EN ISO 15614–1, 2017.
12. Boyer HE (1986) Atlas of fatigue curves. American Society for Metals, Metals Park, Ohio

Study on the Number of Primary and Secondary Fragments Produced by Explosion of Horizontal Vessel

Zijie Li, Dongliang Sun, Jiahui Sun, and Juncheng Jiang

Abstract The fragments produced by explosion of horizontal tank are destructive, which may lead to the damage of adjacent vessels, and then cause the domino effect. Therefore, it is very important to study the primary and secondary fragments respectively produced by source explosion and target destruction. The possible accident types of industrial horizontal vessels were analyzed, and the number of fragments including tank body and accessories were obtained, with the corresponding accident frequency. Based on the maximum entropy principle, the probability distribution models of the number of primary fragments including tank body and accessories were established. Then, the probability model of the number of secondary fragments was obtained by further deducing the probabilistic formula of the volume of fragments formed by the adjacent facilities under the explosion load. Compared to previous work, the fragments including vessel, its accessories and secondary fragments caused by objective crack were comprehensively explored, with the improved results, which had more practical values and significance and were an effective reference for the layout of chemical plants and accident prevention.

Keywords Horizontal vessel explosion · Primary and secondary fragments · Probabilistic distribution model

1 Introduction

Domino accident referred to the phenomenon that the destructive fragments produced by explosion of chemical containers hit adjacent storage tanks and might insert the shell. Therefore, the adjacent storage vessels would suffer from perforation or

Z. Li · D. Sun (✉) · J. Sun
State Environmental Protection Key Laboratory of Environmental Risk Assessment and Control On Chemical Process, School of Resources and Environmental Engineering, East China University of Science and Technology, Shanghai 200237, China
e-mail: dongliangsun@ecust.edu.cn

J. Jiang
School of Environmental and Safety Engineering, Changzhou 213164, China

M. Abdel Wahab (ed.), *Proceedings of 1st International Conference on Structural Damage Modelling and Assessment*, Lecture Notes in Civil Engineering 110,
https://doi.org/10.1007/978-981-15-9121-1_28

plastic failure [1–3]. Explosive fragments were the main factors causing domino effect [4–5]. Countless number of scholars has carried out research on explosive fragments. Gubinelli et al. [6–7] analyzed 143 explosive debris ejection accidents. It was concluded that the number of horizontal tank accidents is the highest accounting to 70.6%, followed by spherical tank accidents accounting to 7.0%. Hauptmanns et al. [8] carried out an analysis of 46 incidents that produced fragments. The results showed that the number of fragments produced by horizontal tank explosion obeyed lognormal distribution, with an average value of 0.85516 and a standard deviation of 0.52448. Mébarki et al. [9–10] used the maximum entropy principle to develop the probabilistic distributions for these fragmented parameters: number, shape, mass, fly off velocity and fly off angles. Djelosevic et al. [11] found that If the mass range of fragments is 10 to 20% of the total mass of the tank, it was most likely to produce two fragments (17%) or three fragments (average 11%).

The conclusion of previous research showed that there were some limitations and shortcomings. For the type of fragments, besides the "large fragments" caused by the cracking of the tank itself, the "small fragments" such as the safety accessories and manholes ejected by the weld fracture, the debris of peripheral facilities damaged by shock wave and other non-tank debris have not been studied.

After the initial explosion accident of the storage tank, the adjacent storage tank or surrounding construction facilities will be affected, and secondary fragments were produced. There were two kinds of secondary debris in the chemical industry park: The secondary debris produced by the initial debris from the tank explosion striking the adjacent tank to make it explode again, and the secondary debris produced by the overpressure generated by the shock wave during the tank explosion to destroy the surrounding facilities. Sun et al. [12] studied the multi domino effect and proposed a model of multi domino scene:

Unit1 (Initial explosion→Fragments)

→(destroy) Unit 2 (Secondary explosion→Fragments)

→(destroy) Unit 3 (Third explosion→Fragments)

→(destroy) Unit 4 (Forth explosion→Fragments)

→……

→(destroy) Unit n-1 ((n-1)-th explosion→Fragments)

→(destroy) Unit n

After the initial explosion, the (n-1)-th explosion caused by debris would also generate debris and lead to the n-th explosion. For the n-th explosion, each n-1 explosion could be regarded as an independent initial explosion, The properties of the fragments were the same as those produced by the initial explosion. Therefore, this kind of secondary debris produced by the adjacent tank was equivalent to the primary debris, so this kind of secondary debris would not be analyzed in this study.

On the basis of summing up the previous research, the possible accident types of industrial vessels were analyzed, and the probabilistic models of primary and secondary debris number were obtained.

2 Number of Primary and Secondary Fragments

2.1 *Number of Primary Fragments Including Tank Body and Accessories*

The types of storage tanks explosion accidents which can produce debris ejection include: BLEVE(Boiling Liquid Expanding Vapor Explosion), ME(Mechanical Explosion), CE(Confined Explosion), RR(Reaction Runaway) [13]. Hauptmanns et al., Holden et al. and Mébarki et al. [8–9, 14–16] studied the fragments produced by BLEVE accidents, and sorted out the accident frequency corresponding to the number of fragments produced. All data was summarized, as shown in Table 1.

Gubinelli and Cozzani [6–7] considered four accident types (BLEVE, ME, CE and RR), and determined the number of fragments according to each type of rupture in each accident type of horizontal tanks, as shown in Table 2.

Many scholars agreed that the number of fragments generated by tank explosion was within the range of [1, 9], they only considered large fragments generated by tank body tearing, but without considering the number of accessories. Accessories would eject due to welding defects and other reasons. The accessories of horizontal storage tank (manhole, safety valve, pressure gauge, level gauge, bracket, joint weld) were shown in Fig. 1. Therefore, the amount of debris produced by tank explosion should be within the range of [1, 15]. At the same time, it should be noted that when large pieces are generated, the pressure inside the tank will be released instantaneously.

Table 1 BLEVE accident data of horizontal tank

Accident types	Fragments number								
	1	2	3	4	5	6	7	8	9
BLEVE	45	42	43	21	3	2	3	0	1

Table 2 Accident data from Gubinelli and Cozzani

Accident types	Fragments number								
	1	2	3	4	5	6	7	8	9
BLEVE	5	56	35	3	0	0	0	0	0
ME	0	6	1	1	0	0	0	0	0
CE	0	9	0	0	1	1	1	1	1
RR	2	3	1	0	1	1	1	1	1

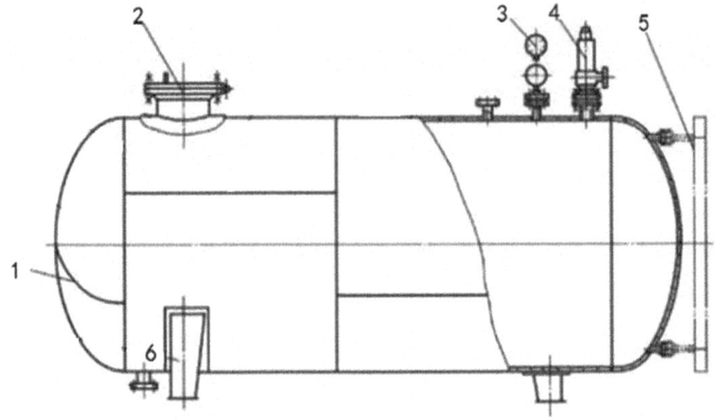

Fig. 1 Accessory of horizontal storage tank

At this time, the possibility of welding fracture and accessories ejecting due to high pressure is very small, that is to say, the possibility of large pieces and accessories pieces appearing at the same time is small. Therefore, it was assumed that the total number of accidents generating [10, 15] fragments under the four accidents types was less than the number of accidents generating 9 fragments. The data in Table 1 and Table 2 were combined, the corresponding number and frequency of explosion and the corresponding number of fragments of the tank body and accessories was obtained, as shown in Table 3.

where 1- Joint weld; 2-Manhole; 3-Pressure gauge; 4- Safety valve; 5-Level gauge; 6- Bracket.

The information available from the data in the table is E1 (average number) and E2 (variance):

$$E_1 = \sum_{i=1}^{15} i\,p(i)_{observed} \tag{1}$$

$$E_2 = \sum_{i=1}^{15} (i - E_1)^2 p(i)_{observed} \tag{2}$$

According to the analysis of the known information, the probabilistic model of the fragments number could be taken as Formula (3):

$$P(N) = e^{-\lambda_0 - \lambda_1 N - \lambda_2 N^2} \tag{3}$$

where N number of primary fragments; λ_0 、 λ_1 、 λ_2 three unknown LaGrange factor values.

Table 3 Summary of types and quantities of horizontal tank explosion accidents, number and proportion of debris including tank body and accessories

Part 1

Accident types	Fragments number						
	1	2	3	4	5	6	7
BLEVE							
Number of events	50	98	78	24	3	2	3
Observed frequency	0.1931	0.3784	0.3012	0.0927	0.0116	0.0077	0.0116
ME							
Number of events	0	6	1	1	0	0	0
Observed frequency	0	0.7500	0.1250	0.1250	0	0	0
CE							
Number of events	0	9	0	0	1		
Observed frequency	0	0.9000	0	0	0.0090	0.0090	0.0090
RR							
Number of events	2	3	1	0	1		
Observed frequency	0.2857	0.4286	0.1429	0	0.0130	0.0130	0.0130

Part 2

Accident types	Fragments number							
	8	9	10	11	12	13	14	15
BLEVE								
number of events	0	1						
observed frequency	0	0.0005	0.0005	0.0005	0.0005	0.0005	0.0005	0.0005
ME								
number of events	0	0	0	0	0	0	0	0
observed frequency	0	0	0	0	0	0	0	0
CE								
number of events	1							
observed frequency	0.0090	0.0090	0.0090	0.0090	0.0090	0.0090	0.0090	0.0090
RR								
number of events	1							
observed frequency	0.0130	0.0130	0.0130	0.0130	0.0130	0.0130	0.0130	0.0130

Table 4 Probability distribution models of the fragments number

Accident types	Probability distribution models
BLEVE	$P(N) = e^{-1.86+0.7N-0.19N^2}$
ME	$P(N) = e^{-6.26+4.8N-1.01N^2}$
CE	$P(N) = e^{-0.25-0.67N+0.02N^2}$
RR	$P(N) = e^{-0.06-0.82N+0.04N^2}$

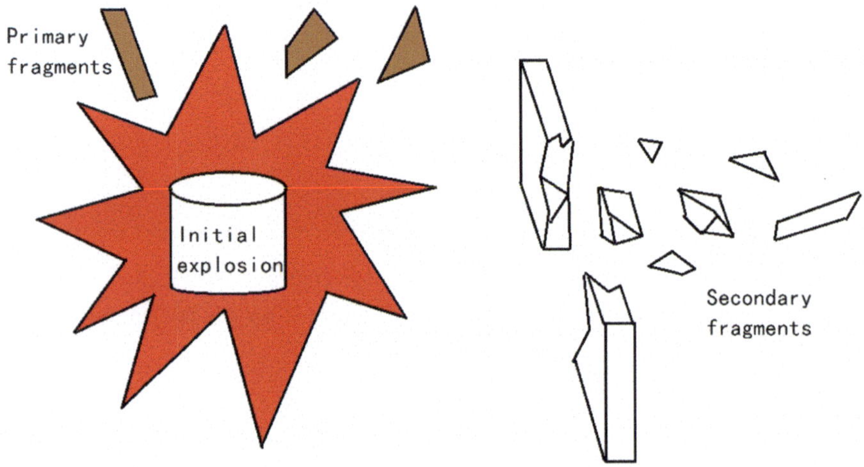

Fig. 2 Failure process of peripheral wall

Therefore, according to the explosion accident data in the table, the Formula (4) was established. The Lagrange factor values were obtained and substituted into Formula (3). Finally, probability distribution models of the fragments number generated by four explosion accidents were obtained, as shown in Table 4.

$$\begin{cases} \sum_{i=1}^{15} e^{-\lambda_0-\lambda_1 i-\lambda_2 i^2} = 1 \\ \sum_{i=1}^{15} i \bullet e^{-\lambda_0-\lambda_1 i-\lambda_2 i^2} = E_1 \\ \sum_{i=1}^{15} (i - E_1)^2 e^{-\lambda_0-\lambda_1 i-\lambda_2 i^2} = E_2 \end{cases} \quad (4)$$

2.2 *Number of Secondary Fragments Generated by Surrounding Facilities*

There may be some construction facilities constructed by brittle materials around chemical tanks. Brittle materials are liable to crack, break and even form debris under explosive load, as shown in Fig. 2. These fragments may impact adjacent facilities again and cause secondary accidents.

Wang et al. [17] used AUTODYN to simulate the process of fragmentation of masonry filling wall under explosive impact, and the generalized limit distribution density function of fragment size formed by masonry filling wall under explosive load was obtained, after further derivation of the formula, the generalized limit distribution density function of fragment number was obtained, as shown in Formula (5).

$$f(n) = \frac{1}{\sigma}\left[1 + k\left(\frac{\sqrt[3]{\frac{6V_{total}}{n\pi}} - \mu}{\sigma}\right)\right]^{-\frac{1}{k}-1} exp\left\{-\left[1 + k\left(\frac{\sqrt[3]{\frac{6V_{total}}{n\pi}} - \mu}{\sigma}\right)\right]^{-\frac{1}{k}}\right\} \tag{4}$$

where n independent variables of probability density function, which is fragment number; μ、σ and k the probability parameters, 0.00897, 0.00412 and 0.56820 respectively.V_{total} total volume of masonry wall.

The probability density function of the primary and secondary fragments number could be summarized in Table 5.

3 Conclusions

Compared to previous researches, fragments such as accessories and secondary fragments were also considered in this study. These were considered more thoroughly than the predecessors, so the research was more accurate and had more practical value and significance. Main results of this study were as follows: (1) Based on the maximum entropy principle, the probability distribution models of the number of primary fragments including tank body and accessories were established. (2) On the basis of previous studies, the generalized limit distribution density function formula for secondary fragment number and mass were further deduced.

Future research trends can focus on the possibility of debris cutting off the support of the vessel to cause the instability of the tank, and the impact of thermal radiation on the debris during ejection to cause flammable combustion around it.

The results of the study can provide an effective reference for the layout of the chemical industry plant and the means of accident prevention of chemical storage tanks.

Table 5 Probability density function of the primary and secondary fragments number

Fragments types	Accident types	Probability density function
Primary fragments	BLEVE	$f(n) = e^{-1.86+0.7n-0.19n^2}(0.7 - 0.38n)$
	ME	$f(n) = e^{-6.26+4.8n-1.01n^2}(4.8 - 2.02n)$
	CE	$f(n) = e^{-0.25-0.67n+0.02n^2}(0.04n - 0.67)$
	RR	$f(n) = e^{-0.06-0.82n+0.04n^2}(0.08n - 0.82)$
Secondary fragments	$f(n) = \frac{1}{\sigma}\left[1 + k\left(\frac{\sqrt[3]{\frac{6V_{total}}{n\pi}} - \mu}{\sigma}\right)\right]^{-\frac{1}{k}-1} \exp\left\{-\left[1 + k\left(\frac{\sqrt[3]{\frac{6V_{total}}{n\pi}} - \mu}{\sigma}\right)\right]^{-\frac{1}{k}}\right\}$	

Acknowledgements The financial support of the Natural Science Foundation of Shanghai (16ZR1408500), the Technology Development Foundation of China Sinopec Qingdao Safety Engineering Institute (313038), the project for Development of Systems of Quantitative Risk Assessment of Domino Effect in the Chemical Industry (B100-81414), the Natural Science Foundation of Jiangsu Province (BK2012824), and the Innovative Practice Program for College Students (201810251073, S19098) are gratefully acknowledged.

References

1. Jia MS, Chen GH, Hu K (2017) Review of risk assessment and pre-control of domino effect in Chemical Industry Park. Chem. Indus. Eng. Progress 36(4):1534–1543
2. Zheng F, Zhang MG, Zuo YW (2019) Study on domino accident in chemical plant area based on dynamic bayesian network. J Nanjing Tech University(Natural Sci Edition) 5:554–560
3. Khakzad N, Landucci G, Reniers G (2017) Application of dynamic bayesian network to performance assessment of fire protection systems during domino effects. Reliab. Eng. Sys. Safety 167:232–247
4. Necci A, Cozzani V, Spadoni G et al (2015) Assessment of domino effect: state of the art and research Needs. Reliab Eng Sys Safety 143:3–18
5. Antonioni G, Spadoni G, Cozzani V (2009) Application of domino effect quantitative risk assessment to an extended industrial area. J Loss Prev Process Ind 22(5):614–624
6. Gubinelli G, Cozzani V (2009a) Assessment of missile hazards: Identification of reference fragmentation patterns. J Hazard Mater 163(2–3):1008–1018
7. Gubinelli G, Cozzani V (2009b) Assessment of missile hazards: evaluation of the fragment number and drag factors. J Hazard Mater 161(1):439–449
8. Hauptmanns U (2001) A procedure for analyzing the flight of missiles from explosions of cylindrical vessels. J Loss Prev Process Ind 14(5):395–402
9. Mébarki A, Mercier F, Nguyen QB et al (2009) Structural fragments and explosions in industrial facilities. Part I: probabilistic description of the source terms. J Loss Prev Process Ind 22(4):408–416
10. Mébarki A, Nguyen QB, Mercier F (2009) Structural fragments and explosions in industrial facilities: Part II—projectile trajectory and probability of impact. J Loss Prev Process Ind 22(4):417–425
11. Djelosevic M, Tepic G (2019) Probabilistic simulation model of fragmentation risk. J Loss Prev Process Ind 60:53–75
12. Sun D, Jiang J, Zhang M et al (2016) Investigation of multiple domino scenarios caused by fragments. J Loss Prev Process Ind 40:591–602
13. Tauseef SM, Abbasi T, Pompapathi V et al (2018) Case studies of 28 major accidents of fires/explosions in storage tank farms in the backdrop of available codes/standards/models for safely configuring such tank farms. Process Saf Environ Prot 120:331–338
14. Nguyen QB, Mebarki A, Saada RA et al (2009) Integrated probabilistic framework for domino effect and risk analysis. Adv Eng Softw 40(9):892–901
15. Holden PL, Reeves A (1985) B: Fragment hazards from failures of pressurized liquefied gas vessels. Inst Chem Eng Symp Ser 93:205–220
16. Abbasi T, Abbasi SA (2007) The boiling liquid expanding vapour explosion (BLEVE): Mechanism, consequence assessment, management. J Hazard Mater 141(3):489–519
17. Wang M (2010) Damage mechanism of steel structures and fragmentation process of masonry walls under blast loading. Tianjin University

A Critical Review on the Structural Health Monitoring Methods of the Composite Wind Turbine Blades

Reza Malekimoghadam, Stefan Krause, and Steffen Czichon

Abstract With increasing turbine size, monitoring of blades becomes increasingly important, in order to prevent catastrophic damages and unnecessary maintenance, minimize the downtime and labor cost and improving the safety issues and reliability. The present work provides a review and classification of various structural health monitoring (SHM) methods as strain measurement utilizing optical fiber sensors and Fiber Bragg Gratings (FBG's), active/passive acoustic emission method, vibration–based method, thermal imaging method and ultrasonic methods, based on the recent investigations and promising novel techniques. Since accuracy, comprehensiveness and cost-effectiveness are the fundamental parameters in selecting the SHM method, a systematically summarized investigation encompassing methods capabilities/limitations and sensors types, is needed. Furthermore, the damages which are included in the present work are fiber breakage, matrix cracking, delamination, fiber debonding, crack opening at leading/trailing edge and ice accretion. Taking into account the types of the sensors relevant to different SHM methods, the advantages/capabilities and disadvantages/limitations of represented methods are nominated and analyzed.

Keywords Structural health monitoring · Wind turbine blades · Sensors · Damage detecting

R. Malekimoghadam (✉)
Department of Mechanical and Aerospace Engineering, Politecnico Di Torino, Corso Duca degli Abruzzi 24, 10129 Torino, Italy
e-mail: reza.malekimoghadam@polito.it

S. Krause · S. Czichon
Fraunhofer Institute for Wind Energy Systems IWES, Bremerhaven, Germany

M. Abdel Wahab (ed.), *Proceedings of 1st International Conference on Structural Damage Modelling and Assessment*, Lecture Notes in Civil Engineering 110,
https://doi.org/10.1007/978-981-15-9121-1_29

1 Introduction

Wind energy as one of the fastest growing renewable energy resources, imparts a significant contribution in the energy market. In accordance with the information supplied, it is appraised that the wind power will provide 12% and 20% of electricity power by 2020 and 2030, respectively [1, 2]. In order to decrease levelized cost of energy (LCoE), reduction of the operations and maintenance cost is an urgently required [3]. Apart from optimizing the design of turbines to ameliorate the availability, another feasible way is exerting reliable and cost-effective condition monitoring system (CMS).

Preventing unexpected catastrophic failures and diminishing unscheduled maintenance, are the prominent purposes that have appeal for employing next generation of wind turbines. Thus, Structural Health Monitoring systems (SHM) can extraordinarily impart reliability and profitability to wind turbines due to detecting the defects at incipient step during operation or testing procedures [4]. Sensing process and the interpretation algorithm are considered as the two crucial parameters which impress SHM growth and development [5, 6]. It should be notified that the blades are considered as one of the most significant components in the turbines [7], since the performance of WT is remarkably dependent on the blades. Furthermore, the blade manufacturing expense is about 15–20% of each wind turbine [8]. Significantly, longer blades are being designed, in order to sweep larger area. Hence, carbon fiber reinforced polymer (CFRP) is increasingly being utilized for very large blades. However, the bigger size of wind turbine blade (WTB) leads to increase the load levels [9].

For sake of eliminating unexpected maintenance, vital failures and minimizing downtime, the wind turbines must be continuously monitored to assure their perfect and appropriate conditions [10]. The situation of health monitoring systems and evaluation techniques for offshore wind turbines (OWTs) were reviewed by Lian et al. [11], considering supervisory control and data acquisition (SCADA) systems and condition monitoring (CMS) as the most conventional types of health monitoring systems of the OWT.

As mentioned by Farrar and Sohn [12], SHM represents a procedure of implementing a damage detection strategy for engineering infrastructures related to aerospace, civil and mechanical engineering, being damaged referring to the variations in material and/or geometric properties of these systems. Some of the most known causes of structural damages are fatigue, wind gusts, moisture absorption [13], thermal stress, corrosion [14], fire and lightning strikes [15]. Moreover, a SHM could be efficient regarding two issues including the fatigue issue and utilization of lighter blade which the former could be useful due to difficulty of predicting the exact life of a wind turbine components, whereas the latter yield higher performance of the wind turbines [16]. Considerably, the advantages of possessing a damage detection system can be classified as an impediment of premature breakdown, reduced maintenance cost, supervision at remote sides and remote diagnosis, improvement of capacity factor and support for further development of a turbine [17]. Schubel et al. [18] presented a review of structural health monitoring methods, encompassing residual cure strain monitoring for the wind turbine blade industry in conjunction with a comparison between different monitoring methods. Thus, a SHM system is required to frequently monitor the condition of the wind turbine blades and warning the failures at incipient level.

The present work furnishes a review and classification considering different damage types and employed sensors in different SHM methods which have been presented in the recent investigations and upcoming promising techniques in both industrial and academic sections; subsequently corresponding pros and cons are introduced.

2 Blade Structural Damages

The damages can occur in any part of the wind turbine while the most prevalent type of damage happens in the blade and tower [19]. Since the blade is the significant part of the wind turbine and its price is about 15–20% of the total cost of the turbine, therefore, tremendous attention has been received by structural health of the blades [20]. In addition, it has been demonstrated that the damage occurred in the blade, is the most expensive type of damage with the longest time for repairing [20]. Likewise, occurring the damage in the blade leads to unbalanced condition of the blade during its rotation which result in secondary failure of wind turbine by collapsing the tower, as well as blade failure [21]. Thus, the present work concentrates on the damages occurring in the rotor blade of the wind turbine. In order to realize the blade damage, the main segments of a blade should be known previously, which are depicted in Fig. 1.

The materials of the conventional blades are often glass/carbon fiber-reinforced composites [23], whereas by increasing the size of the blade nowadays, the utilization of carbon-fiber reinforced composites has recently been augmented in order to fabricate the turbine blades. Regarding the blade structure, there is a main spar tube, and the upwind side and downwind side of the blade are constructed and joined at both the leading edge and the trailing edge using adhesive. Blade damages can occur in different ways which typical damages in the wind turbine blades are represented in Table 1 [24, 25] and schematic illustrations of corresponding damages are delineated in Fig. 2.

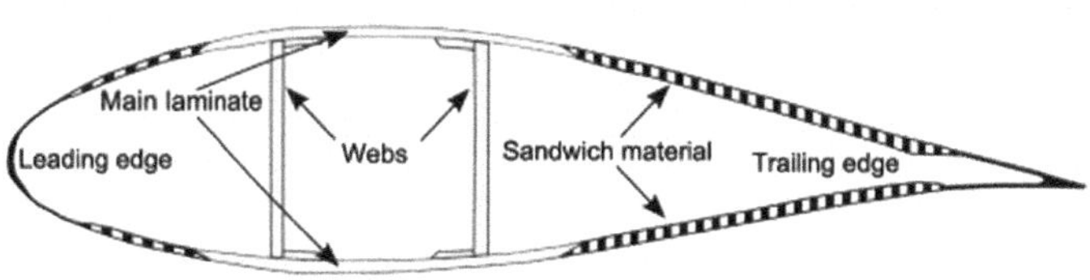

Fig. 1 Sketch of cross-section of a wind turbine blade [22]

Table 1 Typical damages of wind turbine blades [10, 24, 25]

Damages types	Descriptions
Damage type 1	Damage formation/growth and debonding in the adhesive layer joining skin and main spar flanges
Damage type 2	Damage formation and growth in the adhesive layer joining the top and down skins along leading or trailing edges
Damage type 3	Damage formation/growth at the interface between face and core in sandwich panels in skins and main spar web
Damage type 4	Internal damage formation/growth and delamination in laminates in skin or the main spar flanges
Damage type 5	Splitting and fracture of separate fibers in laminates of the blade structure and main spar
Damage type 6	Buckling of the skin due to damage formation and growth in the bond between skin and main spar under compressive load
Damage type 7	Debonding of the gelcoat from the blade surface (gel-coat cracking and gelcoat-skin debonding)

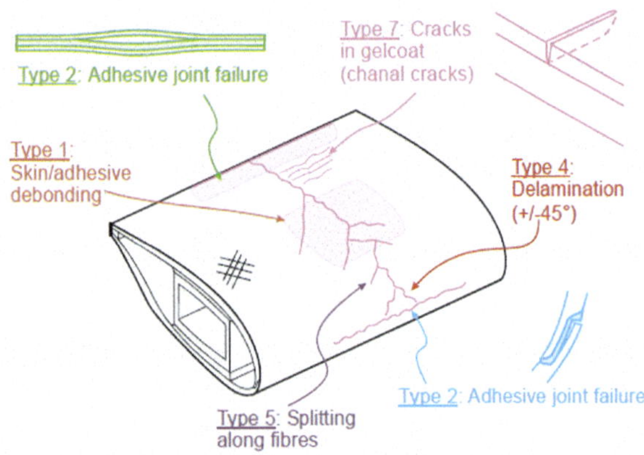

Fig. 2 Types of damage withstood by the wind turbine blade

Below the corresponding damages types in a wind turbine blade are illustrated in accordance with Fig. 2.

A report concerning the identification of different types of damage that developed for Vestas A/S V52 wind turbine blade, tested to failure under quasi-static loading, were provided by Sørensen et al. [24]. Some of the occurred blade failures during the laboratory tests are described in Fig. 3.

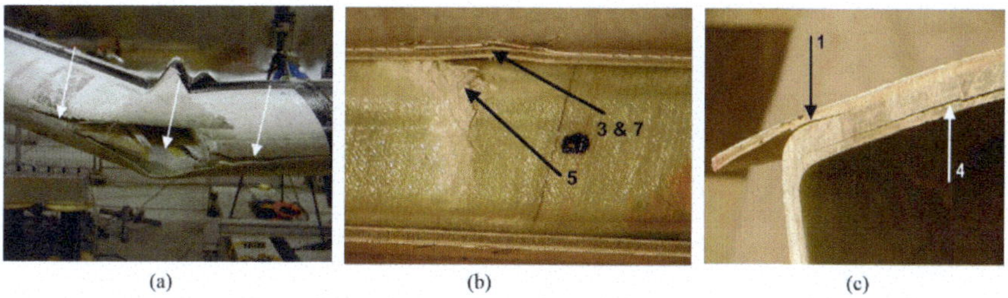

Fig. 3 **a** Adhesive joint failure between skins at the leading edge (Damage type 2), **b** sandwich panel face/core debonding (Damage types 3), laminate failure in compression (Damage types 5) and gel-coat/skin debonding (Damage types 7), **c** Debonding of the main spar flange (Damage type 1) and delamination in the laminated structure (Damage type 4) [22]

3 Structural Health Monitoring Methods (SHM)

In this chapter, a review of recent investigations and novel promising techniques is presented. While it is difficult to provide a quantitative comparison between methods, advantages and drawbacks of presented methods are discussed.

3.1 Acoustic Emission Method (AE)

The acoustic emission phenomenon is based on the release of energy in the form of transitory elastic waves within a material having a dynamic deformation process [26]. Damages such as crack growth, debonding, large deformation, delamination and impacts can stimulate transient alteration in the elastic energy in specific position of a structure which are detected by acoustic emission method. AE is capable of detecting the fault and malfunction in blades, gearboxes, bearings, shafts of the wind turbines [27].

The common AE parameters which are measured during the tests include amplitude, root mean square (RMS) value, energy, counts, and events [28]. Regarding the sensor types, piezoelectric transducers and optical fiber sensors are utilized in AE technique. The signal attenuation should be considered rigorously as a significant limitation of this method. Therefore, the sensors which are employed in AE method must be positioned as close as possible to the damage source [28], which causes an important limitation in exerting AE method in wind turbines. Although the acoustic emission method is based on fast release of localized energy in form of elastic wave in material, the damage process and stress waves also emit airborne sound where the latter doesn't suffer from signal attenuation phenomenon [29]. By analyzing the sensors attenuation, Van Dam et al. [30] demonstrated that the sensors distance should be limited to a maximum of 1 m.

It is noteworthy to mention that the propagation of the acoustic emission waves is influenced by damage inside the material, and Fiber Bragg gratings (FBGs) can

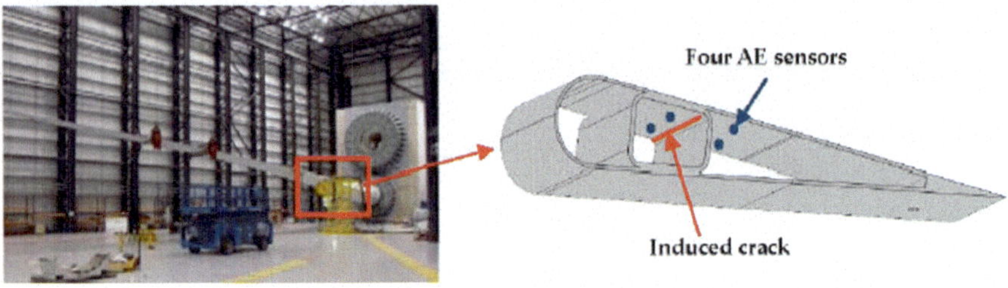

Fig. 4 Testing the turbine blade and AE sensors mounted on the internal side of the blade

be utilized to detect the strain alterations of Lamb waves. Hence, FBGs, with their small diameter, exhibit remarkable performance as Lamb wave sensors for damage detection and debonding in composite structures [31].

The pertinent standards to this method are 'ASTM E2374-15: Standard Guide for Acoustic Emission System Performance Verification' and 'ASTM E976: Standard Guide for Determining the Reproducibility of Acoustic Emission Sensor Response', (pencil-lead breakage test). On the other hand, the high accuracy, the higher number of sensors are required which engenders to increase the amount of output data for signal processing. Thus, to diminish the number of output data, Schulz et al. [16] proposed a structural neural system (SNS) for SHM which possesses several input channels from the sensors and two output channels, reducing the quantity of data acquisition channels for SHM. Utilizing AE method, a turbine blade with length of 45.7 m was tested under fatigue loading by Tang et al. [32], employing piezoelectric sensors as described in the Fig. 4.

Foregoing research was conducted different damage mechanisms through frequency-based methodologies [33–36]. Summarized results are listed in Table 2.

Muñoz and Márquez [37] proposed to detect and locate cracks on the surface of the blades employing three macro-fiber composite (MFC) sensors [38] in a section of a wind turbine blade. The results demonstrated that by employing the three low cost sensors, a fiber breakage occurred in a blade can be detected and located appropriately. The graphical approach of triangulation was applied for the signal processing.

Table 2 Frequency analysis outcomes for different damages [33–36]

Damages types	Range of frequency (kHz)			
	Glass-polyester [35]	Glass-polypropylene [36]	Carbon-epoxy [33]	Carbon-epoxy [34]
Matrix cracking	30–150	–	<100	0–50
Delamination	–	–	–	50–150
Debonding	180–290	100	200–300	200–300
Fiber breakage	300–400	450–550	400–450	400–500
Fiber pull out	180–290	200–300	–	500–600

Fig. 5 **a** Wind turbine blade under test. **b** CRM installed on the wind turbine blade [39]

The method was capable of detecting the position of the acoustic emission with high accuracy with the maximum error of 9 mm.

A laboratory study was reported by Tang et al. [39] regarding fatigue damage growth monitoring through AE in a complete 45.7 m long wind turbine blade. Using compact resonant masses (CRM), the blade was excited for 187.000 fatigue cycles, during which AE monitoring was performed with a 4-sensor array. The CRMs were exerted in order to generate the crack and to simulate the blade vibration. The full-length blade and the CRMs are displayed in Fig. 5.

According to frequency content of AE signals, three failure modes encompassing fiber failure (300 kHz), fiber-matrix debonding (250 kHz) and matrix cracking (110 kHz) were distinguished. Moreover, a vibration test of the blade was implemented before conducting the fatigue test. After about 67,000 fatigue cycles, a crack with dimensions of $0.2 \times 0.05 \times 0.01$ m^3 was intentionally introduced in one of the shear webs and the four sensors were positioned in the foregoing zone internally [39]. Significantly, the results revealed that after the introduction of the defect, there is no visible difference in the modal frequencies, highlighting that these modes are not proper for detecting damages by monitoring of the mode frequency shift which can be seen in Fig. 6b.

In accordance with calculations, one sensor is capable of monitoring with the range of 0.8 m in all directions. Based on the range of the AE signals as an important parameter, 28 sensors could prepare basic monitoring for a 45 m blade. Notably, with changing the threshold, the sensor range could be varied leads to changes in number of the sensors [39].

Audio-level acoustics-based damage detection approach as a non-traditional technique has received little attention up to now, which is based on passively generated acoustic signals, such as noise of cracking, debonding, etc. [28, 40–42].

Utilizing the airborne sound proposed by Krause and Ostermann [42], the damage detection procedure of a 34-m blade was carried out employing three optical microphones. The optical microphones were mounted inside the blade which are depicted in Fig. 7. The microphones were installed orthogonal on the blade spars and all the microphones were considered as omnidirectional.

The characteristics of audible cracking sounds were computed for all microphone signals. According to the abovementioned research, it is difficult to validate this

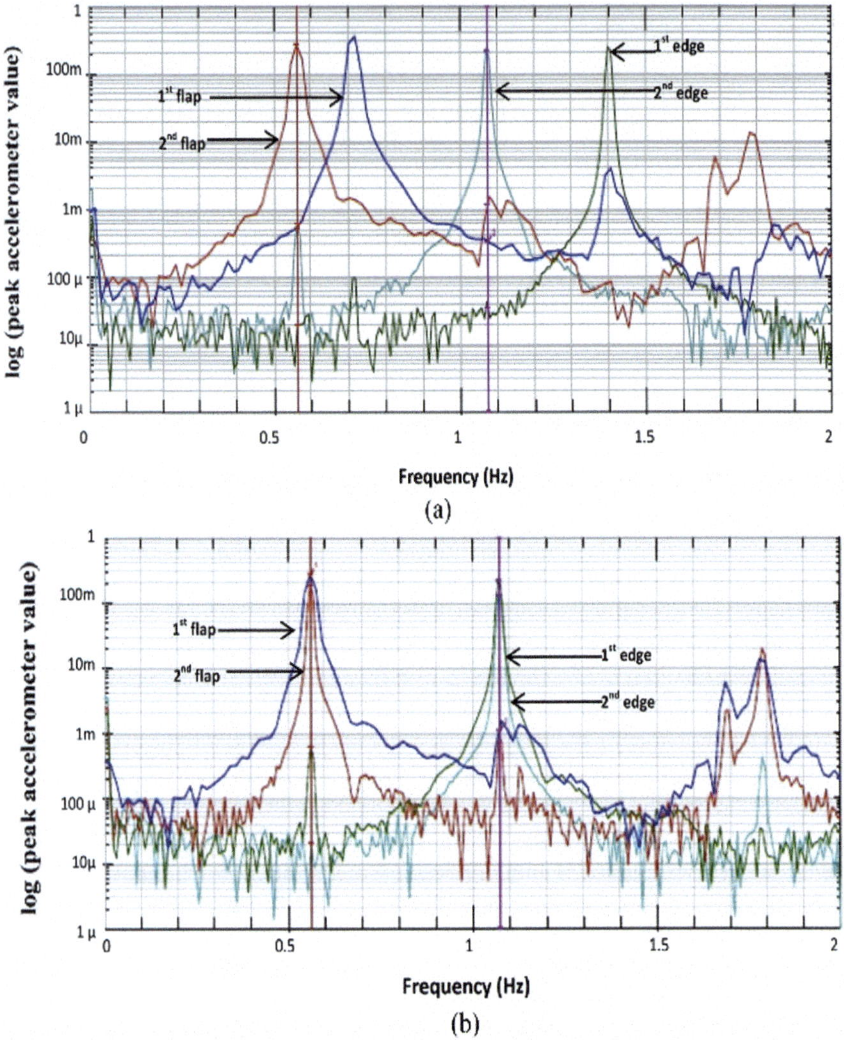

Fig. 6 **a** Frequency shifts after adding mass **b** Frequency spectrums after applying of a 0.2 m defect [39]

method generally, but nonetheless, the estimation exhibits good correlation with the size of damages after visual inspections [42]. Markedly, providing reliable data of blade damage detection using airborne sound method includes several uncertainties which requires further investigations and conducting experimental tests under different operational conditions [42]. Likewise, this method is highly dependent on the environment and background noises which could be considered as one of the limitations of the mentioned technique.

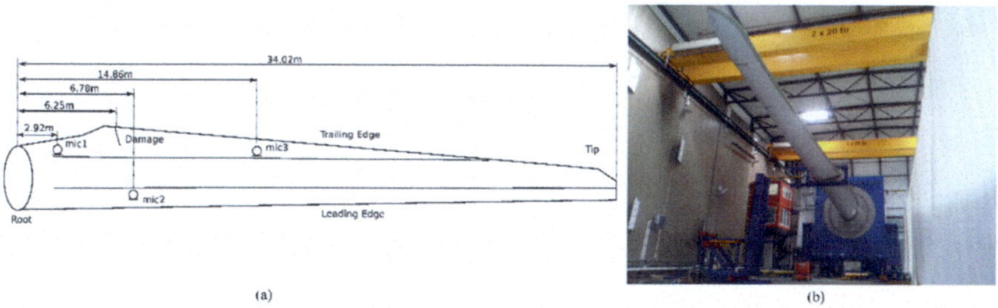

Fig. 7 **a** Drawing of the blade indicating the positions of microphones and damage **b** Blade fatigue test implemented at fraunhofer institute for wind energy and energy system technology [42]

The other significant type of acoustics-based method utilizes the acoustic signatures which are actively generated by an acoustic speaker, for sake of detecting damage, abating the inherent limitations of passive-based techniques by representing flexibility regarding the procedure and duration of structure excitation [43]. The active acoustic damage detection approach, schematically represented in Fig. 8 for a wind turbine blade, damage, speakers and microphones, in which the acoustic energy transmitted from the internal cavities of the structure to its exterior [44].

The acoustic signal alterations during transmitting through boundaries of structural cavities is investigated by Beale et al. [45] as a novel damage detection method. Focusing on active acoustic emission method, the damage was detected by measuring the acoustic pressure responses externally whereas the structure is subjected to internal acoustic excitations. The method was employed for damage detection of 46 m wind turbine blade considering various types of damage and different severity and locations. Extraordinarily, due to the low attenuation of sound in air, very few or even a single microphone could be sparsely located on the tower for capturing the excitation

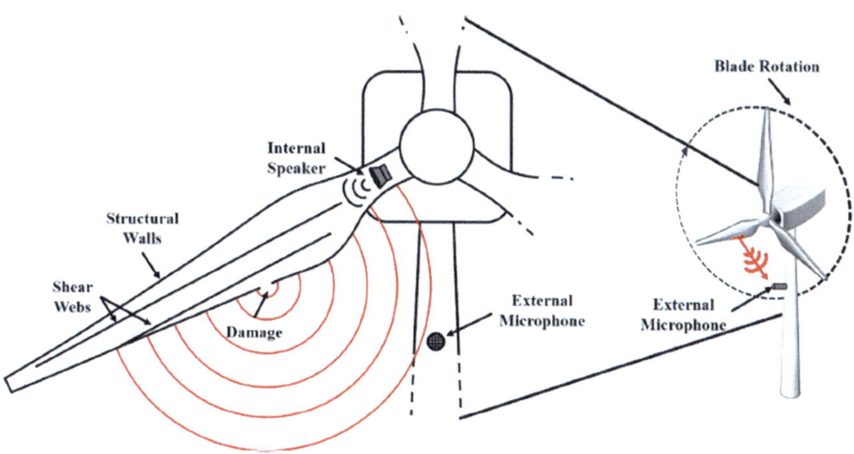

Fig. 8 Schematic illustration of the active acoustic damage detection method [44]

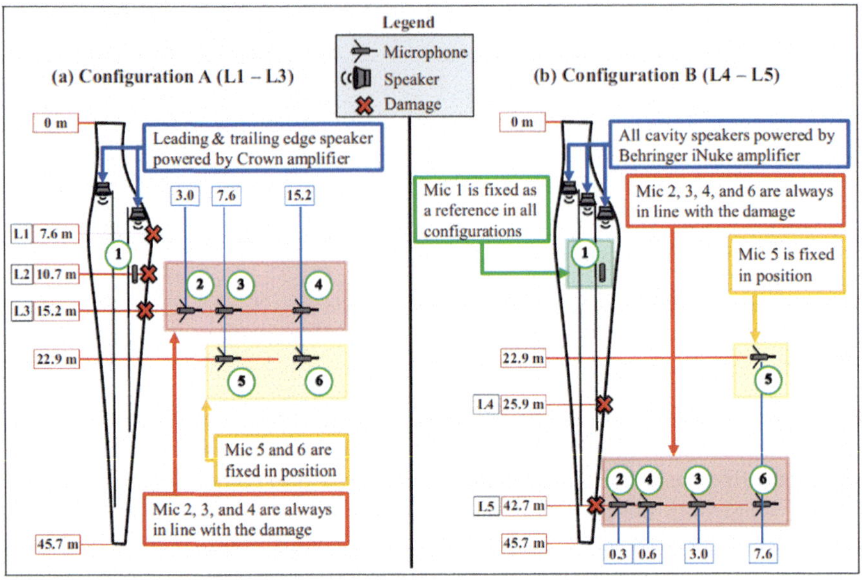

Fig. 9 Schematic of 46 m wind turbine blade, with the positions of the microphone and speaker representing the active acoustic emission method [45]

signals [45]. Moreover, minimum acoustic signal is transmitted through undamaged blade whereas a damaged blade allows high acoustic signal to be transmitted due to sound leakage from a damage position on the blade.

Figure 9 describes two configurations of experimental set up, consisting of microphone and speaker positions.

The results divulged that damages as small as 5.1 cm introduced before the mid-length of the blade were conveniently captured by different sensors ranging between 3 and 17.1 m from the damage position. It is worth mentioning that the active-based method should be limited for detecting edge splits, cracks, holes, and damage types that are extended through the entire skin of the structure.

In addition to available academic researches, various commercial SHM systems are available on the market. Based on damage sound detection, a damage detection system has recently been represented by Ping Services company [46]. As their algorithms are mostly proprietary, the commercial methods are not discussed further.

3.2 Thermal Imaging Method

Thermal imaging method is a detection technique for subsurface defects or abnormalities exerting temperature differences observed on the assessed surface by employing infrared sensors or cameras [47]. So, damages occurred in the structures are detectable due to a change in the thermal diffusivity [48]. Based on thermal excitation method, the thermal imaging is classified as passive or active methods. The passive type is exerted to assess structures that are at a different temperature than ambient [49]. The active approach possesses an external stimulus source such as optical flash lamps, or heat lamps. The passive method is not usual in wind turbine SHM during the tests and more improvements are needed to be flourished as a promising method. Thermoelastic stress method as a specific type of active thermal imaging method, is based on the thermoelastic effect, in which the stress alterations in the materials will cause the change in temperature [49]. This technique has been demonstrated as an efficient approach for fatigue tests of wind turbine blades [50], since stress concentrations could be captured during the test before extension of damage on the surface.

A fatigue test of the 13.4 m sectional blade was accomplished by means of a thermoelastic stress analysis camera (TSA), in order to observe the stress distributions [51]. The detected areas of stress concentration had not previously been predicted by the finite element method. Hence, at those damage-prone locations more strain sensors were exerted. Thus, measuring the strains could be useful to calibrate the signals exhibited by the TSA camera. Figure 10 illustrates the finite element analysis and the blade structure during the test conditions.

Regarding the composite structures, a suggested relationship can be expressed as following [51]:

$$\Delta \mathrm{T} = \frac{-T}{\rho C_p}(L\Delta\sigma_{11} + M\Delta\sigma_{22}) \quad (1)$$

which is proper for observation at a surface, thus $\Delta\sigma_{33}$ is excluded because it is necessarily zero and the directions of 11 and 22 imply on the principal directions of the laminate. The parameters L, M must be obtained from measurement [51].

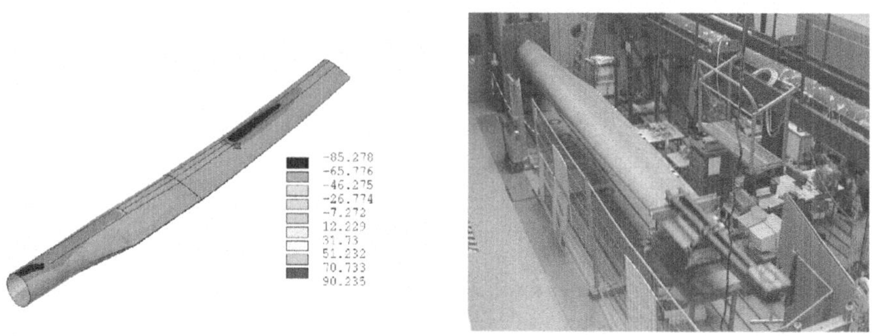

(a) (b)

Fig. 10 **a** Finite element results (strains in lengthwise direction). **b** Blade structure during test [51]

Furthermore, as well as blade damage detection, thermal imaging method is capable of diagnostic and monitoring in a broad range of applications in components such as bearings, gear boxes, electric generators, etc. However, due to slow spread of temperature inside the material, it is not an appropriate method for early damage detection [52, 53].

The vibro-thermographic technique, as a promising branch of thermal imaging, is able to detect the crack, damage, voids and stress concentrations utilizing high power ultrasounds [54], or oscillating stresses with a mechanical shaker [55]. Therefore, this method could prosper as a promising SHM method for wind turbines, while more investigations are required in order to diminish the sensitivity to temperature alterations [50]. More assessments are needed for the application of thermal imaging method to in-service WT's due to its sensitivity to ephemeral temperature alterations [50]. In addition, thermal imaging method is capable of producing a full-field measurement in image form which provides a rapid evaluation possible even for a non-professional operator. However, the cardinal problem of the thermal imaging technique is the thermal excitation procedure. Regarding the passive type, the environmental conditions (such as cloud cover, solar radiation, wind speed) play a crucial role on the outdoor infrared thermographic surveys. Regarding the active excitation, it is not cost-effective and efficient to excite the turbine on site [10]. Regarding the thermoelastic stress excitation method, its shortcoming is the necessity to apply load cyclically, which can lead to damage growth during defining the existing damage [56]. As a specific technique, non-contact ultrasound-induced thermography could be a promising technology for using in SHM of wind turbines, as in aircrafts [57]. As other drawbacks of thermal imaging method, detecting the cracks depends on camera resolution and it is impossible to penetrate in extended depths of the structures. Eventually, it can be concluded that this method could be utilized as a complementary method to other monitoring techniques for acquiring a comprehensive and accurate damage detection.

3.3 Ultrasonic Methods

Ultrasonic method is a technique commonly employed for investigating the inside of structures [58] which is based on propagation and reflection of elastic wave within the material. The basic principle of the technique is that an ultrasonic wave is passed through the material and is then reflected and/or mode converted by a defect. Three different techniques can be taken into account for this method as pulse-echo, through transmission, and pitch-catch [59, 60]. This technique is capable of detecting planar cracks that occur perpendicularly to the direction of wave propagation. This technique is widely employed by the wind energy industry for structural assessment of towers and blades [61].

This method provides an evaluation of the location and nature of the damage through wave propagation which can be used for SHM and determination of material properties of WT components [62]. In addition, ultrasound scanning can be exerted to observe the beneath surface and delamination in composite structures [63]. The transit time and amplitude can be exploited for defining the position and the severity of the damage [25]. Moreover, ultrasonic could be used for inspection of adhesively bonded parts in the composite structures for detecting of delamination and interlaminar weakness [64]. Furthermore, employing a scanning sensing laser beam, ultrasonic images can be constructed by utilizing the ultrasonic excitation at a specific point while measuring corresponding ultrasonic responses from multiple points [8]. Lee et al. [65] propose a portable long-distance ultrasonic propagation imaging (LUPI) system that uses a laser beam targeting and scanning system to excite, from a long distance, while acoustic emission sensors are installed in the blade. It was proposed to a portable health monitoring system for in situ long distance applications in wind turbine blade. The feasibility of the introduced method was verified on a 2 MW commercial wind turbine system. The embedded AE sensors at various damage hot spots acted as the ultrasonic sensors during the damage evaluation and visualization process, as well as the sensing component for damage localization. A schematic diagram of the entire Portable ultrasonic propagation imaging (UPI) system configuration is shown in Fig. 11a. Figure 11b and c represent the configuration and installation of a portable long-distance ultrasonic propagation imaging (LUPI) laser targeting and scanning system in a turbine tower.

A study of complete noncontact laser ultrasonic wavefield imaging technique was proposed by Park et al. [66], to automatically detect and visualize hidden delamination and debonding in composite structures. A blade of 10 kW wind turbine was provided for testing, depicted in Fig. 12a. The blade possesses the dimensions of $3500 \times 500 \times 3\ mm^3$ made of GFRP materials, which comprises 6 plies. Owing to simulate the internal delamination, a 15 mm diameter Teflon tape was embedded between the plies during manufacturing process. Figure 12 b and c compare cumulative total wave energy (CTWE) and cumulative standing wave energy (CSWE) images acquired from the scanning region of the blade. In CSWE image, the incident wave energy and other noise components were eliminated, and only the location of delamination was displayed.

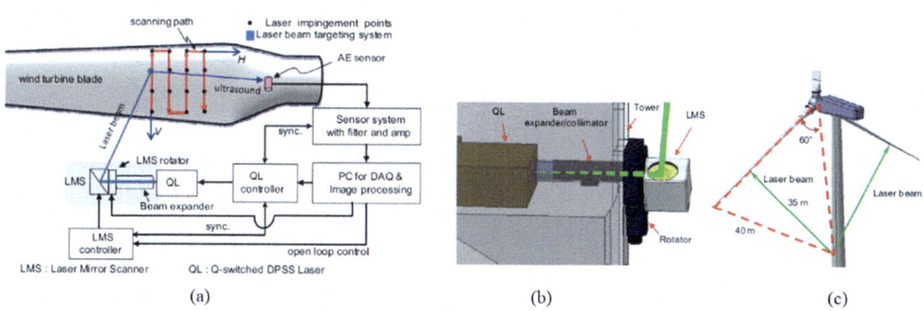

Fig. 11 **a** Portable UPI system configuration for long distance application. **b** System configuration and installation of long-distance laser excitation system. **c** Parked turbine blades at optimal position for scanning [65]

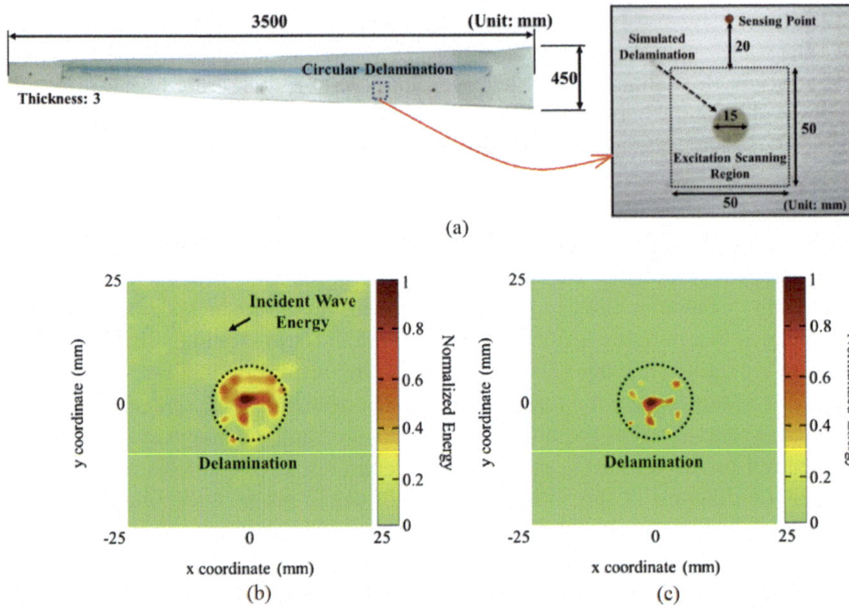

Fig. 12 **a** Composite blade of wind turbine. **b** CTWE image, and **c** CSWE image [66]

The introduced technique possesses following advantages: (1) it does not require any sensor installation; (2) it is noninvasive, rapidly deployable and applicable to harsh environments; and (3) it can visualize damage with high spatial resolution without any baseline data, which enables automated and intuitive damage diagnosis [66].

Ultrasound methods can be exerted for other purposes, such as measuring the thickness of the parts as well as crack detection. However, ultrasound techniques are insensitive to single fiber break in composite structures [10]. As a remarkable limitation, planar defects parallel to the wave propagation may not be detected. Moreover, this method has problem of monitoring irregular, rough, thin, or non-homogeneous materials.

3.4 Strain Measurement Method

Strain monitoring is a method for capturing microscopic length changes in a component [4] and the length variations directly correlate to stresses generated in the structure [18]. Markedly, strain sensors must be located at specific positions, where large deformations are anticipated which restrict their applicability for damage detection in a structure [26]. To this end, prior knowledge of the stress field of structure such as employing finite element analysis is required, thus, the sensors should be installed in vital locations.

Multifarious sensors are exploited in strain measurement method as traditional electrical sensors (e.g. piezoelectric materials, capacitance, inductance, etc.), fiber-optic sensors and recently the most popular types of sensors namely, Fiber Bragg Grating (FBG) sensors. Some main drawbacks of the electrical sensors are presented as below [18, 67]:

1. Influence of temperature on the performance of the piezoelectric sensors.
2. Non-linearity.
3. Hysteresis and zero shift due to cold work.
4. Weakening of the signal over a considerable distance.

Fiber-optics sensors can be mounted on the wind turbine blades for load measurement. By applying the strain on the component, the power of the light source will reduce which enables measurement of the applied load [68]. As a simplest sensing principle, the light power dwindles linearly with increasing of the strain, while decreases tremendously when the crack appears in the structure [68]. Some of the advantages of optical fiber sensors can be indicated as higher fatigue resistance, eliminating the wiring issues, and monitoring more locations with the same cable.

Fiber Bragg gratings (FBGs), as wavelength-based sensors, are made by illuminating the core of a suitable optical fiber with a spatially-varying pattern of intense ultra-violet (UV) laser light [69].

The advantages of FBG's are expressed as following:

1. Minimal size, high strength, very sensitive.
2. Lightening safety and neutrality to Electromagnetic interference (EMI).
3. Capability of measuring various parameters such as strain, temperature, sounds, pressure, accelerometer, etc.
4. Multiplexing: ability to engrave several sensors on a single fiber; for an instance, temperature and strain can be measured with the same sensor.
5. Combining multiple gratings in a single fiber: reading simultaneously large numbers of sensors on a very few fibers, leading to reduced cabling requirements and easier installation.

The Bragg wavelength, λ_B is determined by the below expression [70]:

$$\lambda_B = 2n_{eff}\Lambda \tag{2}$$

where n_{eff} indicates the effective refractive index and Λ the grating pitch.

As the measuring procedure, by applying the strain and altering the mechanical or physical of grating region, Λ and n_{eff} will be changed and consequently the reflected wavelength (λB) will be influenced accordingly, therefore, it can be utilized to measure the applied loads [71]. A schematic illustration of FBG response during imposed strain and its working principle are depicted in Fig. 13.

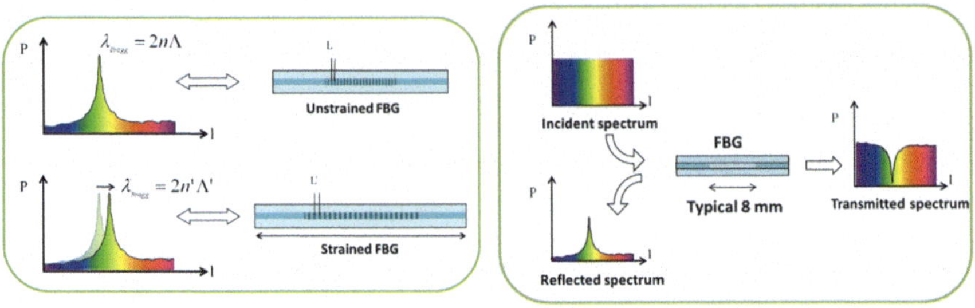

Fig. 13 **a** FBG response as function of strain. **b** Working principle of FBG [72]

An FBG measurement system for strain monitoring in WT blade with length of 45 m, has been mounted and demonstrated by Schroeder et al. [73]. In order to conduct the experiment, three strain sensor pads each have been attached to both the windward and the leeward internal surfaces of the blade, on opposite symmetrical positions which can be seen in Fig. 14.

For sake of obtaining the averaged load transfer from the blade to the relatively short FBG strain sensor, a very long sensor pad with length of 400 mm was selected, in comparison with the FBG sensor length of 5 mm. Finally, the average strain over the pad length of 400 mm was calculated. The photo of the six sensor pads is exhibited in Fig. 15, emphasizing that the third pad comprises a temperature sensor.

FOS4X sensor technology [74], employs fiber-optical strain and accelerometer sensors based on Fiber Bragg grating sensor (FBG) for structural health monitoring of wind turbine blades.

FIBERSAIL [75] is a shape sensing system based on FBG fiber optic sensors to monitor and analyze windmill blades in terms of shape, condition, and behavior.

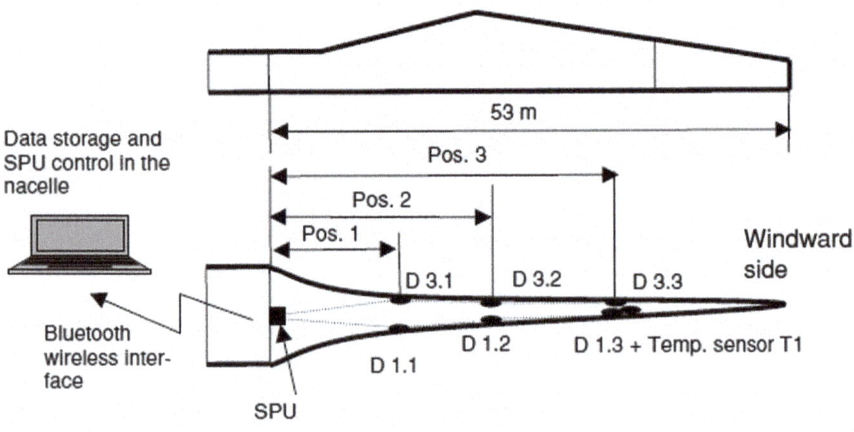

Fig. 14 Scheme of the positions of sensor pads and signal-processing unit (SPU) in the blade [73]

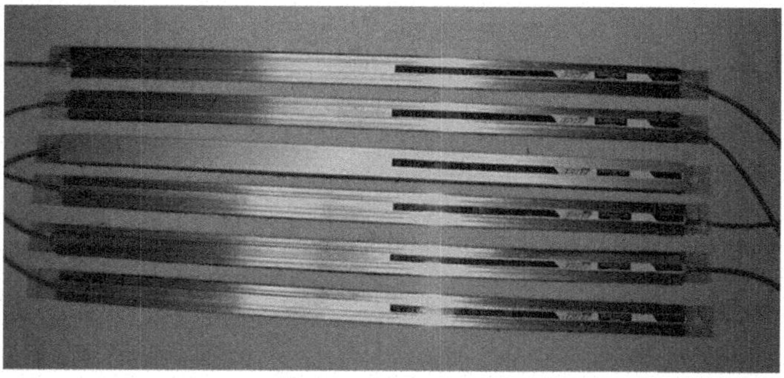

Fig. 15 The sensor pads for strain monitoring in the WT blade [73]

As another strain measurement approach, Moog Insensys [76, 77] has designed a new system for monitoring the blade pitch angle and for measuring the mass increase of the blades due to ice accretion, employing fiber optic strain sensors.

3.5 Vibration–based Method

The vibration-based methods are among the most significant SHM methods for OWT [78, 79, 118, 119]. The method is based on the fact that modal parameters such as frequencies, mode shapes and modal damping, are functions of the physical properties of the structure such as mass, damping and stiffness. Therefore, changes in the physical properties, such as reductions in stiffness originating from the onset of cracks or loosening of a connection, will engender detectable changes in the modal properties [80]. Moreover, the damage detection relies on a comparison between the dynamic response of the structure condition before and after occurring damage. So, changes in modal features, could be utilized as damage indicators [81, 82]. To investigate the dynamic response of the structure by studying the mode shapes, several accelerometers must be mounted [83]. In order to compare the resonant frequency, Ghoshal et al. [13] investigated the application of PZT patches at hot spots. The damage can be detected by measuring the differences in the resonance responses of the healthy and damaged structure.

As another damage detection approach based on vibration methods called Impedance-based damage detection method, various investigations were studied [84, 85]. In this method, by appearing damages in the structure, the electrical impedance of piezoelectric materials will be changed and monitored subsequently.

Pitchford et al. [85], employed foregoing method to detect damage on a wind turbine blade. The technique utilizes small piezoceramic (PZT) patches attached to a structure as self-sensing actuators to both excite the structure with high frequency excitations and monitor any changes in structural mechanical impedance. By monitoring the electrical impedance of the PZT, assessments can be made about the integrity of the mechanical structure. Their outcomes demonstrated that impedance based SHM is capable of detecting damage on the blade section and it is a promising technique for exerting in WT blades.

On the other hand, conducting accurate analyses on a full-scale offshore wind turbine (OWT) during operational condition is tremendously troublesome due to the high number of uncertainties induced by offshore environment such as wind and wave loading. Due to mentioned limitations, Operational Modal Analysis (OMA) is employed for measuring the modal parameters relying on the assumption that the structure is exposed to unknown random loads and the system is linear time-invariant during the analyzed time interval [86–88]. Devriendt et al. [89] evaluated that by utilizing 10 min as time segment for monitoring of the OWT, adequate modal parameters could be identified properly. Regarding the active vibration-based structural health monitoring, Tcherniak and Mølgaard [90] presented a technique which is capable of detecting structural damages of WT blade. The mentioned investigation demonstrated the system on a Vestas V27 [91] wind turbine. It was demonstrated that, for the specific damage type introduced to the blade, even small damage of at least 15 cm size could be captured remotely without stopping the WT. The proposed SHM system relies on the vibration method, in which the plunger frequently hits the blade and the generated vibrations propagated along the blade are detected by installed accelerometers. The actuator was mounted outside the blade, on its upwind side about 1 m from the root as delineated in Fig. 16. 12 monoaxial piezoelectric accelerometers were used in this study and their locations on the blade are described in Fig. 16 (d). However, it is worth mentioning that it is difficult to detect the defect using the frequency method as mentioned by [43].

The influence of the scour on the natural frequency of offshore wind turbine was studied by Weijtjens et al. [92], where it was demonstrated that by increasing the scour, the natural frequencies of the support structure, and therefore the wind turbine diminish. Vibration- based method with a three-tier structural health monitoring framework of a 34 m blade was exploited by Tsiapoki et al. [93], for sake of damage and ice detection. The ice accretion steps on the blade tip are exhibited in Fig. 17.

Results from the modal test shown that structural alterations due to damage at the trailing edge and adding mass could be captured by changes in the condition parameters [93].

The ‘IDD. Blade’ ice detection system is a vibration-based blade condition monitoring system developed by Wölfel [77, 94], which measures vibrations of WT blades with structural noise sensors.

3.6 Employing Carbon Nanotubes (CNTs) as Damage Impediment and Sensor Agents

As mentioned in the previous section, SHM approaches encompass the application of strain gages, accelerometers, piezoelectric sensors, and fiber optic sensors for

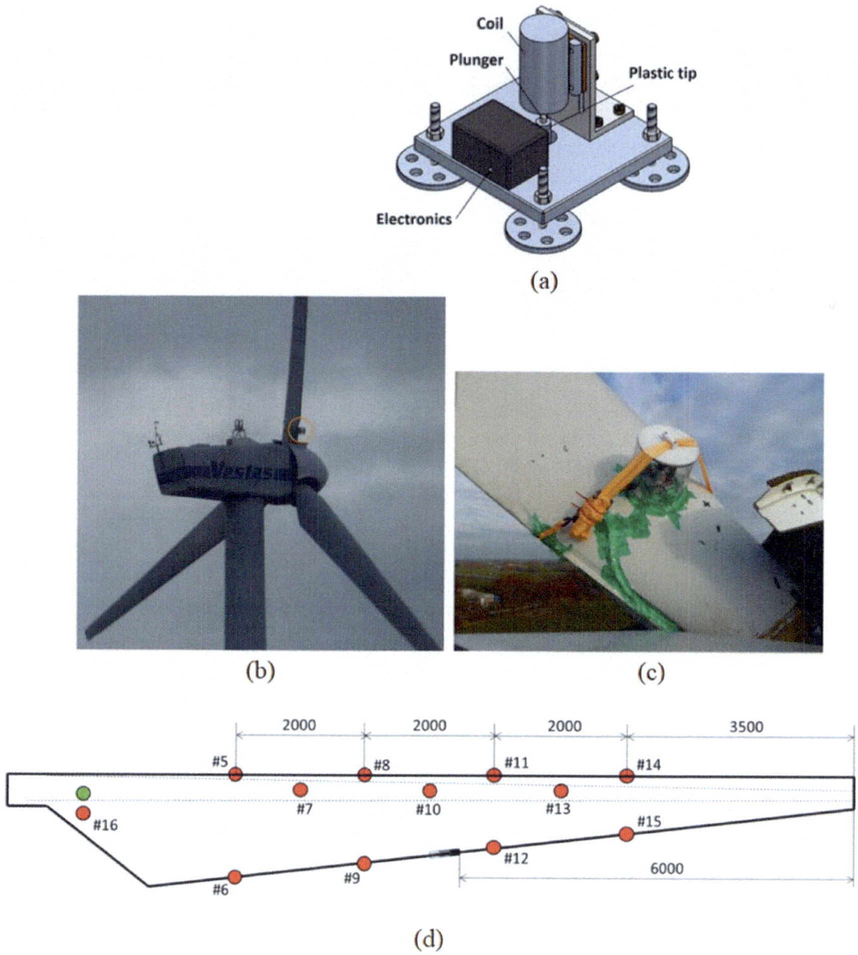

Fig. 16 Actuator **a** design. **b** actuator location. **c** mounting on the blade. **d** blade sketch indicating the red and green circles as the locations of the accelerometers and the location of the actuator, respectively [90]

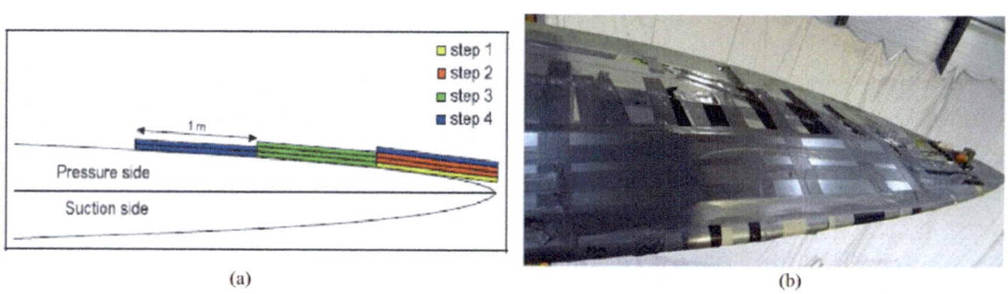

Fig. 17 **a** Ice accumulation steps. **b** Steel sheets mounted on blade surface for simulation of ice accretion [93]

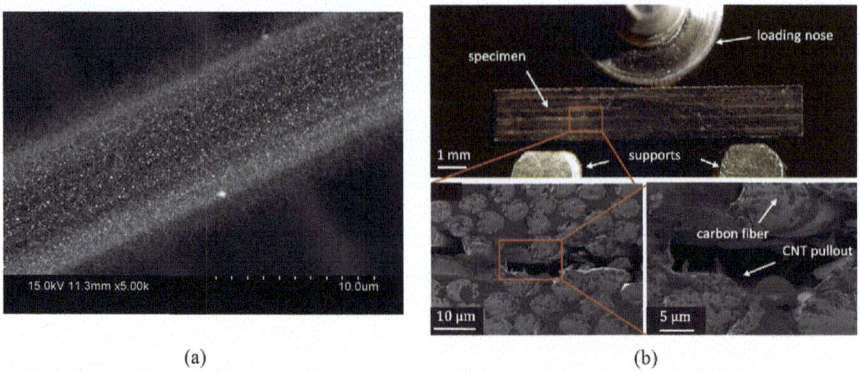

Fig. 18 **a** SEM micrograph of CNT growth on glass fiber (Fuzzy fiber) **b** short-beam shear (SBS) test configuration and the effect of CNTs which are pulled out of the polymer matrix during delamination [113, 114]

measuring strain, vibration, etc. [95–99]. Many of these methods furnish sensing only near the sensor, so, they must be positioned near to critical regions for sake of damage detection. Thus, if damage takes place at other unexpected regions, it may go undetected. Several investigations have concentrated on piezoresistive polymers made by dispersing CNTs into a polymer to produce a conductive matrix [100–106]. Fuzzy fiber sensors exhibit promising features as low cost, low weight, and convenient consolidation within a composite material. Furthermore, they are insensitive to manufacturing methods or alignment. Thus, foregoing remarkable features can provide an extremely high number of sensing elements to be embedded inside a composite structure with the minimum cost and weight. Furthermore, adding CNT to the conventional composite can augment the interfacial and interlaminar strengths to preclude the delamination and debonding damages in the composite structures [107–112]. The SEM image of carbon fiber coated with CNT (fuzzy fiber), and the three-point bending test of hybrid composite (conventional composite with CNTs) are delineated in Fig. 18. As it can be seen, the bridging effect of CNTs at the interface represents the significant improvement of such nano-engineered materials regarding the interlaminar and interfacial strengths of the composite structures.

Sebastian et al. [113], utilized a novel, multi-modal, nanomaterial-based sensor technology that can provide wide area detection of damage in composite structure. The efforts presented in this research serve as a feasibility study into the incorporation of carbon nanomaterials into structural composites as sensors. Surprisingly, the carbon nanotube covered fiber (fuzzy fiber) sensors exhibit similar sensitivity to conventional strain gages and are more easily integrated into composite structures as the sensor itself is a composite. The fuzzy fiber strain gages can be employed to sense strain within composite structures and can be consolidated into the laminated composite to prepare sensing over large sections and in positions which are not accessible to conventional strain gauging approach. The superior properties of the

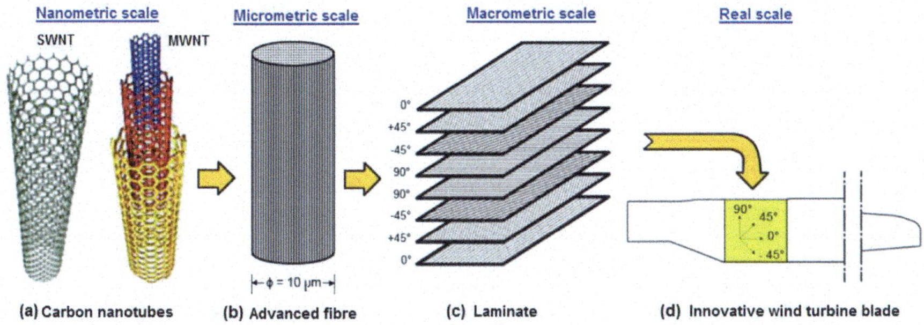

Fig. 19 Application of CNT-based continuous fibers for design of wind turbine blades [117]

fuzzy fiber provide its application in a wide range of sensing tasks within a composite structure including strain, temperature, degradation, etc.

Due to exceptional mechanical properties of CNTs and their potential to be used for monitoring the structural damages, application in wind industry is explored by several research groups [115–117]. These competitive advantages make CNTs attractive candidate for the reinforcement of composite long wind turbine blades with trivial increase in the blade weight, that can produce ultra-mega sources of clean energy [117]. A schematic application of CNTs- based wind turbine blades is illustrated in Fig. 19.

As a classification, all the introduced methods with their advantages and disadvantages are represented in the Appendix 1.

4 Conclusion

The present work is devoted to preparation of a review of disparate structural health monitoring (SHM) methods of wind turbine blades encompassing active/passive acoustic emission method, strain measurement method, vibration–based method, thermal imaging method and ultrasonic method, based on the recent investigations. The damages considered in the present work are fiber breakage, matrix cracking, delamination, fiber debonding, crack opening at leading/trailing edge and ice accretion. Moreover, employing the carbon nanotubes (CNTs) as both reinforcement agent and internal sensors for monitoring the internal health of the blade structure is introduced. Finally, the advantages and disadvantages of the represented methods are nominated and classified.

Appendix 1

(Table 3).

Table 3 Comparison and classification of the introduced SHM methods

SHM Methods	Definitions/Companies/ Employed Sensors	Advantages/Capabilities	Disadvantages/Limitations
Acoustic Emission Method (AE)	1. The conventional method is based on the release of localized energy in form of elastic wave in material 2. The damage process and stress waves also emit Airborne sound (audible crack) 3. Two approaches as active and passive types 4. Active Method: *a. Based on propagation of acoustic pressure in the air* *b. Using fiber optic microphones* *c. Eliminating the wave attenuation inside the material* *d. Optimizing number of speakers* *e. Optimizing the inside design of blade (avoiding narrow zone)* 5. Passive Method: *a. Based on elastic wave propagation inside material* *b. More required sensors than Active method* *c. Conducting attenuation test for reducing the sensor numbers* 6. Capable of detecting: Fiber breakage, Matrix cracking, Delamination, Fiber debonding, Crack opening at leading/trailing; Also capable of fault detection in gearboxes, bearings, shafts 7. Relevant Standards: ASTM E976, ASTM E2374- 15 8. Prior knowledge of stress field using finite element analysis is required for detecting damage-prone locations 9. Damage sound detection used by Ping Services company **Employed Sensors** Piezoelectric sensors, Fiber Optic Sensors, Fiber Brag Grating (FBG), Macro-fiber Composites Sensors (MFC)	1. Powerful in detecting/identifying damages and their locations up to micro-scale 2. Allowing a convenient, fast and cost-effective monitoring of a structure 3. Proper response at low frequencies and sensitive in detecting the incipient defect 4. Detecting widespread types of damages 5. Improving the damage evaluation capability of AE by employing Structural Neural System (SNS) 6. Airborne sound method: *a. there is NO signal attenuation due the inherent feature of this method* *b. Due to use of microphone, the number of sensors is remarkably reduced than conventional passive AE*	1. Depending on the algorithm 2. High sensitivity to background noise; Assessing the normal condition operating, different wind, and environment conditions such as raining condition, hailing, gust 3. Attenuation of the signal during wave propagation inside the material (except Airborne method) 4. Importance of sensor locations study for making cost-effective process 5. Avoiding the installation of metallic components/cables, due to lightning strike 6. Airborne sound method: *a. Dependent on the environment noise* *b. Incapable of detecting small crack or detecting at incipient*

(continued)

Table 3 (continued)

SHM Methods	Definitions/Companies/ Employed Sensors	Advantages/Capabilities	Disadvantages/Limitations
Strain Measurement Method	1. Detecting microscopic length changes in a structure 2. Prior knowledge of stress field using finite element analysis is required for detecting critical areas for mounting the sensors 3. Recently utilizing the optical fiber and FBG's strain sensors are highly attracted 4. Employed by FOS4X GmbH company for damage detection based on FBG strain sensors 5. Employed by Moog Insensys Ltd for ice and load measurement, based on four optical fiber strain sensors **Employed Sensors**: Strain gauge (Capacitance, inductance, semiconductor, and resistance), Fiber Optic Sensors, Fiber Brag Grating (FBG)	1. Convenient installation process once proper training was implemented 2. Mature technology 3. Using fiber- optic sensors due to less prone to fatigue, eliminates wiring issues and allows more points to be monitored with the same cable	1. Mechanical properties limitations 2. Sensitive of installation to misalignments 3. Influenced by EMI noise (except FBG's) 4. Distance between the sensor and the Data Acquisition System impresses accuracy and restricts location of sensor (except FBG's and optical fiber sensors)
Vibration-Based Methods	1. Considering the modal parameters such as resonance frequency, mode shapes and modal damping, as functions of the physical properties of the structure (mass, stiffness & damping) 2. Changes in the physical properties, resulting from the onset of cracks or damage initiation, will lead to detectable changes in the modal properties 3. Detecting the damages in the structure by comparing between the response before and after occurring damage 4. Employed by FOS4X and Wölfel for ice and damage detection 5. Active vibration based SHM used by VESTAS company 6. Deferent analysis types of modal base method: *a. Using PZT patches at the hot spots and exciting actuator, damage is detected using the differences of resonance response of healthy and damaged structure* *b. Impedance-based SHM by monitoring the electrical impedance of the PZT* 7. Operational Modal Analysis (OMA) *a. Considering that the system is linear and doesn't change during the time in the analyzed time interval* *b. Compared with traditional experimental modal analysis (EMA), OMA does not require expensive excitation sources* **Employed Sensors**: Accelerometer, MEMS, optical fibers, FBGs, Velocimeter	1. Reliable and mature technology 2. convenient mounting 3. Existing of many various techniques available based on this method 4. Solving some limitations based on development in Operational Modal Analysis	1. High number of uncertainties which makes difficult to accomplish a detailed dynamic/modal analysis of the entire of the structure 2. Incapable of detecting very small damages or localizing damages at incipient level 3. Difficult analysis in operating conditions 4. Increasing the scour, leads to decrease the natural frequency of the support structure and therefore wind turbine 5. Considering the variation of environmental and operational conditions in the results 6. Due to uncertainties and unmeasurable input forces such as wind and waves, the Classical Experimental Modal Analysis (EMA) methods cannot be employed for offshore WT

(continued)

Table 3 (continued)

SHM Methods	Definitions/Companies/ Employed Sensors	Advantages/Capabilities	Disadvantages/Limitations
Thermal Imaging Method	1. Based on the fact: when a component starts to malfunction, the temperature increases beyond the normal values 2. Two approaches as passive and active method based on excitation method 3. Passive Method: Utilizing to investigate materials that are at a different temperature than ambient (often higher)- NOT common in SHM of wind turbines 4. Active Method: Employing an external stimulus source such as optical flash lamps, heat lamps, etc., for inducing pertinent thermal contrasts on the part 5. 'Thermoelastic stress method' as one type of active thermal imaging method, based on the thermoelastic effect *a. Changing the temperature of material due to the change of stress in the structure* *b. Necessity to apply load periodically on the component being tested, which results in growth of damage* **Employed Sensors**: Impedance tomography, Thermography (infrared cameras)	1. Rapid assessment of the structure 2. Cost effective (Passive method) 3. Capable of producing a full-field measurement in image form	1. Detecting cracks depends on camera resolution 2. Impossible to penetrate in extended depths 3. Passive method: An important role of environmental conditions on outdoor infrared thermographic surveys (cloud cover, solar radiation, wind speed) 4. Active Method: Exciting the turbine on site, is not cost effective 5. Infeasible to use during the operation of a wind turbine, since the detection capability is constrained by the effect of environmental conditions 6. This method could be used as complementary to other SHM method
Ultrasonic Method	1. Utilizing three different techniques for investigation: *a. Pulse-echo Method* *b. Through transmission Method* *c. Pitch-catch Method* 2. Damage is detected by changing in wave reflection or time of flight, comparing with the time that a wave needs to travel a distance through a medium 3. Determining the defect position and defect severity by transit time and the amplitude, respectively 4. Revealing planar cracks (e.g. delamination) oriented perpendicular to the direction of sound wave propagation **Employed Sensors**: Piezoelectric transducers	1. Sensitive to both surface and subsurface discontinuities 2. The depth of penetration for flaw detection or measurement is superior to other NDT methods 3. Superior to other NDT methods 4. Only single-sided access is needed when the pulse-echo technique is used 5. It has other uses, such as thickness measurement, in addition to flaw detection	1. Difficulty in evaluation of irregular, rough, or non-homogeneous materials 2. Surface must be accessible to transmit ultrasound 3. Insensitive to single fiber rupture in fiber composites 4. Linear defects oriented parallel to the sound beam may go undetected 5. Requiring of reference standards for both equipment calibration and flaw characterization 6. Inefficient for full-scale SHM of large structures and it can be used as complementary method

References

1. Soua S, Van Lieshout P, Perera A, Gan TH, Bridge B (2013) Determination of the combined vibrational and acoustic emission signature of a wind turbine gearbox and generator shaft in service as a pre-requisite for effective condition monitoring. Renew Energy 51:175–181
2. Liu WY, Tang BP, Han JG, Lu XN, Hu NN, He ZZ (2015) The structure healthy condition monitoring and fault diagnosis methods in wind turbines: a review. Renew Sustain Energy Rev 44:466–472
3. Simani S, Farsoni S, Castaldi P (2015) Fault diagnosis of a wind turbine benchmark via identified fuzzy models. IEEE Trans Ind Electron 62:3775–3782
4. Martinez-Luengo M, Kolios A, Wang L (2016) Structural health monitoring of offshore wind turbines: a review through the statistical pattern recognition paradigm. Renew Sustain Energy Rev 64:91–105
5. Kim H, Melhem H (2004) Damage detection of structures under earthquake excitation using discrete wavelet analysis. Eng Struct 26:347–362
6. Yang W, Tavner PJ, Crabtree CJ, Wilkinson M (2010) Cost-effective condition monitoring for wind turbines. IEEE Trans Ind Electron 57:263–271
7. Zhou HF, Dou HY, Qin LZ, Chen Y, Ni YQ, Ko JM (2014) A review of full-scale structural testing of wind turbine blades. Renew Sustain Energy Rev 33:177–187
8. Yang R, He Y, Zhang H Progress and trends in nondestructive testing and evaluation for wind turbine composite blade. Renew Sustain Energy Rev
9. Yang B, Sun D (2013) Testing, inspecting and monitoring technologies for wind turbine blades: a survey. Renew Sustain Energy Rev 22:515–526
10. Ciang CC, Lee J-R, Bang H-J (2008) Structural health monitoring for a wind turbine system: a review of damage detection methods. Meas Sci Technol
11. Lian J, Cai O, Dong X, Jiang Q, Zhao Y (2019) Health monitoring and safety evaluation of the offshore wind turbine structure: a review and discussion of future development. Sustainability 11:494
12. Farrar CR, Sohn H (2000) Pattern recognition for structural health monitoring. Workshop on mitigation of earthquake disaster by advanced technologies (Las Vegas, NV, USA, 30 Nov.–1 Dec. 2000)
13. Ghoshal A, Sundaresan MJ, Schulz MJ, Pai PF (2000) Structural health monitoring techniques for wind turbine blades. J Wind Eng Ind Aerodyn 85:309–324
14. Adedipe O, Brennan F, Kolios AJ (2016) Review of corrosion fatigue in offshore structures: present status and challenges in the offshore wind sector. Renew Sustain Energy Rev 61:141–154
15. Larsen FMF, Sorensen T (2003) New lightning qualification test procedure for large wind turbine blades. Int Confe Lightning 36:1–10
16. Schulz MJ, Sundaresan MJ (2006) Smart sensor system for structural condition monitoring of wind turbines. Subcontract report NREL/SR-500-40089, National Renewable Energy Laboratory, CO, USA
17. Caselitz P, Giebhardt J, Mevenkamp M (1994) On-line fault detection and prediction in wind energy converters. Proc. EWEC (Thessaloniki, Greece) pp 623–7
18. Schubel PJ, Crossley RJ, Boateng EKG, Hutchinson JR (2013) Review of structural health and cure monitoring techniques for large wind turbine blades. Renew Energy 51:113–123
19. Caithness Windfarm Information Forum (2005) Wind turbine accident Data to December 31st 2005, https://www.caithnesswindfarms.co.uk/
20. Flemming ML, Troels S (2003) New lightning qualification test procedure for large wind turbine blades. Int. Conf. Lightning and Static Electricity (Blackpool, UK) pp 36.1–10
21. Rosenbloom E (2006) A Problem with Wind Power www.aweo.org
22. Zangenberg J, Brøndsted P, Koefoed M (2014) Design of a fibrous composite preform for wind turbine rotor blades. Mater Des 56:635–641. https://doi.org/10.1016/j.matdes.2013.11.036
23. Jureczko M, Pawlak M, Mezyk A (2005) Optimisation of wind turbine blades. J Mater Process Technol 167:463–471

24. Sørensen BF, Jørgensen E, Debel CP, Jensen FM, Jensen HM, Jacobsen TK, Halling KM (2004) Improved design of large wind turbine blade of fibre composites based on studies of scale effects (Phase 1). Summary report (Risø-R Report) Risø National Laboratory, Denmark
25. Sundaresan MJ, Schulz MJ, Ghoshal A (2002) Structural health monitoring static test of a wind turbine blade. Subcontract report NREL/SR-500-28719, National renewable energy laboratory, CO, USA
26. Balageas D, Fritzen C-P, Guemes A (2006) Structural health monitoring. ISTE Ltd., London, UK
27. Tchakoua P, Wamkeue R, Ouhrouche M, Slaoui-Hasnaoui F, Tameghe TA, Ekemb G (2014) Wind turbine condition monitoring: state-of-the-art review, new trends, and future challenges. Energies 7:2595–2630
28. Tan CK, Irving P, Mba D (2007) A comparative experimental study on the diagnostic and prognostic capabilities of acoustics emission, vibration and spectrometric oil analysis for spur gears. Mech Syst Signal Process 21:208–233
29. Krause T, Preihs S, Ostermann J (2014) Detection of impulse-like airborne sound for damage identification in rotor blades of wind turbines. 7th European workshop on structural health monitoring (EWSHM 2014), July 8–11
30. Van Dam J, Bond LJ (2015) Acoustic emission monitoring of wind turbine blades. In: SPIE smart structures and materials and nondestructive evaluation and health monitoring. International society for optics and photonics
31. Takeda N, Okabe Y, Kuwahara J, Kojima S, Ogisu T (2005) Development of smart composite structures with small diameter fibre Bragg grating sensors for damage detection: quantitative evaluation of delamination length in CFRP laminates using lamb wave sensing. Compos Sci Technol 65:2575–2587
32. Tang J, Soua S, Mares C, Gan TH (2017) A pattern recognition approach to acoustic emission data originating from fatigue of wind turbine blades. Sensors (Basel) 17(11):2507
33. Ni QQ, Iwamoto M (2002) Wavelet transform of acoustic emission signals in failure of model composites. Eng Fract Mech 69:717–728
34. Gutkin R, Green CJ, Vangrattanachai S, Pinho ST, Robinson P, Curtis PT (2011) On acoustic emission for failure investigation in CFRP: pattern recognition and peak frequency analyses. Mech Syst Signal Process 25:1393–1407
35. Suziki M, Nakanishi H, Iwamoto M, Jinen E (27–29 June 1988) Application of static fracture mechanisms to fatigue fracture behaviour of class A-SMC composite. In Proceedings of the 4th Japan-US Conference on composite materials, Washington, DC, USA
36. Ramirez-Jimenez CR, Papadakis N, Reynolds N, Gan TH, Purnell P, Pharaoh M (2004) Identification of failure modes in glass/polypropylene composites by means of the primary frequency content of the acoustic emission event. Compos Sci Technol 64:1819–1827
37. Gómez Muñoz CQ, García Márquez FP (2016) A new fault location approach for acoustic emission techniques in wind turbines. Energies 9:40
38. NASA. https://technology.nasa.gov/patent/LAR-TOPS-209
39. Tang J, Soua S, Mares C, Gan T-H (2016) An experimental study of acoustic emission methodology for in service condition monitoring of wind turbine blades. Renew Energy 99:170–179
40. Stearman RO, Schulze GH, Rohre SM (1997) Aircraft damage detection from acoustic and noise impressed signals found by a cockpit voice recorder. J Acoust Soc Am 101:3085
41. Fazenda B, Comboni Bustos D (12–14 June 2012) Acoustic condition monitoring of wind turbines: Tip faults. In: Proceedings of the 9th international conference on condition monitoring and machinery failure prevention technologies, London
42. Krause T, Ostermann J (2020) Damage detection for wind turbine rotor blades using airborne sound. Struct Control Heal Monit p 27
43. Niezrecki C, Inalpolat M (2015) Structural health monitoring of wind turbine blades using wireless acoustic sensing. Patent PCT/US2014/062329, USA
44. Beale C, Niezrecki C, Inalpolat M (2020) An adaptive wavelet packet denoising algorithm for enhanced active acoustic damage detection from wind turbine blades. Mech Syst Signal Process 142:106754

45. Beale C, Inalpolat M, Niezrecki C (2020) Active acoustic damage detection of structural cavities using internal acoustic excitations. Struct Health Monit 19(1):48–65
46. Ping Services, Australia, https://www.pingmonitor.co/
47. Avdelidis NP, Almond DP, Ibarra-Castanedo C, Bendada A, Kenny S, Maldague X Structural integrity assessment of materials by thermography conference. Damage in composite materials CDCM 2006 (Stuttgart, Germany)
48. Hellier CJ (2003) Handbook of nondestructive evaluation; McGraw-Hill Professional Publishing: New York, NY, USA
49. Krstulovic-Opara L, Klarin B, Neves P, Domazet Z (2011) Thermal imaging and thermoelastic stress analysis of impact damage of composite materials. Eng Fail Anal 18:713–719
50. Dutton AG (2004) Thermoelastic stress measurement and acoustic emission monitoring in wind. European wind energy conference (EWEC) 2004, London
51. Hahn F, Kensche CW, Paynter RJH, Dutton AG, Kildegaard J, Kosgaard C (2002) Design, fatigue test and NDE of a sectional wind turbine rotor blade. J Thermoplast Compos Mater 15:267–277
52. Bodil A, Mats D, Magnus U (2011) Feasibility study of thermal condition monitoring and condition based maintenance in wind turbines. Elforsk Report 11:19; ELFORSK: Stockholm, Sweden
53. Ge Z, Du X, Yang L, Yang Y, Li Y, Jin Y (2011) Performance monitoring of direct air-cooled power generating unit with infrared thermography. Appl Therm Eng 31:418–424
54. Walle G, Abuhamad M, Toma E, Netzelmann U (2004) Defect indications in sonothermography in relation to defect location and structure. In: Proceedings of quantitative infrared thermography conference, Rhode Saint Genèse, Belgium
55. Rantala J, Wu D, Busse G (1996) Amplitude modulated lock-in vibrothermography for NDE of polymers and composites. J Res Nondestruct Eval 7:215–228
56. Wilson DW, Charles JA (1981) Thermographic detection of adhesive-bond and interlaminar flaws in composites. Exp Mech 21:276–280
57. Anastasi RF, Zalameda JN (2004) Damage detection in rotorcraft composite structures using thermography and laser-based ultrasound. NDT.net 9:8
58. Li S, Shi K, Yang K, Xu J (2015) Research on the defect types judgment in wind turbine blades using ultrasonic NDT. IOP Conf Ser Mater Sci Eng 87:012056
59. Rose JL (2004) Ultrasonic guided waves in structural health monitoring. Key Eng Mater 270:14–21
60. Raisutis R, Jasiuniene E, Sliteris R, Vladisauskas A (2008) The review of non-destructive testing techniques suitable for inspection of the wind turbine blades. Ultrasound 63:26–30
61. Hyers R, McGowan J, Sullivan K, Manwell J, Syrett B (2006) Condition monitoring and prognosis of utility scale wind turbines. Energy Mater 1:187–203
62. Drewry MA, Georgiou GA (2007) A review of NDT techniques for wind turbines. Insight-Non-Destr Test Cond Monit 49:137–141
63. Cheng, L., Tian, G.Y.: Comparison of nondestructive testing methods on detection of delaminations in composites. J. Sens. 2012, 2012.
64. Tuzzeo D, Lanza di Scalea F (2001) Noncontact air-coupled guided wave ultrasonics for detection of thinning defects in aluminum plates. Res Nondestr Eval 13:61–78
65. Lee J-R, Shin H-J, Chia CC, Dhital D, Yoon D-J, Huh Y-H (2011) Long distance laser ultrasonic propagation imaging system for damage visualization. Opt Lasers Eng 49:1361–1371
66. Park B, An YK, Sohn H (2014) Visualization of hidden delamination and debonding in composites through noncontact laser ultrasonic scanning. Compos Sci Technol 100:10–18
67. Dally JW, Riley WF (1996) Experimental stress analysis. J Appl Mech 33
68. Takeda N (2002) Characterization of microscopic damage in composite laminates and real-time monitoring by embedded optical fiber sensors. Int J Fatigue 24:281–289
69. Doyle C (2003) Fiber bragg grating sensors—an introduction to bragg gratings and interrogation techniques smart fibers. Ltd Report. www.smartfibres.com

70. Campanella CE, Cuccovillo A, Campanella C, Yurt A, Passaro VMN (2018) Fiber bragg grating based strain sensors: review of technology and applications. Sensors 18:3115
71. Iodice M, Striano V, Cappuccino G, Palumbo A, Cocorullo G Fiber bragg grating sensors-based system for strain measurements. Proc. 2005 IEEE-LEOS workshop on fibres and optical passive components (Palermo, Italy) pp 307–12
72. https://fbgs.com/technology/fbg-principle/
73. Schroeder K, Ecke W, Apitz J, Lembke E, Lenschow G (2006) A fibre bragg grating sensor system monitors operational load in a wind turbine rotor blade. Meas Sci Technol 17:1167–1172
74. FOS4X GmbH, Germany, https://fos4x.com/en/downloads/
75. FIBERSAIL, Netherlands, https://www.fibersail.com/
76. Perovic S, Osborne M, Lloyd G, Bridges P (2010) Intelligent blade ice detection and measurement. (Moog Insensys Ltd., Southampton, UK.) European Wind Energy Conference and Exhibition. Warsaw, Poland
77. Cattin R, Heikkilä U (2015) Evaluation of ice detection systems for wind turbines. Final report
78. Carden EP (2004) Vibration based condition monitoring: a review. Struct Health Monit 3:355–377
79. Bab S, Khadem SE, Shahgholi M, Abbasi A (2017) Vibration attenuation of a continuous rotor-blisk-journal bearing system employing smooth nonlinear energy sinks. Mechan Sys Signal Process 84:128–157
80. Gross E, Simmermacher T, Rumsey M, Zadoks RI (1999) Application of damage detection techniques using wind turbine modal data. American society of mechanical engineers wind energy symposium (Reno, NV, USA) AIAA 99–0047
81. Farrar CR, Doebling SW (1997) An overview of modal-based damage identification methods DAMAS 97. Sheffield, UK
82. Doebling SW, Farrar CR, Prime MB (1998) A summary review of vibration-based damage identification methods. Shock Vib Dig 30:91–105
83. Jang J-H, Yeo I, Shin S, Chang S-P (2002) Experimental investigation of system identification-based damage assessment on structures. J Struct Eng 128:673–682
84. Park G, Sohn H, Farrar CR, Inman DJ (2003) Overview of piezoelectric impedance-based health monitoring and path forward. Shock Vib Dig 35:451–463
85. Pitchford C, Grisso BL, Inman DJ (2007) Impedance-based structural health monitoring of wind turbine blades. Proc.SPIE, 6532. https://doi.org/10.1117/12.715800
86. Hermans L, Van Der Auweraer H (1999) Modal testing and analysis of structures under operational conditions: industrial applications. Mech Syst Signal Process 13:193–216
87. Devriendt C, El-kafafy M, De Sitter G, Guillaume P (2012) Damping estimation of offshore wind turbines on a monopile foundation using state-of-the-art operational modal analysis techniques. International conference on noise and vibration engineering, Leuven 2647–62
88. Devriendt C, Magalhães F, Weijtjens W, De Sitter G, Cunha Á, Guillaume P (2014) Structural health monitoring of offshore wind turbines using automated operational modal analysis. Struct Health Monit 13(6):644–659
89. Devriendt C, Jan Jordaens P, De Sitter G, Guillaume P (2013) Damping estimation of an offshore wind turbine on a monopile foundation. IET Renew Power Gener 7(4):401–412
90. Tcherniak D, Mølgaard LL (2017) Active vibration-based structural health monitoring system for wind turbine blade: demonstration on an operating Vestas V27 wind turbine. Struct Health Monit 16(5):536–550
91. Vestas Denmark, https://www.vestas.com/
92. Weijtjens W, Verbelen T, De Sitter G, Devriendt C (2015) Foundation structural health monitoring of an offshore wind turbine: a full-scale case study. Struct Health Monit
93. Tsiapoki S, Häckell MW, Grießmann T, Rolfes R (2018) Damage and ice detection on wind turbine rotor blades using a three-tier modular structural health monitoring framework. Struct Health Monit 17(5):1289–1312
94. Wölfel Group. Germany, https://www.woelfel.de/en/home.html

95. Bois C, Herzog P, Hochard C (2007) Monitoring a delamination in a laminated composite beam using in-situ measurements and parametric identification. J Sound Vib 299:786–805
96. Zou Y, Tong L, Steven P (2000) Vibration-based model-dependent damage (delamination) identification and health monitoring for composite structures—a review. J Sound Vib 230(2):357–378
97. Leng J, Asundi A (2003) Structural health monitoring of smart composite materials by using EFPI and FBG sensors. Sens Actuators A 103(3):330–340
98. Ignatovich SR, Menou A, Karuskevich MV, Maruschak PO (2013) Fatigue damage and sensor development for aircraft structural health monitoring. Theor Appl Fract Mech 65:23–27
99. Teo YH, Chiu WK, Chang FK, Rajic N (2009) Optimal placement of sensors for sub-surface fatigue crack monitoring. Theor Appl Fract Mech 52:40–49
100. Karimov KS, Khalid FA, Chani MTS (2012) Carbon nanotubes-based strain sensors. Measurement 45:918–921
101. Kang I, Schulz MJ, Kim JH, Shanov V, Shi D (2006) A carbon nanotube strain sensor for structural health monitoring. Smart Mater Struct 15(3):737–748
102. Thostenson ET, Chou T-W (2008) Real-time in situ sensing of damage evolution in advanced fiber composites using carbon nanotube networks. Nanotechnology 19:1–6
103. Alamusi HuN, Fukunaga H, Atobe S, Liu Y, Li J (2011) Piezoresistive strain sensors made from carbon nanotubes-based polymer nanocomposites. Sensors 11:10691–10723
104. Arash A, Jiang JW, Rabczuk T (2015) A review on nanomechanical resonators and their applications in sensors and molecular transportation. Appl Phys Rev 2:021301
105. Grow RJ, Wang Q, Cao J, Wang D, Dai H (2005) Piezoresistance of carbon nanotubes on deformable thin-film membranes. Appl Phys Lett 86:93–104
106. Boehle M, Jiang Q, Li L, Lagounov A, Lafdi K (2012) Carbon nanotubes on glass fiber as a strain sensor for real time structural health monitoring. Int J Smart Nanomater 3(2):162–168
107. Malekimoghadam R, Icardi U (2019) Prediction of mechanical properties of carbon nanotube-carbon fiber reinforced hybrid composites using multi-scale finite element modelling. Compos Part B Eng 177:107405
108. Vu-Bac N, Lahmer T, Zhang Y, Zhuang X, Rabczuk T (2014) Stochastic predictions of interfacial characteristic of polymeric nanocomposites (PNCs). Compos Part B Eng https://doi.org/10.1016/j.compositesb
109. Msekh MA, Cuong NH, Zi G, Areias P, Zhuang X, Rabczuk T Fracture properties prediction of clay/epoxy nanocomposites with interphase zones using a phase field model. Eng Fract Mech 188:287–299
110. Silani M, Ziaei-Rad S, Talebi H, Rabczuk T (2014) A semi-concurrent multiscale approach for modeling damage in nanocomposites. Theor Appl Fract Mech 74:30–38
111. Hosseini SA, Saber-Samandari S, Maleki Moghadam R (2017) Multiscale modeling of interface debonding effect on mechanical properties of nanocomposites. Polym Compos 38:789–796
112. Moghadam RM, Saber-Samandari S, Hosseini SA (2016) On the tensile behavior of clay–epoxy nanocomposite considering interphase debonding damage via mixed-mode cohesive zone material. Compos Part B Eng 89:303–315
113. Sebastian J, Schehl N, Bouchard M, Boehle M, Li L, Lagounov A, Lafdi K (2014) Health monitoring of structural composites with embedded carbon nanotube coated glass fiber sensors. Carbon N Y 66:191–200
114. Ni X, Furtado C, Kalfon-Cohen E, Zhou Y, Valdes GA, Hank TJ, Camanho PP, Wardle BL (2019) Static and fatigue interlaminar shear reinforcement in aligned carbon nanotube-reinforced hierarchical advanced composites. Compos Part A Appl Sci Manuf 120:106–115
115. Dai G, Mishnaevsky L Jr (2015) Carbon nanotube reinforced hybrid composites: computational modeling of environmental fatigue and usability for wind blades. Compos Part B Eng 78:349–360
116. Zhou HW, Mishnaevsky L, Yi HY, Liu YQ, Hu X, Warrier A, Dai GM (2016) Carbon fiber/carbon nanotube reinforced hierarchical composites: effect of CNT distribution on shearing strength. Compos Part B Eng 88:201–211

117. Attaf B (17–22 May 2016) On the application of carbon nanotube-based composite materials for smart design of wind turbine blades, international conference on materials and energy ICOME 16, La Rochelle, France
118. Abbasi A, Khadem SE, Ghandchi-Tehrani M Chaos suppression of a macro electromechanical system via linear feedback control based on Lyapunov exponent theorem, 5th international conference on acoustics and vibration
119. Abbasi A, Khadem S, Bab S (2018) Vibration control of a continuous rotating shaft employing high-static low-dynamic stiffness isolators. J Vib Control 24(4):760–783

Correlation Between Mechanical Properties and Cavitation Erosion Damage of Carbide and Nitride Thick Coatings for Turbomachinery Applications

Vladimir Safonov, Anna Zykova, Janusz Steller, and Tomasz Seramak

Abstract The cavitation erosion resistance of carbide and nitride thick (up to 20 μm) coatings by means of the rotating disk testing method was analysed in the present study. The effect of the coatings composition and mechanical properties on the damage mechanism and material weight loss has been confirmed. Results show that all coated samples under investigation are characterised by lower cavitation wear than that of the uncoated ones. Both nitride TiN, CrN, and carbide CrC coatings significantly improve erosion resistance of the stainless steel samples. The correlation between the evolution of mechanical properties of both nitride and carbide coatings and weight losses has been investigated. The differences in the composition affect the mechanical properties of coatings and, finally, cavitation erosion resistance of tested materials. Furthermore, the CrC coated samples have shown improved elastic behaviour and high resistance to plastic deformation as compared to the CrN and TiN coated ones. Young modulus E and hardness H parameters of coatings exceed those estimated for the reference samples. The higher H/E and H^3/E^2 coefficients in the case of carbide coatings in comparison with nitride ones and reference samples have been established. CrC thick coatings with the ability to form nanocrystalline structure and higher mechanical parameters demonstrate the best protective efficacy for various anti-erosion applications.

V. Safonov · A. Zykova
National Scientific Centre "Kharkov Institute of Physics & Technology", National Academy of Sciences of Ukraine, Kharkiv, Ukraine
e-mail: v_safonov@kipt.kharkov.ua

A. Zykova
e-mail: zykova.anya@gmail.com

J. Steller (✉)
Institute of Fluid-Flow Machinery, Polish Academy of Sciences, Gdansk, Poland
e-mail: steller@imp.gda.pl

T. Seramak
Faculty of Mechanical Engineering, Gdansk University of Technology, Gdansk, Poland
e-mail: tseramak@pg.edu.pl

M. Abdel Wahab (ed.), *Proceedings of 1st International Conference on Structural Damage Modelling and Assessment*, Lecture Notes in Civil Engineering 110,
https://doi.org/10.1007/978-981-15-9121-1_30

Keywords Cavitation · Erosion resistance · Arc-PVD method · Hardness · Elastic modulus · Chromium carbide coatings

1 Introduction

Cavitation erosion of hydraulic turbine blades, elements of shut-off and control valves, pump impellers, etc. is a process of progressive material damage due to continued exposure to cavitation. High impact pressure generated stress waves act on the material surface to produce plastic deformation, crack initiation, crack growth and material removal [1–3]. Cavitation and hydro-abrasive wear may contribute significantly to hydraulic machinery residual lifetime decrease or at least to degradation of streamlined surfaces and deterioration of machinery performance characteristics. The mechanism of erosive wear of steam turbine rotor blades is associated mainly with the blade material fatigue failure due to high mechanical stresses arising in the process of droplet impact on the blade surface [4, 5]. Despite differences in the impingement character, fatigue is also generally considered the main mechanism of cavitation erosion.

An effective way to extend the lifetime of equipment, subject to the abovementioned erosive load, is to use the vacuum ARC PVD coatings based on carbides and nitrides. The protective coatings application allows decreasing the effect of a complex of destructive factors, mainly such as corrosion and erosion. Different types of coatings have been already applied [6–14] for increasing the useful lifetime of such machine elements as fuel injector components, steam turbine blades, diesel engines, as well as pipe valves for the industrial applications.

Cavitation erosion behaviour of the coated materials is significantly influenced by the main substrate and coatings properties [8]. Surface roughness, scratches and pits on the coated substrate surface and defects in the coating structure such as inclusions and droplets are recognised as preferred cavitation initiation sites [15]. Coatings degradation mechanism was previously observed for films deposited by the cathodic arc technique [16, 17] and magnetron sputtered coatings [9, 12]. The effect of coating thickness was analysed by Blowers' impact model using [18]. The decrease of damage for thick coatings was explained in terms of stress wave effect. Droplet impact loads resulted in the deformation of the material by generation of impact stress pulses. The stress waves are partially transmitted and reflected in the process of propagation into coating. The intensity of stress waves reaching the substrate changes depending on the coating structure and properties. Moreover, the increase in coatings thickness induces an increase in stress attenuation in the coating, because of the rising time for the stress wave to reach the coating/substrate interface. Hence, the resultant stress transmitted into the substrate would decrease to result in a smaller deformation of the substrate. The thickness of the anti-erosion coatings proposed for the erosion protection of the rotor blades, should provide minimization of internal stresses at the coating-substrate interface. In the previous study the estimated coating thickness was about 15 $\div$ 20 micrometers for Arc-PVD chromium coatings [19].

Hardness H and the elastic modulus reciprocal $1/E$ represent material mechanical strength parameter and elastic deformation vulnerability. Lower Young's modulus is associated with high wear and cavitation resistance. The higher E value leads to more effective elastic energy transport through the coating to the substrate and results in significant material damage. The resistance of material to failure is dependent on the H/E ratio and its resistance to plastic deformation on the H^3/E^2 parameter. The H/E and H^3/E^2 parameters play important role in controlling wear performance. Higher ratios mean that the coatings exhibit highly elastic behavior and high resistance to plastic deformation [20–22]. The H/E ratios for both the coating and the substrate are key parameters of the R_{cav} cavitation resistance factor as derived in the beginning of the passing decade by Krella [23, 24]. The additional parameters involved in Krella's approach are Poisson coefficients as well as those of thermal conductivity and expansion. The number of phases in the coating, its adhesion strength and thickness are also included in the consideration.

The tests reported in the current paper have been conducted with only one substrate material. The H/E and H^3/E^2 parameters as dependent on coatings composition and microstructure are discussed at the end of Sect. 4. It is shown that at least in case of coatings under investigation higher H^3/E^2 and H/E values lead to increasing material resistance to wear and erosion.

The authors believe that the results are of significance for a wider scope of applications as previous studies indicate a correlation in damage mechanism of the water droplet erosion and cavitation erosion [25, 26]. Cavitation erosion is generally considered to result from the eroded surface impingement by micro-jets formed in course of an asymmetric implosion of cavitation bubbles close to this surface and shock pressure waves generated in the implosion final stage due to sudden deceleration of the nearly radial water movement [27]. The abovementioned mechanical load gives rise to stress wave propagation in the material surface layer [28] and cumulating the internal stress energy which gets released by generating micro-cracks and other structural defects, leading eventually to a major damage and material loss.

For evaluation of the material cavitation resistance, different methods and experimental facilities are applied, such as cavitation tunnels, rotating disks, cavitating jet cells, vibratory rigs etc. From the point of view of industrial application, the rotating disk stand with ability to test simultaneously a larger number of samples under conditions resembling those in a hydraulic machine is of great importance. Moreover, the intensity of erosive load at rotating disk facility as used in the hydraulic lab of the Szewalski Institute of Fluid-Flow Machinery (IMP PAN) is significantly higher than that at the other test facilities involved in the International Cavitation Erosion Test project at the turn of 1980 s and 90 s [29–31]. Furthermore, the advantage of this method is a significant gradient of load distribution on the sample surface which may be helpful for some kinds of accelerated evaluation techniques of the cavitation erosion resistance.

The aim of this study is the analysis of correlation between mechanical properties and cavitation erosion damage of carbide and nitride thick (up to 20 μm) coatings basing on the rotating disk test results.

2 Materials and Technology

Nanocrystalline coatings of TiN, CrN and CrC were deposited by cathodic arc evaporation (ARC PVD) method using the AIR-3 multifunctional installation. An evaporator with magnetic stabilization of a cathode spot was applied [32]. For deposition, cathodes made of commercially pure Ti (99.9%) and a low-alloyed chromium alloy VH-2 K (La 0.3…0.4%, Ta 0.3…0.5%, V 0.3…0.5%) were used. The deposition of coatings was performed on austenitic stainless steel Cr18Ni10Ti samples with a diameter of 30 mm and thickness of 8 mm. Substrates were preliminarily quenched at 1050 °C, and then ground and polished to the roughness of Ra ≤ 0.05 μm.

A typical process of coating deposition was initiated by vacuuming the technological chamber below $2 \div 3\ 10^{-3}$ Pa. Then the surface of the sample was subjected to ion cleaning of the cathode material (titanium or chromium) at a negative bias volt-age of 0.8 kV. After ion cleaning, deposition of a thin metal intermediate layer (~0.5–1 μm thick) followed throughout the period of 3 min. The TiN and CrN coatings were deposited to a thickness of 18–20 μm in a nitrogen atmosphere, while the CrC ones in a toluene (C_7H_8) atmosphere. During the coating formation process, a negative bias voltage was applied to the samples. The temperature of the substrate during coating deposition was controlled by an IR pyrometer and was kept within the 350–400°C range. The distance between the samples and the cathode was 270 mm. The parameters used for deposition of TiN, CrN, and CrC coatings are presented in Table 1. Characteristics of the deposited coatings were studied using various techniques. The thickness of CrC coatings was measured by a Calotest device. The adhesion properties were measured by standard methods with the use of the Revetest scratch tester (CSM Instruments). The phase composition of the coatings was studied using a DRON-4-07 X-ray diffractometer in copper Cu-Kα radiation using a Ni selectively absorbing filter. Qualitative phase analysis was carried out using the card index of the international database of crystallographic compounds ICDDPDF-2.

Table 1 Deposition parameters, thickness, hardness, Young's modulus and adhesion strength of TiN, CrN, CrC coatings

Deposition and coatings parameters	Type of coating		
	TiN	CrN	CrC
Pressure of residual gases, Pa	$2 \cdot 10^{-3}$	$3 \cdot 10^{-3}$	$3 \cdot 10^{-3}$
Working pressure of, N_2, Pa	1	0.4	
Working pressure of C_7H_8, Pa	–	–	0.6
Arc current, A	100	130	130
Substrate bias voltage, V	−150	−100	−80
Coatings thickness, μm	18	16	19
Adhesion strength, N	82	65	74
Hardness, GPa	24,3	22,45	23,1
Young's modulus, GPa	404	390	303

The study of substructural characteristics (the size of coherent scattering regions CSR and the level of micro-deformations of the samples) was based on the integral line width. The analysis was carried out according to the Williamson-Hall method. In the case of CrC coatings, CSR sizes were determined using the Scherrer formula. The coating microstructure and chemical composition were investigated using the JSM-7001F scanning electron microscope (SEM) and the energy dispersive X-ray (EDX) spectroscopy. Hardness and Young's modulus were measured by the *NanoHardness* Tester (*CSM Instruments*) using the Oliver and Farr method. Main results are presented in Table 1.

3 Erosion Tests

Erosion resistance was evaluated basing on the tests conducted at the rotating disk facility in the IMP PAN lab [29–31]. The key component of the rig is the test disk of 330 mm diameter, rotating in the cavitation chamber with 2950 rpm speed and coupled directly to an asynchronous engine of 40 kW output. Before each test run the specimens are inlaid into the disk, in the holes located in implosion zones of cavitating wakes generated by eight cylindrical bolt cavitators (Fig. 1a). The standard static pressure and temperature inside the test chamber are 160 kPa (abs) and 24 ± 2 °C, respectively. Demineralised water is used in order to ensure high repeatability of test conditions and to avoid unwanted depositions at the chamber walls and within

Fig. 1 Test disk with inlaid ceramic samples in an open cavitation chamber (left). 3D view of the test facility—a "photo" taken during an *Autodesk Inventor*® session (right)

the rest of the test loop. As significant amount of heat is generated in course of each test, the test rig water system comprises an open cooling loop with an efficient heat exchanger and cooling water taken directly from the municipal water network. Pressure control is possible thanks to an air cushion in the stilling water tank. A *Labview®* platform based computerized control system takes care to keep both the temperature and pressure at nearly constant level. A 3D overall view of the facility is shown in Fig. 1b.

The opportunity to test simultaneously 8 specimens under heavy cavitation load conditions is a true advantage in case of the need to select material or protective coating of high cavitation resistance out of many options available. In case of more detailed studies (various load conditions) cavitation tunnel tests are generally preferred. An option under consideration is replacing the currently used rotating disk test chamber by that of another design and implementing the rotation speed control.

The cumulative mass loss (average values) curves of TiN, CrN, and CrC coatings as well as the Cr18Ni10Ti and Armco iron reference samples have been determined as a result of 6 h long testing at the facility. As it can be seen from Fig. 2 this test duration was sufficient for substantial damage of protective coatings with erosion penetrating locally deep into the substrate material. At the same differences in the erosion mechanism can be noticed as TiN coating shows more brittle performance whereas undulations typical for higher ductility and/or poorer adhesion can be noticed in case of the CrC coating in the early stage of erosion.

Weight loss measurements were performed after each exposure period. Before the test and after each test interval, the samples were cleaned, dried and re-weighed. At the beginning of the cavitation test, measurements were carried out every 15 min of exposure period (during the first 60 min of the test), then the intervals were increased to 30 and 60 min. Afterwards cavitation erosion damage was analyzed using an optical and scanning electron microscope (SEM) and 3D X-ray micro tomography (*Phoenix* v/tome/x m, *GE Inspection Technologie sGmbH*, Germany). The device was equipped with both a 180 kV/15 W high-power nanofocus X-ray tube and a 240 kV/320 W micro focus tube. Scan data were reconstructed by the implementation of a graphics processing unit (GPU).

4 Results and Discussion

The surface morphology and composition of the coated samples are presented in Fig. 3. The characteristic peaks of Cr, Ti, N, and C correspond to the peaks of the main coatings elements. The diffraction patterns of both nitride (TiN, CrN) and carbide CrC coatings are shown in Fig. 4.

Only one low-intensity diffraction line, which corresponds to chromium with the lattice parameter a = 2.863 Å (less than the values known from literature, which may be due to the macro-stress presence) was observed in the CrC coated sample (Fig. 4a). The size of the CSR of chromium is $D = 14.5$ nm as estimated according to the Scherrer formula. The diffraction pattern is revealed as a two diffraction halo in

Fig. 2 TiN (above) and CrC (below) coatings after 30 and 360 min of test duration

the angles regions $2\theta = 39 \div 46°$ and $2\theta = 72 \div 89°$. This fact indicates the presence of an X-ray amorphous phase in the sample, in addition to the crystalline phase (that is, either amorphous or the CSR size is less than 3 nm). It was established [14] that under ordinary deposition conditions, usually only one type of crystalline structure is formed - the NaCl type. With low carbon content, these materials are mainly single-phase structures, often containing carbide nanocrystallites. With an increase in the total carbon content in the coating, the evolution of the microstructure leads to the formation of an additional phase (amorphous carbon) and a nanocomposite microstructure [14].

It should be noted that some carbide films are X-ray amorphous, but a more thorough study using transmission electron microscopy can reveal the existence of nanosized carbide grains in an amorphous matrix. Magnusson et al. [33] showed that $Cr_{1-x}C_x$ films are amorphous in a wide range of compositions, but at the same time at $x \approx 0.47$, using high-resolution transmission electron microscopy, nanocrystalline

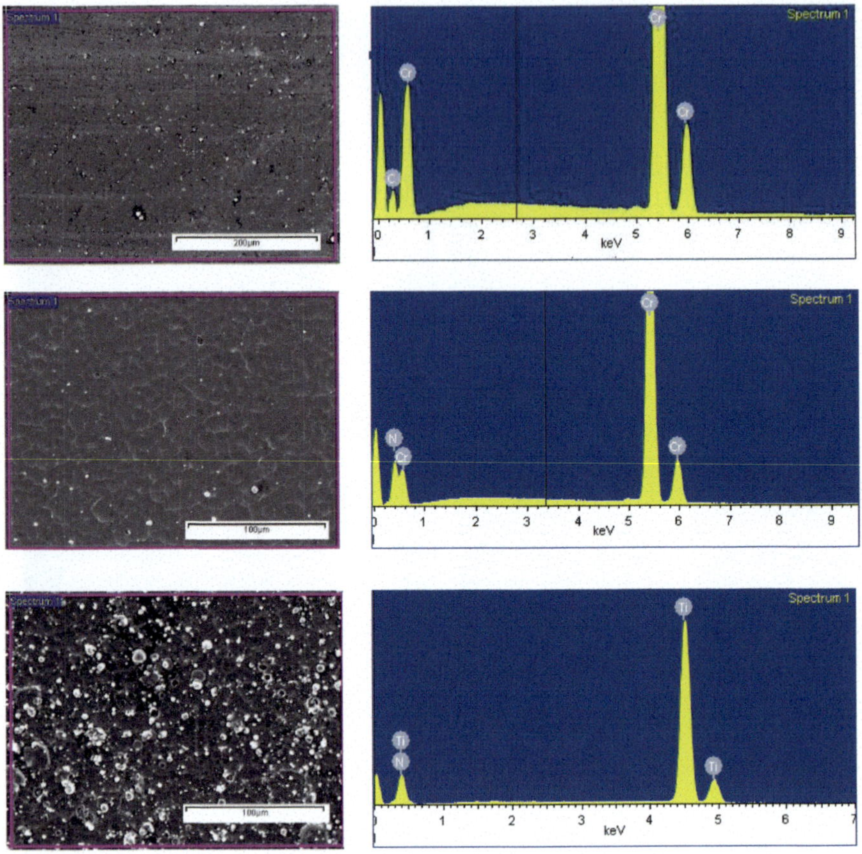

Fig. 3 Results of SEM-EDS surface chemical composition analysis of CrC, CrN and TiN protective coatings as ordered from top to bottom

grains with a diameter of < 3 nm can be detected. An additional analysis of electron micro-diffraction patterns has shown that such nanocrystallites most likely have a cubic CrC structure.

The CrN coating (Fig. 4b) is practically a single-phase one; chromium nitride CrN (cubic syngonium, structural type NaCl) can be easily detected from the diffractogram. The additional line to be identified in the diffractogram probably refers to hexagonal Cr_2N nitride. The intensity distribution of the diffraction lines of CrN nitride indicates the presence of a strong texture phase (220) with a lattice parameter $a = 4.200$ Å. In addition, a slight shift towards large angles for some lines was detected. This fact indicates the presence of macro-stresses in the coating. The CSR size in CrN nitride is $D = 17.3$ nm.

Two phases were detected in the TiN sample (Fig. 4c): titanium nitride TiN (structural type NaCl) and titanium Ti-α (small amount). The weight content of the main phase of the TiN in the sample is 95.1% wt, and the lattice parameter is $a = 4.257$

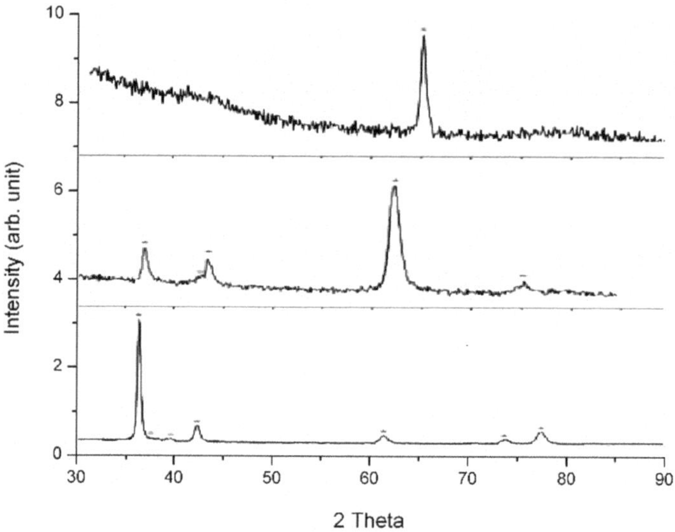

Fig. 4 Diffraction patterns of coatings **a** Cr-C, **b** CrN, **c** TiN

Å. The intensity distribution of the diffraction lines of the nitride indicates the presence of the (111) texture. The significant part of the phase consists of the grains mainly oriented by the crystallographic {111} planes parallel to the sample plane. The calculation of the substructural characteristics showed that the size of the CSR in TiN nitride is $D = 36.9$ nm. The content of Ti-α phase in the sample was about 4.9% wt. The lattice parameters of Ti-α phase were: $a = 2.97$ Å; $c = 4.73$ Å, respectively.

It should be noted that the lattice parameters of titanium are greater than the corresponding literature values. It can be assumed that this phase is a solid solution of nitrogen in titanium.

The cumulative mass loss curves of TiN, CrN, and CrC coated samples, as well as those of Cr18Ni10Ti stainless steel and Armco iron uncoated reference samples are presented in Fig. 5.

Typical damage process in materials subjected to highly concentrated mechanical load due to impact of solid or liquid particles and shock waves or micro-jets generated by collapsing cavitation bubbles can be divided into some characteristic periods:

- incubation period, in which internal stresses are cumulated in the surface layer without visible volume losses;
- acceleration period resulting from exceeding the fatigue strength limit at a large surface area;
- steady state damage period featured by a balance of energy cumulated and lost with removed material particles [30].

The course of the cumulative weight loss curves (Fig. 5) indicates that all studied coatings are characterized by a rather low cavitation wear intensity.

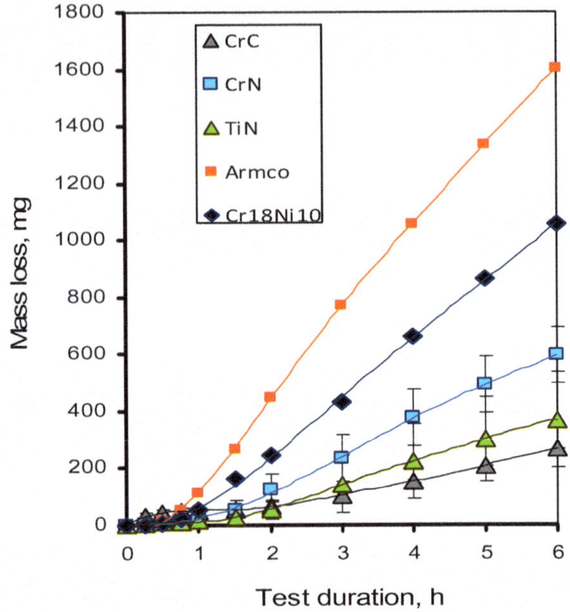

Fig. 5 Cavitation wear kinetic curves of TiN, CrN, and CrC coated samples as well as the uncoated Cr18Ni10Ti stainless steel and Armco iron reference ones

The highest erosion resistance has been shown by the CrC coatings. The CrC coatings demonstrate the lowest wear, 5 times less intense than the wear of the uncoated Cr18Ni10Ti steel. It should be noted that the sample testing lasted up to 6 h, and the maximum depth of erosion defects in the films reached 5…6 mm, i.e. significantly surpassing the thickness of the coatings.

This fact was clearly documented by means of 3D the X-ray micro tomography method. The time interval for the formation of the defects shown in Fig. 6 was about 360 min. Both the nature and area of the eroded surfaces, as well as the rate of cavitation wear are determined by the deposited coatings. The coatings significantly reduce the surface erosion area, and also slow down the rate of erosion propagation into the substrate material.

The erosion of the reference stainless steel sample surface was more extensive than that of the coated samples. In contrary to the coatings, the Armco iron and the stainless steel sample surfaces show developed damage combining plastic deformation and uniform surface erosion. Deep craters are found here after 30 min of cavitation testing. The degree of damage is slightly lower at the CrN and TiN coated sample as compared to the stainless steel one. The minimal damage was detected in the case of CrC coated samples.

Results demonstrate the principal possibility to decrease the cavitation erosion rate by using the nanostructured coatings. After 6 h of test, mass losses of the CrC coated Cr18Ni10Ti samples appeared 4 times smaller than those of the uncoated Cr18Ni10Ti sample and decisively smaller than those of the CrN and TiN coated samples.

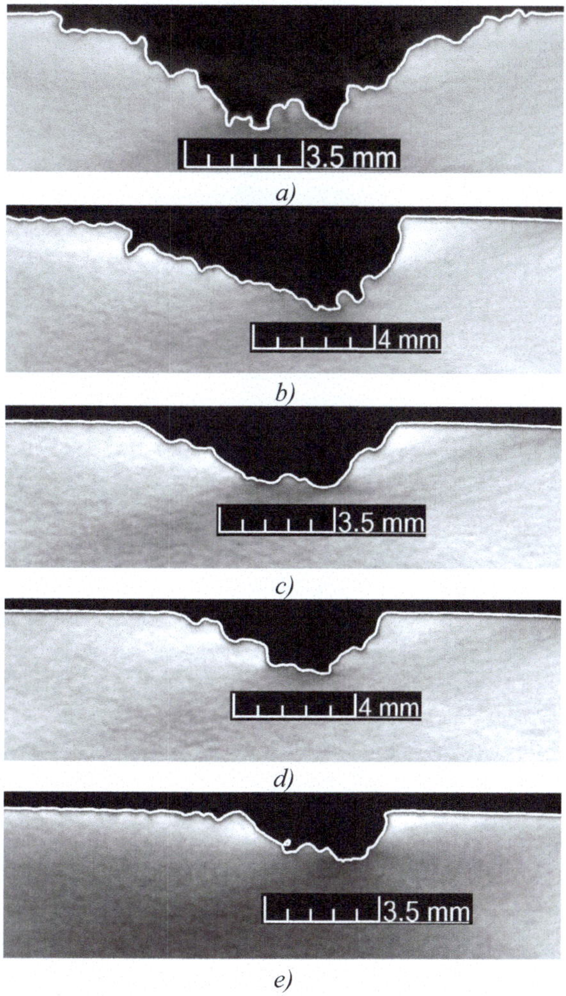

Fig. 6 Samples cross-section after 6 h of cavitation tests. **a** Armco iron, **b** uncoated Cr18Ni10Ti, **c** CrN, **d** TiN and **e** CrC coatings

The presence of coatings on stainless steel slows down further deformation process. However, as erosion developed, the material was removed due to brittle fracture of the coating at the edge of the formed cavity. Brittle fracture is the dominant erosion mechanism of the TiN, CrN, and CrC coated samples.

The CrC coated sample showed less surface damage as compared to the CrN and TiN coated ones (Fig. 6). The presence of micro-droplets at the CrC sample surface also affects the coating properties, but to a lesser extent than in the case of TiN and CrN. Nevertheless, both nitride TiN, CrN, and carbide CrC coatings significantly improved erosion resistance of stainless steel samples.

Figure 7 shows the erosion rate dependence on the exposure time. In case of the samples covered with TiN, and especially CrC coatings, measurable mass loss could

Fig. 7 Erosion rate dependence on exposure duration

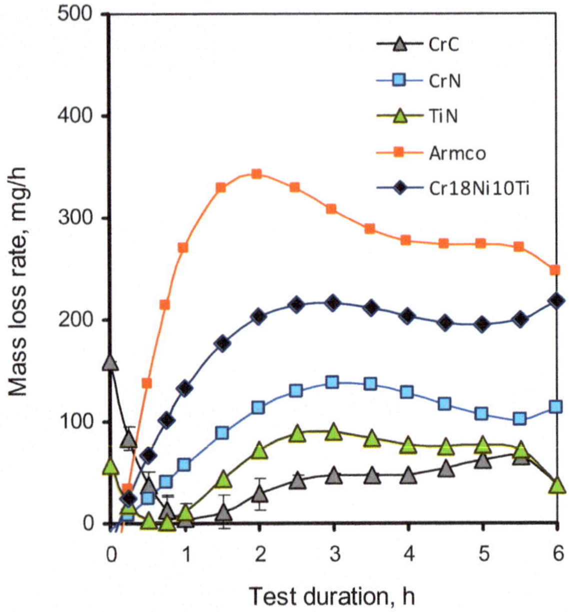

have been observed already in the very beginning of the cavitation test. This short period of intense mass loss (15–30 min) was followed by further damage incubation lasting up to 1 h in case of the TiN coating and about 2 h in case of the CrC one. Such a sequence of events is generally attributed to the presence of weak points at the impinged surface of the otherwise erosion resistant material. The mentioned weak points can be usually identified with micro droplets of condensed substrate vapour. The micro droplet formation is the principal disadvantage of the Arc PVD method. Their number generally rises with the coating thickness rise.

At the beginning of the erosion process of TiN, CrN, and CrC coated samples, weakly bound micro drops were primarily removed from the coating surface. After removal of micro drops, the mass losses demonstrated negligible values until the end of the erosion incubation period. Of course, the phenomenon of erosion incubation meant as cumulation of internal stresses and structural defects in the subsurface layer at "healthy" surface portion continued from the very beginning of the exposure, irrespective of the initially intensified mass loss at the weak sites. Generally, both the nitride (TiN, CrN) and the carbide CrC coated samples demonstrated a lower rate of weight loss than the uncoated stainless steel reference samples.

As the erosion rate rise at the end of the incubation period may follow different patterns, the precise value of the incubation period duration is a matter of stipulation. In addition to the traditional nominal incubation period definition, the ASTM G 32 updates of the passing decade [34] use also the terms of test duration up to a certain erosion depth. Nevertheless, the nominal incubation period parameter seems to be especially well suited for assessing protective coating erosion resistance as

Table 2 Nominal incubation time τ_{inc} and time to reach maximum erosion rate t_{max} of tested coatings and reference materials

Coating/material	CrC	CrN	TiN	Armco	Cr18Ni10Ti
τ_{inc}, min	109*	74	86	43	56
t_{max}, min	210	180	165	120	165

*Calculated with disregard of the initial erosion due to micro droplet presence

it usually represents the time needed to destroy a significant portion of the coating and the initiation of intense damage of the protected material. As Table 2 shows the nominal incubation period duration of the uncoated stainless steel samples was less than 56 min, such period for the TiN coated samples lasted less than 86 min and for CrN coated samples −74 min. The analogous period for CrC coated samples demonstrated the largest value up to 109 min. The second row in Table 2 shows the time to reach the maximum erosion rate of the tested coatings and reference materials. Again, the CrC coating shows the most favourable results.

The fracture images and morphology of uncoated stainless steel as well as the TiN, CrN and CrC coated samples were observed by means of the SEM technique. Both the nitride TiN, CrN and the carbide CrC coated samples are characterized by fine crystalline microstructure. As a result of cavitation, the sample surfaces are randomly degraded due to the impact of micro-jets or shock pressure waves [35]. SEM observations have revealed the damage character resulting from cavitation tests (Fig. 8). The cyclic load caused by exposure to cavitation impingement has led to deformation of the substrate, and occurrence of microcracks in the coatings. In addition, cracks were formed as a result of the post-micro drops surface failure. The coated samples were plastically deformed, and cracks appeared mainly on top of the bumps.

With present study, the correlation between the evolution of mechanical properties of nitride and carbide coatings and their weight losses was investigated. The differences in the chemical composition affect the mechanical properties of coatings, as well as the coating/substrate system performance and, finally, cavitation erosion resistance of tested materials.

Young's modulus E and hardness H parameters of coatings exceed those estimated for the reference samples. The hardness of stainless steel Cr18Ni10Ti was 3.7 GPa, Armco iron—1.36 GPa, Young's modulus—190 GPa, and 156 GPa, respectively.

The CrC coatings demonstrate improved elastic behavior and high resistance to plastic deformation (Fig. 9).

The H/E and H^3/E^2 coefficients of carbide coatings were evaluated as higher than those of chromium and titanium nitrides. The effect of chemical composition and mechanical properties on the cavitation erosion resistance of coated materials has been confirmed. High coatings thickness and higher H/E, H^3/E^2 ratios are apparently preferential for the cavitation erosion resistance of coated materials. The increase in the coatings thickness and the aforementioned mechanical parameters result in increasing of cavitation resistance of the coating/substrate material systems. The

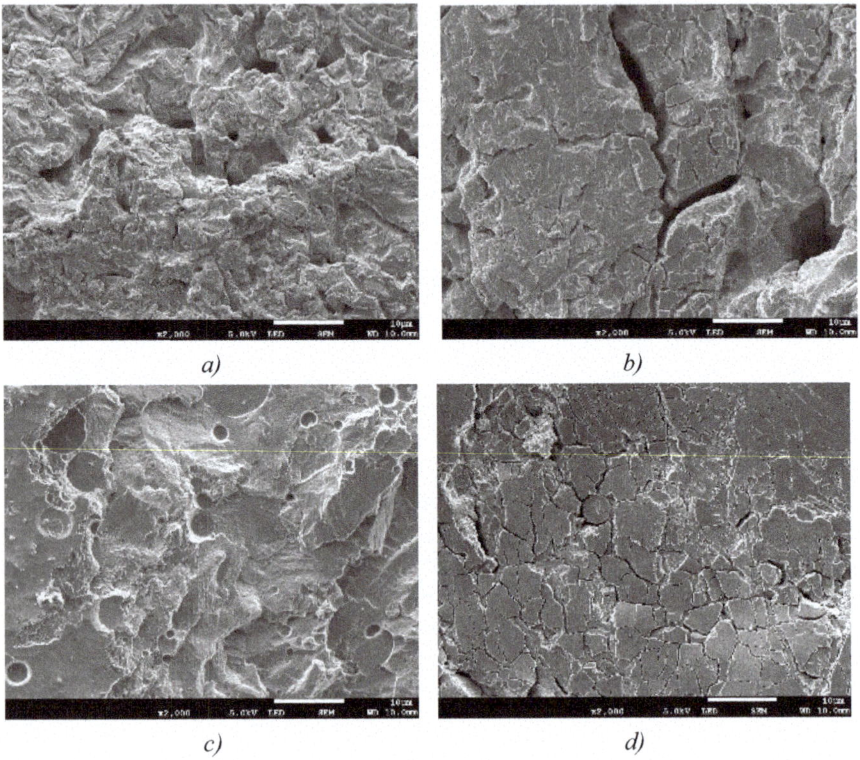

Fig. 8 SEM images of samples surface after 6 h of cavitation exposure. **a** uncoated stainless steel Cr18Ni10Ti, **b** CrN, **c** TiN, and **d** CrC coatings

Fig. 9 Mass losses after 6 h of specimen exposure versus H/E and H^3/E^2 ratios

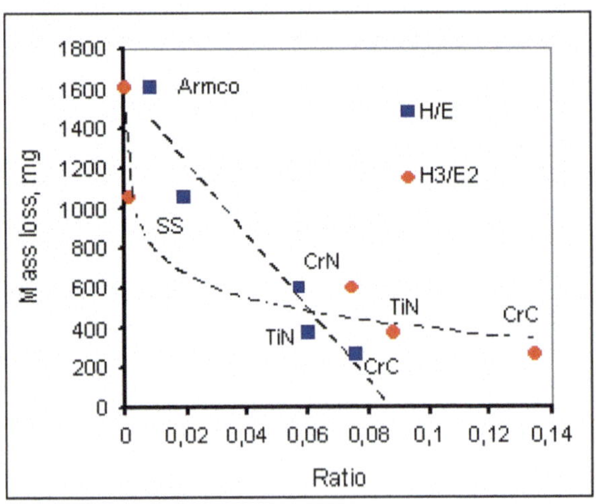

obtained results demonstrate that cavitation erosion resistance of various turbomachinery components made of stainless steel can be significantly improved by both nitride and carbide coatings deposition.

5 Conclusion

With present study a correlation between some mechanical properties and cavitation erosion resistance of the TiN, CrN, and CrC coatings with thicknesses up to 20 μm, deposited on Cr18Ni10Ti steel by Arc-PVD method, has been shown. Kinetic curves of cavitation wear of TiN, CrN, and CrC coatings as well as the stainless steel reference samples have been derived. It has been confirmed that cavitation resistance is strongly related to the coatings type. Results show that all studied coatings are characterized by cavitation wear lower than that of the uncoated samples. Both nitride (TiN, CrN) and carbide CrC coatings significantly have improved the erosion resistance of stainless steel samples. The results confirm that cavitation erosion of elements made of stainless steel can be prevented by nitride TiN, CrN and carbide CrC coatings deposition.

Furthermore, the CrC coated samples have shown improved elastic behavior and high resistance to plastic deformation as compared to the CrN and TiN coated ones. Higher H/E and H^3/E^2 coefficients have been stated in the case of carbide coatings than in the case of nitride ones and reference samples. The effect of the coatings composition and mechanical properties on the damage mechanism and material weight losses has been confirmed. Thus, the CrC thick coatings with their ability to form amorphous and nanocrystalline structures and higher mechanical H/E, H3/E2 ratios are promising candidates for various erosion protective applications.

Acknowledgements The present study has been conducted under a collaboration project between the Polish Academy of Sciences and the National Academy of Sciences of Ukraine (#46: Enhancement of turbomachinery blade erosion resistance by means of carbide and nitride protective coatings, 2018–2020). The authors of this paper find it their pleasure to thank Professor Alicja Krella for her SEM analysis and valuable consultations in the initial stage of research. Thanks for conducting the rotating disk tests are due to Mr P. Żółtko.

References

1. Heymann FJ (1969) High-speed impact between a liquid drop and a solid surface. J Appl Phys 40:5113–5122
2. Hamed A, Tabakoff W, Wenglarz R (2006) Erosion and deposition in turbomachinery. J Propul Power 22(2):350–360
3. Szala M, Walczak M, Pasierbiewicz K, Kamiński M (2019) Cavitation Erosion and Sliding Wear Mechanisms of AlTiN and TiAlN Films Deposited on Stainless Steel Substrate. Coatings 9:340–350

4. Wu Y, Hong S, Zhang J, He Z, Guo W, Wang Q, Li G (2012) Microstructure and cavitation erosion behavior of WC–Co–Cr coating on 1Cr18Ni9Ti stainless steel by HVOF thermal spraying. Int J Refract Metal Hard Metal 32:21–26
5. Maurer C, Schulz U (2013) Erosion resistant titanium based PVD coatings on CFRP. Wear 302:937–945
6. Azar GT, Yelkarasi C, Urgen M (2017) The role of droplets on the cavitation erosion damage of TiN coatings produced with cathodic arc physical vapor deposition. Surf Coat Technol 322:211–217
7. Borawski B, Todd JA, Singh J, Wolfe DE (2011) The influence of ductile interlayer material on the particle erosion resistance of multilayered TiN based coatings. Wear 271(11–12):2890–2898
8. Krella A (2009) The influence of TiN coatings properties on cavitation erosion resistance. Surf Coat Technol 356:263–270
9. Yang Q, Seo DY, Zhao LR, Zeng XT (2004) Erosion resistance performance of magnetron sputtering deposited TiAlN coatings. Surf Coat Technol 188–189:168–173
10. Branco JRT, Gansert R, Sampath S, Berndt CC, Herman H (2002) Solid particle erosion of plasma sprayed ceramic coatings. Mater Res 7(1):147–153
11. Cai F, Huang X, Yang Q, Wei R, Nagy D (2010) Microstructure and tribological properties of CrN and CrSiCN coatings. Surf Coat Technol 205:182–188
12. Cai F, Huang X, Yang Q (2015) Mechanical properties, sliding wear and solid particle erosion behaviors of plasma enhanced magnetron sputtering CrSiCN coating systems. Wear 324–325:27–35
13. Deng J, Wu F, Lian Y, Xing Y, Li S (2012) Erosion wear of CrN, TiN, CrAlN and TiAlN nitride coatings. Int J Refract Metal Hard Metal 35:10–16
14. Jansson U, Lewin E (2013) Sputter deposition of transition-metal carbide films: A critical review from a chemical perspective. Thin Solid Films 536:1
15. Krella A (2013) Cavitation erosion of TiN and CrN coatings deposited on different substrates. Wear 297:992–997
16. Sebastiani M, Piccoli M, Bemporad E (2013) Effect of micro-droplets on the local residual stress field in CAE-PVD thin coatings. Surf Coat Technol 215:407–412
17. Tkadletz M, Mitterer C, Sartory B, Letofsky-Papst I, Czettl C, Michotte C (2014) The effect of droplets in arc evaporated TiAlTaN hard coatings on the wear behavior. Surf Coat Technol 257:95–101
18. Lee MK, Kim WW, Rhee CK, Lee WJ (1999) Liquid impact erosion mechanism and theoretical impact stress analysis in TiN-coated steam turbine blade materials. Metall Mater Trans 30A:961–968
19. Safonov V, Zykova A, Krella A, Steller J, Gajowiec G, Szkodo M, Ghaemi H (2016) Cavitation erosion resistance of thick CrC coating deposition by the Arc-PVD method. In: Polish hydropower conference RENEXPO, Warsaw, pp 52–54
20. Musil J, Kunc F, Zeman H, Poláková H (2002) Relationships between hardness, Young's modulus and elastic recovery in hard nanocomposite coatings. Surf Coat Technol 154:304–313
21. Leyland A, Matthews A (2000) On the significance of the H/E ratio in wear control: A nanocomposite coating approach to optimised tribological behaviour. Wear 246:1–11
22. Leyland A, Matthews A (2004) Design criteria for wear-resistant nanostructured and glassy metal coatings. Surf Coat Technol 177–178:317–324
23. Krella A (2011) The new parameter to assess cavitation erosion resistance of hard PVD coatings. Eng Fail Anal 18:855–867
24. Krella A (2012) Degradation of the *nanocrystallite coating/ steel substrate* systems under cavitation damage conditions. Zeszyty Naukowe IMP PAN 556/1515/2012; D.Sc. Thesis (in Polish)
25. Hammitt, F.G.: Cavitation and Multiphase Flow Phenomena. Mc Graw-Hill Inc. (1980)
26. Hattori S, Takinami M (2010) Comparison of cavitation erosion rate with liquid impingement erosion rate. Wear 269:310–316

27. Hsiao ChT, Jayaprakash A, Choi JK, Chahine GL (2014) Modeling of material pitting from cavitation bubble collapse. J Fluid Mech 755:142–175
28. Ma D, Harvey TJ, Wellman RG, Ehiasarian AP, Hovsepian PEh, Sugumaran AA, Purandare YP, Wood RJK (2019) Cavitation erosion performance of CrAlYN/CrN nanoscale multilayer coatings deposited on Ti6Al4V by HIPIMS. J Alloy Compd 788:719–728
29. Steller J (1999) International cavitation erosion test and quantitative assessment of material resistance to cavitation. Wear 233–235:51–64
30. Steller, J., Gireń, B.G.: International Cavitation Erosion Test. Final Report. Zeszyty Naukowe IMP PAN 560/1519/2015 (2016), https://doi.org/10.13140/rg.2.1.4330.8883
31. Steller J (2018) The concept of cavitation resistance of materials by means of a fractional approach. Zeszyty Naukowe IMP PAN 561/1520/2015; D.Sc. Thesis (in Polish)
32. Safonov V (2017) Plasma activated EB-deposition: different modes of arc discharge and plasma characteristics. Probl At Sci Technol 5:65–71
33. Magnuson M, Andersson M, Lu J, Hultman L, Jansson U (2012) Electronic structure and chemical. Bonding of amorphous chromium carbide thin films. J Phys Condens Matter 24:225004
34. ASTM G32-16 (2016) Standard test method for cavitation erosion using vibratory apparatus
35. Mesa DH, Garzón CM, Tschiptschin AP (2011) Influence of cold-work on the cavitation erosion resistance and on the damage mechanisms in high-nitrogen austenitic stainless steels. Wear 271:1372–1377

Numerical and Experimental Investigation of the Thermal Propagation Inside the Carbon Fiber Composites

Jan Novosád and Norbert Pomp

Abstract The infra-red non-destructive testing is a contactless method suitable for thin shells. The method is based on the non-uniform thermal excitement propagation through the materials of different thermal characteristics, as caverns, delamination or inserts. The knowledge of the thermal impulse propagation through the material velocity could be very helpful for the phase shift diagrams evaluations. Some investigations on problems of determination of the thermal wave velocity are described on this article. Several samples was made of the carbon fiber fabric saturated with polymer resin by vacuum infusion. The thickness of samples was various. The thermal propagation was measured on these samples. The heat load was applied by two methods, using the halogen lamp or a warm object of high thermal capacity and conductivity. The time delay of thermal wave was evaluated using thermograms. The velocity of propagation is evaluated as the ratio of time delay to thickness. Then the numerical simulation of heat conduction inside the samples was performed using similar conditions as during the experiments. ANSYS software have been used for numerical simulation. Finally the experimental and numerical results were compared. The limitations and usage of all the methods have been discussed.

Keywords Non-destructive testing · Thermography · Thermal propagation · Thermal simulation · Carbon fiber

1 Introduction

1.1 Infrared Non-destructive Testing

As the computation methods and modern materials production abilities grows, products are more and more precisely designed and the failures are more probable. The non-destruction testing and controlling is thus increasingly important.

J. Novosád (✉) · N. Pomp
VÚTS, a.s., Svárovská, 619 Liberec, Czech Republic
e-mail: jan.novosad@vuts.cz

M. Abdel Wahab (ed.), *Proceedings of 1st International Conference on Structural Damage Modelling and Assessment*, Lecture Notes in Civil Engineering 110,
https://doi.org/10.1007/978-981-15-9121-1_31

The infra-red non-destructive testing is contactless method suitable for (not only) large-area thin-walled shells. The method is based on the non-uniform thermal excitement propagation throw the materials of different thermal characteristics, as caverns, delaminations or inserts [1]. The publications of Cernuschi et al. [2] and Pieczonka et al. [3] describe the possibilities of different measurement techniques for damage detection. Additionally, Mayr et al. [4] dealt with the carbon fiber polymers.

While testing, one surface of tested object is periodically exciting by infrared emitter (there are much more ways of exciting, let's discuss the here mentioned only) and the temperature distribution on the same surface is measured by infrared camera at the same time. Then the exciting and the surface temperature Fourier transformation is compared. It can be profitable to detect the material irregularity on phase shift picture, see, as the phase shift is more independent to the absolute values of exiting.

Part of our company research is aimed to the investigation of fiber composites properties [5]. Following this, the internal research of the IR NDT abilities was made. The obtained diagrams evaluation was find to be not elementary. The several areas of different phase shift can be seen in Fig. 1. The phase difference, compared with the surrounding material, is positive for all of them. Despite of this the character and also the distance from the surface of these failures is different. The circular areas are various diameter aluminium inserts and the longish ones are delaminations. To recognize the irregularities character can be tricky apart from the irregularity depth [6].

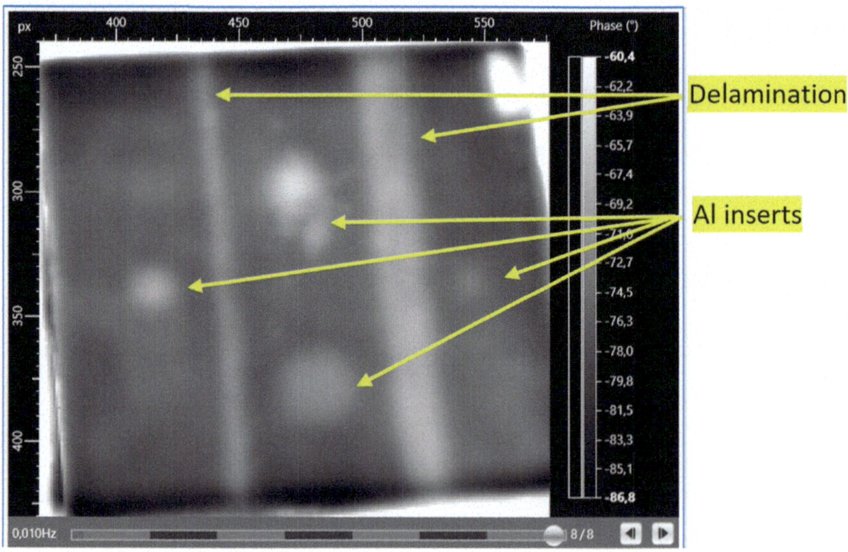

Fig. 1 The phase shift diagram with the material irregularities depicted

The knowledge of the thermal impulse propagation through the material velocity could be very helpful for the phase shift diagrams evaluations. Some investigations on problems of determination of the thermal wave velocity are described on this article.

1.2 Conduction Heat Transfer and Thermal Waves

The heat transfer inside the solid bulk could be described as a conduction problem only. These topic is well described as a fundamentals of heat transfer, e.g. by Lienhard [7]. In general the solution of temperature field is based on the Fourier's law:

$$\dot{q} = -\lambda \nabla T, \tag{1}$$

where q ($W{\cdot}m^{-2}$) is the heat flux density, ($W{\cdot}m^{-1}{\cdot}K^{-1}$) is the thermal conductivity and ∇T ($K{\cdot}m^{-1}$) is temperature gradient. Hence the time dependent variation of temperature and the anisotropic material properties are not covered by the fundamental theory. Based on Eq. (1) the heat conduction equation in three dimensions [7] was derived as

$$\rho c_p \frac{dt}{d\tau} - \nabla \cdot \lambda \nabla T = \dot{q} \tag{2}$$

where ρ ($kg{\cdot}m^{-3}$) is the material density and c_p ($J{\cdot}kg^{-1}{\cdot}K^{-1}$) the specific heat. Both properties are considered as constant for bulk material. Anisotropy of thermal conductivity can have a significant effect on the solution, especially for composite materials. Therefore the substitution of Laplacian $\nabla{\cdot}\nabla T$ have to be applied, e.g. for Cartesian coordinate system x, y, z by equation:

$$\nabla \cdot \lambda \nabla T = \frac{\partial}{\partial x}\left(\lambda_x \frac{\partial T}{\partial x}\right) + \frac{\partial}{\partial y}\left(\lambda_y \frac{\partial T}{\partial y}\right) + \frac{\partial}{\partial z}\left(\lambda_z \frac{\partial T}{\partial z}\right) \tag{3}$$

where λ_x, λ_x, λ_x are the directional thermal conductivity. Following these rules the composite material could be described as anisotropic. The anisotropy is caused by the presence of various materials with different thermal conductivity values, e.g. fibers and filler. In addition, the change in thermal conductivity may indicate changes in structure, i.e. different kind of material, cracks, cavities, etc.

Experimental determination of anisotropic thermal conductivity can be accomplished with commercially available devices, e.g. C-Therm's TCi thermal conductivity analyzer, a heat flow meter like HFM 446 Lambda etc. The testing sample preparation is required, the minimum and recommended sizes and thickness of the sample are also defined. Thus the troubles in manufacturing of that samples could be a limitation for use of the measurement of thermal conductivity. Furthermore, it is important to note that the thermal conductivity varies with temperature.

In general, the IR NDT is based on the thermal wave theory. Contrary, Salazar [8] had published the study of heat conduction in solid samples that are excited by periodic heat source. He had shown that the resulting temperature oscillations inside the sample have the same mathematical expression as a thermal wave, but as they do not transport energy they cannot be considered as real travelling waves. In conclusion, Salazar formulated the argument that, taking into account the time delay between temperature gradients and heat flow, temperature oscillations do not transmit energy, although they are a solution to the wave equation. Nevertheless, in our work we accept this fact, but the term thermal wave is still used.

2 Tested Samples

IR NDT is suitable for fiber composite shells well. So four carbon fiber composites sheets of different thickness were made internally. The samples were flat of dimensions about 15 × 15 cm. The sample final thickness was obtained by secondary bonding of the several layers of one millimeter thick semiproduct within the high pressure. The semiproduct consist of four layers the biaxial carbon fiber fabric type KC Industry 200 tex [9] saturated with polymer resin LH385 [10] with hardener H286 [11] using the Vacuum Infusion Process. The fabric specific weight was 200 g/m^2, the polymer resin specific weight was 1200 kg/m^3. Physical properties are listed in Table 1.

The fiber-to-resin volume ratio was computed from the measured samples weight, materials specific weight and volume, when the volume was assessed on the basis of sample dimension measurement. The gauged fiber-to-resin volume ratios are listed in the Table 1. The differences in the fiber-to-resin ratios can be explained by the semiproduct unevenness. The secondary bonded samples hold some extra resin layers.

Table 1 Properties of carbon fiber fabric and polymer resin [12]

Material	Density (kg·m^{-3})	Specific heat (J·kg^{-1}·K^{-1})	Thermal conductivity s(W·m^{-1}·K^{-1})
Carbon	1750	800	200
Polymer	1200	1000	0.2

3 Thermal Wave Velocity Measurement

The thermal wave velocity measuring methods were designed to be able to be carried out on site with equipment used for IR NDT. Both methods were based on the time delay between the heating start and temperature arise measurement. The sample was heated on one its side and the temperature was measured on the opposite side of the flat sample. The temperature course was measured by infrared camera which detected the heating start as well.

The halogen lamp as the heating source was applied in the first measurement case, see Fig. 2. The infrared camera was covered in the closed box to avoid the reflected temperature influence. The cutout little bit smaller then sample dimension was made on the box wall and this cutout was fully enveloped by measured sample. Near the sample (within the camera visual angle) the small hole was carried out. The hole allowed to detect heating start and had no effect to recording thermograms. The experiments were carried out with three heating intensity levels: 25, 50 and 75% of full lamp input. The absolute values were not to care, relative comparison was concern in only. Heating as well as the measuring distance was 0.5 m. The camera frame rate was 300 Hz [1].

The flexuous warm object of high thermal capacity and conductivity was set as the second heating source. The temperature of this object was about 30 °C and the temperature fall was compensated by the subject inner heating partly. The sample one side was heated and the opposite side was observed by infrared camera again. As the heating start was considered the time when the sample slightly moved due to pressing by heating object. The camera frame rate was 90 Hz only because the window reduction could not be used and the camera could not achieve so high frame frequency.

The measurement were carried out on the laboratory where the conditions were constant. Particularly it was attended to avoid the reflected temperature influence.

The thermal wave velocity is evaluated as the time delay of temperature arise. When the every energy loss are neglected and the heating sources power is constant the temperature increase is constant too. The infrared heating suits this assumption

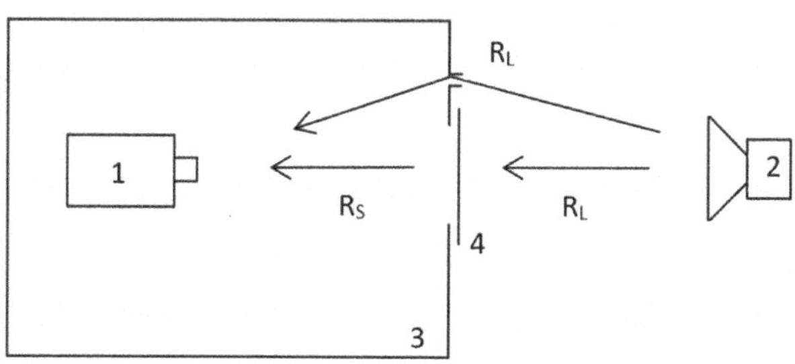

1 - IR camera, 2 - Halogen lamp, 3 – Closed box, 4 - Measured sample.

Fig. 2 The thermal propagation velocity measurement scheme

Table 2 Particular samples fiber-to-resin volume ratios

Thickness (mm)	Sample weight (g)	Matrix weight (g)	Sample volume (mm^3)	Matrix volume (mm^3)	Matrix volume (%)
1	29,8	18	21998	10251	46.6
2	59,2	36	87538	20571	47.0
4	124,7	72	47181	41143	43.6
5	156,1	90	122159	51429	42.1

well and in the case of heating object the best to keep its constant temperature was done. The temperature arise first moment was considered to be the intersection of sample temperature linear regression and initial sample temperature level. In the case of IR heating the temperature linear development is relatively long, 3–5 s, in the case of contact heating significantly sorter (about 0.8 s).

The temperature arise delay for performed experiments is listed in Table 2. The sample thickness divided by found temperature arise delay is the thermal excitation velocity, see Table 3. The velocity data is shown in Fig. 2 also. The points 1, 2 and 3 on horizontal axis correspond to the IR heating of level 25, 50 and 75%, points 4 and 5 came of the contact heating experiments. Missing point on the 1 mm thick sample is not shown in the tables and graph because is out of graph range. The velocity values variances measured on the 1 mm thick sample are so high the contact heating method is pronounce invalid and one point is shown only.

Both methods seem to be very stable (except of 1 mm sample). When IR heating method and contact heating method are compared, the measurement results are almost equal for sample of thickness 4 and 5 mm. However the contact method provides the velocity values more than twice as high as IR method for the sample of thickness 2 mm (Table 4).

Table 3 The temperature arise delay for performed experiments

Thickness [mm]	IR heating			Contact heating	
	25%	50%	75%	M.No 1	M.No 2
1	0.69	0.45	0.20	0.25	
2	2.44	2.30	2.29	0.85	0.83
4	5.02	5.14	5.20	3.77	3.50
5	5.26	5.26	5.98	5.51	5.35

Table 4 Measured thermal excitation velocity

Thickness [mm]	IR heating			Contact heating	
	25%	50%	75%	M.No 1	M.No 2
1	1.51	2.33	5.25	4.36	
2	0.82	0.87	0.87	2.13	2.25
4	0.78	0.76	0.75	1.03	1.11
5	0.93	0.93	0.82	0.89	0.92

4 Heat Transfer Simulation

Numerical simulations of heat transfer are a supporting tool in our research. The main goal of these simulations is to add some parameters necessary for further setup of experiments, which would be tedious or unrealistic to obtain experimentally due to the large number of samples tested and their modifications.

4.1 *Model Idea*

The so-called thermal wave is the base of NDT. Thus the numerical simulation have to reach the time delay between temperature gradients and heat flow as a result. These results should be similar to the reality i.e. similar to experimental results. A simple composite plate is used as a testing sample following the experimental setup.

4.2 *Material Properties*

For composite materials it is important to determine the material properties correctly. The properties of pure components of the composite are well known. Then the density and specific heat are predicted as a weighted mean using the matrix volume. To predict the anisotropic thermal conductivity the electro–thermal analogy is used. The thermal resistance R_{Ti} ($K \cdot W^{-1}$) of each component i is calculated as

$$R_{Ti} = \frac{d_j}{\lambda_i \cdot A}, \tag{4}$$

where d_j (m) is the thickness of the component layer in the j direction, λ_i ($W \cdot m^{-1} \cdot K^{-1}$) is the thermal conductivity of component i and A (m^2) is the section area in the normal direction to j. After that, the whole bulk of each material could be represented by one thermal resistance. For example the composite made of the carbon fiber fabric saturated with polymer resin could be represented by only two thermal

Table 5 Properties of carbon fibre composite

Material	Density (kg·m^{-3})	Specific heat (J·kg^{-1}·K^{-1})	Thermal conductivity (W·m^{-1}·K^{-1})
Composite	1475	900	100 (along fibers) 0.4 (perpendicular to fibers)

resistance values. The thermal conductivity values are determined by connecting the thermal resistances in serial or parallel based on the composite structure in various directions. The property values of composite (50% of carbon fiber) are evaluated based on the properties of pure materials (see Table 1). The evaluated values are listed in Table 5.

4.3 Computational Model and Boundary Conditions

A computational model is based on the experimental setup. The sample is a plate 15 × 15 cm, the thickness of plate is various in the range {2, 3, 4 and 5} mm. Material properties have been set as the properties of composite listed in Table 5.

Boundary faces (see Fig. 3) were taken into account to have similar effects as during the experiment. *Side walls* were considered as insulated, therefore no heat could pass through. The heat load was defined on *wall-heated* surface by defined heat flux q as 25, 50 and 75% of nominal heat flux $q_0 = 1000$ W·m^{-2}. The opposite side (*wall_camera*) is cooled by natural convection of air. Convection is described by the heat transfer coefficient (HTC) $\alpha = 15$ W·m^{-2}·K^{-1} and the free stream temperature of air $t_\infty = 20$°C. The initial temperature for a whole bulk was set to $t_0 = 20$°C.

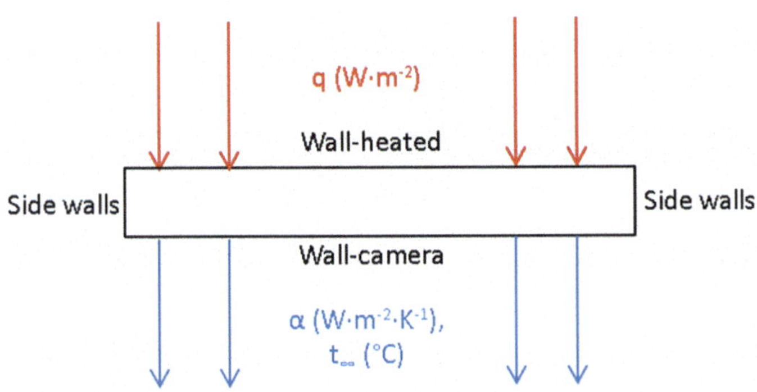

Fig. 3 Computational model—boundary conditions

4.4 Solver Setup

Computation have been realized using ANSYS Fluent software [13]. The second order discretization scheme was used for solving energy equation. The time step size have been set to be 0.5 s. Totally 200 time steps were used for each computation. All the combination of sample thickness and heat load create 12 cases, totally.

4.5 Results Assessment

The temperature was monitored at several surface points. The average temperature on *wall-heated* and *wall-camera* surface have been obtained at each time step. An example of temperature over time is shown in Fig. 4. The curves represent the temperature arise on both sides of the tested sample after the heat load was applied. The Assesment of a propagation velocity is based on the time delay required for temperature arise on the sample surface closer to camera. The linear regression of *wall-camera* temperature values was performed for a part of the significant temperature rise. The time delay was determined as the intersection of the linear regression curve with the horizontal axisvalue at $t_0 = 20$ °C. The sample thickness/delay ratio represents the thermal propagation velocity.

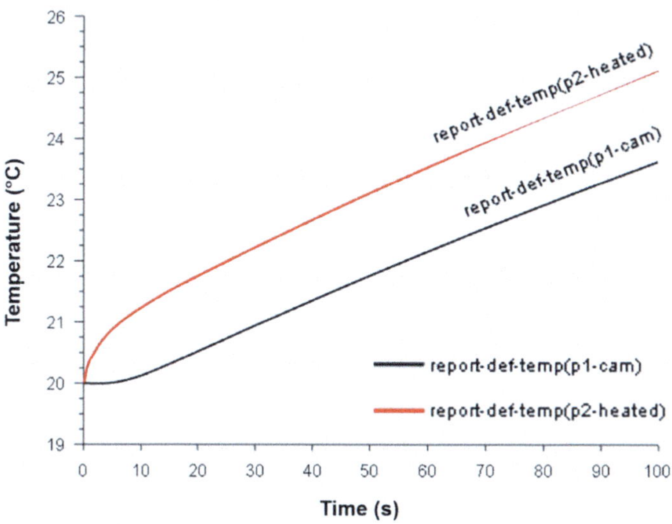

Fig. 4 Temperature over time—example: thickness 4 mm, 50% of q_0

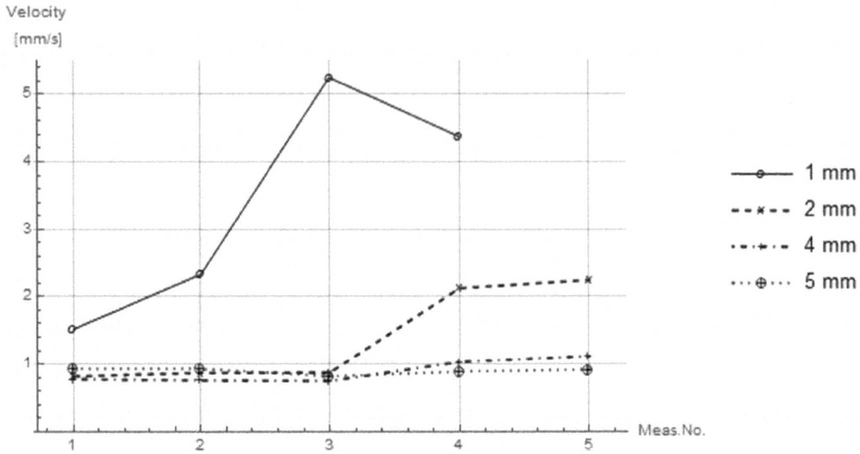

Fig. 5 Thermal propagation velocity—experimental results

4.6 *Results and Discussion*

The experimental part results are shown in Fig. 5. The points 1, 2 and 3 on horizontal axis correspond to the IR heating of level 25, 50 and 75%. Then points 4 and 5 were evaluated during the contact heating experiments. Curves in Fig. 5 show the significant difference between thin 1 mm sample and the other samples. The propagation velocity values are higher and the course of the curve is different for 1 mm sample. Other curves are very close together for the cases of heat load by halogen lamp. The propagation velocity could be considered as a constant with very small deviation (within less than 5%). However the heat load applied by the contact heating causes the increase of propagation velocity at measured area. Thus the high thermal capacity of contact aluminium solids together with inhomogeneous heat load are probably causing this increase. The increase is dominant for the 2 mm sample where the ratio of thermal capacities between composite sample and contact solid is very low. Then with increasing thickness, i.e. increasing the thermal capacities ratio, the values of propagation velocity are closer to the values obtained for experiments using the homogeneous heat load (by halogen lamp).

Figure 6 shows the evaluated values of velocity of the thermal propagation obtained from numerical simulation. Simulations were provided for four different thickness of samples {2, 3, 4 and 5} mm. Each sample was tested using a heat load at three levels {25, 50 and 75%} of the nominal heat load $q_0 = 1000\ \mathrm{W/m^2}$. All samples were set with same material properties that correspond to a 50% carbon fiber material (see Table 5). The propagation velocity values for different thickness and heat load regimes show similarity to the experimental data. The values could be considered as a constant again, the deviation is less than 3%. The absolute values are slightly higher than received from experiments. This could be caused by the

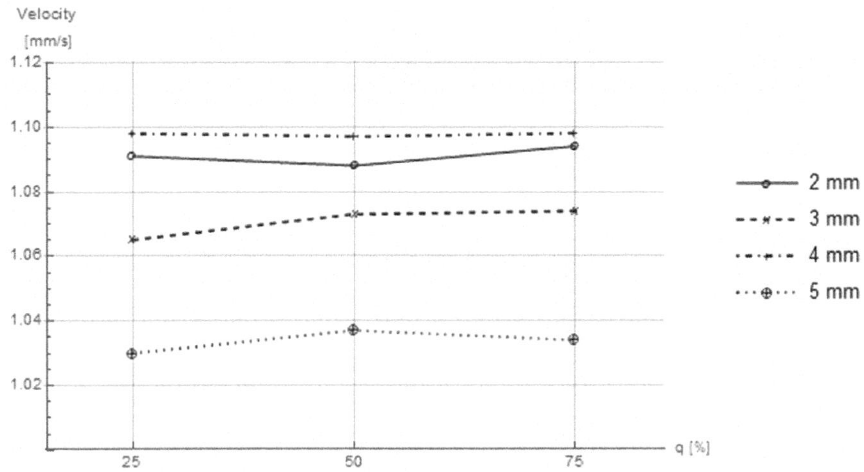

Fig. 6 Thermal propagation velocity—simulation results

simplification of the numerical model. To compare the experimental and numerical setup the material and thermal properties were set for 50% of matrix volume for the simulation. Otherwise the experimentally tested samples consists of smaller volume matrix. Consequently, larger amount of carbon causes the higher thermal conductivity which leads to higher thermal propagation velocity. Thus the experimental and numerical results correspond to each other.

5 Conclusion

Three methods which are focused to simplify the IR NDT parameters setting and its results interpretation are described and compared in the paper.

The experimental methods do not need any more equipment and results evaluation is fast. The method based on the infrared heating appears to be suitable for wider range of material thickness, while the contact heating method is much simpler in respect to the measurement set-up. The both sides of tested shell have to be accessible but infrared lamp installation on the inner part of tested shell is much complicated.

Heat transfer simulations allows to find the thermal wave velocity for wide range of the fiber-to-resin ratios in advance. The results of velocity are independent on the heat load, hence the solver setup seems to be usable for this problem. The standard deviation of velocity for different thickness is around 2%, the value could be therefore considered as a constant.

Comparing the measurement results to each other and with simulation results as well, very good agreement can be found for 4 and 5 mm thin samples. Computed values are slightly higher. The reason of the model and measurement differences is in the higher fiber-to-resin ratio taken into the model assumptions. The results of infrared heating and the computation fit to each other even in case of 2 mm thin samples. Contrary to this findings the result of contact measurement on 2 mm thin plate and mainly on 1 mm thin plate (not only contact method) are. It should be caused by time delay measurement uncertainty.

The future work should continue with validation of simulation. The simulation results could be used as a prediction of IR NDT settings for measurement of fiber composite shells.

Acknowledgements The publication was created with the support of the project LO1213—Excellent Engineering Research, the grantor - MŠMT, beneficiary—VÚTS, a.s., in support of the National Sustainability Program, with the financial contribution of the Ministry of Education, Youth and Sports of the Czech Republic.

References

1. Cernuschi F, Bison PG, Marinetti S, Figari A, Lorenzoni L, Grinzato E (2002) Comparison of thermal diffusivity measurement techniques, 6th international conference on QIRT, Dubrovnik, Croatia
2. Pieczonka L, Szwedo M, Uhl T (2013) Investigation of the effectiveness of different thermographic testing modalities in damage detection. Key Eng Mater 558:349–356
3. Mayr G, Plank B, Sekelja J, Hendorfer G (2011) Active thermography as a quantitative method for non-destructive evaluation of porous carbon fiber reinforced polymers. NDT & EInternational 44:537–543
4. Wolf W, Zissis G (1985) The infrared handbook. General dynamics
5. Zbončák R (2018) Elastic properties prediction models of continuous fibres composites. VÚTS, a.s. ISBN 978-80-87184-81-3
6. Maldague X, Moore P (2001) Nondestructive testing handbook, vol 3. American Society for nondestructive testing
7. Lienhard JH (2019) A heat transfer textbook. 5th edn. Mineola, Dover Publications, New York, ISBN 978-0486837352
8. Salazar A (2006) Energy propagation of thermal waves. Eur J Phys 27:1349–1355
9. Havel Composites CZ s.r.o.: KC Industry 200 tex (datasheet) [online]. [Cited 2020-05-10]. Available at: www.havel-composites.com
10. Havel Composites CZ s.r.o.: Epoxy resin LH 385 (datasheet) [online]. [Cited 2020-05-10]. Available at: www.havel-composites.com
11. Havel Composites CZ s.r.o.: Hardener 286 (MGS) (datasheet) [online]. [Cited 2020-05-10]. Available at: www.havel-composites.com
12. Gay D (2015) Composite materials, design and application, 3rd edn. CRC Press, Taylor & Francis, Boca Raton. ISBN 9781466584877
13. ASYS Fluent User's guide, Release 2020R1. Ansys Inc

Experimental Research Methods of Mechanical Behavior of Structural Materials Under Complex Thermomechanical Influences

Valery Wildemann

Abstract Improving the methods for solving the problems of structural strength and safety of critical structures and structures with the aim of predicting and mitigating the consequences of industrial accidents. This direction does not lose its relevance in the context of the constant development and complication of technical objects and the widespread use of modern constructional, including composite, materials. The questions of a comprehensive study of the properties of structural materials using modern testing systems and measuring instruments on the basis of the Center of Experimental Mechanics at the Perm National Research Polytechnic University, obtaining new data on the laws of the processes of deformation and fracture under complex unsteady and combined thermomechanical effects are considered. Some relevant areas of research on the properties of materials (metals and composites) related to the effects of damage accumulation, supercritical deformation, multi- and low-cycle fatigue under complex stress conditions, thermomechanical fatigue, crack resistance, and cyclic crack resistance are considered. Based on a set of new experimental data, it is possible to: develop non-linear models of continuum mechanics, study the possibilities of ensuring conditions for the stable accumulation of damage to materials in structures and the realization of their deformation reserves.

Keywords Structural materials · Experimental mechanics · Test systems · Non-Destructive testing methods · Deformation · Complex stress state · Fracture · Damage accumulation · Survivability · Safety

1 Introduction

The creation of modern and promising technical systems, especially in the aviation industry, requires an in-depth study of the laws and features of the mechanical behavior of structural materials in complex, close to operational, conditions of external influences. This creates the conditions for improving the technology of

V. Wildemann (✉)
Perm National Research Polytechnic University, 29 Komsomolsky Prospekt, Perm, Krai, Russia
e-mail: wildemann@pstu.ru

M. Abdel Wahab (ed.), *Proceedings of 1st International Conference on Structural Damage Modelling and Assessment*, Lecture Notes in Civil Engineering 110,
https://doi.org/10.1007/978-981-15-9121-1_32

designing and creating materials and products, developing mathematical models of their mechanical behavior and methods of strength analysis, including safety assessment, critical structures. In connection with these tasks, the development of the methodology and the conduct of experimental research of materials using modern testing and measuring systems are of particular relevance.

For this purpose, the Perm National Research Polytechnic University operates the Center for Experimental Mechanics, equipped with servo-hydraulic, electromechanical, electrodynamic, shock testing machines, means for recording deformation and temperature fields, and acoustic emission signals. Complex studies of the properties of materials in a wide range of temperatures (from −100° to 1500°C), displacement velocities (from 0.001 mm/min to 24 m/s), complex stress state and complex modes of external influences seem promising. Important information on the behavior of materials under conditions close to operational or technological is provided by tests during combined stage quasi-static, cyclic, shock and temperature influences. Very useful is the simultaneous use of various registration and diagnostic tools in synchronization with the testing machine. Some methodological features and the results of experimental studies of the properties of metal, polymer and composite materials under complex thermomechanical effects are described in [1–4].

2 Actual and Promising Areas of Study of the Mechanical Behavior of Structural Materials for Aviation Purposes

2.1 Evidence-Based Development of Approaches to Assess the Survivability and Safety of Aerospace Products

The issues of survivability and safety analysis of critical products are related to the development of a rating system for a whole range of indicators [4–9]. The importance of which is determined by the purpose, working conditions of structures, engineering practice and other factors. Significant experience in studying the laws of supercritical deformation of materials and structures.

Creation of appropriate mathematical models. Development of a methodology for a comprehensive strength analysis of the safety of load-bearing objects. All this contributes to the development of approaches to assessing the reserves of bearing capacity, increasing the survivability and safety of deformable systems (see Fig. 1).

It is proposed to consider the behavior of bodies with cracks by analogy with the supercritical stage of deformation of materials. That in turn allows us to predict the moment of destruction at the stage of softening, including taking into account the rigidity of the loading system. It is advisable to conduct a comparative analysis of structural materials in terms of ensuring greater survivability of critical parts [4, 10].

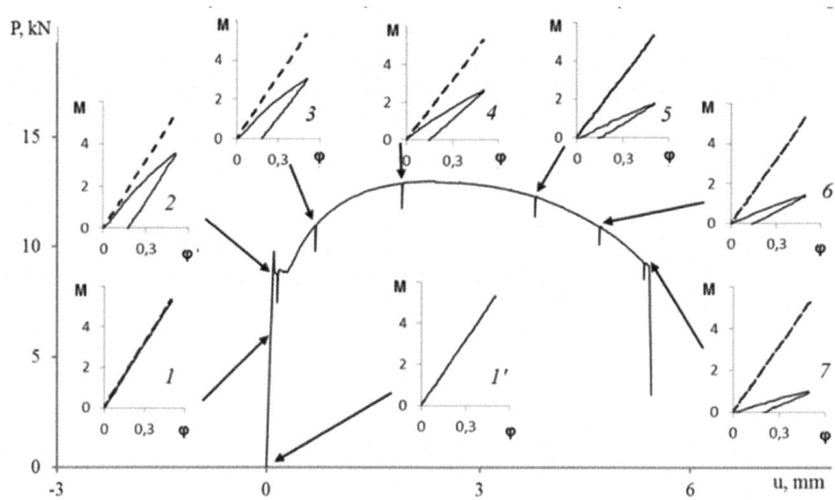

Fig. 1 Change in torsional stiffness of a specimen (1–7) at various stages of elastoplastic and supercritical deformation under tension (20X)

2.2 *The Mechanical Behavior of Materials Under Cyclic Stresses in Complex Stress Conditions*

The need for research in the field of multiaxial fatigue of materials is due to the possible destruction of structural elements that function for a long time with a complex of unsteady, usually cyclically changing loads. It is known that the type of stress-strain state can have a significant effect on the durability, and in some cases significantly reduce the service life of structural elements.

Work in this current area includes an experimental study of the fatigue strength of modern and advanced metals and alloys for aviation purposes in difficult stress conditions. Namely, the effects of constant and variable stress components on fatigue life under bimodal loading (see Fig. 2). Study of multi-cycle fatigue of alloys under disproportionate loading under conditions of single- and dual-frequency bimodal effects. Analysis of the influence of stress concentrators on the durability of multiaxial fatigue.

The practical use of the obtained results provides for the development and testing of mathematical models for predicting fatigue life under difficult stress conditions with the aim of more accurately calculating the resource of parts and assemblies.

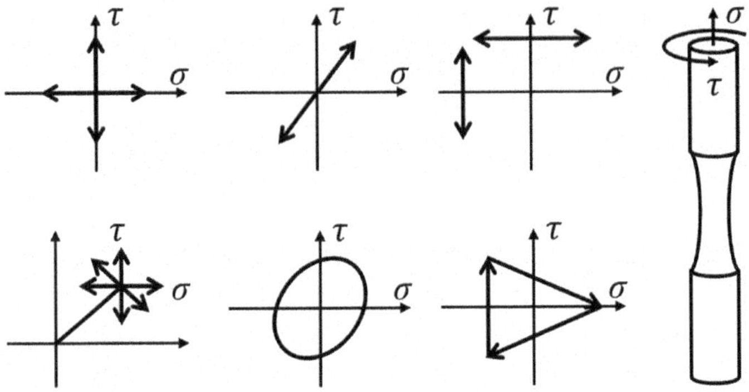

Fig. 2 Loading schemes

2.3 Determination of the Characteristics of Static and Cyclic Fracture Toughness

From the point of view of safety of structural elements and products in the presence of developing cracks, the most important is the stage of their equilibrium growth and determination of threshold values corresponding to the transition to nonequilibrium modes of crack propagation. The stability of this process can be associated both with the stiffness parameters of the loading system with respect to the crack propagation region, and with the parameters of the stress-strain state in the vicinity of its tip (see Fig. 3).

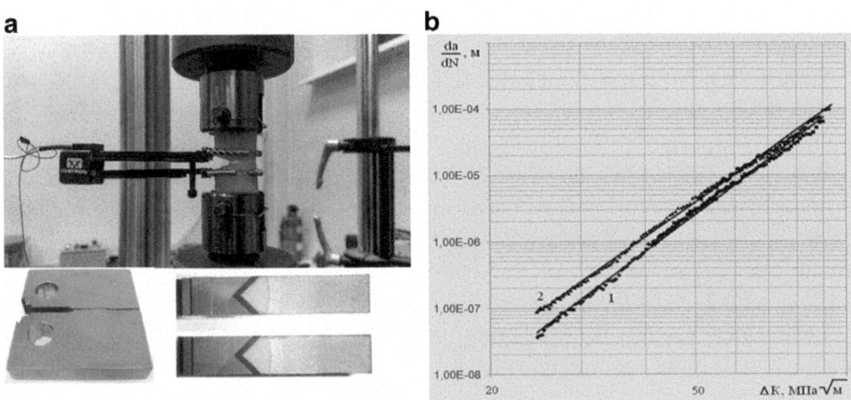

Fig. 3 An experimental study of the characteristics of cyclic crack resistance in simple and complex cycle forms; **a**—samples for studies of crack resistance of the material and sample in the grips of a testing machine with a high temperature sensor installed; **b**—kinetic diagrams of fatigue crack growth in the Paris section for a simple (1) and complex (2) cycle shape

The survivability assessment of critical aviation structures requires testing for static crack resistance with determination of crack growth rate under cyclic loading according to standard GOST and ASTM methods. Also, complex experimental studies of the behavior of materials and structural elements in the presence of fatigue cracks. With various types of stress-strain state and loading modes under conditions of elevated and low temperatures. All studies with the additional use of acoustic emission systems, infrared thermal scanning and non-contact construction of deformation fields in the crack growth zone [11].

2.4 An Experimental Study of the Characteristics of Low-Cycle Fatigue Resistance and with Simple and Complex Types of Stress-Strain State in a Wide Temperature Range

Low-cycle fatigue tests are carried out under hard or soft loading conditions in a wide range of elevated temperatures (see Fig. 4). Complex programs are being implemented for experimental studies of cyclic durability with complex shapes and parameters of the cycle, different types of stress-strain state (uniaxial, plane stress state) and loading sequences (tension, torsion), combinations of which allow us to take into account the real operating conditions of materials in structural elements [12].

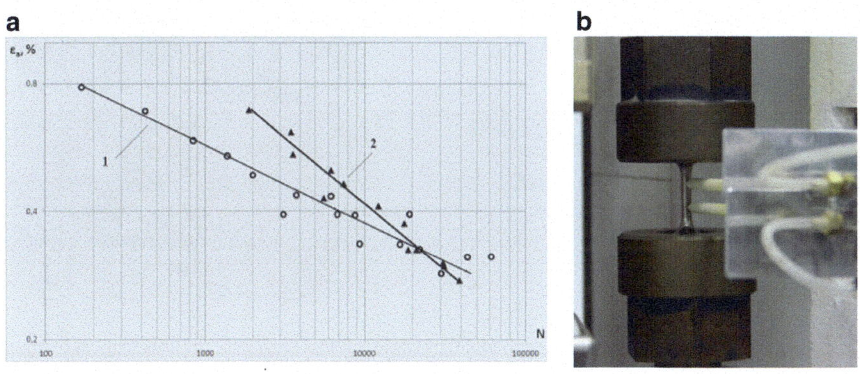

Fig. 4 Experimental data on low-cycle fatigue of a nickel alloy at: **a**—a complex cycle shape and different temperatures: 650 °C (1) and 22 °C (2); **b**—sample in the grips of a testing machine

2.5 An Experimental Study of the Thermomechanical Fatigue of Structural Steels and Alloys Used in Aircraft Engine Manufacturing at Elevated Heating and Cooling Rates

The tests are aimed at studying the characteristics of low-cycle thermomechanical fatigue resistance of structural steels and alloys used in aircraft engine manufacturing under conditions of cyclic heating to 1000 °C in air (see Fig. 5).

Available testing equipment allows independent loading and heating, which makes it possible to experimentally simulate the actual operating conditions of highly loaded parts and structural elements under conditions of cyclically varying temperatures. The system of high-frequency induction heating of conductive structural steels and alloys makes it possible to realize high heating and cooling rates of samples reaching 25 °C/s. It is possible to conduct tests at various phases and laws of variation of stresses, axial strains and temperature in a cycle.

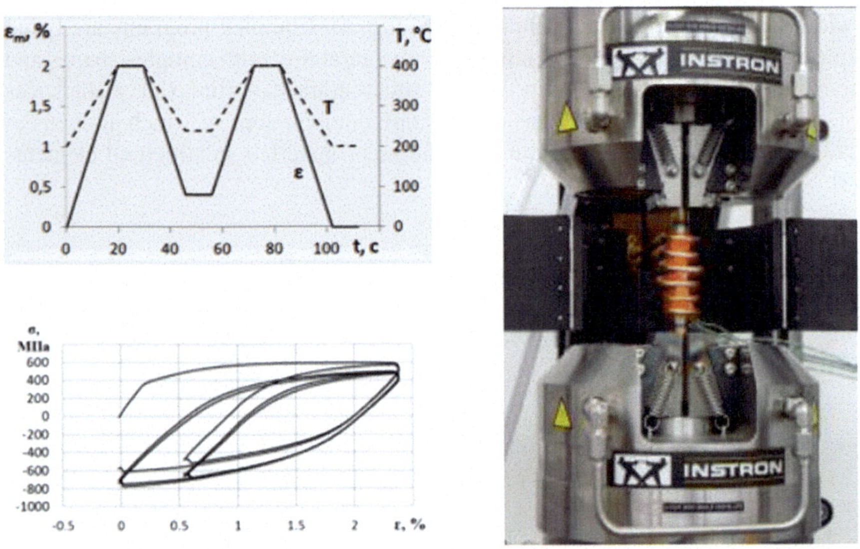

Fig. 5 The loading program and the mechanical hysteresis loops during thermomechanical loading

2.6 Comprehensive Study of the Implementation of the Deformation and Strength Properties of Materials in Products Obtained Using Modern Additive Technologies

Actual and promising direction is associated with the determination of the complex of mechanical characteristics of materials obtained using additive technologies. Such characteristics of materials as deformation, fatigue, strength, resistance to growth of fatigue cracks [13]. Also, a study of the degree of implementation of the required mechanical characteristics in finished products for aviation purposes (see Fig. 6).

The most relevant area of research in the field of additive manufacturing methods for complex parts is the study of the influence of technological parameters of manufacturing. Namely the characteristics of the laser, layer scanning parameters, properties and temperature of powders, arrangement of parts in the chamber. As well as the effect of additional high-temperature processing methods on the structure, porosity, physical and mechanical properties and mechanical behavior of SLS alloys (see Fig. 7).

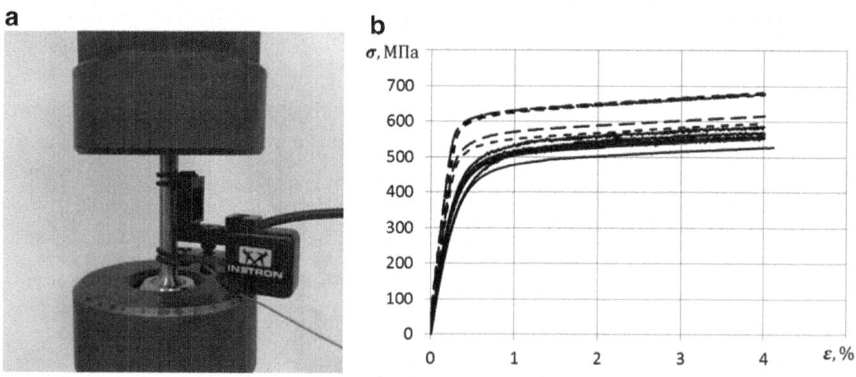

Fig. 6 An experimental study of the characteristics of low-cycle fatigue resistance: **a**—fixing the axial strain sensor on the sample; **b**—the influence of the manufacturing technology of the samples on the mechanical behavior of the SLS-12X18H10T under tension under layer-by-layer formation of the sample in the transverse (solid line) and longitudinal (dashed line) directions

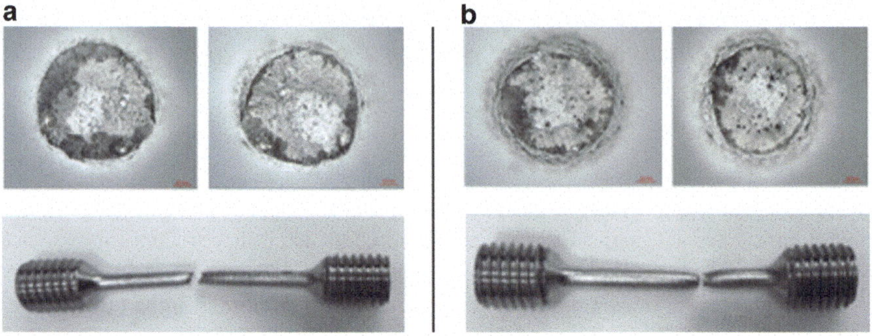

Fig. 7 Experimental data on low-cycle fatigue of a nickel alloy at: **a**—a complex cycle shape and different temperatures: 650°C (1) and 22°C (2); **b**—sample in the grips of a testing machine

2.7 Study of the Deformation and Strength Characteristics of Composite Materials, Structurally Similar Elements and Prototypes

The desire to expand the use of new materials requires comprehensive studies of the mechanical characteristics of composite materials under tension, compression, shear using special tools, as well as using non-standard methods (see Fig. 8) [14].

Since materials are created simultaneously with structures, it is extremely important to conduct mechanical tests of structurally similar elements (contours, bends, bulges, joints, flange elements, etc.) in order to identify the influence of structural and technological features on the strength and deformation characteristics. Important information is provided by temperature tests [15, 16], experiments under condi-

Fig. 8 Studies of the bearing capacity of composite structural elements under different types of loads using specialized equipment

tions of moisture saturation and exposure to aggressive media using climate and temperature chambers, and studies of the thermomechanical fatigue of composites [17–19].

2.8 Comprehensive Experimental and Theoretical Studies of Fatigue and Residual Strength of Aircraft-Grade Structural Composites for Various Types of Stress-Strain State and the Presence of Concentrators

Experimental studies of the laws of fatigue accumulation of damage and the destruction processes of layered fibrous and spatially reinforced structural composites under various types of stress-strain state and loading conditions are carried out (see Fig. 9).

Also in conditions of field heterogeneity caused by stress concentrators, with control of acoustic emission signals and heating of samples during cyclic loading. The methods used make it possible to evaluate the residual strength and stiffness of structural composites after cyclic loading under various types of stress-strain state.

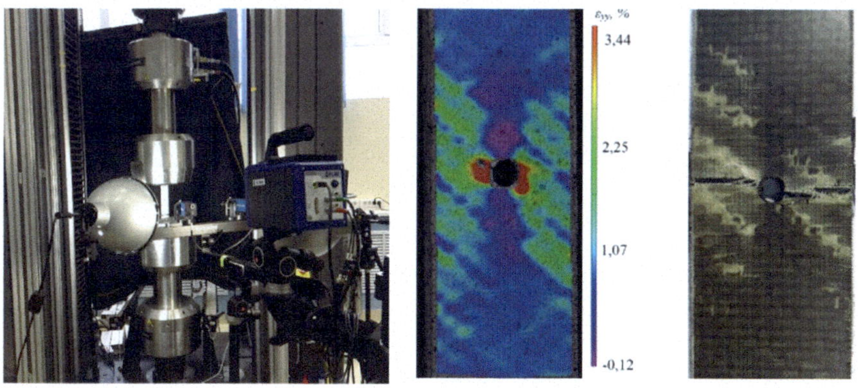

Fig. 9 Studies of the laws of inelastic deformation and fracture of composite samples in the zone of stress concentrators

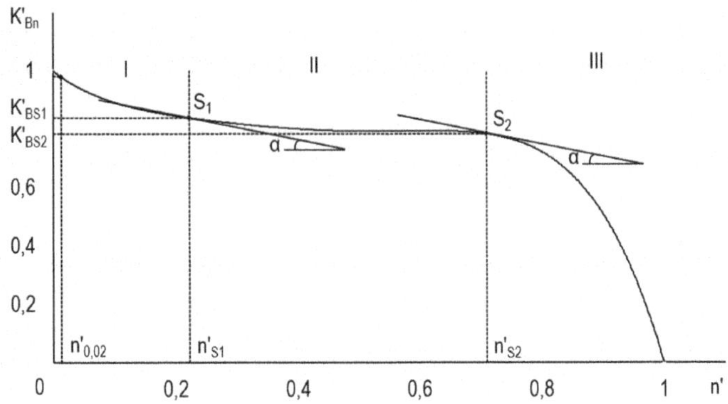

Fig. 10 Fatigue sensitivity diagram I—stage of initial fatigue sensitivity; II—stage of stabilization; III—stage of exacerbation [20]

2.9 *Studies of the Patterns of Deformation Processes Under Complex Unsteady and Combined Quasistatic Effects in a Wide Temperature Range*

Another relevant area is the study of the laws of deformation processes, the mechanisms of damage accumulation and destruction of composites under complex influences. Namely, complex unsteady and combined quasi-static, dynamic, cyclic and shock effects in a wide temperature range.

Assessment of the survivability of composite products ("loyalty" to damage) is carried out during testing of products with technological or operational defects, as well as after repair operations.

Of great importance is the study of the behavior of composite materials for aviation purposes under combined effects, in particular, the analysis of the dependence of the residual static strength on the preliminary cyclic loading, the dependence of the fatigue life on the preliminary quasistatic loading, and the residual compressive strength after punching the composite panels (see Figs. 10 and 11) [20, 21].

The peculiarity of the approach is the joint use of modern loading systems, data recording tools and non-destructive testing methods (non-contact three-dimensional video system for analyzing deformation fields, acoustic emission, thermal scanning). The research results are reflected in [1–23].

3 Conclusions

The questions of the development of methods of experimental mechanics are examined in terms of studying the inelastic behavior of materials under complex stress states, combined and complex thermomechanical influences. Attention is paid to the

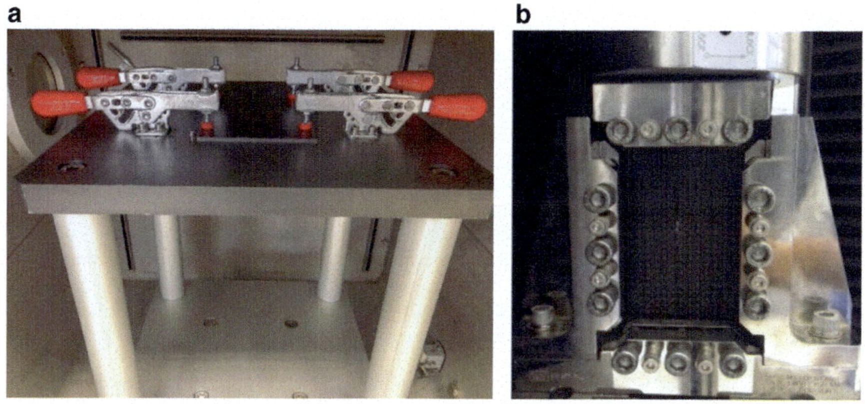

Fig. 11 Studies of the patterns of deformation processes under impacts in a wide temperature range: **a**—photograph of samples installed in devices for striking; **b**—compression after impact [21]

features of the joint use of modern test systems and means for recording deformation and temperature fields, as well as acoustic emission signals. The data on supercritical deformation of steels under plane stress state, multi- and low-cycle fatigue of alloys under biaxial impacts, as well as the implementation of complex cyclic loading conditions are presented. The experimental data obtained on damage accumulation and destruction of polymer layered fiber composites under complex quasistatic, cyclic and shock impacts are analyzed.

Acknowledgements Experimental studies were carried out using the large-scale research facilities « Complex of testing and diagnostic equipment for studying properties of structural and functional materials under complex thermomechanical loading » PNRPU modernized with funds by the Ministry of Science and Higher Education of the Russian Federation, Unique project identifier RFMEFI61920X0017.

The Experimental studies in the framework of the description of the mechanical behavior of structurally inhomogeneous materials were conducted within the State Assignment of the Ministry of Science and Higher Education of the Russian Federation (№ FSNM-2020-0027).

References

1. Wildemann VE, Babushkin AV, Tretyakov MP, Ilinykh AV, Tretyakova TV, Ipatova AV, Slovikov SV, Lobanov DS (2011) The mechanics of materials. Methods and means of experimental research. Publishers Inc. PNRPU, Perm 165
2. Wildemann VE (2012) Experimental studies of the properties of materials under complex thermomechanical effects. M.: FIZMATLIT 204
3. Tretyakova TV, Wildemann VE (2016) Spatial-temporal heterogeneity of inelastic deformation of metals. M.: FIZMATLIT 120
4. Wildemann VE, Lomakin EV, Tretyakov MP, Tretyakova TV, Lobanov DS (2018) Experimental studies of supercritical deformation and fracture of structural materials. Publishers Inc. PNRPU, Perm 155

5. Tretyakova TV, Mugatarov AI, Feklistova EV (2020) Experimental study of macroscopic localization of plastic flow in al-mg alloy under complicated stress-strain state. IOP Conf. Series: Materials Science and Engineering 747:012132
6. Vildeman VE, Lomakin EV, Tret'yakova TV, Tret'yakov MP (2016) Development of inhomogeneous fields under postcritical deformation of steel specimens in extension. Mech Solid 51(5):612–618
7. Vildeman VE, Lomakin EV, Tretiakova TV (2015) Yield delay and space-time inhomogeneity of plastic deformation of carbon steel. Mech Solid 50(4):412–420
8. Tretyakova TV, Zubova EM (2018) Influence of additional vibration impact on kinetics of strain bands due to the Chernov-Lüders deformation and Portevin-Le Chatelier effect in metals. Struct Integrity Procedia 13:1739–1744
9. Lobanov DS, Staroverov OA, Tretyakov MP (2020) Realization effects of postcritical deformation the regularities for GFRP in tensile tests. IOP Conf. Series: Materials Science and Engineering 747:012120
10. Tretyakov MP, Feklistova EV, Lykova AV (2020) Experimental study of creep and relaxation of steels at the postcritical stage of deformation. IOP Conf. Series: Materials Science and Engineering 747:012032
11. Lomakin EV, Tretyakov MP (2016) Fracture properties of graphite materials and analysis of crack growth under bending conditions. Meccanica 51(2):2353–2364
12. Lomakin EV, Tretyakov MP, Ilinykh AV, Lykova AV (2019) Mechanical behavior of structural steel X15CrNi12-2 under biaxial low-cycle fatigue at normal and elevated temperatures. PNRPU Mech Bull 1:77–86
13. Ilinykh AV (2018) Mechanical properties of 12Kh18N10T steel obtained by selective laser fusion. PNRPU Aerosp Eng Bull 55:103–109
14. Strungar EM, Feklistova EV, Babushkin AV, Lobanov DS (2019) Experimental studies of 3D woven composites interweaving types effect on the mechanical properties of a polymer composite material. Procedia Struct Integrity 17:965–970
15. Lobanov DS, Babushkin AV (2017) Experimental studies of the high temperature influence on strength and deformation properties of combined glass organoplastics. PNRPU Mech Bull 1:104–117
16. Strungar EM, Yankin AS, Zubova EM, Babushkin AV, Dushko AN (2020) Experimental study of shear properties of 3D woven composite using digital image correlation and acoustic emission. Acta Mech Sin 36(2):448–459
17. Lobanov DS, Staroverov OA (2019) The fatigue durability GFRP under increased temperatures. Procedia Struct Integrity 17:651–657
18. Lobanov DS, Slovikov SV (2017) Mechanical properties of a unidirectional basalt-fiber-reinforced plastic under a loading simulating operation conditions. Mech Compos Mater 52(1):767–772
19. Zubova EM, Lobanov DS, Strungar EM, Temerova MS (2020) Research of the carbon/epoxy laminated curved beams strength in four-point bending test. IOP Conf. Series: Materials Science and Engineering 747:012139
20. Wil'deman VE, Staroverov OA, Lobanov DS (2018) Diagram and parameters of fatigue sensitivity for evaluating the residual strength of layered GFRP composites after preliminary cyclic loadings. Mech Compos Mater 54(3):313–320
21. Staroverov OA, Babushkin AV, Gorbunov SM (2019) Evaluation of the damage degree to carbon-fiber composite materials under impact. PNRPU Mech Bull 1:161–172
22. Lobanov DS, Zubova EM (2019) Research of temperature aging effects on mechanical behaviour and properties of composite material by tensile tests with used system of registration acoustic emission signal. Procedia Struct Integrity 18:347–352
23. Tretyakova TV, Dushko AN, Strungar EM, Zubova EM, Lobanov DS (2019) Comprehensive analysis of mechanical behavior and fracture processes of specimens of three-dimensional reinforced carbon fiber in tensile tests. PNRPU Mech Bull 1:173–183

A Comparative Study on the Evolution of Plastic Zone Between Indentation and Flattening Contact

Qingming Deng, Xiaochun Yin, and Magd Abdel Wahab

Abstract The peak of the ratio between the average pressure to yield strength is called hardness. The average contact pressure of elastic–plastic flattening and indentation contacts is studied by using numerical simulation. The similarity and difference between the two kinds of single asperity contacts are investigated. This paper chooses five kinds of elastic–plastic materials as the deformable-body. The yield strength of these materials covers the typical steel materials range used in the engineering project. And the effect of the evolution of the plastic zone under the contact surface on the contact pressure distribution is analyzed. Before the elastic zone under the contact surface disappears completely, the evolution of the plastic zone of the two types of contact is similar. However, the average pressure of indentation contact continues to increase after this, until it reaches the maximum. The average pressure of flattening contact reached the plateau since elastic core has disappeared, and after a while, it began to decrease.

Keywords Hardness · Flattening contact · Indentation contact · Average pressure · Finite element analysis

1 Introduction

Both flattening and indentation are fundamental problems in contact mechanics and have direct applications in a broad range of engineering fields [1–8]. Hertz [9] gave the prediction formulas for the pure elastic spherical contact. The Hertzian purely

Q. Deng (✉) · X. Yin
Department of Mechanics and Engineering Science, Nanjing University of Science and Technology, Nanjing 210094, China
e-mail: dqm2017@qq.com

X. Yin
e-mail: yinxiaochun2000@aliyun.com

Q. Deng · M. Abdel Wahab
Soete Laboratory, Department of Electrical Energy, Metals, Mechanical Constructions and Systems, Faculty of Engineering and Architecture, Ghent University, Ghent, Belgium

M. Abdel Wahab (ed.), *Proceedings of 1st International Conference on Structural Damage Modelling and Assessment*, Lecture Notes in Civil Engineering 110,
https://doi.org/10.1007/978-981-15-9121-1_33

elastic contact theory [10] is shortly reviewed, here. The contact area and force of Hertz theory are given by:

$$A_E = \pi R'\omega \tag{1}$$

$$F_E = \frac{4}{3} \times E'\sqrt{R'} \times \omega^{\frac{3}{2}} \tag{2}$$

where ω is the interference between two contact bodies. And the equivalent elastic modulus, E', and an equivalent radius, R', are expressed as:

$$\frac{1}{E'} = \frac{1 - v_1^2}{E_1} + \frac{1 - v_2^2}{E_2} \tag{3}$$

$$\frac{1}{R'} = \frac{1}{R_1} + \frac{1}{R_2} \tag{4}$$

where E_1, E_2, v_1, v_2, R_1, and R_2 are Young's modulus, Poisson's ratio, and the radius of curvature of sphere and half-space, respectively. Later, many researchers have proposed many empirical models to predict contact response [9–17]. Ghaednia et al. divided those contact models into two categories: indentation and flattening model [11].

Figure 1 shows a schematic of a frictionless elastic–plastic half-space indented by a rigid sphere of radius R. The parameters a and a' are real contact radius and truncated contact radius, respectively. A normal force F acts on top of a rigid sphere. The ω is interference between two contact bodies. Figure 2 shows a contact of an elastic–plastic hemisphere with a rigid half space. The hardness test is implemented to evaluate the material's resistance to plastic deformation. Indentation contact is an important research content about hardness test [11, 12]. The hardness H is equal

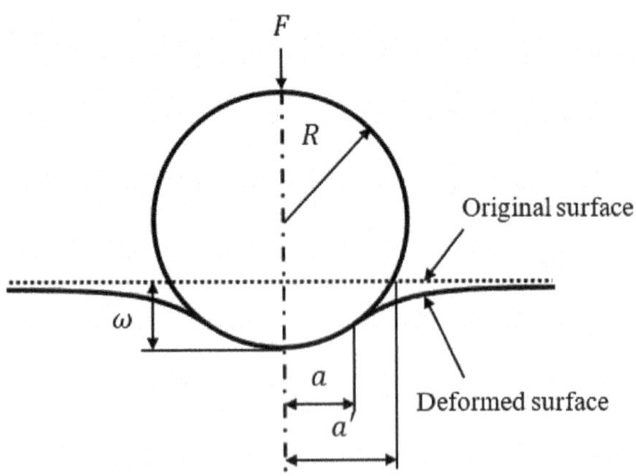

Fig. 1 An indentation model, in which the sphere is rigid and the flat is deformable

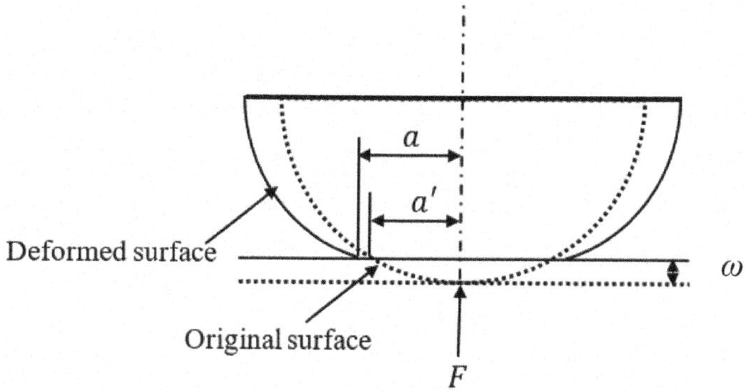

Fig. 2 A flattening model, in which the flat is rigid and the sphere is deformable

to the average pressure when contact surface under fully plastic deformation. So the evolution of plastic zone under the contact surface is important for predict the hardness. Johnson [12] defined as $H = 2.8S_y$ using a constant multiple of yield strength. However, by using numerical simulation, Mesarovic and Fleck [13] found that the hardness depended on the material property. Other numerical research also had similar conclusions [11, 14–16].

And by observing the evolution of the plastic zone some new contact models were proposed [22–24]. The flattening contact is also a research hotspot. Many models suitable for flattening contact were proposed [17–21]. Kogut and Etsion also defined the hardness as a constant multiple of its yield strength ($H = 2.8S_y$). However, Jackson and Green found that the hardness of sphere was similar like half-space which depends on the material properties [14]. And they obtained a new critical interference ω_c resulting in the initial yield as:

$$\omega_c = \left(\frac{\pi \times c \times S_y}{2E'}\right)^2 R \tag{5}$$

where S_y and v are the yield limit and Poisson's ratio of the elastic–plastic body, respectively. And parameter c is expressed as:

$$c = 1.295e^{0.736v} \tag{6}$$

Jackson and Kogut considered that the indentation contact had big difference compared with flattening contact [22]. However, when predicting hardness, there were cases where these two different contact models were mixed [14–16]. This abuse and mixing phenomenon also appeared in the prediction of collision problems [23]. The main purpose of this article is to compare the average pressure of two types of contact by studying the evolution of the plastic zone below the contact surface.

2 Finite Element Model

In this paper, for flattening and indentation contact problems, an axisymmetric 2-D model is used similar to other researches in the literature [14, 16, 18, 24–26]. The commercial program ANSYS™ is used. For indentation FEM model, as shown in Fig. 3a, the left side of the half-space is fixed horizontally and can move freely in the vertical direction. The bottom of half-space is completely fixed and a displacement ω is applied on the top of rigid sphere. The thickness T and width L are gradually increased, until the simulation result difference less than 1% between iterations as shown in Fig. 3. **Finite** element model of (a) indentation and (b) flattening contact, $L = 10R$ and $T = 6R$. For flattening model, the central axis of the hemisphere is fixed horizontally and is free vertically. The quadrilateral, eight node element solid183 is used here. The mesh size satisfies the study of convergence and the number of the elements under the contact surface should be more than 30 to ensure the accuracy of the contact radius.

For all cases, the elastic modulus and Poisson's ratio of deformable body are 200 GPa and 0.32, respectively. And yield strengths are 0.210, 0.5608, 0.9115, 1.2653, and 1.619 GPa. In the following, those five kinds of material models are referred to as Mat 1, Mat 2, Mat 3, Mat 4, and Mat 5. The deformation considered in this paper is before the contact radius reaches 0.8 *R*.

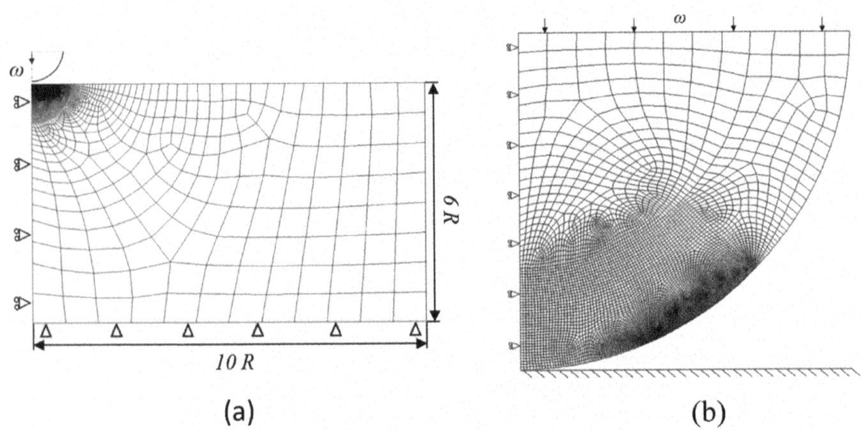

Fig. 3 Finite element model of **a** indentation and **b** flattening contact

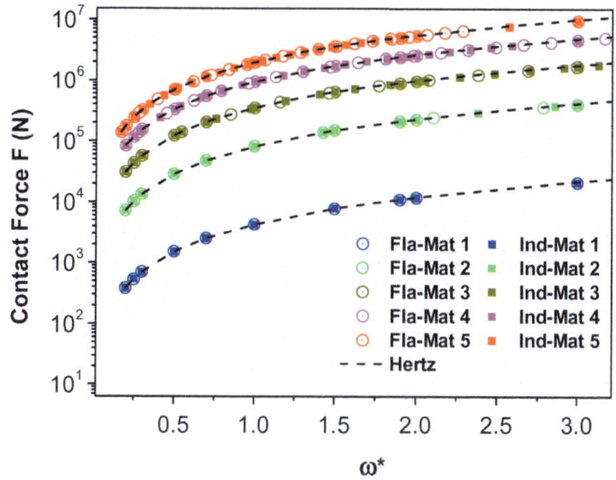

Fig. 4 Comparison of contact force predicted by simulation and Hertz theory in the elastic deformation regime

3 Numerical Results and Discussion

3.1 *FEM Model Validation*

Normally, the verification of the simulation model is verified by hertz theory. Figure 4 shows the comparison of contact force predicted by FEM and Hertz theory in the elastic phase. It shows that the maximum difference between FEM and Hertz results is less than 0.5%. This verifies the suitability of the finite element mesh shown in Fig. 3.

3.2 *The Ratio of Average Contact Pressure to Yield Strength*

As interference increasing, the average contact pressure also increases. Until reaching a maximum, it will decrease with the increase of interference. This peak value is always considered to be the hardness of the material. Researchers use this to measure the ability of materials to resist plastic deformation.

The average contact pressure to yield strength ratio of all cases are shows in Fig. 5a. With the same deformable-body material, the two different contacts show similarity. It can be seen from Fig. 5b that when the deformation is small, the difference between the two types of contact is close to zero. Mat 5 case has larger similarity interference range than Mat 1 case.

In order to consider the different initial yield interference ω_c of different materials, the interference can be normalized through Eq. (5). And the variation of average contact pressure to yield strength ratio with dimensionless interference is shown in Fig. 6a. Compared with Fig. 5a, it can be seen that the different material cases have same change trend. As the interference increases, Mat 1 first plastically yields. In

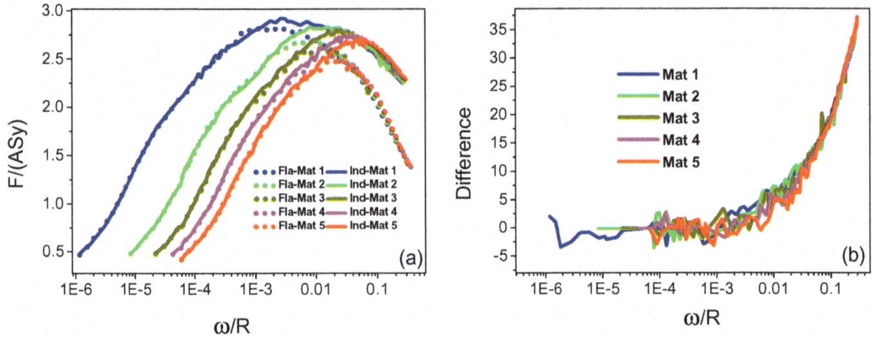

Fig. 5 **a** The variation of average contact pressure to yield strength ratio with interference. **b** The variation of difference with interference

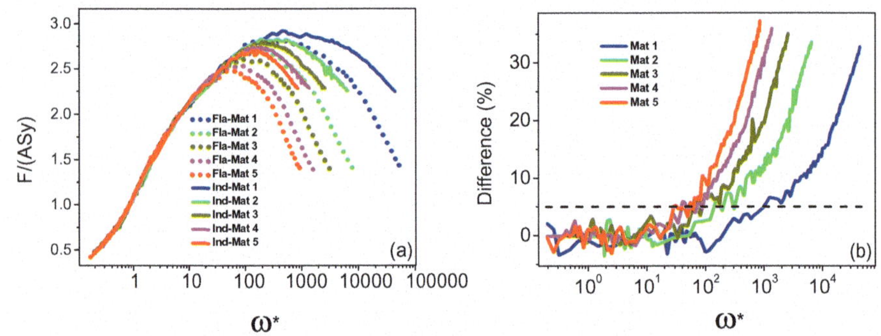

Fig. 6 **a** The variation of average contact pressure to yield strength ratio with dimensionless interference and **b** The variation of difference with dimensionless interference

order to facilitate the evaluation of the difference between the two types of contact in engineering applications, the dimensionless interference that makes this difference reaches 5% is defined as the upper limit interference ω_{lh}^{*}. When the dimensionless interference is less than ω_{lh}^{*}, the average contact pressure difference is less than 5%, so the results are reliable when predicted with different contact models.

The ω_{lh}^{*} of different material cases predicted by the simulation are list in Table 1. And a prediction formula is given as:

Table 1 Upper limit interference for different material

Material	Mat 1	Mat 2	Mat 3	Mat 4	Mat 5
E^{*}/S_{y}	1061.03	397.32	244.45	176.1	137.626
ω_{lh}^{*}	1443.3	240.8158	116.0272	78.3215	57.6173

$$\omega_{lh}^{*} = 34.669 + 0.00141 \times \left(\frac{E^{*}}{S_{y}}\right)^{1.9827} \tag{7}$$

3.3 Plastic Zone Evolution

The change in contact response has a great relationship with the deformation near the contact surface. In order to study the effect of plastic zone evolution on average contact pressure, the indentation and flattening contact of Mat 3 are chosen as an example and are considered in this paper. According to the evolution characteristics of the plastic zone, the whole deformation process is divided into several stages as shown in Fig. 7. The plastic yield first appears ($\omega^{*} = 1$) as shown as point 1. And when $\omega^{*} = 1.9$, the plastic zone is in the subsurface of contact according to point 2. Plastic zone is for the first time reached and covers the whole surface according to point 3 and 4, respectively. During the deformation process, the average pressure will remain at the peak for a while, like a plateau as shown in Fig. 7 between point 5 and 6. It is worth noting that the flattening contact, point 4 coincides with point 5.

The plastic zone evolution of indentation and flattening contacts are shown in Figs. 8 and 9. As plotted in Figs. 8a and 9a, the initial plastic yield of both types of contact occurs when $\omega^{*} = 1$. And the initial yield point of indentation and flattening contact is located at 0.00513 *R* and 0.0052 *R* below the contact surface, respectively. The stress distribution on the contact surface complies with hertz theory.

And in the work of Jackson and Green [14], they considered that the dominance elastic contact phase will continue until $\omega^{*} = 1.9$ as shown in Figs. 8b and 9b. Although the plastic zone is large enough, the contact surface is still purely elastically deformed. Later on, the plastic zone will extend to the contact surface when $\omega^{*} =$

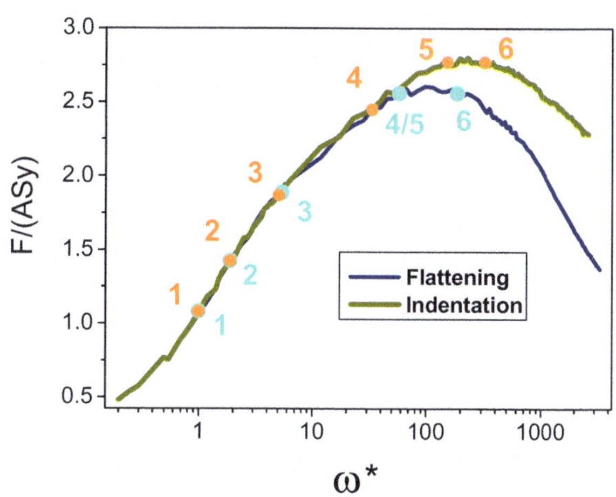

Fig. 7 The average contact pressure to yield strength ratio of Mat 3

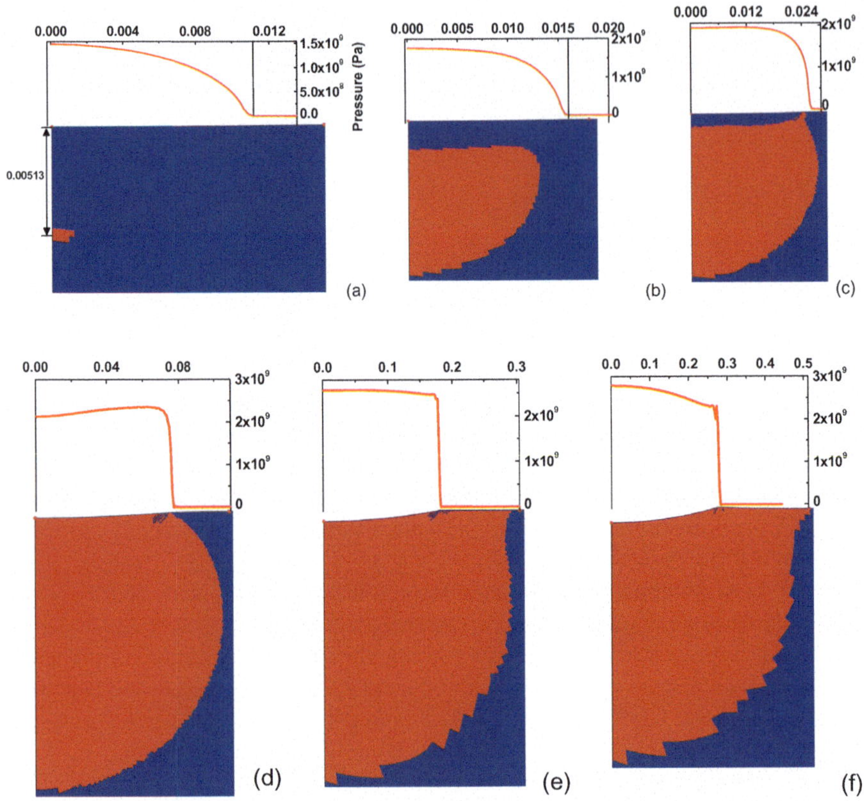

Fig. 8 The plastic zone evolution of indentation contact

5.115 and $\omega^* = 5.44$ for indentation and flattening contact as shown in Figs. 8c and 9c. The stress distribution on the contact surface becomes flatter than elastic phase. At this stage, there is still a decreasing elastic core at the contact surface. The plastic zone, when this elastic core completely disappears ($\omega^* = 33.474$), and the two critical points of the average pressure plateau ($\omega^* = 153$ and $\omega^* = 321.317$) for indentation contact are shown in Fig. 8d–f. For flattening contact, when the elastic core disappear ($\omega^* = 57.7$), the average contact pressure will close to the peak value as shown in Fig. 9d. The distribution of the plastic zone when the average pressure starts to decrease ($\omega^* = 186.89$) is shown in Fig. 9e. It can be seen that before the elastic zone under the contact surface disappears completely, the evolution of the plastic zone of the two types of contact is very close. However, the average pressure of indentation contact continues to increase after this, until it reaches the maximum. The average pressure of flattening contact has reached the plateau since then, and after a while, it begins to decrease. So the maximum average pressure of the indentation contact is greater than the flattening contact.

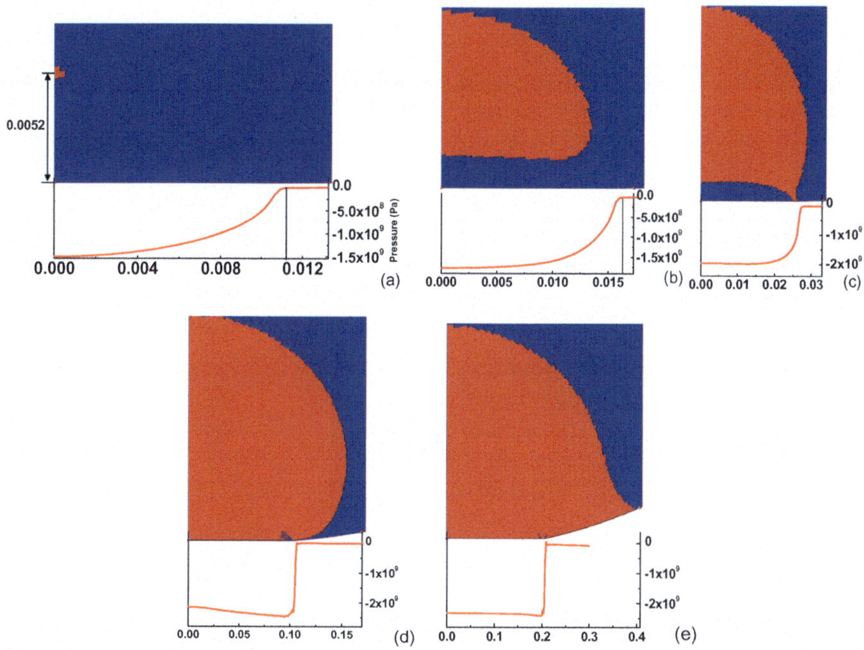

Fig. 9 The plastic zone evolution of flattening contact

4 Conclusions

With the same deformable-body material cases, the average contact pressure of two different contacts show similarity. The larger yield strength case has a larger similarity interference range than the smaller cases. With dimensionless interference, smaller yield strength case has a larger similarity interference range than larger cases. And before the elastic zone under the contact surface disappears completely, the evolution of the plastic zone of the two types of contact is similar. However, the average pressure of indentation contact continues to increase after this, until it reaches the maximum. The average pressure of flattening contact has reached the plateau since elastic core disappear, and after a while, it begins to decrease. So the maximum average pressure of the indentation contact is greater than the flattening contact for the same deformation body cases.

Acknowledgements The authors would like to acknowledge the financial support of the grants from the China Scholarship Council (201806840127), the National Natural Science Foundation of China (Grant No. 11372138 and No.11572157).

References

1. Mccarthy B et al (2002) A dynamic model, including contact bounce, of an electrostatically actuated microswitch. J Microelectromech Syst 11(3):276–283
2. Pal RK, Awasthi AP, Geubelle PH (2013) Wave propagation in elasto-plastic granular systems. Granular Matter 15(6):747–758
3. Komvopoulos K (2000) Head–disk interface contact mechanics for ultrahigh density magnetic recording. Wear 238(1):1–11
4. Cowles BA (1996) High cycle fatigue in aircraft gas turbines—an industry perspective. Int J Fract 80(2–3):147–163
5. Polycarpou AA, Etsion I (2000) A Model for the Static Sealing Performance of Compliant Metallic Gas Seals Including Surface Roughness and Rarefaction Effects. Tribol Trans 43(2):237–244
6. Wang Y et al (2002) A mixed-TEHD analysis and experiment of journal bearings under severe operating conditions. Tribol Int 35(6):395–407
7. Jiang X, Cheng HS, Hua DY (2000) A theoretical analysis of mixed lubrication by macro micro approach: part I—results in a gear surface contact. Tribol Trans 43(4):689–699
8. Kweh CC et al (1992) Simulation of elastohydrodynamic contacts between rough surfaces. J Tribol 114(3):412–419
9. Hertz H (1882) Ueber die Berührung fester elastischer Körper.: Journal für die reine und angewandte Mathematik (Crelle's Journal). Journal Für Die Reine Und Angewandte Mathematik
10. Hertz H (1882) Uber die beruhrung fester elastische korper und uber die harte (On the contact of rigid elastic solids and on hardness). Verhandlungen des Vereins zur Beforderung des Gewerbefleisses, Leipzig
11. Ghaednia H et al (2016) A comprehensive study of the elasto-plastic contact of a sphere and a flat. Tribol Int 93:78–90
12. Johnson KL (1987) Contact mechanics. Cambridge University Press
13. Mesarovic SD, Fleck NA (2000) Frictionless indentation of dissimilar elastic–plastic spheres. Int J Solids Struct 37(46–47):7071–7091
14. Jackson RL, Green I (2005) A finite element study of elasto-plastic hemispherical contact against a rigid flat. J Trib 127(2):343–354
15. Komvopoulos K, Ye N (2000) Three-dimensional contact analysis of elastic-plastic layered media with fractal surface topographies. J Trib 123(3):632–640
16. Song Z, Komvopoulos K (2013) Elastic–plastic spherical indentation: deformation regimes, evolution of plasticity, and hardening effect. Mech Mater 61:91–100
17. Chang W, Etsion I, Bogy DB (1987) An elastic-plastic model for the contact of rough surfaces. J Tribol 109(2):257–263
18. Kogut L, Etsion I (2002) Elastic-plastic contact analysis of a sphere and a rigid flat. J Appl Mech 69(5):657–662
19. Zhao Y, Maietta DM, Chang L (2000) An asperity microcontact model incorporating the transition from elastic deformation to fully plastic flow. J Trib 122(1):86–93
20. Jacq C et al (2002) Development of a three-dimensional semi-analytical elastic-plastic contact code. J Trib 124(4):653–667
21. Vu-Quoc L, Zhang X, Lesburg L (2000) A normal force-displacement model for contacting spheres accounting for plastic deformation: force-driven formulation. J Appl Mech 67(2):363–371
22. Jackson RL, Kogut L (2005) A comparison of flattening and indentation approaches for contact mechanics modeling of single asperity contacts. J Tribol 128(1):209–212
23. Wang H et al (2017) Experimental and theoretical analysis of the elastic-plastic normal repeated impacts of a sphere on a beam. Int J Solids Struct 109:131–142
24. Kogut L, Komvopoulos K (2004) Analysis of the spherical indentation cycle for elastic–perfectly plastic solids. J Mater Res 19(12):3641–3653

25. Larsson PL, Olsson E (2015) A numerical study of the mechanical behavior at contact between particles of dissimilar elastic–ideally plastic materials. J Phys Chem Solids 77:92–100
26. Olsson E, Larsson P-L (2016) A unified model for the contact behaviour between equal and dissimilar elastic–plastic spherical bodies. Int J Solids Struct 81:23–32

An Advanced Elasto-Plastic Approach to Design Notched Metals Against Variable Amplitude Multiaxial Fatigue Loading

Namiq Zuhair Faruq and Luca Susmel

Abstract The present paper summarises an attempt to devise and validate a novel, advanced approach suitable for designing notched components against multiaxial fatigue. This fatigue assessment technique was formulated by combining together the Modified Manson-Coffin Curve Method, the Shear Strain-Maximum Variance Method, and the elasto-plastic Point Method. The accuracy of this design method was checked against numerous experimental results generated by testing notched cylindrical samples of carbon steel En8. The validation exercise being carried out allowed us to show that the systematic usage of the fatigue design approach being proposed returned predictions falling within an error factor of 2. This level of accuracy is certainly satisfactory especially because this fatigue assessment technique was applied by directly using a commercial Finite Element (FE) software to determine the cyclic elasto-plastic local stress/strain fields in the vicinity the notches being investigated.

Keywords Multiaxial fatigue · Notch · Variable amplitude loading

1 Introduction

Real engineering components not only have complex geometries that result in localised stress/strain concentration phenomena, but also are subjected to in-service multiaxial load histories that vary randomly. Accordingly, systematic research work has been carried out since the beginning of the 1900s to formulate and validate reliable design techniques capable of addressing such complex engineering problem, with this being done by always reaching an adequate level of accuracy.

As to the different fatigue design approaches that have been formulated so far, examination of the state of the art shows that stress-based methods are usually employed to assess damage in the medium/high-cycle fatigue regime—i.e.,

N. Z. Faruq · L. Susmel (✉)
Department of Civil and Structural Engineering, The University of Sheffield, Mapping Street, Sheffield S1 3JD, UK
e-mail: l.susmel@sheffield.ac.uk

M. Abdel Wahab (ed.), *Proceedings of 1st International Conference on Structural Damage Modelling and Assessment*, Lecture Notes in Civil Engineering 110,
https://doi.org/10.1007/978-981-15-9121-1_34

when cyclic plasticity can be disregarded with little loss of accuracy. In contrast, strain-based methods are usually used in those applications where metals undergo large-scale cyclic plastic deformations [1–3].

As far as low/medium-cycle fatigue assessment of notched components is concerned, root stresses and root strains post-processed according to the critical plane concept are seen to return accurate estimates [4–6], with this holding true provided that the notches being designed are relatively blunt [5]. In contrast, in the presence of sharp notches strain-based approaches are seen to return predictions that are somehow too conservative, with the degree of conservatism increasing as the sharpness of the notch being assessed increases [7, 8].

In this challenging scenario, this paper aims at extending the use of the strain-based approach also to those situations involving stress/strain gradients arising not only from blunt, but also from sharp notches. This will be done via the formulation and validation of a novel fatigue lifetime estimation technique making use of the following three key ingredients: (i) the Modified Manson-Coffin Curve Method (MMCCM) [3, 5, 9], (ii) the Shear Strain-Maximum Variance Method (γ-MVM) [3, 10, 11], and (iii) the elasto-plastic Point Method (PM) [7, 8].

2 The Modified Manson-Coffin Curve Method (MMCCM)

The MMCCM postulates that fatigue lifetime can be predicted in a very accurate way by using non-conventional bi-parametrical Manson-Coffin curves [9]. The formulation of this design method is based on the assumption that fatigue damage is quantified via log–log diagrams that plot the shear strain amplitude, γ_a, relative to that plane experiencing the maximum shear strain amplitude (i.e., the so-called critical plane) against the number of reversals to failure, $2N_f$ (Fig. 1). As per the schematic modified

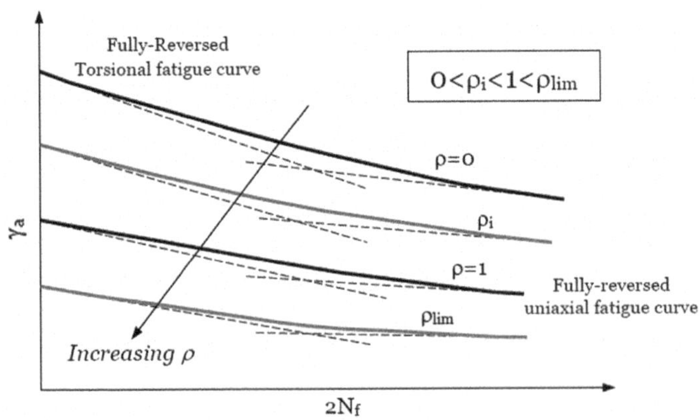

Fig. 1 The modified Manson-Coffin diagram

Manson-Coffin diagram shown in Fig. 1, the equation describing any fatigue curve is directly written as [3, 9]

$$\gamma_a = \frac{\tau'_f(\rho)}{G} \cdot \left(2 \cdot N_f\right)^{b(\rho)} + \gamma'_f(\rho) \cdot \left(2 \cdot N_f\right)^{c(\rho)} \quad (1)$$

where critical plane stress ratio ρ is defined as:

$$\rho = \frac{\sigma_{n,m} + \sigma_{n,a}}{\tau_a} = \frac{\sigma_{n,\max}}{\tau_a} \quad (2)$$

In Eq. (2) τ_a is the shear stress amplitude relative to the plane of maximum shear strain amplitude. $\sigma_{n,m}$, $\sigma_{n,a}$ and $\sigma_{n,max}$ are instead the mean value, the amplitude and the maximum value of the stress perpendicular to the critical plane, respectively. In this context, thanks to the way it is defined, ratio ρ is sensitive not only to the degree of multiaxiality and non-proportionality of the applied loading path, but also to the presence of superimposed static stresses [3, 9, 11].

Turning back to Eq. (1), functions $\tau'_f(\rho)$, $\gamma'_f(\rho)$, $b(\rho)$ and $c(\rho)$ can directly be estimated using the following relationships [9]:

$$\frac{\tau'_f(\rho)}{G} = \rho(1+\nu_e)\frac{\sigma'_f}{E} + (1-\rho)\frac{\tau'_f}{G}; \quad \gamma'_f(\rho) = \rho\left(1+\nu_p\right)\varepsilon'_f + (1-\rho)\gamma'_f \quad (3)$$

$$b(\rho) = \frac{b \cdot b_0}{(b_0 - b)\rho + b}; \quad c(\rho) = \frac{c \cdot c_0}{(c_0 - c)\rho + c} \quad (4)$$

The above calibration functions were derived by directly using the conventional fully-reversed uniaxial and torsional Manson-Coffin fatigue curves, with these two curves being both expressed in terms of maximum shear strain amplitude γ_a, i.e.[3, 9]:

$$\gamma_a = (1+\nu_e)\frac{\sigma'_f}{E}\left(2 \cdot N_f\right)^b + (1+\nu_e)\varepsilon'_f\left(2 \cdot N_f\right)^c \text{ (Uniaxial case—}\rho = 1\text{)} \quad (5)$$

$$\gamma_a = \frac{\tau'_f}{G}\left(2N_f\right)^{b_0} + \gamma'_f\left(2 \cdot N_f\right)^{c_0} \text{ (Torsional case—}\rho = 0\text{)} \quad (6)$$

In the above equations ν_e and ν_p are Poisson's ratio for elastic and plastic strain, respectively.

To determine the orientation of the material planes of maximum shear strain amplitude, the MMCCM is suggested to be applied in conjunction with the γ-MVM [10, 11]. According to this method, the orientation of the critical plane is estimated via that direction, MV, that is characterised by the maximum variance of the resolved shear strain, $\gamma_{MV}(t)$—see Fig. 2a, b. As far as variable amplitude (VA) load histories are concerned, the key feature of the γ-MVM is that the shear stresses and shear

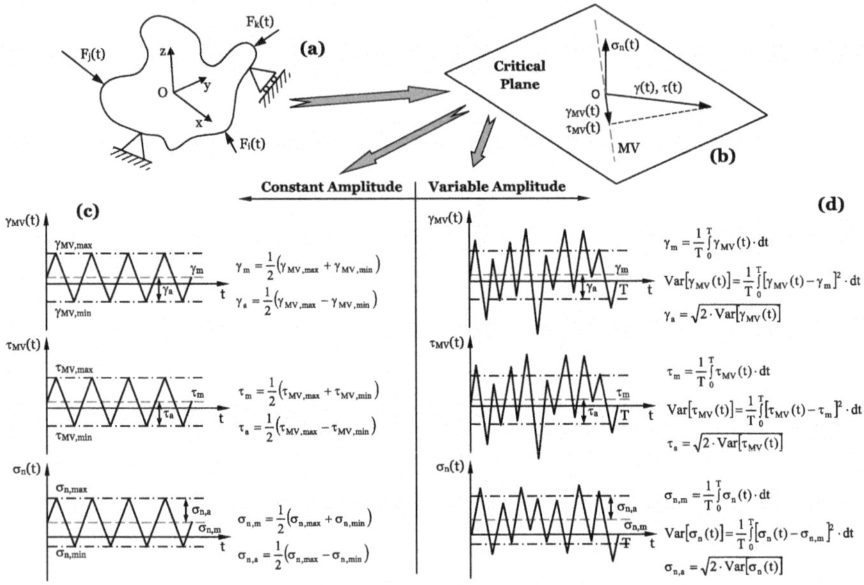

Fig. 2 Stress/strain components relative to the critical plane determined via the γ-MVM

strains relative to the critical plane are calculated via resolved quantities $\tau_{MV}(t)$ and $\gamma_{MV}(t)$, respectively (Fig. 2d). Since, by definition, these stress/strain quantities are monodimensional signals, fatigue cycles under variable amplitude multiaxial fatigue loading are counted unambiguously by directly using the standard Rain-Flow counting method that was originally devised to specifically address uniaxial fatigue situations.

3 Fatigue Assessment of Notched Components Under Multiaxial Variable Amplitude Fatigue Loading

The multiaxial fatigue life estimation technique discussed in the present paper makes use of the MMCCM, the γ-MVM, and the elasto-plastic Point Method (PM) [7, 8]. The key features of this design approach [12] will be explained below by addressing separately the constant amplitude (CA) and variable amplitude (VA) case. Further, the fatigue design approach being explained will be formulated by taking as a starting point the assumption that the PM critical distance, L, is known a priori. The procedure for the determination of L will be explained in Sect. 3.3.

3.1 Constant Amplitude Load Histories

Figure 3 summarises the design procedure being recommended to follow to estimate CA fatigue lifetime of notched components. Consider then the notched structural detail sketched in Fig. 3a and assume that, during in-service operations, it is subjected to a complex system of CA time-variable external forces and moments.

To use the MMCCM along with the PM, initially the elasto-plastic cyclic stress/strain fields along the so-called focus path must be determined. The focus path is defined as a straight line that emanates from the crack initiation point (i.e., point A in Fig. 3a) and is normal to the component surface at the initiation point itself [3]. According to the PM, the stress/strain quantities used to estimate fatigue lifetime are determined by post-processing the cyclic stress and cyclic strain tensors determined, along the focus path, at a distance from the assumed crack initiation point equal to L/2 (i.e., at point O in Fig. 3a). The orientation of the critical plane is determined from the time-variable strain tensor at point O via the γ-MVM. As soon as direction MV is known (Fig. 3b), γ_a (Fig. 3c) and τ_a (Fig. 3d) can be calculated directly according to the standard definitions reported in Fig. 2c. Similarly, by post-processing stress $\sigma_n(t)$, the mean value and the amplitude of the stress normal to the critical plane can be determined unambiguously (Fig. 3e) according to the definitions reported in Fig. 2c.

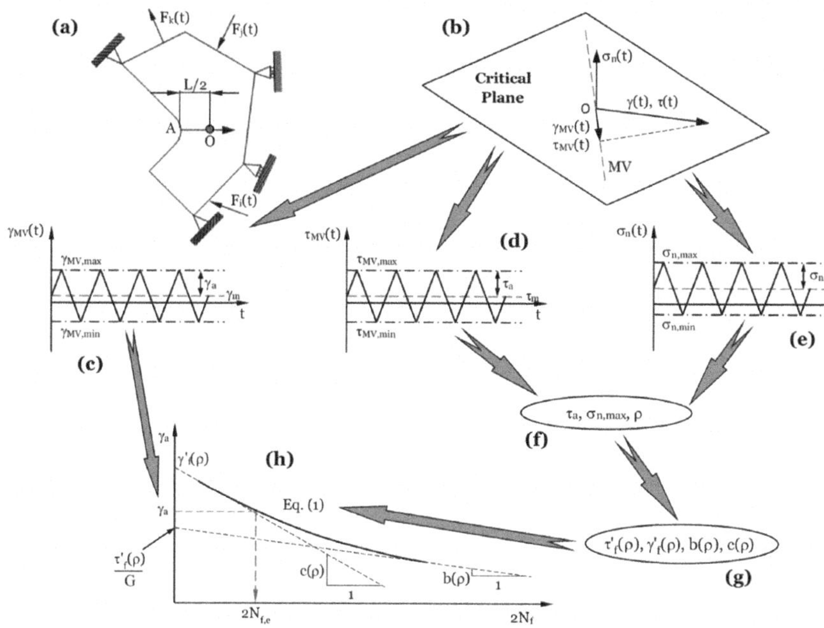

Fig. 3 Designing notched components against CA uniaxial/multiaxial fatigue loading

The values for τ_a, $\sigma_{n,m}$, and $\sigma_{n,a}$ being determined via the γ-MVM are then used to estimate ratio ρ (Fig. 3f). Using the calculated value for ρ, the corresponding modified Manson-Coffin curve, Eq. (1), can then be derived from calibration functions (3) and (4)—see Fig. 3g. Finally, as per Fig. 3h, it is straightforward to estimate the number of reversals to failure, $2N_f$, by using the modified Manson-Coffin curve determined according to Eq. (1) and calibrated via relationships (3) and (4).

3.2 Variable Amplitude Load Histories

Consider the notched component shown in Fig. 4a and assume that it is subjected to a complex system of external VA forces/moments. Analogous to the CA case, under VA load histories as well the first step is determining the elasto-plastic cyclic stress/strain fields along the focus path. By so doing, the relevant stress/strain tensors can determined at a distance from the assumed crack initiation point equal to L/2 (Fig. 4a). Subsequently, this strain tensor is re-analysed according to the γ-MVM in order to determine direction MV (Fig. 4b).

Given the orientation of the critical plane (Fig. 4b), the equivalent amplitude of the shear stress, τ_a, as well as the mean value, $\sigma_{n,m}$, and the equivalent amplitude, $\sigma_{n,a}$, of the normal stress are determined by making use of the definitions explained in

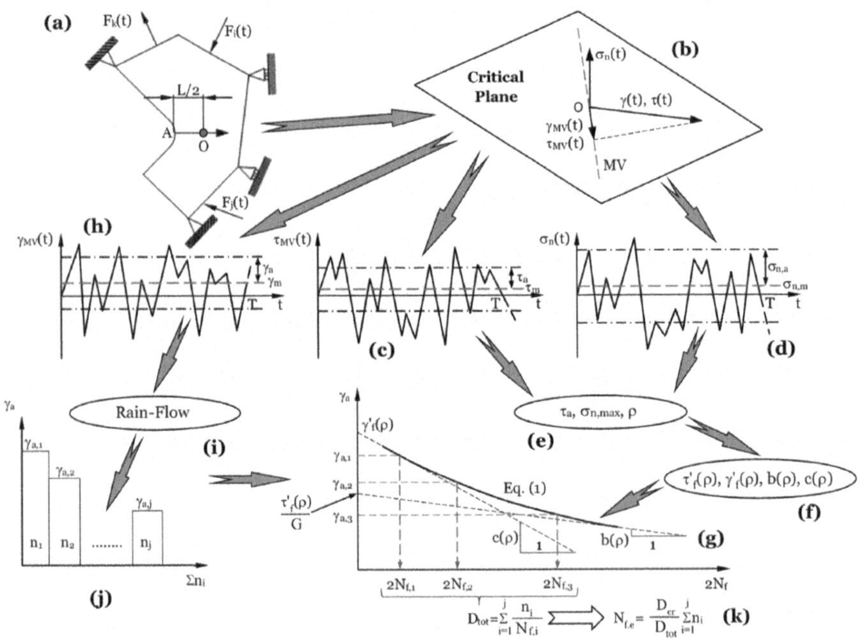

Fig. 4 Designing notched components against VA uniaxial/multiaxial fatigue loading

Fig. 2d—see also Figs. 4c and 4d. Equivalent stress components τ_a and $\sigma_{n,max}$ [11] can then be used to estimate the value of the critical plane stress ratio, ρ, characterising the VA fatigue load history being assessed (Fig. 4e). Stress ratio ρ together with functions (10) allow then the MMCCM to be calibrated correctly (Fig. 4f).

As soon as the modified Manson-Coffin curve to be used to assess fatigue damage is known (Fig. 4g), the next step is using the Rain-Flow method to count the cycles from signal $\gamma_{MV}(t)$. This allows the shear strain spectrum associated with the load history under investigation to be determined unambiguously (Figs. 4h, i). The obtained shear strain spectrum (Fig. 4j) together with the modified Manson-Coffin curve estimated via Eq. (1) are then used to quantify the extent of damage associated with the shear strain cycles being counted (Fig. 4g), with the number of cycles to failure being estimated as follows (Fig. 4k):

$$D_{tot} = \sum_{i=1}^{j} \frac{n_i}{N_{f,i}} \Rightarrow N_{f,e} = \frac{D_{cr}}{D_{tot}} \sum_{i=1}^{j} n_i \tag{7}$$

To conclude, it is worth point out that in the above relationships, D_{tot} is the total value of the damage sum, whereas D_{cr} is the critical value for D_{tot} that results in the initiation of a fatigue crack in the notched component being assessed.

3.3 Determination of Critical Distance L

Figure 5 summarises the procedure being recommended for the experimental determination of critical distance L. To this end, the uniaxial and torsional fatigue properties of the un-notched material under investigation are supposed to be known from the experiments.

Consider now the notched specimen of Fig. 5 and assume that this sample is subjected to a fully-reversed nominal uniaxial loading. Under the applied uniaxial CA load history, this notched specimen is seen to fail at a number of cycles to failure equal to $N_{f,c}$.

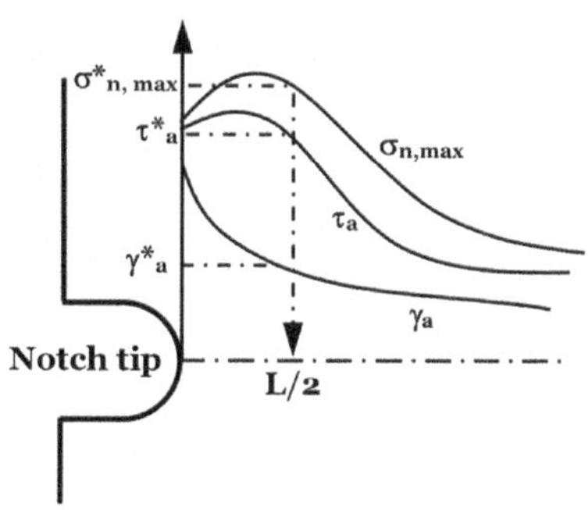

Fig. 5 Critical distance L determined according to the Point Method

The stress/strain-distance curves along the focus path associated with the investigated stress/strain concentrator are determined by directly post-processing the results from an elasto-plastic Finite Element (FE) model. Subsequently, the stress and strain components along the focus path are re-analysed according to the γ-MVM so that the distribution of γ_a, τ_a, $\sigma_{n,max}$, and ρ along the focus path itself can be determined unambiguously (Fig. 2c). Finally, according to the PM [7, 8], L/2 is the distance from the notch tip where the shear strain amplitude, γ_a^*, is equal to the shear strain amplitude estimated via Eq. (1) that causes failure in the plain material at $N_f = N_{f,c}$ (Fig. 5).

4 Experimental Validation of the Fatigue Design Technique

To check the accuracy of the multiaxial notch fatigue design approach discussed in the present paper, a systematic experimental investigation was carried out in the Structural Integrity Laboratory of the University of Sheffield, UK [12]. The plain and notched samples being tested were made of medium-carbon steel En8 having ultimate tensile stress equal to 701 MPa, yield stress to 453 MPa, Young's modulus to 210 GPa, and shear elastic modulus equal to 80 GPa.

The notched cylindrical samples under investigation had net diameter equal to 18 mm, gross diameter to 38 mm and notch root radius, r, equal to 1.5 mm, to 3 mm, and to 6 mm.

The fatigue experiments were run by using a SCHENCK axial-torsion fatigue machine, with the cylindrical specimens being clamped using two MTS 646.25S hydraulic collet grips. The strain-controlled tests as well as the force/torque-controlled tests were run under CA/VA sinusoidal loading signals and the frequency was set in the range 0.5–2 Hz. The experimental results generated from the smooth samples were run by controlling the axial and the shear strains by using an Epsilon 3550-025 M bi-axial extensometer having gauge length equal to 25 mm. The number of cycles/blocks to failure was defined both for the plain specimens and the notched samples by 5% axial/torsional stiffness drop, with the stiffness decrease being estimated under the maximum amplitude of the force applied during testing.

By adopting the concave upwards spectrum with sequence length, S_L, of 50 cycles that is shown in Fig. 6a, the VA fatigue behaviour of the plain and notched specimens being tested was investigated by controlling the axial/shear strain and the axial/torsional loading, respectively. In the spectrum of Fig. 6a, given either the axial force, the torque, the axial strain, or the shear strain, $\Omega_{a,i}$ denotes the amplitude of the reference strain/loading quantity characterising the i-th cycles, whereas $\Omega_{a,max}$ denotes the maximum value in the spectrum of the amplitude of the strain/loading quantity of interest. The tests under VA loading were run by applying the cycles in random order.

The experimental values of the fully-reversed uniaxial cyclic mechanical properties of the carbon steel being tested were as follows: K' = 971.5 MPa, n' = 0.188, $\sigma'_f = 852.3$ MPa, b = −0.105, $\varepsilon'_f = 0.477$, and c = -0.554. The corresponding

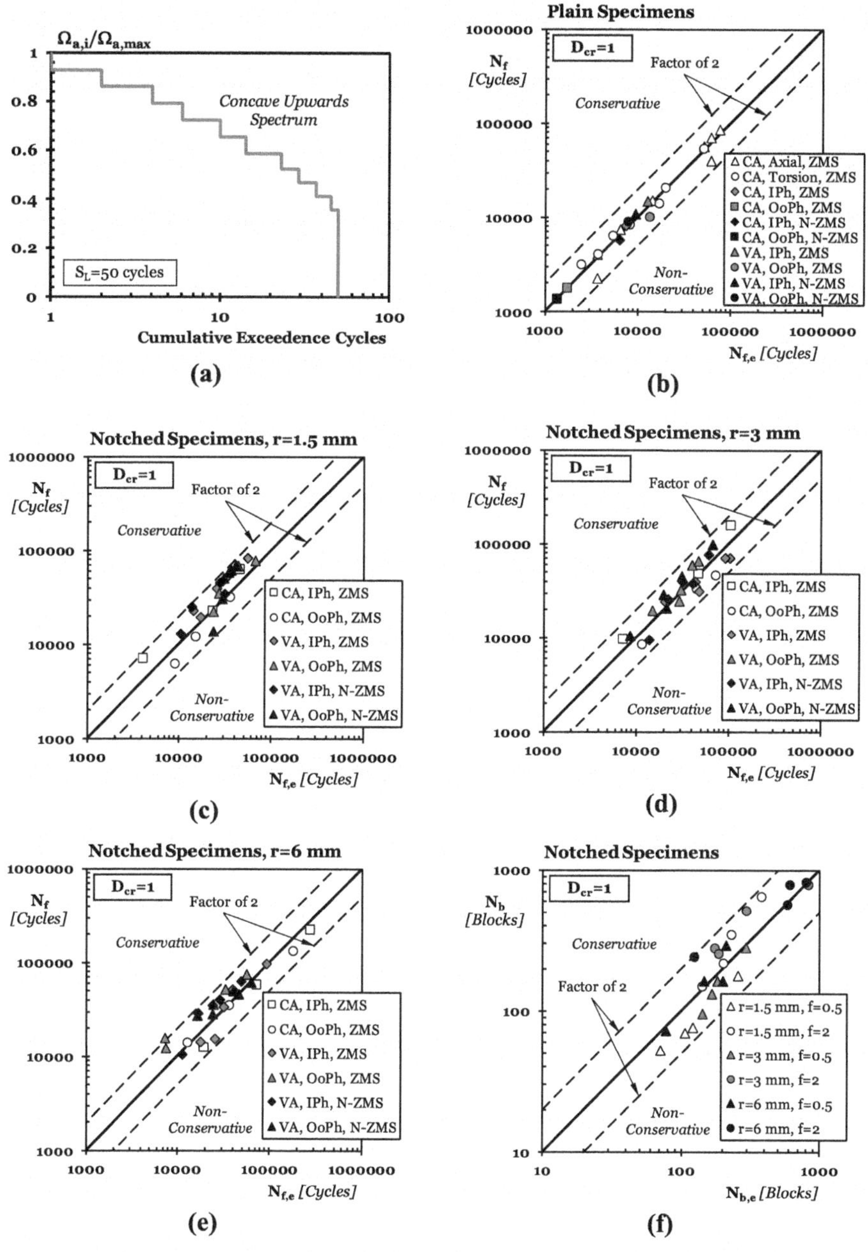

Fig. 6 Overall accuracy of the proposed multiaxial fatigue life estimation technique (CA = Constant Amplitude; VA = Variable Amplitude; IPh = In-Phase; OoPh = 90° Out-of-Phase; ZMS = Zero Mean Stress; N-ZMS = Non-Zero Mean Stress; f = ratio between the frequencies of the axial and torsional loading channels)

mechanical properties under fully-reversed torsion were instead as follows: $K'_0 = 442.4$ MPa, $n'_0 = 0.107$, $\tau'_f = 460.6$ MPa, $b_0 = -0.068$, $\gamma'_f = 1.55$, and $c_0 = -0.648$. These material constants were used to calibrate the MMCCM —i.e., relationships (3) and (4)—for the specific carbon steel being investigated.

Having calibrated the MMCCM, initially we checked its accuracy in estimating fatigue lifetime of un-notched components when the required stress/strain quantities relative to the critical plane are determined according to the γ-MVM. The results from this initial validation exercise are summarised in the experimental, N_f, vs. estimated, $N_{f,e}$, number of cycles to failure diagram reported in Fig. 6b. According to Palmgren and Miner [1, 2], the predictions under VA loading shown in Fig. 6b were obtained by taking the critical value of the damage sum, D_{cr}, invariably equal to unity. This error chart demonstrates that the use of this design approach resulted in estimates falling within an error factor of 2, i.e., within a scatter band as wide as the one containing the fatigue results generated under axial and torsional CA fatigue loading and used to calibrate the MMCCM itself.

To re-analyse the results generated by testing notched specimens, FE code ANSYS® was employed to derive the necessary time-variable elasto-plastic local stress/strain fields. To this end, the notched specimens were modelled using three-dimensional 8-node structural element SOLID185, with the elasto-plastic behaviour of the tested steel being simulated by using a multi-linear kinematic rule. After performing conventional convergence analyses, a very refined mesh was used to model the material in the vicinity of the tips of the notches being investigated.

The critical distance value for the tested carbon steel was determined experimentally by applying the PM along with the MMCCM according to the procedure shown in Fig. 5. By so doing, the results generated by testing six sharply notched specimens under uniaxial fatigue loading returned an average value for L/2 of 0.78 mm.

The accuracy of the design procedure schematically described in Figs. 3 and 4 is seen in Figs. 6c to 6f, where fatigue lifetime was estimated by taking the critical value of the damage sum, D_{cr}, invariably equal to unity [1, 2]. According to the error diagrams of Fig. 6c, f, the use of the proposed multiaxial fatigue life estimation technique was seen to result in a very high level of accuracy, with all the experimental points falling within an error factor of 2.

5 Conclusions

The multiaxial fatigue assessment methodology investigated in the present paper is successful in estimating lifetime of notched metallic materials subjected to both CA and VA multiaxial load histories. In particular, it is important to point out that a very high level of accuracy was reached independently of sharpness of the notch, degree of non-proportionality of the load history, mean stress level, and ratio between the frequencies of the applied loading components. Accordingly, it is possible to conclude by saying that this approach can be treated as a powerful tool suitable for performing the multiaxial fatigue assessment of real notched structural components, the key advantage being that it can be applied by directly post-processing the relevant stress/strain fields determined from elasto-plastic numerical models solved using commercial FE codes.

References

1. Stephens RI, Fatemi A, Stephens RR, Fuchs HO (2001) Metal fatigue in engineering, 2nd edn. John Wiley & Sons, USA
2. Socie DF, Marquis GB (2000) Multiaxial fatigue. SAE, Warrendale, PA, USA
3. Susmel L (2009) Multiaxial Notch Fatigue: from nominal to local stress-strain quantities. Woodhead & CRC, Cambridge, UK
4. Tipton SM, Fash JW (1989) Multiaxial fatigue life prediction for the SAE specimen using strain based approaches. In: Leese GE, Socie DF (eds) Multiaxial fatigue: analysis and experiments, SAE AE-14, pp 67–80
5. Susmel L, Meneghetti G, Atzori B (2009) A novel multiaxial fatigue criterion to predict lifetime in the low/medium-cycle fatigue regime. Part II: Notches. Trans ASME, J Eng Mater Technol 131, 021010–1/8
6. Gates N, Fatemi A (2014) Notched fatigue behavior and stress analysis under multiaxial states of stress. Int J Fatigue 67:2–14
7. Susmel L, Taylor D (2010) An elasto-plastic reformulation of the Theory of Critical Distances to estimate lifetime of notched components failing in the low/medium-cycle fatigue regime. Trans ASME, J Eng Mater Technol, 132, 021002–1/8
8. Susmel L, Taylor D (2015) Estimating lifetime of notched components subjected to variable amplitude fatigue loading according to the elasto-plastic Theory of Critical Distances. Trans ASME, J Eng Mater Technol 137, 011008–1/15 (2015).
9. Susmel L, Meneghetti G, Atzori B (2009) A simple and efficient reformulation of the classical Manson-Coffin curve to predict lifetime under multiaxial fatigue loading. Part I: plain materials. Trans ASME, J Eng Mater Technol 131: 021009–1/9
10. Susmel L (2010) A simple and efficient numerical algorithm to determine the orientation of the critical plane in multiaxial fatigue problems. Int J Fatigue 32:1875–1883
11. Wang Y, Susmel L (2016) The Modified Manson-Coffin Curve Method to estimate fatigue lifetime under complex constant and variable amplitude multiaxial fatigue loading. Int J Fatigue 83:135–149
12. Namiq ZF, Susmel L (2019) Proportional/non-proportional constant/variable amplitude multiaxial notch fatigue: cyclic plasticity, non-zero mean stresses, and critical distance/plane. Fatigue Fract Eng Mater Struct 42(9):1849–1873

The manufacturer's authorised representative in the EU is Springer Nature Customer Service Centre GmbH, Europaplatz 3, 69115 Heidelberg, Germany. If you have any concerns regarding our products, please contact ProductSafety@springernature.com

Printed and bound by CPI Group (UK) Ltd, Croydon, CR0 4YY
30/09/2026
02225232-0007